ENVIRONMENTAL AND SAFETY CONCERNS IN UNDERGROUND CONSTRUCTION
VOLUME 2

PROCEEDINGS OF THE 1ST ASIAN ROCK MECHANICS SYMPOSIUM: ARMS '97
A REGIONAL CONFERENCE OF ISRM/SEOUL/KOREA/13-15 OCTOBER 1997

Environmental and Safety Concerns in Underground Construction

Edited by
HI-KEUN LEE
Seoul National University, Korea

HYUNG-SIK YANG
Chonnam National University, Kwangju, Korea

SO-KEUL CHUNG
Korea Institute of Geology, Mining and Materials, Taejon, Korea

VOLUME 2

A.A.BALKEMA / ROTTERDAM / BROOKFIELD / 1997

The texts of the various papers in this volume were set individually by typists under the supervision of each of the authors concerned.

Authorization to photocopy items for internal or personal use, or the internal or personal use of specific clients, is granted by A.A.Balkema, Rotterdam, provided that the base fee of US$1.50 per copy, plus US$0.10 per page is paid directly to Copyright Clearance Center, 222 Rosewood Drive, Danvers, MA 01923, USA. For those organizations that have been granted a photocopy license by CCC, a separate system of payment has been arranged. The fee code for users of the Transactional Reporting Service is: 90 5410 910 6/97 US$1.50 + US$0.10.

Published by
A.A.Balkema, P.O.Box 1675, 3000 BR Rotterdam, Netherlands (Fax: +31.10.413.5947)
A.A.Balkema Publishers, Old Post Road, Brookfield, VT 05036-9704, USA (Fax: 802.276.3837)

For the complete set of two volumes, ISBN 90 5410 910 6
For Volume 1, ISBN 90 5410 911 4
For Volume 2, ISBN 90 5410 912 2

© 1997 A.A.Balkema, Rotterdam
Printed in the Netherlands

Table of contents

Modelling techniques for safety evaluation: Evaluation of rock mass properties

Fiber reinforced shotcrete simulation using the discrete element method P.Chryssanthakis, N.Barton, L.Lorig, M.Christianson & Y.H.Suh	547
Numerical study on reinforcement effect of cable bolt in discontinuous rock mass H.Kinashi, S.Amano, H.Tsuchihara, H.Yoshioka & K.Michihiro	553
Reliability research of surrounding rock based on elasto-plastic analysis of stochastic field finite element method D.Liu & Y.Zheng	559
Numerical simulation and experimental study of excavation and anchoring support of underground chambers in jointed rockmass W.Zhu, W.Chen & B.Wang	565
A finite element method of probabilistic keyblock analysis D.S.Young	571
Viscoplastic-damage analysis of underground structures in rock A.Nanda & S.Sengupta	577
Design of the reinforced concrete lining in Bakun Diversion Tunnels W.R.Jee, Y.K.Sim & T.J.Lim	581
Criterion of energy catastrophe for rock project engineering system failure in underground engineering M.Cai, G.Kong & J.Liang	587
A study of the frictional-slide behaviour in the post-failure region of soft rock X.Li & S.Wang	591
The effect of the number of joints on the failure pattern of discontinuous rock mass around circular hole E.K.Cho & Y.S.Kim	595
Fractal characteristic of width of branching cracks in coal mass J.Shen, Y.Zhao & K.Duan	601

A practical approach for obtaining ground reaction curve for tunnels under squeezing ground conditions 607
R.K.Goel, J.L.Jethwa & B.B.Dhar

AE characteristics on progressing fracture within solid media in multiple-stage rock shear tests 613
T.Shiotani, K.Matsumoto, M.Tsutsui & H.Chikahisa

Evaluation of in situ hydromechanical properties of rock fractures at Laxemar in Sweden 619
J.Rutqvist, C.F.Tsang, D.Ekman & O.Stephansson

Physical simulation of energy exchange aimed at detection of dangerous forms of rock mass failure 625
E.V.Lodus

Evaluation of JRC from image processing of borehole wall 631
J.Kim & O.Stephansson

Fractal geometry and mechanical behavior of rock fracture surfaces 635
J.A.Wang & M.A.Kwaśniewski

Tests on mechanical properties of model fracture zones 643
Ö.Aydan, Y.Shimizu, T.Akagi & T.Kawamoto

Physical model and numerical model analyses of jointed rocks 649
B.Wang, S.Kwon, H.D.S.Miller, D.H.Lee & H.K.Lee

Effects of the thermal cracks on the physical and mechanical properties of the Pocheon granite 655
Y.K.Yoon

Assessment of parameters for rock weathering classification: A case study on Malanjkhand Copper Project, India 661
K.S.Rao & A.S.Gupta

Virgin state of stresses and discontinuities of rock masses 667
M.El Tani

Influence of water saturation on the dynamic response of rock mass surrounding a circular tunnel 673
S.H.Kim, K.M.Chang & K.J.Kim

In situ stress measurement and its application to mining design in Jinchuan Nickel Mine, China 679
M.Cai, T.Liu & C.Zhou

Effect of residual tectonic stresses on the TKI-ELI Soma Işıklar Decline stability, Turkey 683
Y.Özçelik

Application of acoustic emission technique to determination of in situ stress 691
M.Seto, M.Utagawa, K.Katsuyama & T.Kiyama

Investigation on the influence of montmorillonite on the stability of the surrounding rock of gallery 697
S.Zhang, H.Wan, Z.Feng, C.Xu, X.Xu, J.Yin, B.Wang & T.Cao

In situ stress measurement using hydraulic fracturing for shallow tunnel in Korea — 703
S.O.Choi, H.S.Shin & K.S.Kwon

Modelling techniques for safety evaluation: Simulation of the coupled behavior

Thermo-hydro-mechanical coupling analysis for underground heat storage: Application of several codes — 709
H.S.Lee, M.H.Kim & H.K.Lee

Experimental and numerical study on the thermo-mechanical behavior of granite — 715
M.H.Jang, H.S.Yang & K.O.Park

An analytical solution and application of the coupled flow of fluids around a single well — 721
Z.Xu, X.Xu & H.Li

Numerical study on thermo-hydro-mechanical coupling analysis in rock with variable properties induced by temperature — 727
H.J.Ahn & H.K.Lee

Analysis of thermo-mechanical behavior of underground cold storage cavern by monitoring and numerical prediction — 731
Y.Park, J.H.Synn, C.Park & H.Y.Kim

Modelling techniques for safety evaluation: Monitoring and interpretation

The automation of MPBX monitoring and the numerical analysis of MPBX displacement — 739
Y.B.Jung, H.K.Lee, H.K.Jung, S.K.Chung & D.H.Kim

Determination of in situ stress using DRA and AE techniques — 745
M.Utagawa, M.Seto & K.Katsuyama

Pillar deformation response delay effect in underground mining — 751
A.S.Voznesensky

Application of time domain reflectometry to the deformation characterization of rock mass — 757
S.L.Jung, S.K.Chung & H.K.Lee

Loosening rock region estimated by field measurements during large underground cavern excavation — 763
Y.Uchita, Y.Hirakawa & A.Mochizuki

Visualization of three-dimensional structure of rocks using X-ray CT method — 769
K.Sugawara, Y.Obara, K.Kaneko, K.Koike, M.Ohmi & T.Aoi

A study on the application of electrical resistivity monitoring technique to detection of seawater intrusion — 775
T.Kang, I.Y.Han, J.Lee & S.J.Hong

A.E. source location considering the stress induced velocity anisotropy in rock — 779
K.S.Lee & C.I.Lee

Crack monitoring in underground construction — 785
S.K.Tewatia

Compact VSP probe for inspection of rock mass quality 789
A. Hirata, S. Baba, T. Inaba & K. Kaneko

Geodynamic safety – The main factor in the exploration of mineral resources 793
and the earth surface
A. N. Shabarov, V. V. Zoubkov & N. V. Krotov

Modelling techniques for safety evaluation: Back analysis and others

Displacement back analysis of tunnels in viscoelastic rock masses 801
F. Zhu & L. Xue

Stochastic finite element analysis of underground rock structure using Latin Hypercube 805
Sampling technique
K. S. Choi, B. Y. Park & H. K. Lee

Swellex® in weak/soft rock 813
U. Håkansson & C. Li

Development of an expert system for safety analysis of structures adjacent to tunnel 819
excavation sites
G. J. Bae, C. Y. Kim, H. S. Shin & S. W. Hong

Safety aspects in tunnelling and salt cavern design 825
R. Rokahr & K. Staudtmeister

A viscoelastic plastic displacement back analysis model for basic parameters of rock mass 831
Z. Shen & Z. Xu

Full scale tests of steel arch supports 835
J. W. Kim & H. K. Lee

Back calculation of initial stress state from incremental displacement measurements 841
Y. K. Lee & C. I. Lee

An integrated back-analysis system for monitoring underground openings 847
N. Shimizu, K. Nakagawa & S. Sakurai

Back analysis of linear and nonlinear deformational behaviors for multiple cross sections 853
in discontinuous rock masses of a large underground power house cavern
S. Akutagawa & S. Sakurai

Theoretical study on deformational behavior and reinforcing effect by bolting for tunnels 857
in soft rock
Y. J. Jiang, T. Esaki & Y. Yokota

Detection of underground cavity by inverse calculation 863
B. S. Suh, K. I. Sohn, B. D. Kwon & H. K. Jung

Back analysis of subsidence above old open stopes in an Indian hard rock mine 869
by numerical simulation
A. K. Ghosh, A. Sinha & D. G. Rao

Evaluation of long cable tendon load distribution using Computer Aided Bolt Load Estimation (CABLE™) *W.F. Bawden, M. Moosavi & A.J. Hyett*	875
The S.M.A.R.T. cable bolt: An instrument for the determination of tension in 7-wire strand cable bolts *A.J. Hyett, W.F. Bawden, P. Lausch, M. Ruest, J. Henning & M. Baillargeon*	883
Application of boundary element method for displacement back analysis of tunnels in viscoelastic rock mass *L. Xue & R. Luo*	891
Limitations of ubiquitous joint models *E.M. Dawson & Y.J. Park*	895
Complex behaviour of the rock mass around the excavation face in large rock caverns *Y.N. Lee, Y.H. Suh, D.Y. Kim & K.S. Jue*	901

Modelling techniques for safety evaluation: Slope stability and landslides

Landslide hazards and stability analysis of coastal cliff regions of Bangladesh *Md. H. Rahman*	909
Safe and economical rockfall protection barriers *W. Gerber & B. Haller*	915
Study on prediction of landslide of rock slopes *X. Ge, D. Xu, X. Gu, C. Chen & Y. Shi*	921
Reliability-based analysis for rock slopes considering multi-failure modes *I.M. Lee & M.J. Lee*	927
Statnamic test for estimating the bearing capacities of rock socketed piles *J.H. Kim, S.H. Lee & M.M. Kim*	933
Putting forward a new concept of grouting engineering – Groutable period *X. Wang & Q. Gao*	937
Implementation of safety measure by two dimensional and three dimensional stability analysis for a highly bedded rock slope *S.K. Chung, K.C. Han, S.O. Choi, C. Sunwoo, H.S. Shin & Y. Park*	941
Evaluation of rock slope stability by analysis of discontinuities in Boryung damsite *C. Sunwoo, H.S. Shin, K.C. Han & S.K. Chung*	951
Author index	955

Modelling techniques for safety evaluation: Evaluation of rock mass properties

Fiber reinforced shotcrete simulation using the discrete element method

P.Chryssanthakis & N. Barton
Norwegian Geotechnical Institute, Oslo, Norway

L. Lorig & M. Christianson
Itasca Consulting Group, Minneapolis, Minn., USA

Y. H. Suh
Hyundai Institute of Construction Technology, Seoul, Korea

ABSTRACT: Fiber reinforced shotcrete has been widely used as part of permanent tunnel support during the last 15 years especially in connection with the application of the Norwegian Method of Tunnelling (NMT). The interaction of the fiber reinforced shotcrete and the rock bolt reinforcement can now be numerically modelled with the Distinct element method (DEM). The discontinuous code UDEC (Universal Distinct Element Code) is used to investigate the overall stability of an excavation, to predict the expected stresses and deformations caused by the excavation and to investigate the optimal excavation sequence to be followed. The jointed rock geometry of Hyundai's shallow test tunnel in jointed biotite gneiss has been considered for demonstrating the fiber reinforced shotcrete, S(fr), subroutine. The results have shown that by using S(fr) and subsequently rock bolts as primary support in the tunnel, the load attained by some of the rock bolts is reduced by approximately half compared to the case were only rock bolts were used.

1 INTRODUCTION

The Norwegian Geotechnical Institute (NGI) of Oslo has been involved in a joint effort with Itasca Consulting Group for establishing an algorithm for improved simulation of the behaviour of fiber reinforced shotcrete S(fr) in multiple layers in underground structures. A special S(fr) subroutine that was developed by Itasca and financed by NGI has been incorporated in UDEC (the two dimensional Universal Distinct Element Code). In NGI's modelling work the UDEC-BB version is generally used. This is a special version of UDEC that includes the Barton - Bandis joint constitutive model (Barton and Bandis 1990).

A project that NGI and Hyundai Institute of Construction Technology (HICT) were involved in 1996 in Seoul has been chosen as an example to demonstrate the use of S(fr) in UDEC-BB. Modelling work was performed simultaneously in NGI and Hyundai and in situ measurements have been taken to be compared with the numerical results. The work involved a tunnel in Hyundai's test station (span 5.4, height 6.4 m) in biotite gneiss.

2 THEORETICAL BACKGROUND FOR THE FIBER REINFORCED SHOTCRETE

The structural elements in UDEC can be used to model the effect of fiber reinforced shotcrete on any rock surface. The area of application of the shotcrete is specified and UDEC automatically creates the elements necessary to represent a uniformly applied layer. The material behaviour model associated with the structural element formulation in UDEC simulates the inelastic behaviour representative of many common surface-lining materials. This includes non-reinforced and reinforced cementitious materials, such as concrete and fiber-reinforced shotcrete, that can exhibit either brittle or ductile behaviour as well as materials such as steel, that behave in a ductile manner. The behaviour of the material model used for S(fr) can be shown on a moment-thrust interaction diagram, see Figure 1. Moment-thrust diagrams are commonly used in the design of concrete columns. These diagrams illustrate the maximum force that can be applied to a typical section for various eccentricities (e). The ultimate failure envelopes for non-reinforced and reinforced cementitious materials are similar. However, reinforced materials have a residual capacity that remains after failure at the ultimate load. Non-reinforced cementitious materials have no residual capacity.

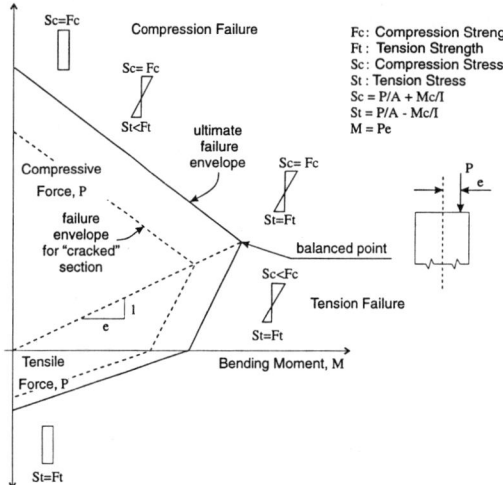

Figure 1. Behaviour of S(fr) as a moment - thrust interaction diagram.

3 BRIEF DESCRIPTION OF THE FIBER REINFORCED SHOTCRETE IN UDEC

The numerical code used at NGI, is a UDEC-BB version 3.0 dated November 1996 with the newly developed S (fr) subroutine implemented. This new version 3.0 is a further developed version of Cundall's original distinct element two-dimensional code (Cundall 1980). Examples of application of this code can be found in (Makurat et al. 1990, Barton et al. 1992). The main characteristics of the fiber reinforced subroutine in UDEC are as follows:

• Possibility to apply S(fr) not only on idealised (geometrical shape) tunnel peripheries but also in uneven peripheries (i.e. after a blasting operation with uneven overbreak).

• Possibility to model the variation in adhesion between the S(fr) and rock interface (e.g., model the difference in adhesion to schist and granite).

• Possibility to model the bolt reinforcement piercing the S(fr). The last feature has a limitation since the S(fr) and bolts are fixed in one single point only.

• Possibility to model the fiber reinforced shotcrete in multiple layers.

Seven different types of graphs can be produced in connection with the S(fr) subroutine in UDEC-BB. These are: axial and shear forces on the S(fr), normal and shear forces on the S(fr)/rock interface, moments on the S(fr), failure plot of the S(fr), and tensile failure plot of the S(fr)/rock bond.

The necessary choice of suitable input data to represent bond strength has resulted in the discovery of potentially very high frictional strength between the shotcrete and rock surface, when the latter is a fresh blasted (or road-header excavated) surface with normal high roughness. Figure 2 shows the principles of the method for selecting relevant values of cohesion and friction in the rock-shotcrete interface, once a designed bond strength (of say 0.5 or 1.0 MPa) has been chosen. The method is based on the non-linear stress dependent BB model (Barton and Bandis, 1990) and on the linear Mohr-Coulomb model of shear strength. The latter is used in describing the bond strength in UDEC-S(fr) using the parameters JTENS which signifies the bond strength, JCOH for the cohesion of the interface, and JFRIC for the friction angle.

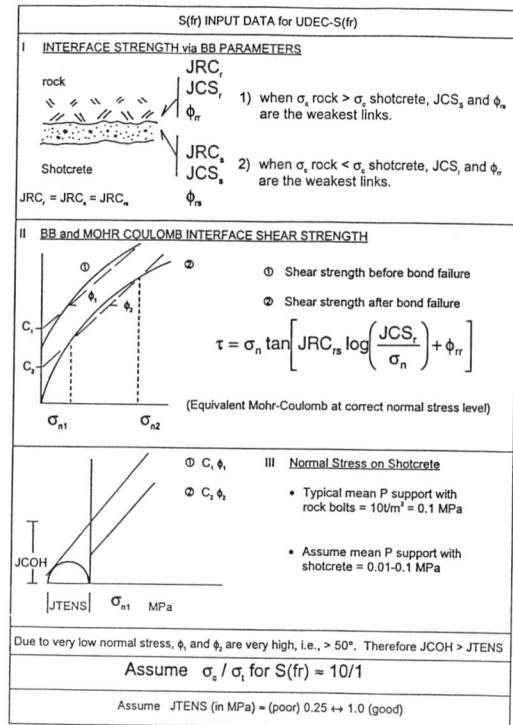

Figure 2. Estimating rock/S(fr) interface strength.

4 EVALUATION OF ROCK QUALITY, TUNNEL SUPPORT AND STRESS SITUATION

4.1 *Rock mass characterisation*

Rock mass classification systems provide guidelines for the estimation of support pressure and for the design of tunnel reinforcement. The Q-System of Barton et al., 1974, Grimstad and Barton 1993, developed at the Norwegian Geotechnical Institute (NGI), has also been applied extensively to derive the geotechnical parameters needed for predicting the performance of rock masses. As mentioned earlier the modelled tunnel section illustrated in this paper is located in biotite gneiss with varying degree of weathering. The upper 10 to 15 m of the modelled section are assumed to be weathered (shaded area in Figure 3) with the lower part strongly altered. The following points can be summarised for the gneiss.

- Two to three joint sets plus random joints, with mostly non-continuous joints in the upper and lower part of the section are observed. Equal weighting regarding frequency and extent, for sub-vertical joints.
- The mean spacing of all joint sets is 0.20 to 0.25 m.
- The continuity of joints is less than the tunnel span.
- Joint roughness is described as smooth to rough undulating.
- Joint weathering varying from nearly fresh with surface stains only to small amounts of clay for the foliation joint set.

Detailed engineering geological mapping of the rock and core logging has been carried out. The same geotechnical logging chart used in this project has been used extensively as an aid in data collection and presentation for the design of underground caverns for radioactive wastes in England (Barton et al. 1992) and for the mapping of large underground openings (Bhasin et al. 1993, Barton et. al. 1994, Grimstad and Barton 1995). The Q-values, range from 0.4 to 3.1 for the different weathering degrees of gneiss. The gneiss has sufficient joint sets for kinematics block release (three or more on average at one location) and will require systematic bolting after the application of fiber reinforced shotcrete. The joint structure of the gneiss in the model contains a wedge at the tunnel crown. A weak zone runs diagonally though the tunnel demonstrating the worst case scenario.

4.2 *Support requirements, fiber reinforced shotcrete*

The estimated support requirements for this tunnel which were derived by the Q-system suggest a total thickness of S(fr) about 10 cm. The S(fr) can be applied in two layers of 5 cm each. The first layer which will be applied immediately after the excavation will be followed by systematic bolting in a 1.5 × 1.5 m pattern, 25 mm in diameter and 2.5 m in length followed by the second 5 cm S(fr) layer.

5 DESCRIPTION OF THE NUMERICAL MODELS

Four numerical models were run and compared in an attempt to get a better understanding of the performance of the fiber reinforced shotcrete in the tunnel.

The analysis of the results in this paper focuses on the behaviour of the S(fr) and the rock bolts. All four models have exactly the same joint geometry (Figure 3), intact rock, joint properties, boundary conditions (roller boundaries) and in-situ rock stresses. The tunnel was excavated in a single excavation step. The S(fr) was applied on the models when approximately 50% of the total expected deformation had occurred to be followed by the installation of bolts at about 60% of the total expected deformation. This was done in an attempt to allow for the elastic deformation that had already occurred at the face of the tunnel. In model 2 where only rock bolts were applied, these were applied at about 50% of the total deformation. The differences between the numerical models are:

- The 1st model (Model 1) was run unsupported, no S(fr), no bolts
- The 2nd model (Model 2) has no S(fr) applied only bolts.
- The 3rd model (Model 3) has total S(fr) thickness of 10 cm applied in two layers of 5 cm each with bolting between these stages.
- The 4th model (Model 4) has a S(fr) thickness 10 cm applied in a single layer followed by rock bolts.

For the numerical modelling a rather conservative ratio of $\sigma_h/\sigma_v = 0.75$ has been used. The deformation modulus for the intact rock is assumed to vary between 0.4 (weak zone) and between 1 and 4 GPa for the upper and lower part of the model, Poisson's ratio varies between 0.27 (weak zone) and 0.3 for the intact rock and density varies between 2300 kg/m^3 (weak zone) and 2500 kg/m^3 for the intact rock.

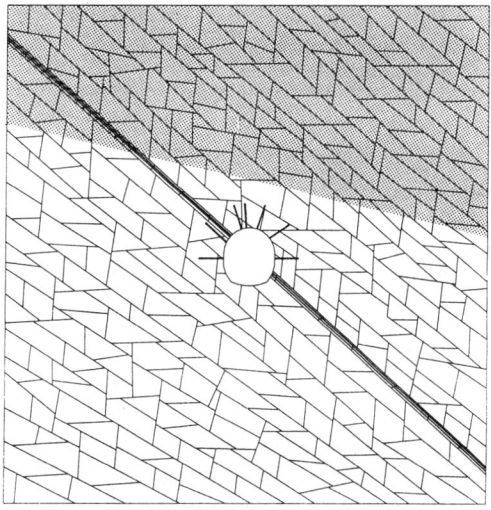

Figure 3. Jointed rock mass geometry, and bolt pattern used in the models.

Table 1. Jointed rock properties for the sub-horizontal joint set and joint sets 2 & 3.

Parameter		Joint set 1 sub-horizontal 0°-40°	Joint set 2 Sub vertical I 90°-140°	Joint set 3 Sub vertical II 40°-90°
JRC_0		5.4	2.3	3.8
JCS_0	MPa	98.0	114.0	136.0
ϕ_r	deg.	28.5	30.5	30.5
σ_c	MPa	190.0	190.0	190.0
L_0	m	0.1	0.1	0.1
L_n	m	0.4	0.3	0.3
Aper	mm	0.192	0.110	0.137

Table 2. Fiber reinforced shotcrete parameters used in the modelling work.

Parameter	All Models
Modulus of elasticity, E (GPa)	15
Poisson's ratio, ν	0.15
Density, ρ (kg/m^3)	2,500
Compressive yield strength, (MPa)	30
Tensile yield strength, (MPa)	3
Residual tensile yield strength, (MPa)	2
Friction in S(fr)/rock interface, (deg.)	60
Cohesion in S(fr)/rock interface,(MPa)	0.86
Tension in S(fr)/rock interface, (MPa)	0.50

Table 3. Summary of the numerical results for model 1, 2, 3 and 4.

Parameters	Model 1	Model 2	Model 3	Model 4
Maximum principal stress,(MPa)	1.41	1.39	1.42	1.53
Maximum displacement (mm) arch crown	4.59	4.17	3.92	4.06
Maximum shear displacement (mm)	2.90	2.56	1.46	2.05
Maximum axial forces on bolts (tnf)		14.76	12.48	13.09
Maximum axial forces on S(fr) (tnf)			32.30	32.60
Maximum moment on S(fr) (tnf)x m			0.18	0.54
Maximum shear forces on S(fr) (tnf)			3.54	4.34
Maximum normal forces on S(fr)/rock (tnf)			7.72	5.51
Maximum shear forces on S(fr)/rock (tnf)			6.77	9.43

The jointed rock properties for all joint sets in the model are shown in Table 1. The necessary UDEC - S(fr) properties for modelling the S(fr) and their values used in this modelling work are listed in Table 2. The rock bolt pattern (bolts of 25 mm diameter) that was applied in reality in the crown and the walls of the tunnel was also modelled numerically (bolt spacing 1.5 m; length 2.5 m). The rock bolt pattern is also shown in Figure 3, UDEC results in Table 3.

6 NUMERICAL RESULTS - UDEC-BB

6.1 *Rock mechanics effects*

There is a little change between different models in the magnitude and direction of principal stresses. In order to study the effects of a falling wedge on the tunnel support system, a wedge has been formed numerically on the tunnel crown. When the tunnel was run unsupported, this wedge was loosened and eventually fell. When the tunnel was run with S(fr) and steel bolts the wedge remained in place. For practical reasons, a single "temporary" bolt was used to keep the falling wedge in place immediately after the numerical excavation and before the application of the S(fr) and the ordinary bolt pattern on the model.

The shearing associated with the active wedge in the tunnel crown for model 1 is significant. S(fr) which was modelled in three of the models will effectively secure smaller blocks from falling, something that will very likely happen in reality.

6.2 *Development of deformation vectors during excavation*

The unsupported tunnel in model 1 exhibits the highest deformation values (Table 3). There is a difference in the deformation magnitudes of about 15% between model 1 and model 3. The maximum deformation value occurs in the invert arch of the tunnel. The application of S(fr) on the tunnel in layers of 5 cm with reinforcing bolts in between (model 3) reduces the maximum shear displacement on the joints by approximately half. Due to the presence of massive blocks around the opening the S(fr) has little effect on the overall stability of the tunnel.

6.3 *Axial forces on bolts*

As expected the use of S(fr) lessens the load on the rock bolts. One of the model shows axial bolt forces approaching or even exceeding the scaled bolt yield limit (14.7 tnf). It is clear that, due to the presence of the unstable wedge in the arch, this particular bolt is heavily loaded, with some others bolts approaching yield limit. It is worth mentioning the 15% decrease of the maximum bolt forces in model 3 where two layers of S(fr) 5 cm each were applied, and maximum bolt load reached 12.5 tnf, (Figure 4, left) and in model 2 where no S(fr) was applied. The yield limit in the bolts is derived from the 22 tnf yield limit for the 25 mm bolts times the reduction factor of 0.67 (bolt pattern 1.5 x 1.5 m) for the UDEC model of 1 m thickness.

It is interesting to note also the development of bolt forces in the models with different thickness of S(fr), models 3 and 4. It is obvious that the application of 2 layers of S(fr) 5 cm each, instead of a single layer of 10cm, contributes to a better distribution of the bolt forces in the rock mass.

6.4 Forces and moments on the S(fr), failure mode

There is a substantial difference between the results obtained between model 3 and 4. The area of application of the forces on the S(fr) is 1 m (depth of model) x 10 cm (e.g. for S(fr) thickness) = 0.1 m^2. The thinner S(fr) layer in model 3 (1st layer) shows a rather uniform distribution of shear forces around the arch and the tunnel walls, while the thicker S(fr) layer in model 4 attains shear forces mainly on the tunnel arch. This can also be observed from the shear plot in Figure 5. The second S(fr) layer of 5 cm on the arch and walls in Model 3 attains about half of the shear forces compared to the 1st applied layer.

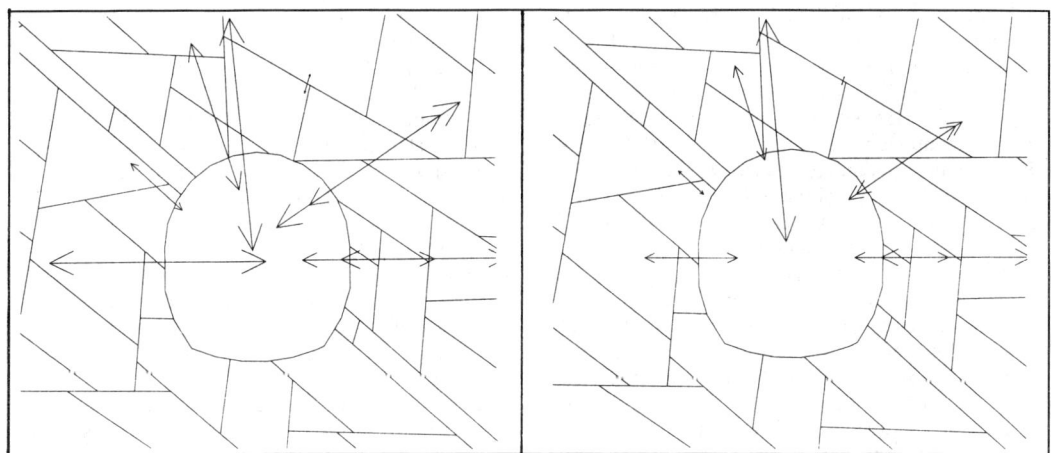

Figure 4 Axial bolt forces for models 2 (left) max. value 14.8 tnf and model 3 (right) max. value 12.5 tnf.

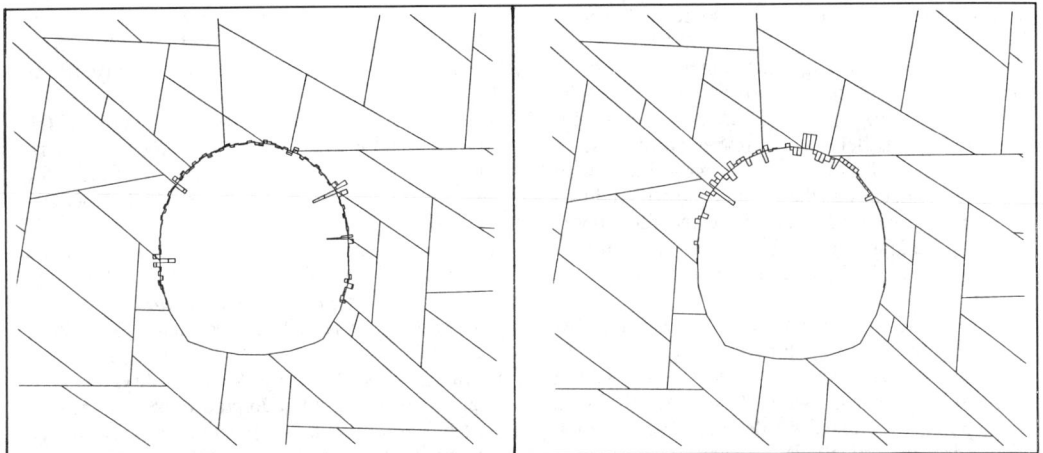

Figure 5. Shear forces plot for model 3 (left) max. value 3.5 tnf and model 4 (right) max. value 4.3 tnf.

7 CONCLUSIONS

- Design and construction of Hyundai's shallow test tunnel in relative weak rock has been carried out by using principles from the Norwegian Method of Tunnelling (NMT). The rock quality varies between very poor, to poor. The geomechanical properties of these rocks have been assessed based on laboratory and field investigations for input to numerical modelling studies. The rock mass characterisation approach (Q-system) has been applied extensively to predict and evaluate appropriate rock reinforcement requirements tunnels. The estimated Q-values measured ranged between 0.4 and 3.1.

- The input data for the UDEC-BB models have been derived from rock joint and rock mass characterisation. The four numerical models that are presented in this article were similar, with variations mainly in the S(fr) thickness. Models 2, 3 and 4 have also been numerically reinforced by systematic bolting. The bolt properties and bolt pattern were derived by means of the Q-system. The discontinuum code UDEC-BB (Barton - Bandis joint constitutive model) was used for the two-dimensional modelling of the tunnel. This is a rather conservative approach since several features of the in situ rock behaviour cannot be modelled in 2 D (e.g. only the joints parallel or sub-parallel to the tunnel axis have been represented etc.).

- The numerical models 3 and 4 have shown that there is little effect of the S(fr) thickness on the overall deformation of the tunnel. The effect of S(fr) is more evident in the maximum shear displacement values where shear displacement is almost reduced by half in the model where S(fr) was applied prior to bolting. This is mainly due to the fact that the tunnel is lying in a jointed biotite gneiss with rather low JRC values especially in the foliation joint set. The load on some of the rock bolts was reduced by about 15% to 50% when S(fr) was used prior to bolting. This is mainly due to the fact that the rock mass deforms rather little. The S(fr) thickness applied on the tunnel was 5 + 5 cm and 10 cm for models 3 and 4 respectively.

- The use of S(fr) in relatively weak rocks with unstable wedges on the tunnel crown reduces significantly the load attained by the rock bolts. The implications for the use of S(fr) as tunnel support in intensively jointed tunnel conditions are obvious.

- The modelled values of axial and shear forces in the shotcrete and forces along the rock-shotcrete interfaces as well as bolt capacity, can each be compared with Q-system designed assumptions, Hyundai's detailed deformation monitoring of tunnel sections and modelling of the tunnel by using ITASCA's FLAC code (Fast Langragian Analysis of Continua). These comparisons will be a subject of a future article.

8 REFERENCES

Barton, N., R. Lien, and J. Lunde 1974. Engineering classification of rock masses for the design of tunnel support. *Rock Mech.*, 6: 189-236.

Barton, N., F. Løset, R. Lien and J. Lunde 1980. Application of the Q-system in design decisions concerning dimensions and appropriate support for underground installations. *Int. Conf. on Subsurface Space, Rockstore, Stockholm, Sub-surface Space,* Vol. 2, pp. 553-561.

Barton, N., F. Løset, A. Smallwood, G. Vik, C.P. Rawlings, P. Chryssanthakis, H. Hansteen and T. Ireland 1992. Geotechnical core characterization for the UK radioactive waste repository design. *Proc. ISRM Symp. EUROCK*, Chester, UK.

Barton, N., T.L. By, P. Chryssanthakis, L. Tunbridge, J. Kristiansen, F. Løset, R.K. Bhasin, H. Westerdahl and G. Vik 1994. Predicted and measured performance of the 62m span Norwegian Olympic Ice Hockey Cavern at Gjøvik.*Int. J. Rock Mech. Min. Sci. & Geomech. Abstr.* Vol. 31, No. 6, pp. 617-641.

Barton, N. and S.C. Bandis 1990. *Review of predictive capabilities of JRC-JCS model in engineering practice. International Symposium on Rock Joints.* Loen 1990. Proceedings, pp. 603-610.

Bhasin, R., N. Barton and F. Løset 1993. Engineering geological investigations and the application of rock mass classification approach in the construction of Norway's underground Olympic stadium. *Eng. Geol.*, 35:93- 101.

Cundall, P.A. 1980. *A generalized distinct element program for modelling jointed rock.* Report PCAR-1-80, Contract DAJA37-79-C-0548, European Research Office, US Army. Peter Cundall Associates.

Grimstad, E. and N. Barton 1993. Updating of the Q-System for NMT. *Proc. Int. Symp. Modern use of wet mix sprayed concrete for underground support,* Fagernes 1993, Norway,: 46-66.

Grimstad, E. and N. Barton 1995. Rock mass classification and the use of NMT in India. *Proc. Conf. on design and construction of underground structures,* 23-25 February 1995, New Delhi, India.

Makurat, A., N. Barton, G. Vik., P. Chryssanthakis and K. Monsen 1990. Jointed rock mass modelling. *International Symposium on Rock Joints.* Loen 1990. Proceedings, pp. 647-656.

Numerical study on reinforcement effect of cable bolt in discontinuous rock mass

H. Kinashi, S. Amano, H. Tsuchihara & H. Yoshioka
Obayashi Corporation, Tokyo, Japan

K. Michihiro
Department of Civil Engineering, Setsunan University, Osaka, Japan

ABSTRACT: This paper will describe a numerical study on the application of cable bolt in a large-scale underground cavern to restrain displacement and deterioration of discontinuous rock mass. The cable bolt is fully bonded and non-pretensioned reinforcement method, which is expected to reduce the installation cost and the construction period in comparison with pretensioned rock anchor. Numerical experiments were carried out to evaluate the reinforcement effect in the stabilization of the large-scale underground cavern. The Distinct Element Method was used to compare the reinforcement effect between the cable bolt and the rock anchor. The result of the numerical experiments indicated that the reinforcement effect of the two methods were nearly equivalent, although the support mechanism of the methods are quite different.

1. INTRODUCTION

Recently, underground caverns such as underground hydraulic-power house or energy storage facility, tend to be large and to be constructed deeply. In these underground caverns, the important subject is to secure the mechanical stability and to reduce the construction cost.

One of the most important factors to affect the stability of the cavern is the presence of the joints in a rock mass. According to the recent result of field measurements conducted in an underground hydraulic-power house (Akagi et al. 1995), it is pointed out that joints have a great influence on the mechanical and hydraulic behavior of the rock mass. The properties of joint such as orientation, frequency, and roughness also are regarded as important parameters for the rock mass classification conducted in tunnels. Therefore, the research of discontinuous rock mass is one of the most important subjects in the recent rock engineering.

A private underground research laboratory was constructed by Obayashi Corporation in Kamioka Mine to study various issues related to the design and construction of underground caverns in discontinuous rock mass. The field experiments were executed with five-year plan starting from 1991 (Tamai et al. 1994). The authors have already reported the site investigation results concerning the joint system of the Kamioka experiment site (Amano et al. 1995a), and the numerical simulation result of a drift excavation considering the mechanical and geometrical properties of joints (Kinashi et al. 1995a). In the numerical simulation, a Distinct Element Method (Cundall 1980) was applied, and the rock mass was modeled with rock blocks divided by joints. The result of the numerical simulation indicated that the joint deformation and convergence observed in the field were satisfactorily simulated.

In this study, numerical experiments were carried out to estimate the stability of the large-scale underground cavern in discontinuous rock mass. In large-scale underground caverns such as underground hydraulic-power house, rock anchor has been commonly used as one of the main supports. Rock anchor is a rock reinforcement method which is installed with pretension. On the other hand, the authors have been studying the applicability of the cable bolt which is fully bonded and non-pretensioned rock reinforcement in order to reduce construction period and cost. In the following, results of numerical experiments will be described concerning the reinforcement effect of the cable bolt and the rock anchor.

2. SUPPORT MECHANISM OF CABLE BOLT AND ROCK ANCHOR

Basically, the same material is used in both the cable bolt and the rock anchor. The difference between both reinforcement methods exists in the presence of pretension and the bonding process. Figure 1 shows the concept of support mechanism of both reinforcement methods. In the case of the cable bolt, the grouting material and the cable strand are installed after the borehole is drilled in the rock mass (Figure 1-a). When the cavern is excavated and displacement occurs in the rock mass, the axial force arises in the cable bolt and the cable bolt exhibits the reinforcement effect. Since joints are generally more deformable than rock matrix, larger axial force on the cable bolt is found at the location of intersection with joints. On the other hand, in the case of rock anchor, the cable strand is installed after the borehole is drilled, and the end of the cable strand is bonded with grouting material. Then, the pretension force is applied on the cable strand. As a result, deformation of joints is restrained by this pretension force.

3. PROCEDURE OF NUMERICAL EXPERIMENTS

3.1 Discontinuous Analysis by Distinct Element Method

In order to consider the mechanical and geometrical properties of joints, UDEC (Universal Distinct Element Cord) was used in the numerical simulation. In UDEC, rock mass is represented as an assemblage of discrete blocks divided by joints. Joints are treated as boundary conditions between blocks. Hence, the mechanical behavior of both rock matrix and joints are considered. The interaction between the rock mass and the cable bolt through the grouting material can be also considered in UDEC so that the shear stress acting on the cable bolt arises in proportion to the relative displacement between the rock mass and the cable bolt. It enables us to simulate the reinforcement effect of the cable bolt in conformity with the actual field behavior. Since UDEC adopts the finite difference approach with successive approximation, axial force is calculated at every time-step from the relative displacement between the rock mass and the cable bolt. The algorithm generally used in finite element method would not be suitable to incorporate the concept of the relative displacement between the rock mass and the cable bolt.

3.2 Modeling of Joint Geometry

Modeling of joint geometry is based on the field investigation conducted in the Kamioka experiment site. Joint geometry which affects on the rock mass behavior is frequency, orientation and size. Procedure and assumptions which are employed to construct the model in this study are as follows.

1) Joints are represented as traces on the two dimensional plane. The geometry of traces is described by the location of the center point, the dip angle, and the trace length. Each center point of the joint is generated based on the mean spacing, which is assumed as 10m in this study.

2) Joint orientation is generated stochastically according to the results of in-situ joint survey at the Kamioka experiment site. Based on the observed distribution of the dip angle, each dip angle is generated using random numbers.

3) Each trace length is assumed as 20 m, which is roughly equal to the width of the cavern.

4) A discrete block model is constructed by managing end points and intersection points of the traces so that the joint trace map can be used in the numerical simulation. Figure 2 shows the discrete block model constructed according to the procedure mentioned

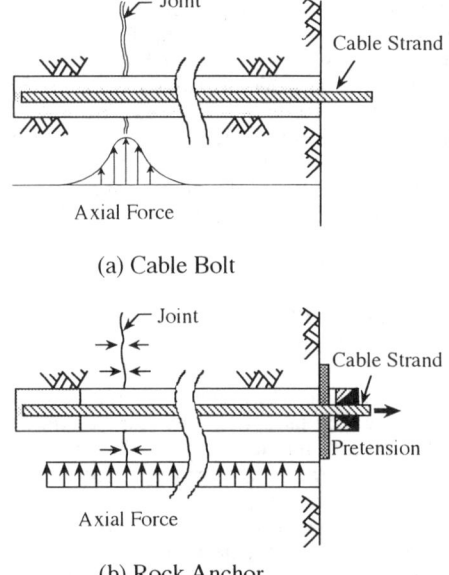

Figure 1. Concept of support mechanism

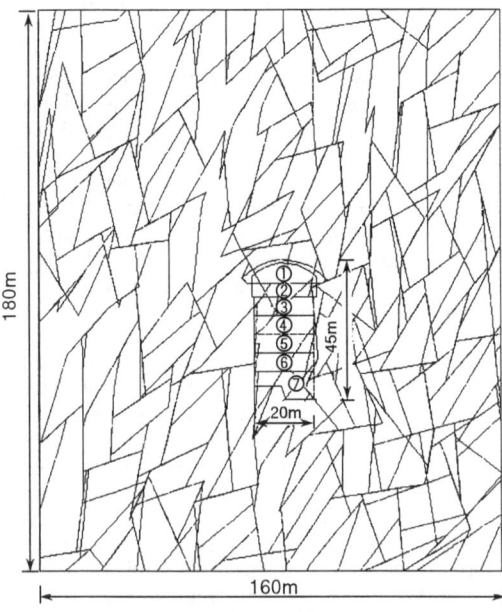

Figure 2. Discrete block model

above. Each block is subdivided into triangular finite difference elements and is treated as elastic material.

3.3 Modeling of Cable Bolt and Rock Anchor

To simulate the support mechanism as shown in Figure 1, both cable bolt and rock anchor are modeled as shown in Figure 3. Since the cable bolt is fully bonded, multiple nodes are set on the cable bolt. The spring which represents the axial stiffness of the cable bolt is also set between each node. The spring and the slider, which represent the stiffness and the strength of grouting material, respectively, are set so as to connect each node with the rock mass element. The interval length of each node is determined as 0.2 m based on the preliminary studies.

On the other hand, in the case of rock anchor, nodes are set only both ends of the cable strand. The pretension force is given between the two nodes. The pretension force of each rock anchor is supposed to be 90 tf. However, actually applied pretension force in the numerical simulation was 60tf on the upper 4 stages and 30 tf on the lower 7 stages, considering the installation pitch of the rock anchor in the depth direction, because the numerical simulation is performed under two dimensional plane strain condition.

3.4 Case and Step of Numerical Simulation

Three cases of numerical simulation were carried out in this study. Those are the cases of using the cable bolt, the rock anchor, and no reinforcement. In the three cases, shotcrete was considered, but rock bolt and final lining concrete were not considered.

The numerical simulation was executed at 8 steps, including the initial stress analysis and excavation of top heading and benches. Excavation steps are shown in Figure 2.

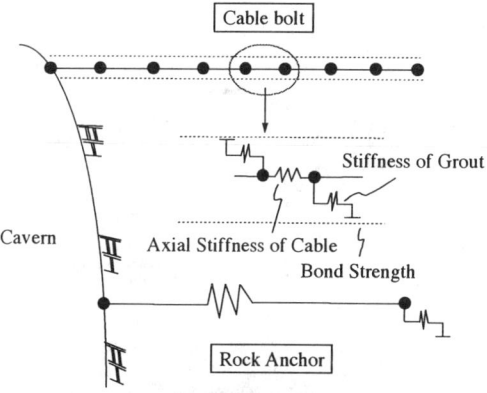

Figure 3. Modeling of cable bolt and rock anchor

3.5 Determination of Input Parameters

The elastic modulus and Poisson's ratio of the rock matrix were determined from the results of the borehole dilatometer test and unconfined compression test, respectively. The cohesion and friction angle of the rock matrix, which are used to estimate the local safety factor, were determined from the results of triaxial compression test. The properties of the rock matrix are summarized in Table 1.

Mechanical properties of joint were determined from direct shear test using jointed samples collected in the Kamioka experiment site. Input parameters used in Barton & Bandis joint model (Barton & Bandis 1990) were obtained as shown in Table 2.

To estimate the bond properties of the cable bolt, in-situ pullout tests were carried out (Tsuchihara et al. 1995), and input parameters of the cable element were determined as shown in Table 3.

Roof arch concrete and shotcrete were also considered in the numerical simulation. Table 4 indicates their input parameters.

The initial stresses in the rock mass were determined from the result of in-situ stress measurement by means of the conical-ended borehole technique conducted in the Kamioka experiment site (Sakaguchi 1994).

Table 1. Parameters of rock matrix

Density	2.600 kg/m³
Elastic modulus : E	10,000 MPa
Poisson's ratio : ν	0.25
Cohesion : c	3 MPa
Friction angle : ϕ	46 deg.

Table 2. Parameters of joint

Normal stiffness : K_n	10.2 MPa
Shear stiffness : Ks	2.2 MPa
Joint compressive strength : JCS	51.0 MPa
Joint roughness coefficient : JRC	9.4
Residual friction angle : ϕ_r	27 deg.

Table 3. Parameters of cable element

CABLE STRAND :	
Density	6,083 kg/m³
Young's modulus	190,000 MPa
Tensile strength	1,550 kN
Failure strain	0.035
Compressive strength	1,550 kN
GROUT :	
Shear stiffness	180 MN/m/m
Bond strength	600 kN/m

Table 4. Parameters of roof concrete and shotcrete

ROOF CONCRETE :	
Density	2,200 kg/m^3
Elastic modulus	19,600 MPa
Poisson's ratio	0.2
SHOTCRETE :	
Density	2,200 kg/m^3
Elastic modulus	17,000 MPa
Poisson's ratio	0.2

Table 5. Initial stress components

σ_{xx}	σ_{zz}	τ_{zx}
2.14 MPa	4.13 MPa	0.77 MPa

Table 5 shows the initial stress components used in the numerical simulation.

4. RESULTS OF NUMERICAL SIMULATION

The restrain of displacement and the improvement of the stability around underground caverns are expected as the result of the reinforcement effect of the cable bolt and the rock anchor. In the following, results of the three cases of numerical simulation at the final stage of excavation will be described.

4.1 Axial Force on Cable Bolt and Rock Anchor

Figure 4 (a) and (b) show the distribution of the axial force on the cable bolt and the rock anchor. As shown in Figure 4 (a), the axial force on the cable bolt has the peak value at the intersection points with joints, and the axial force arises through grouting material in proportion to the relative displacement between the rock mass and the cable bolt. This type of the axial force distribution is a characteristic feature of the cable bolt as fully bonded reinforcement.

As shown in figure 4 (b), the axial force on the rock anchor has uniform value along the cable strand, whose value is equal to the sum of the initial pretension force and additional load caused by the excavation.

4.2 Joint Displacement

Figure 5 shows the joint shear displacement comparing with the three cases. In Figure 5, thickness of the line along each joint indicates the magnitude of joint shear displacement. As indicated in Figure 5, the maximum value of the joint shear displacement is 28.2mm, 29.0mm, and 37.5mm, respectively. The maximum value of the joint opening is 2.5 mm, 2.2mm, and 34.4mm, respectively. The result of numerical simulation clearly indicates that the joint displacement is restrained by the cable bolt or the rock anchor. The reinforcement effect is remarkable in the region of the right side of the cavern, because the orientation of the joint is unstably inclined in the region of the right side of the cavern.

4.3 Rock Mass Displacement

Figure 6 shows the contour map of the horizontal displacement around the cavern. In the case of no reinforcement, displacement at the right side of the cavern is fairly larger than those of the cases of using the cable bolt or the rock anchor. On the other hand, the horizontal displacement at the left side of the cavern is nearly equal in three cases. As indicated in Figure 6, the maximum horizontal displacement is 41.4mm, 40.8mm, and 124.3mm, respectively. The reinforcement effect of the cable bolt can be said nearly equal to that of the rock anchor in terms of the restrain of the horizontal displacement.

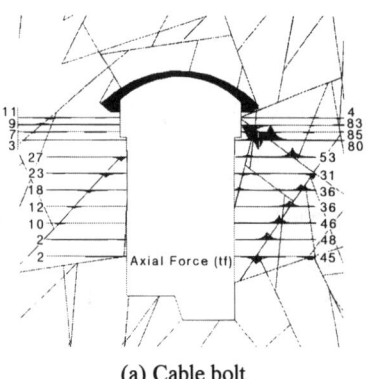
(a) Cable bolt

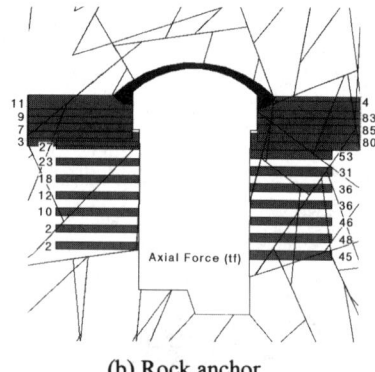
(b) Rock anchor

Figure 4. Axial force of cable bolt and rock anchor

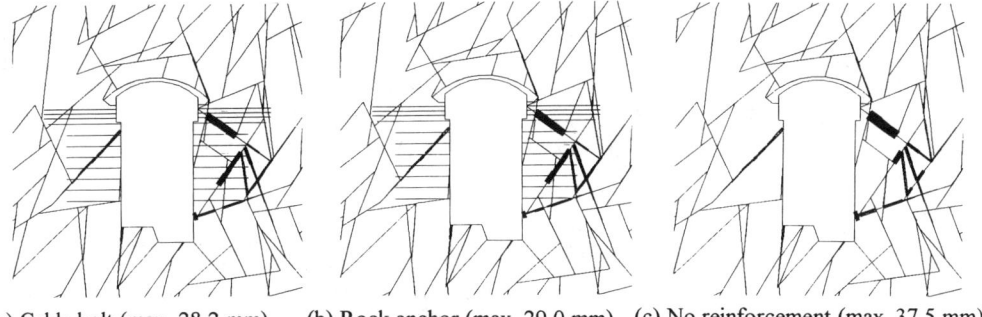

(a) Cable bolt (max. 28.2 mm) (b) Rock anchor (max. 29.0 mm) (c) No reinforcement (max. 37.5 mm)

Figure 5. Shear displacement of joint

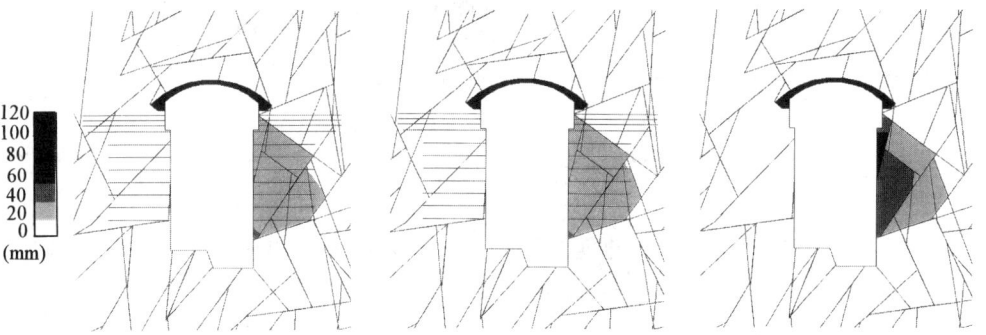

(a) Cable bolt (max. 41.4 mm) (b) Rock anchor (max. 40.8 mm) (c) No reinforcement (max. 124.3 mm)

Figure 6. Contour map of Horizontal displacement

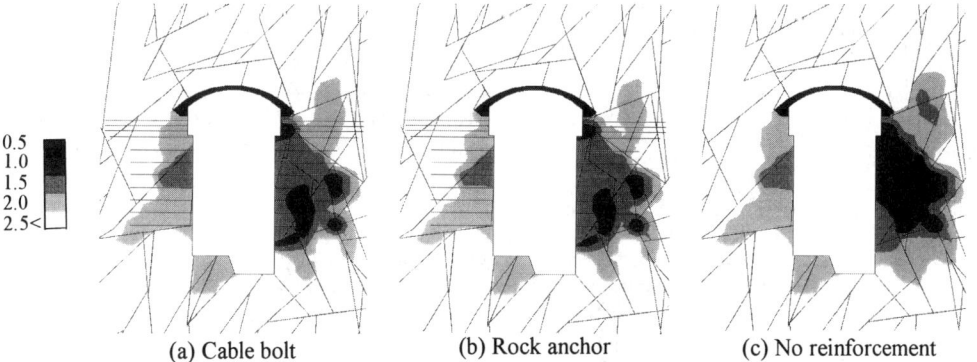

(a) Cable bolt (b) Rock anchor (c) No reinforcement

Figure 7. Contour map of local safety factor around cavern

4.4 Mechanical Stability of Cavern

Figure 7 shows the contour map of the local safety factor (L.S.F) in the rock mass around the cavern. L.S.F was estimated using Mohr-Coulomb's criterion. In the case of no reinforcement, the region where L.S.F is below 1.0 is fairly larger than that of the cases of using the cable bolt or the rock anchor. This result indicates that the stability of the cavern is improved by the cable bolt or the rock anchor. This can be attributed to the fact that the stress state around the cavern is improved from the uniaxial to the biaxial condition so that the confining pressure is added to the rock mass due to the cable bolt or the rock anchor.

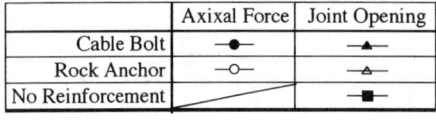
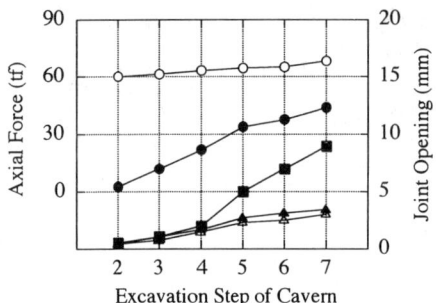

Figure 8. Progress of axial force and joint opening with excavation

4.5 Axial Force and Joint Behavior during Bench Excavation

Figure 8 shows the progress of the joint opening and the axial force at a location on the upper right of the cavern during the bench excavation. In the case of no reinforcement, joint opening increases with the bench excavation, and finally amounts to approximately 9mm. On the other hand, in the case of using the cable bolt, the joint opening at the final excavation stage is restrained to 3.5mm in response to the increased axial force during the bench excavation. In the case of using the rock anchor, the joint opening is 3.1mm, which is roughly equal to the case of using the cable bolt.

5. CONCLUSIONS

The result of numerical experiments indicates that the reinforcement effect of the cable bolt and the rock anchor is almost equivalent under the assumed condition in this study, although the support mechanism of both reinforcement methods are clearly different. In the case of the cable bolt, the axial force increases with the joint displacement. The displacement and the deterioration of the rock mass strength are restrained because the joint displacement is restrained by the axial force generated on the cable bolt. In the case of rock anchor, the joint displacement and the deterioration of the rock mass strength are restrained because the pretension force actively gives the confining pressure to the joint surface. Though the pretension is more effective in some instances, it is expected that the reinforcement effect of the cable bolt and the rock anchor is almost equivalent in accordance with the result of the numerical simulation.

The cable bolt is expected to be used as the reinforcement method for the large-scale underground caverns instead of the rock anchor. However, it will be still necessary to precisely estimate the bonding properties of the grouting material on the occasion of the field application of the cable bolt.

REFERENCES

Akagi T., H. Horii, T. Tanaka, T. Yamabe, Y. Uchida : Investigation, test and measurement of discontinuous rock mass, *Proc. of 26th. Sympo. of Rock Mech., Committee on Rock Mech., Japan Society of Civil Engineers*, pp. 574-583, 1995 (in Japanese)

Amano S. & H. Shirahata : Discrimination of conductive fractures based on the information from geological investigations, *Proc. of 26th. Sympo. of Rock Mech., Committee on Rock Mech., Japan Society of Civil Engineers*, pp. 126-130, 1995a (in Japanese)

Amano S., H. Kinashi & T. Tanaka : Numerical experiments on support effect of cable bolts in fractured rock mass, *Proc. of 5th. Tunnel Engineering, Japan Society of Civil Engineers*, pp. 155-160, 1995b (in Japanese)

Barton N. & S. Bandis : Review of predictive capabilities of JRC-JCS model in engineering practice, *Proc. of Int. Sympo. on Rock Joints*, pp.603-610,1990

Cundall P.A. : UDEC - A generalized distinct element program for modelling jointed rock, *Report PCAR-1-80, European Research Office, U.S Army*, Contact DAJA37-79-C-0548, 1980

Kinashi H., H. Shirahata, K. Nagahisa, T. Tamano & T. Tanaka: Estimation of joint characteristics based on joint survey and application of DEM analysis to the discontinuous rock mass, *Proc. of 26th. Sympo. of Rock Mech., Committee on Rock Mech., Japan Society of Civil Engineers*, pp. 126-130, 1995a (in Japanese)

Kinashi H., S. Amano & T. Tanaka : Stochastic modelling of rock joint system and numerical experiments on stability of underground openings, *Proc. of 26th. Sympo. of Rock Mech., Committee on Rock Mech., Japan Society of Civil Engineers*, pp. 16-20, 1995b (in Japanese)

Sakaguchi K., K. Sugawara, K. Nagahisa & T. Kaneda: Application of conical ended borehole technique to inhomogenious rock and consideration, *Proc. of 9th. Japan Sympo. on rock Mech.*, pp.229-234, 1994 (in Japanese)

Tamai A., T. Mikami & K. Akiyoshi : Study on three-dimensional mechanical behaviors by tunnel excavation in discontinuous rock mass, *Proc. of 9th. Japan Sympo. on rock Mech.*, pp.605-610, 1994 (in Japanese)

Tsuchihara H., T. Ninomiya, M. Inoue & K. Nagahisa : Application test of cable bolts to in-situ rock mass, *Proc. of 5th. Tunnel Engineering, Japan Society of Civil Engineers*, pp. 149-154, 1995 (in Japanese)

Reliability research of surrounding rock based on elasto-plastic analysis of stochastic field finite element method

Dongsheng Liu & Yingren Zheng
Department of Civil Engineering, Logistical Engineering University, Chongqing, People's Republic of China

ABSTRACT: In this paper, strength and deformation parameters of surrounding rock are simulated as two-dimensional stable normal stochastic fields. By use of discretization, each stochastic field can be divided into a group of stochastic variables on discreted elements. A stochastic field finite element method (SFFEM) for elasto-plastic silution of surrounding rock is proposed. Besed on a probabilistic yield criterion and the concept of target yield probability, some probabilistic plastic zones of surrounding rock corresponding to different levels of target yield probabilitiy can be obtained by SFFEM and the system reliability of surrounding rock can be properly appreciated and estimated.

1. INTRODUCTION

Stability appreciation of surrounding rock, to great extent, depends upon the correct realization of parameters of surrounding rock as well as the loads that is applied to it. From many years engineering practical experiance, it is deeply realized that the traditional constant simulation to parameters of surrounding rock may cause unreasonable appreciation to safety of underground engineering. It is proved by theories and practices that geomaterials always have inherent spatial variations and stochastic characteristics to some extent because of the influences of many objective nature factors. Both the traditional constant and single random variable simulation can not reflect the spatial variations and the stochastic characteristics of geomaterials as good as possible. So, a better approach is to simulate the rock parameters as stochastic fields which vary with the spatial coordinate. For simplification, both strength parameters c, φ and deformation parameters E, μ of surrounding rock are simulated as four two-dimensional stable normal stochastic solar fields respectively.

Considering the spatial variation and stochastic characteristics of geomaterials, a probabilistic yield criterion for geomaterials is proposed and a stochastic field finite element (SFFEM) is developed to calculate yield probabilities of elements and determine probabilistic plastic zone. Based on the results of SFFEM, the system reliability of surroumding rock can be approximately estimated.

2. DISCRETIZATION OF STOCHASTIC FIELD

In order to analyze the stability of surrounding rock by SFFEM, it is necessary to discrete the stochastic fields of rock parameters into corresponding stochastic variables on every finite element. There are many ways of discretization, of them the local average theory proposed by E. H. Vanmarcke of MIT is the most efficient method. This method is extensively used in prctical engineering for its lower requirement to origin data, faster converge speed and higher accuracy. To overcome the disadvantage of the traditional local average discretization that requires the discrete element to be rectangular and to expend the range of application of local average theory, a discretization method based on the isoparametric local average theory is developed for the purpose of dealing with the discrete element with irregular element shapes. This method can be used for both arbitrary guadrilateral element and rectangular element.

Let $S(x,y)$ to be a two-dimensional stable normal stochastic field with mean m and variance σ^2,

The local average of the sochastic field on a domain Ω can be difined as

$$S_\Omega = \frac{1}{A} \int_\Omega S(x,y) dx dy \qquad (1)$$

The local average mean on the domain is

$$E(S_\Omega) = E\left(\frac{1}{A} \int_\Omega S(x,y) dx dy\right) = m \qquad (2)$$

where A is area of the domain (or area of discrete element). The covariance between two arbitrary discrete elements is

$$Cov(S_\Omega, S_{\Omega'}) = \frac{\sigma^2}{AA'} \iint_{\Omega\Omega'} \rho(\Delta x, \Delta y) dx dy dx' dy' \qquad (3)$$

where $\rho(\Delta x, \Delta y)$ is the standard auto-correlation founction of $S(x,y)$.

It is not difficult to obtain Eq. (4) from Eq. (3) by use of isoparametric transformation.

$$Cov(S_\Omega, S_{\Omega'}) = \frac{\sigma^2}{AA'} \int_{-1}^{1}\int_{-1}^{1}\int_{-1}^{1}\int_{-1}^{1} \rho(r,s)$$

$$|J||J'| d\xi d\eta d\xi' d\eta' \qquad (4)$$

where $r = \sum_{i=1}^{4}(N_\Omega^i x_\Omega^i - N_{\Omega'}^i x_{\Omega'}^i)$
$s = \sum_{i=1}^{4}(N_\Omega^i y_\Omega^i - N_{\Omega'}^i y_{\Omega'}^i)$

$|J|$ $|J'|$ are Jacobian determinants of element Ω and Ω' respectioely.

Eq. (4) can be easily calculated by Gaussian integration formula.

Thus it can be seen that for a stable normal stochastic field, both the local average mean on a discrete element and the covariance of the means between two arbitrary discrete elements can be calculated by Eq. (2) and Eq. (4) once the mean, the variance and the correlation distance of the stochastic field are known.

3. PROCESS OF ELASTO-PLASTIC ANALYSIS OF SFFEM

The key problem of stochastic field finite element method (SFFEM) is to get the means and variances of displacement and stress field caused by that of rock parameters and loads.

Just as the traditional deterministic elasto-plastic analysis of FEM, 'initial stress' approach with load increments is used in elasto-plastic analysis of SF-FEM.

Because of the spatial variations and stochastic characteristics, deformation and strength parameters of surrounding rock are looked upon as four two-dimensional stable normal stochastic fields, i.e.

$$E(x,y), \ \mu(x,y), \ \varphi(x,y), \ c(x,y)$$

After discretization, each stochastic field will be divided into a group of random variables on corresponding discrete elements, that is E_e, μ_e, φ_e and c_e (where e is the number of discrete element). The four discrete random variables may be independent or correlated to each other to some extant and this kind of correlation will be taken into account in the calculation of element yield reliability. But the autocorrelation between the same kind of random variables on different discrete elements always exists and the degree of auto-correlation will depend on the correlation distance of the stochastic field as well as the distance between the two discrete elements.

Elasto-plastic analysis of SFFEM based on 'initial stress' computational process can be summarized as follows.

1. Apply load increment ΔP_i and determine elastic increment of stress $\{\Delta\sigma_i\}_j$ and strain $\{\Delta\varepsilon_i\}_j$ as well as covariance between components of elastic increment of stress $Cov(\Delta\sigma_i^k, \Delta\sigma_i^l)_j$, (where i is the number of load increment and j is the number of times of iteration).

2. Add $\{\Delta\sigma_i\}_j$ to stress existing to start of increment $\{\sigma_i\}_{j-1}$ to obtain current stress $\{\sigma_i\}_j$, so does $Cov(\Delta\sigma_i^k, \Delta\sigma_i^l)_j$ to obtain covariances of current stress $Cov(\sigma_i^k, \sigma_i^l)_j$, i.e.

$$\{\sigma_i\}_j = \{\sigma_i\}_{j-1} + \{\Delta\sigma_i\}_j \qquad (5)$$

$$Cov(\sigma_i^k, \sigma_i^l)_j = Cov(\sigma_i^k, \sigma_i^l)_{j-1} + Cov(\Delta\sigma_i^k, \Delta\sigma_i^l)_j \qquad (6)$$

3. Check whether $Prob(F \geq 0) < P_{f0}$, in which F is the yield criterion choosen and P_0 is the element yield probability given. If above staisfied only elastic strain charges occur and process is stoped, if not proceed to 4.

4. If $Prob(F \geq 0) \geq P_{f0}$, i.e. element is in yield, increment of stress $\{\Delta\sigma'_i\}_j$ and 'initial stress'

$\{\Delta\sigma_0\}_j$ should be

$$\{\Delta\sigma'_i\}_j = [D]_{ep}\{\Delta\varepsilon_i\}_j \quad (7)$$

$$\{\Delta\sigma_0\}_j = \{\Delta\sigma_i\}_j - \{\Delta\sigma'_i\}_j \quad (8)$$

corresponding covariance of $\{\Delta\sigma'_i\}_j$ can be calculated and noted as $Cov\{\Delta\sigma'^l_i, \Delta\sigma'^k_i\}_j$.

5. Store current stress and covariances as fellows

$$\{\sigma_i\}_j = \{\sigma_i\}_{j-1} + \{\Delta\sigma'_i\}_j \quad (9)$$

$$Cov\{\sigma'^k_i, \sigma'^l_i\}_j = Cov\{\sigma^k_i, \sigma^l_i\}_{j-1} + Cov(\Delta\sigma'^k_i, \Delta\sigma'^l_i)_j \quad (10)$$

6. Calculate nodel forces of element corresponding to the equilibrating body force. These are given for any element by

$$\{f_0\} = \int_\Omega [B]^T \{\Delta\sigma_0\} dA \quad (11)$$

Assemble $\{f_0\}$ according to nodel number of element to form total nodel forces $\{\Delta F_0\}$ corresponding to the equilibrating body force.

7. Repeat steps 1 to 6 under the action of $\{\Delta F_0\}$ until all element are converged to the disired precision.

8. Apply next increment of load $\{\Delta P_{i+1}\}$ and repeat steps 1 to 7 until all increments of load are applied.

It should be noted that because of variations of parameters and stresses, the traditional deterministic yield criterion shouldn't be used in elastio-plastic analysis of SFFEM. So, a probabilistic yield criterion based on reliability theory is developed and the concept of target yield probability and probabilistic plastic zone are introduced by the authors.

4. SYSTEM RELIABILITY OF SURROUNDING ROCK

In finite element analysis, surrounding rock can be looked as an overall system that is consisted of many elements. It is obvious that yield or failure of an element doesn't mean the failure of the whole surrounding rock system, but yield or failure of certain number of finite elements will lead the failure of the system. Based on this point of view, system reliability of surrounding rock should have a close relation to the yield probability of every element in the system. On the other hand, failure of surrounding rock always takes place in a partial field of the system, this implies that failure of whole system is always controled by only a number of elements in the system, not all of them. So, it is important to find out the partial field of surrounding rock and the elements located in the field before calculating the reliability of the whole system.

Suppose a partial field corresponding to a certain failure model of surrounding rock has been determined and yield probabilities of the elements in the field have been calculated be SFEME, the system failure probability of the partial field can be approximately calculated by the following formula

$$P_{fs} = \frac{\prod\limits^{N_\Omega-m} P_{fi}}{(P_{f0})^{N_\Omega-m}} \quad 100\% \quad (12)$$

in which P_{fi} is the failure (yield) probability of element i in the field, P_{f0} is the target yield probability of element, N_Ω is the number of elements included in the field and m is the number of element which are in yield.

From the formula above, it can be seen that the system failure probability of a failure model will be increased with the increase of yielded elements in the field. When all elements in the field are in yield, m equals to N_Ω and P_{fs} equals to 100%; when all elements in the field are not in yield, m equals to zero and P_{fs} is almost zero.

For a given underground engineering problem, there are several possible main failure models of surrounding rock, system failure probability for each failure model P_{fs} can be abtained by used of Eq. (12). As a result, the system failure probabitity for the whole surroumding rock is the maximum of P^i_{sf} and the system reliability P_{fs} is the minimum of P^i_{rs}, that is

$$P_{fs} = Max(P^1_{sf} P^2_{sf} \cdots\cdots P^n_{sf}) \quad (13)$$

$$or \ P_{rs} = Min(P^1_{sR} P^2_{sR} \cdots\cdots P^n_{sR}) \quad (14)$$

where n is the number of failure models.

5. NUMERICAL EXAMPLE

Figure 1 shows a SFFEM mesh of a quarter of a deeply located round tunnel, the in-situ stresses are

$\sigma_{0x} = \sigma_{0y} = 10$MPa and $\tau_{xy} = 0$. Deformation and strength parameters of surrounding rock are simulated as four stable normal stochastic fields and their statistical characteristics are listed in table 1. Probabilistic plastic zones with different levels of target yield probabilisties are obtained by SFFEM and shown in Figure 3 to Figure 7 and table 2 to table 4.

Table 1. Statistical Characteristics of Parameter Stochastic Fields

Parameter	E	μ	c	φ
Mean	2×10^4MPa	0.3	0.5MPa	35°
Cofficients of varition	0.3	0.3	0.3	0.3
Correlation distance	1m	1m	1m	1m

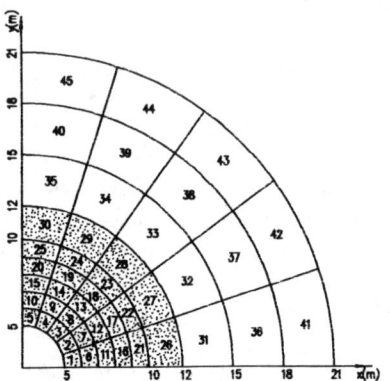

Figure 3. Probabilistic plastic zone given by SFFEM with target yield probability $P_{f0}=15.8\%(\beta=-1)$

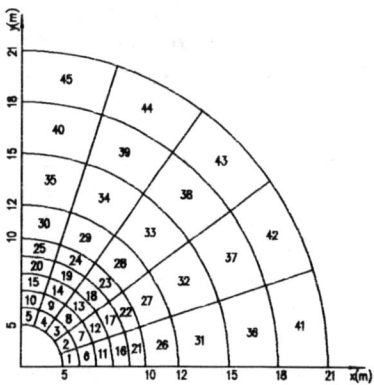

Figure 1. SFFEM mesh

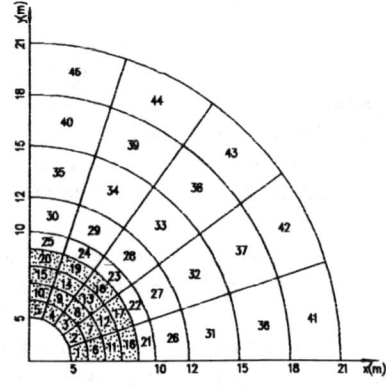

Figure 4. Probabilistic plastic zone given by SFFEM with target yield probability $P_{f0}=30.9\%(\beta=-0.5)$

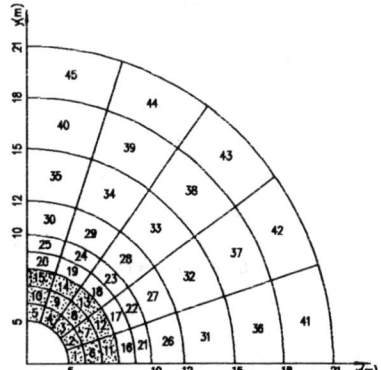

Figure 2. Deterministic plastic zone given by traditional FEM

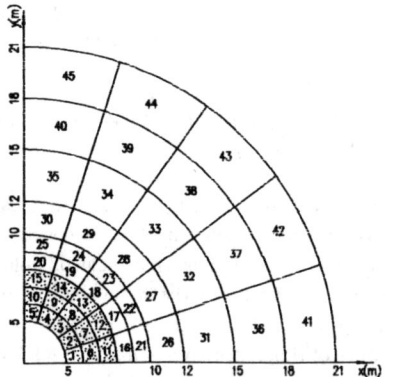

Figure 5. Probabilistic plastic zone given by SFFEM with target yield probability $P_{f0}=50\%(\beta=0)$

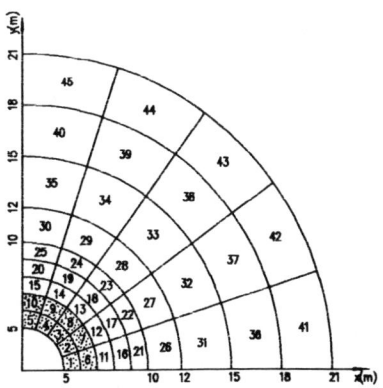

Figure 6. Probabilistic plastic zone given by SFFEM with target yield probability $P_{f0}=69.1\%(\beta=0.5)$

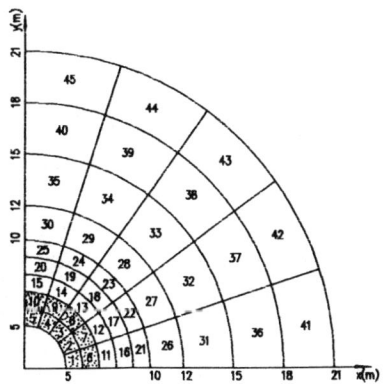

Figure 7. Probabilistic plastic zone given by SFFEM with target yield probability $P_{f0}=84.2\%(\beta=1)$

Table 2. System Failure Probabilites ($P_{f0}=15.8\%$)

Number of element	Yield probability of element (%)	Failure model	System failure probability (%)
1~5	15.8	1~5	100
6~10	15.8	1~10	100
11~15	15.8	1~15	100
16~20	15.8	1~20	100
21~25	15.8	1~25	100
26~30	15.8	1~30	100
31~35	12.1	1~35	76.58
36~40	7.2	1~40	35.39
41~45	2.1	1~45	4.77

Table 3. System Failure Probabilites ($P_{f0}=5.0\%$)

Number of element	Yield probability of element (%)	Failure model	System failure probability (%)
1~5	50.0	1~5	100
6~10	50.0	1~10	100
11~15	50.0	1~15	100
16~20	41.3	1~20	82.60
21~25	22.8	1~25	37.67
26~30	9.2	1~30	6.93
31~35	0.1	1~35	0.01
36~40	0.0	1~40	0.00
41~45	0.0	1~45	0.00

Table 4. System Failure Probabilites ($P_{f0}=84.2\%$)

Number of element	Yield probability of element (%)	Failure model	System failure probability (%)
1~5	82.4	1~5	100
6~10	82.4	1~10	100
11~15	75.9	1~15	90.14
16~20	60.4	1~20	64.66
21~25	31.6	1~25	24.66
26~30	12.1	1~30	3.48
31~35	0.0	1~35	0.00
36~40	0.0	1~40	0.00
41~45	0.0	1~45	0.00

6. CONCLUSION

Because of the stochastic characteristics and spatial variations, parameters of surrounding rock should be simnulated as stochastic fields rather than constants or single random variables, so, the traditional deterministic yield criterion should be replaced by the probabilistic yield critorion based on the yield reliability of element. As a result, the plastic zone of surrounding rock should not be a certain one but a kind of probabilistic plastic zone contralled by the means and the coifficients of variation of the parameters as well as the target yield probability given.

By comparison the probabilistic plastic zones obtained by SFFEM, if the parameters are simulated as stable normal stochastic fields, the traditional deterministic plastic zone based on the means of these fields by FEM and the probabilistic plastic zone with target yield probability $P_0 = 50\%$ by SFFEM have the exactly same result, this implies that under this situation the daterministic plastic zone is a special case of the probabilistic plastic zones and SFFEM is a more general numerical method in which the traditional FEM is involved.

System reliability of surrounding rock is a comprehensive index to judge the safty of an underground engineering, it depends not only on the yield probabitlities of some elements and target yield probability choosen, but also on the failure model of surrounding rock.

Target yield probability P_{f0} is also an important index in determining the range of plastic zone by SFFEM and it should be determined according to practical situation of an engineering. It is obvious that the range of plastic zone and the system reliability for the same failure model of surroundign will be increased with the decrease of target yield probability P_{f0}. For an important engineering, a lower target yield probability will be choosen and as a result, a lower system reliability will be obtained.

REFERENCE

Isaac Eltshakoff, Probabilistic Method in the Theroy of Structure, John Wiley & Sons, USA.

Vanmarcke, E. H, Probabilistic modeling of soil profiles, J. Geotech. Engig, ASCE, 103, 11(1977), 1035—1053

Vanmarcke, E., H, Random field: New concepts and engineering applications, Structural Safety Studies (Ed. James, Yao. T. P) ASCE, New York (1985), 7—17.

Vanmarcke, E. H. etal, Random field and stochastic finite elements, Journal of Sturctual Safety, 3(1986), pp. 143—166.

Hasofer, A. M. Simulation of random field, Probabilistic Methods in Geotechnical Engineering, (Li & Lo eds), 1993, 45—61.

Beacher, G. B. and Ingra, T. S. Stochastic FEM in settlement prodictions, J. Geotech. Engrg, Div, 107 (4), (1981), 449—463.

W. Q. Zhu, Y. J. Ren and W. Q. Wu, Stochastic FEM based on local averages of random vector field, J, Engrg, Mech, Vol. n8, No, 3, 3(1992).

D. S. Liu and Y. R. Zheng, A probabilistic yield criterion of geomaterials, Proc, Int. Conf. on Rock Mechanics and Environmental Geotechnology, Chongqing, P. R. China, April 1997.

D. S. Liu, Reliability Research of Surrounding Rock Based on Elasto-plastic Analysis of Stochastic Field Finite Element Method, Ph. D thesis, Chongqing Jianzhu University, 1996.

Numerical simulation and experimental study of excavation and anchoring support of underground chambers in jointed rockmass

Weishen Zhu & Weizhong Chen
Institute of Rock and Soil Mechanics, The Chinese Academy of Sciences, Wuhan, People's Republic of China

Baolin Wang
Mining Research Laboratories, CANMET, Ottawa, Ont., Canada

ABSTRACT: Based upon characters of a jointed rockmass, this paper simulates the property of large deformation of a large-size cavern excavated in a jointed rockmass and bolting-supported afterwards by Block-Spring method. Then the paper describes the simulation experimental study for the deformability of jointed surrounding rockmass under excavation and anchoring effect, which reveals the varying law of the mechanical behavior of stepwise excavated and bolting supported rockmass surrounding large underground caverns under different initial geostresses. The testing results and the numerical simulating results show basic coincidence in the top failure zone and the motional law of the jointed rockmass during excavation, thus to provide a reliable basis for rational guiding engineering practice.

1 INTRODUCTION

Having experienced a very long time geological structural motion, a rockmass becomes geological medium with certain structures, mostly, the rockmass is cut by such continuities as bedding planes and joints. In the case of water conservancy and mining engineerings, many a large underground cavern is built in jointed rockmasses, so, more often than not, it is unsuitable to simply use a method based upon continuum mechanics for the numerical analysis of the caverns' stability but the emphasis should be stressed on the discontinuous character of the medium.

Large underground engineerings often use pre-stressed bolts or bonding bolts as a means for supporting. Nevertheless, man has not yet had enough knowledge of motion and failure law and anchoring effect of jointed rockmasses under excavation to establish a valid supporting scheme, which may often result in unrealistic engineering design and construction.

Numerical simulation technique of the mechanical behavior of jointed rockmass can be traced back to 1960'. The earlier numerical analysis most introduced the finite element method and made special treatment to joints (Zienkiwicz et al. (1969); Goodman (1968); Best (1970); Ghaboussi et al. (1973); Heuze and Barbour (1982). Kawai (1977), Plesha (1983) developed rigid blocks to simulate discontinuous medium. Lorig (1984) developed the coupled program of discrete and boundary elements, Cundall (1985) successfully developed a discrete element program of UDEC, which can be used to analyse both deformability and failure criterion of rock blocks simultaneously.

This paper applies the Block-spring model (Baolin W· and V. K. Garga; V. K. Garga and Baolin W· 1993) to simulate the deformability and the failure law of underground caverns' surrounding rockmass as well as the effect of supporting parameters of bolts on anchoring effectiveness with different initial geostress field and at different excavation stages. This model is obviously characterized by (1) simulating large deformation and (2) part by part simulation of reinforcing action of bolts. This paper also compares the computing results and the large-sized simulation testing results of the excavation and bolt-supporting of underground caverns in jointed rockmass. Finally, the paper summarizes the corresponding relationship of motional law of surrounding rockmass with geostress field and prestress of bolts.

2 NUMERICAL ANALYSING FOR BLOCK-SPRING MODEL

This paper only gives a brief introduction to the basic principle of Block-spring model, the readers refer to literature (Baolin W. and V. K. Garga 1993; V. K. Garga and Baolin W. 1993) for details.

2.1 Basic formulation

2.1.1 Physical equations

Assume that the center of block i. has two translations of U_i and V_i and a rotation of θ_i the translations are positive if they are coincident with coordinate direction and the rotation θ_i is positive if the block rotates counterclockwise, then the displacement of any point on the block can be determined by the displacements of the centroid of the block as follows:

$$\{U\}_i = [B]_i \cdot \{U\}_i \quad (1)$$

where $\{U\}_i = \{U_i, V_i\}^T$ (U_i and V_i are the horizontal and vertical components of the displacement respectively, shown in Fig. 1.

$$\{U\}_i = \{U_i, V_i, \theta_i\}^T$$

$$[B]_i = \begin{bmatrix} 1 & 0 & -y'_i \\ 0 & 1 & -x'_i \end{bmatrix} \quad (2)$$

Where $x_i = x - x_{ci}$; $y_i = y - y_{ci}$ (x, y are the coordinates of the contact point; x_{ci}, y_{ci} are the coordinate of the centroid of block i.

Assume that blocks i and j contact at point Q, the relative displacement of the contact point Δu can be expressed as:

$$\{\Delta u\}_{ij} = [D]_{ij} \cdot \{U\}_{ij} \quad (3)$$

where $\{\Delta U\}_{ij} = \{\Delta U_{ij}, \Delta V_{ij}\}^T$ is the relative displacement of the two contact points; $\Delta u_{ij} = u_j - u_i$; $\Delta V_{ij} = V_j - V_i$; $\{\Delta u\}_{ij} = \{U_i\ V_i\ \theta_i\ U_j\ V_j\ \theta_j\}^T$.

$$[D]_{ij} = [-B_i | B_j] \quad (4)$$

The relative tangential and normal displacements of the contact point is:

$$\{\delta\}_{ij} = [T]_i \{\Delta U\}_{ij} \quad (5)$$

where $\{\delta\}_{ij} = \{\delta\}_j - \{\delta\}_i$; $\{\delta_k\} = \{\delta_{nk}, \delta_{sk}\}^T$ ($k=i,j$) and δ_{nk} ($k=i,j$)

are the tangential and normal components of the point; $[T]_i$ is the matrix of the rotating angle.

$$[T]_i = \begin{bmatrix} cos\alpha_i & sin\alpha_i \\ -sin\alpha_i & cos\alpha_i \end{bmatrix} \quad (6)$$

where α_i is the angle of the outward normal N of the contact surface on block i from coordinate axis of x.

Assume that the normal and the tangential forces between two blocks, F_n and F_s, are properly proportional to the corresponding relative normal and the tangential displacements respectively, then,

$$\{F_n\}_{ij} = [K]_{ij} \cdot \{\delta\}_{ij} \quad (7)$$

where

$$\{K\}_{ij} = \begin{bmatrix} K_n & 0 \\ 0 & K_s \end{bmatrix} \quad (8)$$

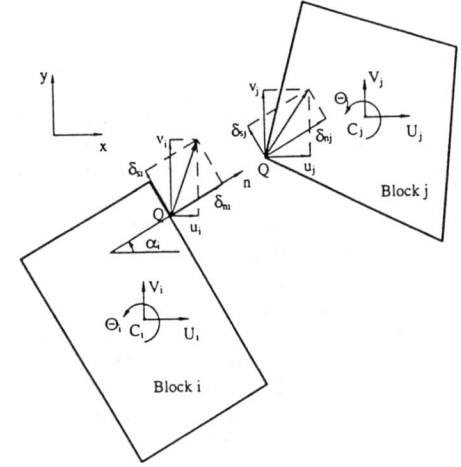

Fig. 1 Relative displacements between two blocks of C_i and C_j

The contact stiffnesses of K_n and K_s are determined according to the stiffnesses of the rock block, K'_n, K'_s, and those of the joint, K''_n and K''_s:

$$K_n = 1/(\frac{1}{K'_n} + \frac{1}{K''_n}), \quad K_s = 1/(\frac{1}{K'_s} + \frac{1}{K''_s}), \quad (9)$$

From eqs · (3),(5) and (7), we have

$$\{F_n\} = [K]_{ij} \cdot [T]_i \cdot [D]_{ij} \cdot \{U\}_{ij} \quad (10)$$

2.1.2 Equilibrium equation

Assume that the block is in a state of static equilibrium under the co-action of the gravity force, external force, boundary reaction force and contact force, the equilibrium state should satisfy the following:

$$\Sigma\{X\} = 0 \quad (11)$$

where $\{X\} = \{X, Y, M\}^T$ (X, Y are respectively the horizontal and the vertical components of the force; M is the moment of that force about the centroid of block).

2.2 Boundary Condition of Displacement

The known displacement boundary can be applied by one-dimention bar element. Providing the angle between the known displacement, w_0, and the axis of x is β(Fig2(a)), it can be assumed that the directions of the element's outer end (B'), and BB' can be used to describe the displacement boundary, w_0. If the stiffness of the bar element is specified as a large number (10^{20}, for example), and the displacement of point B is determined from eq. (1), the reaction force of point B is (Fig. 2(b)):

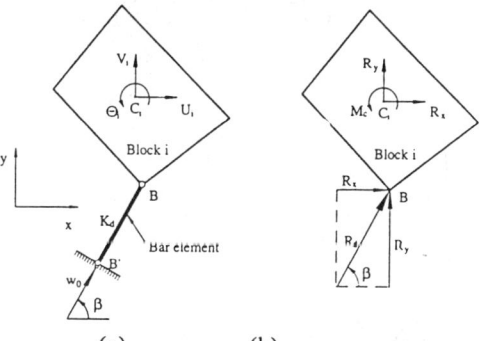

Fig. 2 Simplification of displacement boundany
(a) bar element (b) boundary reaction force

$$R_d = K_d W_0 - K_d \{C\}_d^T [B]_i \{U\}_i \quad (12)$$

where K_d is the stiffness of the bar element;

$$\{C\}_d = \{\cos\beta \quad \sin\beta\}^T$$

substitution (10) and (12) into (11), we have:

$$[KK] \cdot \{U\} = \{P\} \quad (13)$$

where $[KK] = [K][T][D]$ is a $3N \times 3N$ stiffness matrix, $\{U\}$ is a column vector of unknown displacement components: U, V, θ; $\{P\}$ is a column vector of $3N$ known prescribed boundary conditions and the gravity force.

2.3 Simulation of bolts

An end anchored bolt can be simulated by one-dimention spring element (Fig. 3(a)) whereas a fully-bonded bolt can be simulated by several one-dimention spring elements (Fig. 3(b)), depending upon the situation of the penetrated rock block.

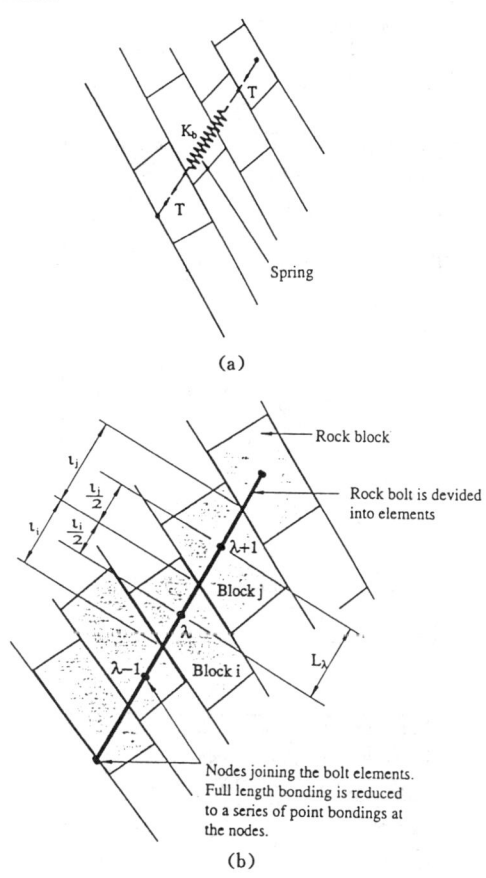

Fig. 3 Simulation of bolts
(a) end-anchored bolt (b) fully-bonded bolt

3 COMPUTATIONAL CASES

The computation model measures (140×180(cm $\times$ cm) as shown in Fig. 4, having two orthogonal joint sets (dipping angle of 45°). The simulated target is a three-span high sidewall opening with sizes of 32cm $\times$ 16cm. The excavation is carried out in three stages. After excavation, eleven bolts are installed along the openings periphery, each having a certain prestress. Different combinations of horizontal and vertical pressures are

Table 1 regimes for computation

regime	confining pressure (MPa)		bolt-supporting and prestress
1	$\sigma_h=0.5$	$\sigma_v=0.5$	no
2	$\sigma_h=0.5$	$\sigma_v=1.0$	no
3	$\sigma_h=0.5$	$\sigma_v=0.5$	no
4	$\sigma_h=0.5$	$\sigma_v=0.5$	$\sigma_{ps}=0.01$
5	$\sigma_h=0.5$	$\sigma_v=0.5$	$\sigma_{ps}=0$
6	$\sigma_h=0.5$	$\sigma_v=0.1$	no
7	$\sigma_h=0.5$	$\sigma_v=0.1$	$\sigma_{ps}=0.01$
8	$\sigma_h=0.5$	$\sigma_v=0.1$	$\sigma_{ps}=0$

selected for simulating various initial geostresses. the computation has eight rigimes which are listed in Table 1.

The computing results show the followings. When the initial geostress ratio $\sigma_h/\sigma_v=1$, inward displacements of the surrounding rockmass will result with the excavation. When $\sigma_h/\sigma_v<1$, inward displacements will result at the opening top whereas outward displacement will result along the sidewall. When $\sigma_h/\sigma_v>1$, with the excava-

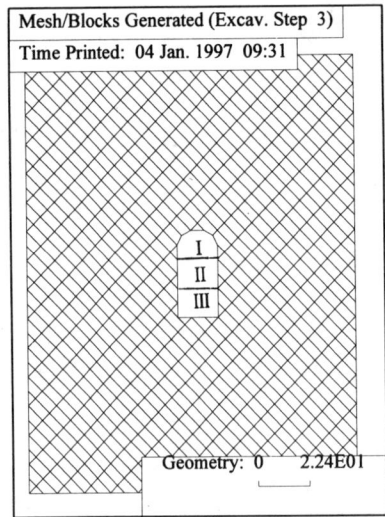

Fig. 4 Computation model

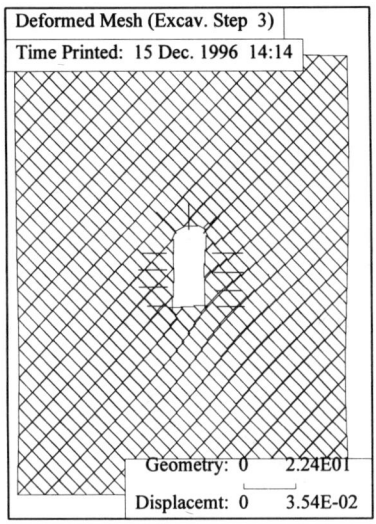

Fig. 6 Diagram of opening deformation after excavation step Ⅲ (regime 7)

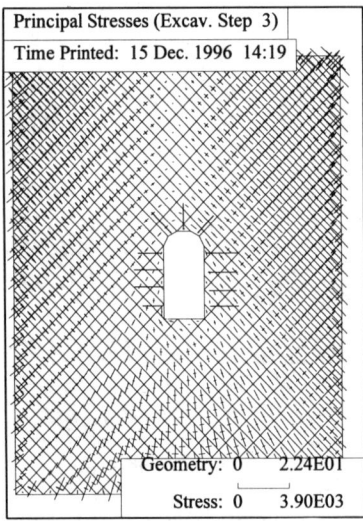

Fig. 5 Diagram of principal stress vectors after excavation step Ⅲ (regime 7)

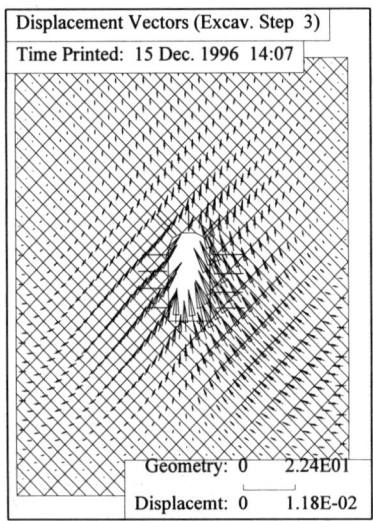

Fig. 7 Diagram of displacement vectors of surrounding rockmass after excavation step Ⅲ (regime 7)

Table 2 Physico-mechnical indexes of material

inner friction angle of rock(°)	density (Kg/m³)	compression strength (MPa)	Young's modulus	Poisson's ratio	K_n (MPa/m)	K_s (MPa/m)	friction angle of joint(°)	bonding stress (MPa)
47°	1.45×10^3	7.9	1.58×10^3	0.15	2.3×10^4	$1.1 \sim 2.9 \times 10^3$	42.8	0.002

tion, the vault of the opening will deform upward owing to the acting of crushing and squeezing; whereas the sidewall will deform inward with an upward displacement component, large displacements take place at the middle of the sidewall.

The physico-mechanical indexes are listed in Table 2.

Bolt-supporting results in a reinforced area around the excavating face of the opening surrounding rockmass, which improves the global stability of the surrounding rockmass. The overall displacement of the reinforced rockmass is reduced but the displacement around the bolts ends is increased. When $\sigma_h/\sigma_v < 1$, the bolts in the vault are generally in a tensile state and the tension will increase monotonouly with the excavation; whereas the bolts in the sidewall are in a state of compression and the bolts in the lower sidewall are heavily compressed. When $\sigma_h/\sigma_v > 1$, with the excavation, the bolts in the vault bear monotonously reduced tension and force-bearing state is locally turned from tension into compression, the bolts in the sidewall bear increasing tension with the excavation.

Shown in Fig 5 to 7 are the computed results for regime 7.

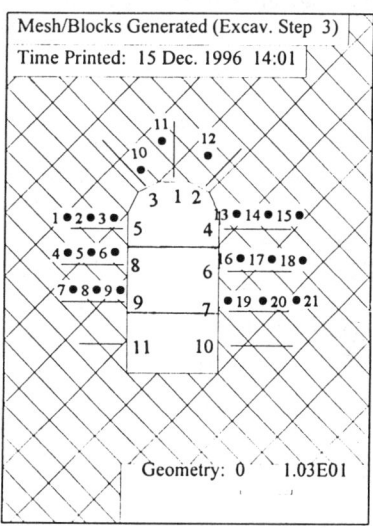

Fig. 8 measuring points and bolts arrangement

Layout of measuring points and bolts arrangement are shown in Fig. 8.

4 SIMULATING TESTS

The Institute of Rock and Soil Mechanics, Academia Sinica and Technical Research Institute, AOKI Corporation have in cooperation conducted large sized lab simulating tests under the loading condition of planar strain to study the stability of the surrounding rockmass excavated for openings. The size and the material mechanics indexes for the simulating test are the same as those for computation. The bolts are made of brass with the diameter D=3mm, prestress exerted on the reinforced area is $\sigma_{ps} = 0.01$MPa and the regimes in testing are the same as those in computing.

The observation of displacements of the surrounding rockmass are made using 21 dial meters (0.01mm) installed on the model surface (Fig. 8). The bolt stress is measured by micro strain gages and finally obtain through conversion. In testing, CCD technique is also used.

5 TESTING AND COMPUTING RESULTS

The rockmass deformability, failure law and reinforcing effect of bolts obtained from simulating tests are similar to the numerical analysing re-

Table 3 Displacement at measuring points after excavating step Ⅲ and supporting (regime 7)

No.	measured displacements (mm)		calculated displacements (mm)	
	U	V	U	V
3	0.26	0.04	0.034	0.044
6	0.31	0.06	0.033	0.040
9	0.23	0.02	0.037	0.031
11	0.17	0.06	0.064	0.034
14	0.11	0.06	0.087	0.057
15	0.07	0.06	0.037	0.037
18	0.08	0.07	0.02	0.04
19	0.09	0.05	0.034	0.074
20	0.07	0.01	0.025	0.055
21	0.07	0.05	0.018	0.046

sults. The state of bearing-force of bolts and the vertical displacement at the measuring points are relatively close between measured and calculated results, but the horizontal displacemants are quite different from each other. Table 3 gives the comparison of the measured and the calculated results of measuring point displacement.

Table 4 comparison of bolt-bearing axial forces at different excavation steps (regime 7).

Table 4 Axial forces of bolts for different excavation steps (regime 7)

No.	measured stress (MPa)			calculated stress (MPa)		
	I	II	III	I	II	III
2	1.13	−1.21	−1.43	0.53	−1.28	−1.95
4	5.99	6.16	9.90	6.10	6.87	7.60
9	/	7.30	11.94	/	6.51	10.37
10	/	/	5.83	/	/	6.52

6 CONCLUSION

During excavation, the force-bearing state and the motional law of the surrounding rockmass of the opening are constrained by the initial geostress field. By reinforcing the rockmass with bolts, the formation of top collapse can be avoided and the prestressed bolt can further improve the stability of the rockmass.

The numerical computing model of Block spring shows to a certain extent advantages in analysing the stability of jointed rockmass. Both numerical analysing and the simulation testing results provide geotechnical engineerings with referential but important conclusions.

REFERENCE

Baolin Wang, Vinod K. Gargo. A Numerical method for modelling large displacements of jointed rocks (I). *Can. Geotech. J.* 30. 96~108, 1993.

Vinod. K. Garga, Baolin Wang. A Numerical method for modelling large displacements of jointed rocks (II). Modelling of rock bolt and ground water and application. *Can. Geotech. J.* 30. 109~123. 1993.

Zhu W. et al. Finite element analysis of jointed rockmass and engineering application, *Int. J. Rock Mech. Min. Sci Geomech. Abstr.*, Vol. 30, No. 5, 537−544. 1993.

Nagai T et al. Behavoir of jointed rockmass around an underground opening under excavation using large scale physical model tests. *Proc. of 27th Symposium on Rock Mechanics of Japan*, 1996.

Zhu W., Ren W. and Zhang Y.. Research on stability and anchoring effect of opening excavation in jointed rocks by model testing. *Proceedings of the Korea Japan Joint Symposium on Rock Engineering. Seoul, korea*, 1996.

A finite element method of probabilistic keyblock analysis

D.S.Young
Department of Mining Engineering, Michigan Technological University, Mich., USA

ABSTRACT: A finite element method was introduced for probabilistic keyblock analysis, which is comparable to that of deformation analysis in continuum mechanics. In this method the fundamental numerical algorithm was developed based on the connectivity matrix, which characterizes the rock mass discontinuity and corresponds to the stiffness matrix. The connectivity matrix completely defines the connection/disconnection conditions between two nodal points and the number of connections at a nodal point. By defining the keyblock failure at a node, whenever the node was included within a keyblock, the keyblock analysis can be easily done on the global connectivity matrix in probabilistic terms. Various important engineering criteria can be obtained from the probabilistic nodal point failure such as the frequency of keyblock formations, their sizes, and locations in full probabilistic terms. Its superiority over the original deterministic keyblock theorem was demonstrated through a case study on a subway tunnel.

1 INTRODUCTION

Rock joints play an important role in rock mechanics (particularly for structural stability analysis) and geohydrology (particularly for fluid-flow in fractured rocks). Site characterization requires the characterization of both the rock mass and the joint systems within the rock mass.

Considering many characteristics of rock joints that are best described in probabilistic terms, there are intrinsic advantages in geotechnical approaches that directly employ the relevant statistical distributions (Warburton 1980).

Consequently, an appropriate model of joint systems in a rock mass is the localized probabilistic model, and a realistic geotechnical analysis of rock structures is a probabilistic approach made on the model, which will yield local structural stability in terms of the probability of failure.

In this paper, a numerical method was developed to identify blocks (i.e. blocks formed by the joints in a rock mass) and calculate their sizes (or volumes), shapes, and locations, as well as their stability. The connectivity matrix was introduced in this numerical approach, which is equivalent to the stiffness matrix of the finite element method of stress analysis. Then, keyblock analysis was extended for probabilistic structural analysis based on the connectivity matrix. Finally, the key question, the localized probabilistic structural analysis was achieved by applying the finite element approach for the keyblock analysis onto the discrete cell-block model of the joint system.

2 GEOSTATISTICAL JOINT MODELS

Rock joint systems surveyed in the field and characterized statistically are often incorporated in the geotechnical analysis through the joint model. Because of the complexity of joint geometries and their characteristic nature, as well as limited accessibility in the field for joint surveys, various degrees of simplification or assumptions are made in joint modeling. Thus, corresponding joint models are developed depending on the field geology, modeling purposes and the model's end usage.

Recently geostatistics has been applied to joint system modeling, joint network simulations (Chiles 1988) and discrete block models of the equivalent continuum media (Young 1987a, 1987b).

2.1 Discrete cell-block model

In most engineering analyses in rock mechanics and geohydrology, the network geometry of joint systems can be replaced with the discrete cell-block model, because the joint parameters can be used directly as input data or they can be effectively converted into the equivalent continuum media that represents the joint systems. In this discrete model, the entire area (or rock mass) to be modeled is divided into uniform cell-blocks and the characteristic parameters are inferred for each cell-block from the sparse sample data measured in the field (Young 1987a, 1987b).

2.2 Local stochastic model

Considering the dispersion of joint parameters over their means and the complexity of the characteristic nature of joints, the mean value does not carry much meaning nor is it close to the reality, and neither is the deterministic engineering analysis. This difficulty was corrected in the stochastic model, which provides the full statistical distribution of joint parameters for each local cell-block. Thus, probabilistic engineering is applicable to the model at the early stages of site exploration and engineering design.

The local probability distribution was estimated by indicator kriging (IK), more precisely Mononodal IK, which is a non-parametric approach (Lemmer 1984). The original IK approach was rederived for vectorial variables (pole vectors in this case) and indicator variables were defined on the two-dimensional area of class intervals, which were projected on Grossman's tangent plane for pole histograms. The accuracy of this stochastic model was then cross-validated by comparing the model with the actual field data in an open pit case (Young 1987b, Young and Hoerger 1988a).

3 FINITE ELEMENT APPROACH FOR KEYBLOCK FAILURE

The traditional keyblock theorem for stability analysis on the structures excavated in a jointed rock mass is a deterministic method based on deterministic infinite joint planes. This means that the location and frequency of joints and size of joint planes are excluded from the block failure analysis and it provides a worse case analysis.

Consequently, a numerical approach was developed, which is general for both joint system models and any structure (their size and shape). Also, it can be combined easily with the local stochastic model of joint systems to achieve the localized probabilistic analysis of block failures.

The numerical algorithm was developed based on the connectivity matrix, which is comparable to the stiffness matrix of the finite element method of stress analysis in continuum mechanics.

3.1 Connectivity Matrix Approach

The local area, where the joint systems were simulated and the keyblock analysis desired, was replaced with the discrete finite element model as used in the finite element method of engineering mechanics. However, the elements were constructed by two-force bars as in truss structures rather than by solid elements. Then, the local area can be represented as a large truss structure with bars connected at nodal points, whose continuity and immobility were secured.

When the rock mass in the local area is cut by joints, some elements will be cut, as well as the connection bars within those elements. Also, many independent small truss structures will be formed when the rock mass is cut into many rock blocks by joints; that is, the whole truss structure is cut into many parts corresponding to those rock blocks. The connectivity matrix was introduced to define the connecting condition of bars (or their continuity conditions), and the whole truss structure was represented by the global connectivity matrix. Then, the independent small size structure representing a block formed by joints can be searched and identified as an independent block matrix system in the global connectivity matrix. Each of the independent matrix elemental blocks in the whole system matrix has its own size, shape, and location. So, the complete information for a block geometry is known and available from the nodal numbers of the matrix elemental block.

3.2 Elemental and Global Connectivity Matrices

The element constructing the whole system of the rock mass is replaced with a truss formed by simple bars connected between two nodes. For simplicity in this paper, a rectangular element with 8 nodal points was used for the rock block calculations. Then, the

8-point equal parameter truss element appears like the usual 8-point solid element in the finite element method, but it consists of 28 two-force bars as shown in Figure 1.

The global continuous truss structure for the entire rock mass was developed by constructing this type of truss element on every element in the model. The inner nodal points will have 26 bars connecting to adjacent nodes around it.

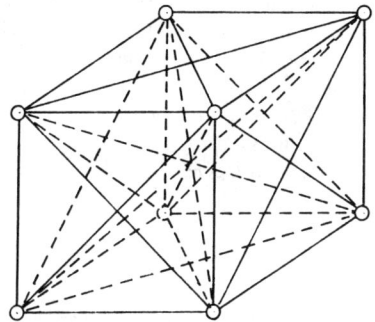

Figure 1. 28 bar truss element model.

The connectivity of the truss element is an 8x8 matrix, since the element has 8 nodal points, and only one freedom at each node is needed for the purpose of block calculations. Consequently, two indicator numbers are enough to define the continuity condition of the connection bars between any two nodes, such as 0 for no connection or the connection cut by a joint, and 1 for the positive connection, so there is a bar to connect them. By applying this indicator system, the connectivity matrix of a truss element can be written simply as follows:

$$[K] = \begin{vmatrix} 7 & 1 & 1 & 1 & 1 & 1 & 1 & 1 \\ 1 & 7 & 1 & 1 & 1 & 1 & 1 & 1 \\ 1 & 1 & 7 & 1 & 1 & 1 & 1 & 1 \\ 1 & 1 & 1 & 7 & 1 & 1 & 1 & 1 \\ 1 & 1 & 1 & 1 & 7 & 1 & 1 & 1 \\ 1 & 1 & 1 & 1 & 1 & 7 & 1 & 1 \\ 1 & 1 & 1 & 1 & 1 & 1 & 7 & 1 \\ 1 & 1 & 1 & 1 & 1 & 1 & 1 & 7 \end{vmatrix}$$

Compared with the exact stiffness matrix of the finite element method, the freedom of [K] was reduced from 24 to 8 and the matrix element of [K] does not express exactly the mechanical behavior of the element. However, the connectivity matrix [K] remains symmetric to follow the principles of mechanics.

Also, the global connectivity matrix can be assembled from the element connectivity matrix [K] by following the same procedure of assembling the global stiffness matrix in the finite element method. The global connectivity matrix of the entire truss structure in the rock mass, [M], is a (nxn) matrix, where n is the number of nodal points in the whole structural matrix. Therefore, the information of the continuity or connection between two nodes can be stored at each matrix element in [M], because each node has one freedom. The [M] matrix is symmetric and banded along the diagonal direction as in the finite element method.

3.3 *Block calculation*

Simulated rock joints were introduced into the whole structural matrix one by one. Whenever one joint plane was introduced, the elements were searched that may be cut by the joint, and which of the 26 bars within the element to be cut were checked. Whenever a bar is cut by a joint, the continuity of the global truss structure will be weakened. This weakness was reflected on the global connectivity matrix by modifying its matrix element corresponding to the bar that was cut by the joint. Actually, the matrix element was subtracted by one from the original number of 26. Consequently, the global connectivity matrix was modified constantly while all of the joints were introduced and all the connection bars were tested for their continuities. The final global connectivity matrix must be singular, and will consist of as many independent substructures as blocks formed by the joint systems.

It is obvious that when a rock block is isolated by joints from the rock mass, its corresponding small substructure will be separated from the global structure. These substructures are independent from each other, i.e. no connection exists among them. This means that the global connectivity matrix, which has been modified completely after the whole joint was introduced, can be transformed into a block matrix by elemental transformation. Then, each elemental matrix block in the final block matrix, transformed from the global connectivity matrix, represents an independent substructure, which is nothing but the rock block isolated by the joints and the block searched for the block calculations.

In this way, the problem of searching blocks in the rock mass was replaced with the problem of identifying the independent matrix blocks on the global connectivity matrix. The knowledge of nodal numbers within the independent matrix block is enough to identify the shape, size or volume, and location of a rock block formed by the joint systems. The volume of a rock block was calculated simply by counting the nodes within its substructural matrix block and summing up their volumes.

For this connectivity matrix method, there are no limits on the sizes and shapes of joint planes as well as the geometry of rock blocks formed by the joint systems, including any random aggregation of elements in the three-dimensional space.

3.4 *Key block failure*

Once blocks were identified, keyblocks were sorted out by following three steps:
1. Collect the joint and excavation plane geometry associated with a particular block.
2. Evaluate kinematic stability of the block with an algorithm based on Shi's theorem (Goodman and Shi 1984) and if the block is kinematically unstable.
3. Evaluate mechanical stability with Warburton's algorithm (Warburton 1980).

Then, the probabilistic analysis of keyblock failures was presented by the positional probability of failure, which was defined by the number of times the position (or node) was evaluated as being contained in a keyblock. In this way the probabilistic keyblock analysis can be effectively combined with the stochastic joint system simulation, and the realistic structural stability can be obtained in probabilistic terms.

One of the interesting aspects of this type of analysis is that it provides for the evaluation of progressive keyblock failures. If it is assumed that a keyblock displaces into the excavation, the next level of blocks exposed to the excavation surface modified by the initial keyblock failure become potential keyblocks. The process may continue over many levels of failure. With this type of analysis it is very simple to evaluate the successive levels of keyblock failure around an excavation surface.

This has a specific application in mining engineering: cavability analysis for the caving method of mining.

3.5 *A case study on a subway tunnel*

A metropolitan subway tunnel was studied to illustrate the difference between the traditional keyblock theorem and the positional probability of keyblock failures by the finite element approach for the block failure. The joint systems and their statistical details were published by Cording and Mahar (1974).

An unit length of the tunnel was isolated based on the discrete cell-block model. The joint systems within this cell-block were modeled and simulated for the stability analysis.

When the frequency of positional block failure was projected along the unit length of the tunnel, the cumulative probability of positional failure can be plotted around the tunnel as shown in Figure 2. Compared with the worst case type of analysis by the traditional keyblock theorem, the positional probability analysis shows clearly the size and frequency of keyblock occurrences in this projection.

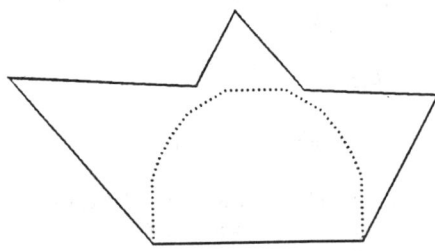

a) Maximum removable area

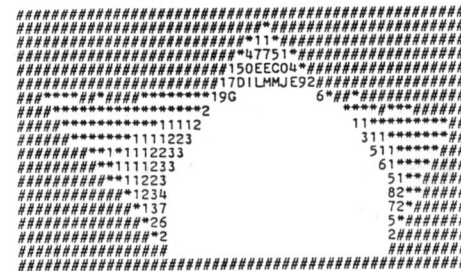

b) Positional probabilities of failure

Figure 2. Comparison of keyblock failures (a) with the positional probability of block failures (b) for a tunnel.

The positional probability of keyblock failure carries important features that can be simply implemented in the geotechnical design of excavations. For the design of a roof bolting system, the anchor should be located in the area where the positional probability is zero or low. The positional probability is the probability of the bolt not being anchored effectively to a stable portion of the rock mass. Also, the parts of the excavation requiring the most support can be identified easily from this.

The other measures of keyblock statistics included here are:
1. The distribution (or histogram) of keyblock sizes and its mean and standard deviation.
2. The total volume of keyblocks which summarizes the susceptibility to keyblock failure.
3. The frequency of different sizes of keyblocks.

4 CONCLUSIONS

1. The most important conclusion, in general, which could be drawn from this work is that localized probabilistic stability analysis for geotechnical structures can be made at the early stages of engineering design and construction, when only sparse sample data is available. This leads to the optimum design of geotechnical structures, optimum in their relative locations and orientations with other peripheral structures, and their shapes and sizes. This is achievable through the geostatistical model of characteristic parameters of rock masses. Then, it can be said that this is an ideal model of joint systems for many engineering analyses in both rock mechanics and geohydrology.

2. The deterministic approach based on block theorems treats joint orientations as constant and requires "engineering judgement" to qualitatively incorporate the quantitatively ignored factors of joint sizes and spacings, and their variabilities. Because of the fixed joint orientations and assumptions of infinite sizes of joints, the maximum removable area approach of the deterministic keyblock analysis provides an upper bound to the keyblock size identified. When the probabilistic analysis is coupled with the localized stochastic model of joint systems in geological formations, a significant amount of engineering judgement required to optimize the size, shape and orientation of an excavation could be replaced by quantitative solutions.

3. The positional probability of keyblock failure carries important features; the parts of excavation requiring the most support, probability of roof bolts not being anchored in stable zones, distributions of keyblock volumes, and keyblock sizes and frequency.

4. In-situ block size distribution should be a part of geotechnical and geohydrological site characterizations. It is a pertinent parameter to be included in keyblock failures, and it is related directly to the transmissibility of fluid flow through the fractured rock as in granular materials. Further study is warranted for this hydrological application of the block size distribution.

REFERENCES

Cording, E. J. and Mahar, J. W. 1974. The Effects of Natural Geologic Discontinuities on Behavior of Rock in Tunnels. *Proceedings of 1974 Rapid Excavation and Tunneling Conference.* San Francisco, CA, pp. 107-138.

Goodman, R. E. and Shi, G. H. 1984. *Block Theory and Its Application to Rock Mechanics.* Prentice-Hall, Englewood Cliffs.

Lemmer, I. C. 1984. Estimating Local Recoverable Reserves via Indicator Kriging. *Proceedings of Geostatistics for Natural Resources Characterization* (ed. by G. Verlys). D. Reidel, Dordrecht, pp. 349-364.

Warburton, P. M. 1980. Stereological Interpretation of Joint Trace Data: Influence of Joint Shape and Implications for Geological Surveys. *International Journal of Rock Mechanics & Mining Science.* 17:305-316.

Young, D. S. 1987a. Random Vectors and Spatial Analysis by Geostatistics for Geotechnical Applications. *Mathematical Geology.* 19:467-479.

Young, D. S. 1987b. Indicator Kriging for Unit Vectors; Rock Joint Orientations. *Mathematical Geology.* 19:481-502.

Young, D. S. and Hoerger, S. F. 1988a. Non-Parametric Approach for Localized Stochastic Model of Rock Joint Systems. *Geostatistical, Sensitivity, and Uncertainty Methods for Ground-Water Flow and Radionuclide Transport Modeling* (ed. by B. Buxton). Battelle Press, Columbus, OH, pp. 361-385.

Young, D. S. and Hoerger, S. F. 1988b. Geostatistics Applications to Rock Mechanics. *Proceedings of 29th U.S. Symposium on Rock Mechanics.* Balkema, Brookfield, pp. 271-282.

Viscoplastic-damage analysis of underground structures in rock

A. Nanda & S. Sengupta
Engineers India Limited, New Delhi, India

ABSTRACT: The time dependent behavior of underground openings in rock is well established and a number of methods are available for design and analysis. However most analytical and numerical methods are restricted to moderate stresses and hence cannot account for creep-rupture type behavior. In this paper a new viscoplastic formulation including damage is presented for the analysis of creep and creep-rupture type problems, in particular the time dependent response of underground openings in rock. The model including its implementation in a finite element code and some preliminary results are presented.

1 INTRODUCTION

It is well known that rock exhibits time dependent behavior. Rock samples under moderate stresses experience primary creep that is followed by secondary creep, during which the strain rate is either constant or decreases with time. Under high stresses, the secondary creep is followed by tertiary creep, during which the strain rate increases with time and may lead to rupture. The behavior of rock mass is a function of the rock quality and the rock mass may experience significant creep under lower stresses. The time dependent behavior of underground openings is also well established and emprical relationships have been established between underground structure dimensions, rock quality and standup time (Bieniawski 1976).

A limited number of analytical solutions for the time dependent response of underground openings are available (Lo and Yuen 1981, Ladayni and Gill 1984). Most of the analytical methods consider circular openings, homogeneous isotropic rock and simple spring and dashpot type models. Numerical techniques including classical plasticity, viscoplasticity and emprical creep laws are also available in the literature. However most of these models consider only secondary creep. Recently a number of models which include both secondary as well as tertiary creep leading to failure have been developed for the analysis of underground structures in rock (Cividini et al 1991, Yamatomi and Mogi 1993, Cristescu 1993).

There has been considerable interest in application of damage continuum theory in modelling material response, in particular material response at high stresses and temperatures. Applications of damage mechanics include materials as diverse as metals, polymers, concrete, rock and soils (Kachnaov 1987, Mroz & Angelillo 1982, Lemaitre & Chaboche 1990). This paper considers the application of a viscoplastic-damage formulation for application of the time dependent response of underground openings in rock. Starting from one dimensional conditions a multiaxial viscoplastic formulation including damage is presented for creep and creep rupture problems. The implementation of this model in a finite element program and some preliminary results are also presented.

2 ONE DIMENSIONAL FORMULATION

In this section starting from Nortons law for creep, a one dimensional viscoplastic formulation for creep and creep-rupture is presented.

2.1 Perfect Viscoplasticity

Figure 1 illustrates a typical strain versus time plot for a constant stress creep test, which is typical for many materials. For low to medium stresses, the initial or primary creep is followed by a steady state or secondary creep in which the strain rate is either constant or reduces with time. Under high stresses, the secondary creep is followed by tertiary creep in which the strain rate increases with time finally leading to rupture. For many materials the secondary creep may be approximated;

$$\epsilon'_{vp} = [(\sigma-\sigma_0)/\mu]^n \quad (2.1)$$

Where ϵ'_{vp} is the viscoplastic strain rate, σ is the applied stress, σ_0 is a yield stress, and μ and n are constants. When the yield stress is zero, the above equation reduces to the well known Nortons law for creep. The equation can be integrated directly and gives a linear strain versus time response.

2.2 Viscoplasticity with hardening

Many materials exhibit a reduction of strain rate with time under low to medium stresses. This may be approximated by introducing linear hardening in the earlier formulation.

$$\epsilon'_{vp} = [(\sigma-(\sigma_0+H\epsilon_{vp}))/\mu]^n \quad (2.2)$$

Where H is a hardening modulus. For n equal to 1.0, the above equation can be integrated. With hardening, the strain rate reduces with time and reaches a limiting strain.

2.3 Perfect viscoplasticity with damage

To model tertiary creep and creep rupture, the above formulation will be modified to include damage.

$$\epsilon'_{vp}=[(\sigma-\sigma_0(1-D))/\mu(1-D)]^n \quad \sigma > \sigma_0 \quad (2.3)$$

$$D' = [(\sigma-\sigma_0(1-D))/A(1-D)]^r \quad (2.4)$$

Where D is the damage which varies between 0 and 1.0 and A and r are constants. When the yield stress is set to zero, the above reduce to the Nortons law for creep and Kachanovs law for damage. The above equations can be integrated exactly to obtain the time to rupture t_r.

2.4 Viscoplasticity with hardening and damage

In the previous model, rupture will occour for all stress states. Most materials exhibit a stable response at low to medium stresses. To model this, combined hardening and damage will have to be considered.

$$\epsilon'_{vp} = [(\sigma-(\sigma_0+H\epsilon_{vp})(1-D))/\mu(1-D)]^n$$
for $\sigma > \sigma_0$ (2.5)

$$D' = [(\sigma-(\sigma_0+H\epsilon_{vp})(1-D))/A(1-D)]^r \quad (2.6)$$

Depending on the stress level, the material will either reach a stable condition with a zero strain rate or rupture will occour. For the special case of n=r, for a stable condition the applied stress should be less than;

$$\sigma < (\sigma_0+H\beta)^2/4H\beta \quad (2.7)$$

where $\beta = (A/\mu)^n$.

3 MULTIAXIAL FORMULATION

The one dimensional formulation can be extended for the multiaxial case;

$$(\epsilon'_{vp})_{ij} = [F/\mu(1-D)]^n(\delta F/\delta\sigma_{ij})$$

for $F > 0$ (3.1)

$$F = (3J_2)^{1/2} + mJ_1 - (\sigma_0+H\epsilon_{eq})(1-D) \quad (3.2)$$

$$\epsilon_{eq} = ((2/3)\epsilon_{ij}\epsilon_{ij})^{1/2} \quad (3.3)$$

$$D' = [F/A(1-D)]^r \quad (3.4)$$

Where J_i are the stress invariants and m is a constant to account for the dependence on the mean stress. Under a uniaxial state of stress, the above formulation reduces to the previous one dimensional formulation.

4 FINITE ELEMENT FORMULATION

A simplified explicit finite element formulation for visco plasticity and damage for creep and creep-rupture is presented. In this formulation the elastic modulus is constant and all nonlinear response including damage is included in the viscoplastic strain.

1. A t=0, solve for the elastic displacements, strains and stresses.

2. At any other time, solve the following equations;

$$d\mathbf{u}^n = \mathbf{K}^{-1}d\mathbf{V}^n \quad (4.1)$$

$$d\mathbf{V}^n = f_v (\mathbf{B}^T \mathbf{C} \boldsymbol{\epsilon}'_{vp} dt) dv \quad (4.2)$$

Where $\mathbf{u}$ are the nodal displacements, $\mathbf{K}$ is the elastic stiffness matrix, $\mathbf{V}$ is a load vector, $\mathbf{B}$ is the strain displacement matrix, $\mathbf{C}$ is the elastic stress-strain matrix and $\boldsymbol{\epsilon}'_{vp}$ is the viscoplastic strain rate vector evalauted at the current stress and damage levels and dt is the time step. The variables are updated a every time step;

$$\mathbf{u}^{n+1} = \mathbf{u}^n + d\mathbf{u}^n \quad (4.3)$$

$$\sigma^{n+1} = \sigma^n + \mathbf{C}(\mathbf{B}d\mathbf{u}^n - \boldsymbol{\epsilon}'_{vp}dt) \quad (4.4)$$

$$D^{n+1} = D^n + D'dt \quad (4.5)$$

The procedure is repeated for every time step till the strain rate is below a specified level. As an explicit formulation has been used, the solution will be conditionally stable. The above formulation has been implemented in a two dimensional finite element program with four noded isoparametric elements.

5 RESULTS

Some results of the finite element analysis are presented in this section. Figure 2 compares the finite element predictions with the exact solution for perfect viscoplasticity, viscoplasticity with hardening and viscoplasticity with damage under one dimensional conditions with the following parameters. (σ=100, A/μ=10, n=r=1, E/A=10, σ_0=0). It can be observed that the comparision is excellent, except very close to rupture.

Figure 3 illustrates the strain-time response for combined hardening and damage for various stress levels. A Hardening modulus E/H=10 was used along with the above parameters. Upto a stress of 250, the response is stable, above which, the material tends to rupture. This is in agreement with the predicted value of H/4 as β=1 and σ_0=0. Note that time scale has been normalized with respect to the time to rupture for a stress of 100.

The above results establish the valaidity of the numerical implementation. Analysis of two dimensional problems is currently underway. This model with a few parameters can represent a wide range of behaviors ranging from elastic, secondary creep with constant or reducing strain rate and creep-rupture type behavior.

6 CONCLUSIONS

A Viscoplastic formulation including damage was presented for creep and creep-rupture type problems with particular reference to the time dependent behavior of underground openings in rock. Starting from a one dimensional formulation a multiaxial formulation was developed and implemented in a finite element code. Some preliminary results were presented for both one and two dimensional cases. The model is simple and has the potential for application to the analysis of the time dependent response of underground structures in rock.

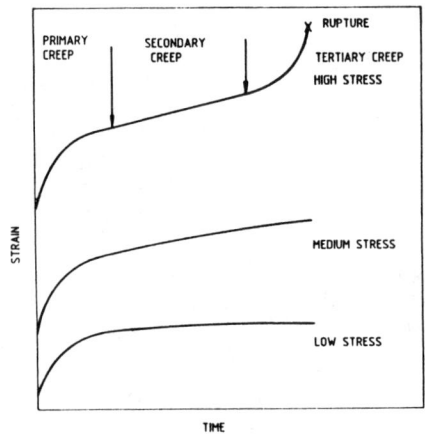

Fig.1 TYPICAL CREEP RESPONSE

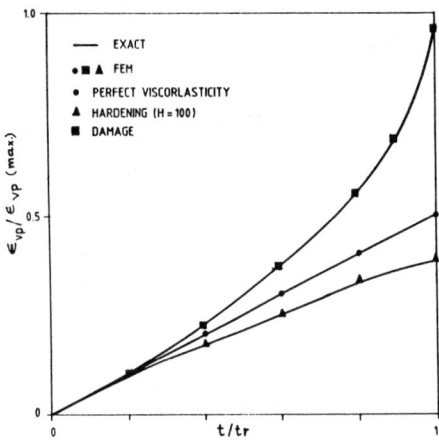

Fig.2 COMPARISON OF FEM WITH EXACT SOLUTION

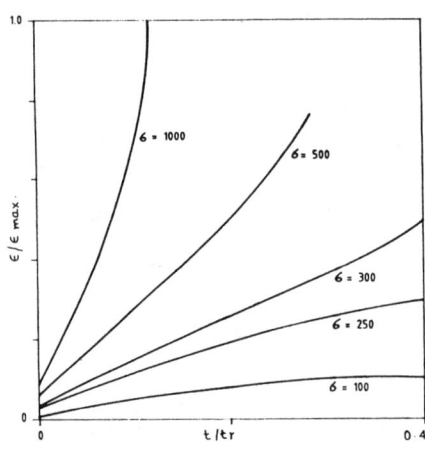

Fig.3 COMBINED HARDENING AND DAMAGE

7 ACKNOWLEDGEMENTS

The authors would like to thank the management of Engineers India Limited.

REFRENCES

Bieniawski, Z.T. 1989. Engineering rock mass classifications. John Wiley, New York.

Cividini, A., Gioda, G. and Carini, A. 1991. A finite element analysis of the time dependent behavior of underground openings. Proc. Int. Conf. Comp. Adv. Geomech., Cairins.

Cristescu, N.D. 1993. "Failure and creep failure around an underground opening. Int. Conf. Preventing Rock failure, Instanbul.

Kachanov, L.M. 1986. Introduction to continuum damage mechanics. Martinus Nijhoff Publishers, Netherlands.

Ladayni, B. and Gill, D.E. 1984. Tunnel lining design in a creeping rock. ISRM Symp., Cambridge.

Lemaitre, J. and Chaboche, J.L. 1990. Mechanics of solid materials. Cambridge university press, Cambridge.

Lo, K.Y. and Yeun, C.M.I. 1981. Design of tunnel lining in rock for long term effects. Can. Geotech. J., 18, 24-34.

Mroz, Z. and Angelillo, M. 1982. Rate dependent degradation model for concrete and rock. Int. Symp. Num. Models Geomech., Zurich, pp. 13-17.

Yamatomi, J. and Mogi, G. 1993. A FEM study on viscoplastic behavior of weak rock. Proc. Int. Conf. Preventing rock failures, Instanbul.

Design of the reinforced concrete lining in Bakun Diversion Tunnels

Wang-Ruel Jee
Bakun Tunnel Design Project, Dong-Ah Const., Seoul, Korea

Young-Kyung Sim & Tae-Jung Lim
Dong-Ah Const., Seoul, Korea

ABSTRACT: The completion of the Bakun Diversion Tunnel is subsequently to the Main Dam construction. Therefore, the completion date is very important for the Bakun Hydroelectric Project. Generally, the tunnel lining work as a finishing phase of the tunnelling project occupies a important portion as well as an excavation and a support work of the tunnels in respect to the construction cost and period. Internal section of Bakun Diversion Tunnel is designed circular shape to reduce the roughness of the water flow with 12 meters in diameter of total length 4314.6 meters of 3 tunnels. The Lining thickness is varied between 50cm and 70cm depending on the structural condition.[1] From the original Tender design of the Bakun Tunnels, the required quantity of steel bars was 5,985 ton designed by Reinforced Concrete(RC) through the entire tunnel Linings. During the detail design stage by the consideration of the rock conditions and various load conditions, we could suggest five kinds of RC lining type including plain concrete lining type. Through the detail design modification, we could reduce the required amount of steel bars to 2,178ton, as a half of original Bill of Quantity.
Finally, this design modification give us the time and cost saving effect to catch up the construction progress in time.

1. Computer modelling and analysis of the Reinforced Concrete Lining

1.1 Modelling work

The structural system of the Concrete Lining is a rigid circular frame elastically supported by the surrounding rock mass. This system is represented by a sequence of 64 beams with nonlinear(no tensile forces to transmit) springs in every node acting perpendicular on the Lining.

For the computer calculation of this frame is used a professional software program of GTST-RUDL. In order to iterate better and faster, some of these springs are left out from the beginning.

Modulus of subgrade reaction in radial direction $K_{S.R}$ can be estimated to [3]

$$K_{S.R} = C_R \cdot \frac{E_D}{R_0}$$

where,

$K_{S.R}$; modulus of subgrade reaction in radial
E_D ; deformation modulus of rock mass
C_R ; factor in the literature given to 0.67

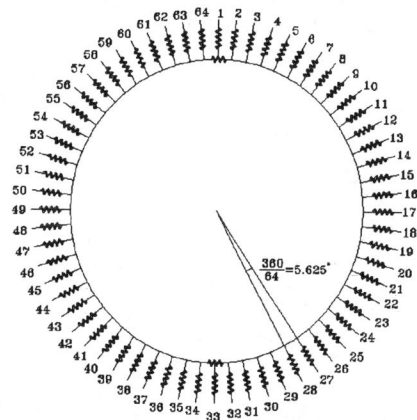

Fig 1. Modelling of the Reinforced Concrete Lining in Diversion Tunnels

Table 1. Rock Pressure applied to Reinforced Concrete Lining of the Diversion Tunnels

Item (RST) \ Rock Support Type	Abbrev.	Unit	A	B	C	D	E	F
Rock Mass Type	RMT	-	I	II	III		IV	
Deformation modulus	E_D	MPa	12,000	8,000	6,000		2,500	
Tunnel Radius	R_0	m	6.5	6.5	6.5		6.5	
Factor(radial)	C_R	-	0.67	0.67	0.67		0.67	
Mod. of Subgrade	$K_{S.R}$	MN/m^3	1236.9	824.6	618.5		257.7	
Vertical Weight	P_V	MPa	0.026	0.039	0.039	0.065	0.078	0.104
Horizontal Weight	P_H	MPa	0.013	0.020	0.020	0.033	0.039	0.052

1.2 *Loading for diversion Tunnel cross sections.*

Conducting the design of Bakun Diversion Tunnel Lining, applied loads are in belows.
- Dead load of the Lining structure.
- Vertical and horizontal rock pressures;
  Released rock pressure (dead weight of yield zone) from excavation phase apply to the model in vertically and horizontally according to Table 1.
- External water pressure;
  External water pressure is distinguished three cases by construction area related Plug position. Plug downstream area is not applied water pressure because of the weephole installation.
  ① 25m height (related to top height of Tunnel Lining) : downstream area of the plugs.
  ② 68m height (related to top height of Tunnel Lining) : according to the water level of +125m asl. for Tunnel I upstream area of plug.
  ③ 152m height (related to top height of Tunnel Lining) : according to the water level of +209m asl. for Tunnel II,III upstream area of plug.
- Internal water pressure;
  Internal water pressure will create cracks in the Concrete Lining, which consequently create equilibrium conditions of water pressure between Lining and rock surface; therefore, internal water pressure will not be considered for further structural design.
- Temperature loads;
  According to principal design criteria, the parameters relevant for the calculation of thermal loads are +25℃ for rock temperature and +20℃ /+30℃ for min/max water temperatures $\Delta T = \pm 5℃$.
- Seismic loads;
  ah=0.05 g, horizontally
  av=0.10 g, vertically

Applied load combinations are fixed according to "ACI code to design reinforced Concrete structures", but we ignore the effect of the temperature load and the earthquake load because that it decrease the safety factor.[2]

Load combnation Case 1;
  U=1.4D+1.7L+1.4F+1.7H
Load combnation Case 2;
  U=0.75(1.4D+1.7L+1.4F+1.7W) or
  0.75(1.4D+1.7L+1.4F+1.4T)
Load combnation Case 3;
  U=0.75(1.4D+1.7L+1.4F+1.7H+1.87E)

where,

U ; Required ultimate strength to resist factored loads.
D ; Dead loads including equipment.
L ; Live loads.
F ; Loads due to weight and pressure of fluids including waves and hydrodynamic pressure.
H ; Load due to weight and pressure of soil (backfill) or rock,
W ; Wind load.
T ; Cumulative effects of reduced temperature, creep and shrinkage.
E ; Seismic loads.

1.3 Remove the Spring elements and Analysis Concrete Lining

When we analyze Concrete Lining, it is necessary to perform the modelling of surrounding rock as a Spring elements and we are analyzing only the compressed Spring elements and removing the tensioned Spring elements.

This process has two methods due to the ability of analysis program.

The frist way of removing the tensional spring elements is modelling of all surrounding rock by the Lining as a Spring elements and appling the load and analyzing them to eliminate the tensioned Spring elements according to the designer's opinion. And then repeat the process until no tensioned Spring elements left in the model.

The second way is to use the program which is the analysis program to remove the tensioned Spring automatically. In this way, we can expect the more accurate calculation and faster analysis because of program's objective standard.

When we use the first method, the position of Spring elements might be alternative results depending on the designer's opinion. Therefore, the different results could be achieved by the individual design.

In this reason, we analyzed the Lining structure of Bakun Diversion Tunnel using the automatic calculation program to reduce the effect of the subjective opinion while calculating manually.

1.4 Safety factors

The structural design of the Reinforced Concrete Lining is executed according to German Standard DIN 1045.[2]

Direct action-effects in Reinforced Concrete,

γ = 1.75 (concrete section is expected to fail after prior warning)

γ = 2.10 (in the case of brittle failure)

Indirect action-effect in Reinforced Concrete,

γ = 1.00

2. Output and Reinforcement Design

Fig 2 illustrates the area of the removed Spring elements and the displacement of the Lining in case of none external water pressure, while Fig 3 shows the analysis results with water pressure.[4]

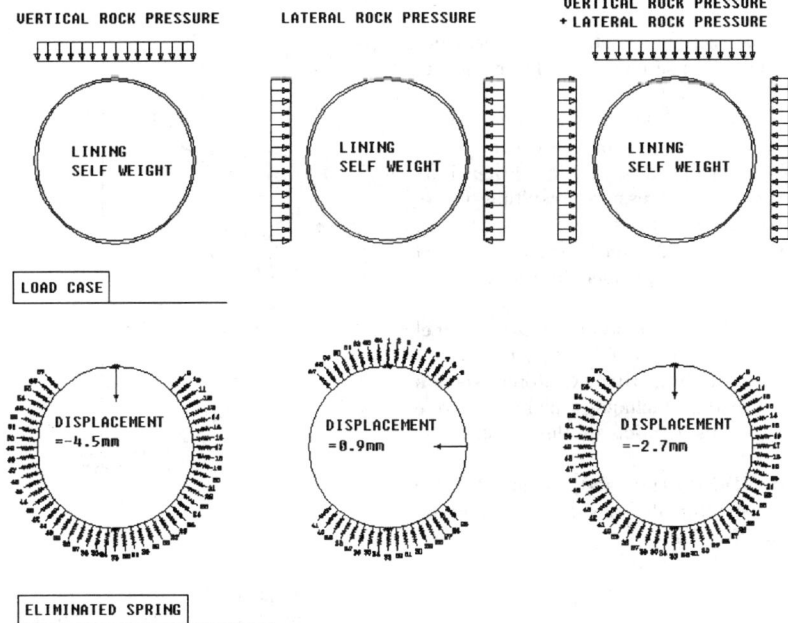

Fig 2. Load combination cases without water pressure (Rock TYPE F)

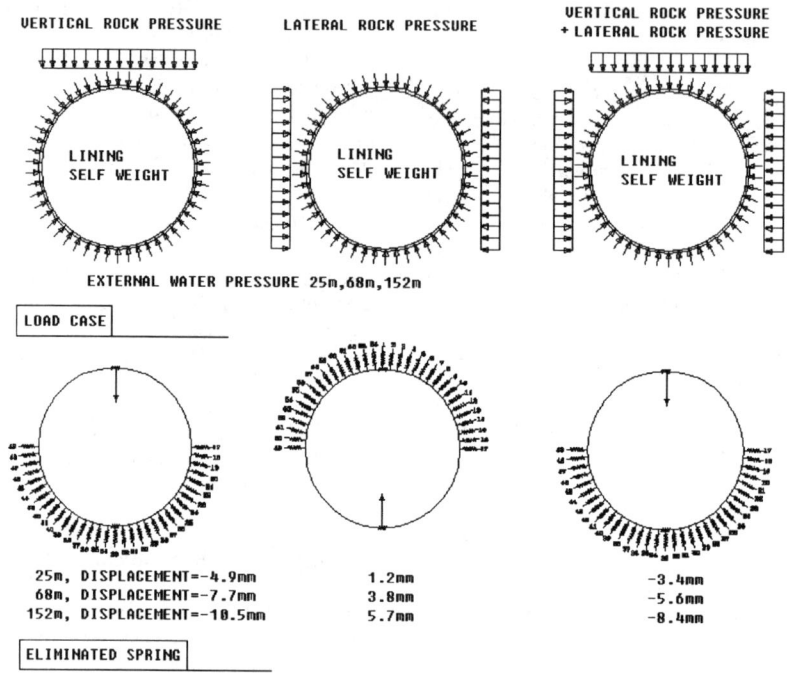

Fig 3. Load combination cases with the water pressure(25m,68m,152m) (Rock TYPE F)

In both cases, when the vertical rock pressure is applied only, maximum displacement is observed.

When the vertical and lateral rock pressure is applied together, lateral rock pressure constraints the lateral displacement around the Spring Line from the vertical rock pressure, resulting in its displacement could be reduced.

Fig 4 describes the calculated structural tunnel section consideration of member forces from the analyzed result.

Although we should subdivide entire Tunnel relying on the structural condition and rock types for the optimum design, five Reinforce section types were selected including plain concrete lining type with consideration of the construction ability.

Concrete B35(DIN 1045) and Rebar BSB460 BS(4449) are used for the design depending on the Tender Specification.

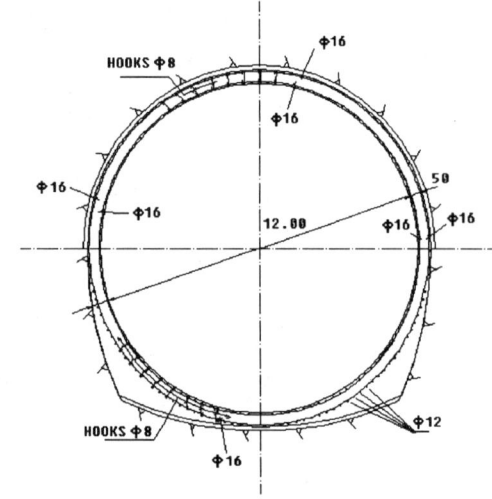

Fig 4. Typical section of the Reinforced Concrete Lining of Bakun Diversion Tunnel

Table 2. Result of structural analysis of the Tunnel Lining

External water Pressure	Rock Type	Rock Pressure	(mm) Displacement	Max (kN.m) Moment	Max (kN) Shear force	Max (kN) Axial force
non external water pressure	A	Vertical	2.7(01)	107.42(01)	42.36(06)	1011.73(33)
		V+L	0.9(01)	28.62(01)	26.43(06)	433.07(33)
	F	Vertical	4.5(01)	148.97(01)	106.94(60)	983.58(33)
		Lateral	0.9(17)	45.42(15)	41.13(19)	495.55(33)
		V+L	3.7(01)	100.18(01)	82.36(06)	1072.52(33)
25m related to top of lining	A	Vertical	3.7(33)	72.29(21)	145.77(18)	2203.29(33)
		V+L	3.3(33)	55.96(20)	125.52(18)	2228.08(33)
	F	Vertical	4.9(01)	139.97(12)	194.85(16)	2234.23(33)
		Lateral	1.2(33)	54.69(33)	131.25(27)	2316.71(33)
		V+L	3.4(01)	57.56(13)	122.62(18)	2326.14(33)
25m related to top of lining	A	Vertical	7.2(33)	113.10(20)	281.38(16)	4875.48(36)
		V+L	6.6(33)	96.28(19)	272.26(16)	4902.05(33)
	F	Vertical	7.7(01)	163.00(13)	318.13(16)	4909.90(33)
		Lateral	3.8(33)	74.50(13)	262.03(28)	4998.86(33)
		V+L	5.6(01)	72.68(13)	255.41(18)	5008.91(33)
25m related to top of lining	A	Vertical	8.8(33)	209.58(17)	566.07(13)	10249.18(35)
		V+L	8.2(33)	176.20(16)	550.23(11)	10286.90(33)
	F	Vertical	10.5(01)	312.45(14)	595.77(08)	10362.70(33)
		Lateral	5.7(33)	141.77(10)	537.88(15)	10409.00(33)
		V+L	8.4(01)	158.84(15)	540.30(24)	10477.60(33)

( ). Nod number (see Fig 1)

3. Conclusion

Rock Tunnel excavation was analyzed using the program FLAC before structural analysis of the Concrete Lining.

In case of the Rock type A(table 1), there is no yield zone found just above the Tunnel Crown by the result of rock support analysis.

According to the result of the structural analysis of FLAC, some of the existing Reinforced Concrete Lining could be modified as a plain Concrete Lining at the downstream area of the Plug because of the sound rock condition of the tunnel.

Required amount of steel bar for the each lining section(type 1,2,3,4) is designed as; 1.209ton/m, 0.990ton/m, 1.004ton/m, 1.676ton/m.

Through the cautious control of main Rebar spacing to minimize the required steel quantity, we could reduce the total required steelbars quantity of the Diversion Tunnels from BOQ amount 5,985ton to changed 2,178ton and this design modification save us time and construction cost.

Reference

1) "BAKUN HYDROELECTRIC PROJECT Tender documents" 1995. EKRAN BERHAD
2) DIN 1045 "Structural Use of Concrete Design and Construction" 1988. Beuth Verlag
3) "Handbuch des Tunnel- und Stollenbaus (2)" ,Dr. Bernhard Maidl. Verlag Glückauf GmbH
4) "Calcuation Report of the BAKUN DIVERSION TUNNEL" Dong-Ah const.

Criterion of energy catastrophe for rock project engineering system failure in underground engineering

M. Cai, G. Kong & J. Liang
College of Resources Engineering, University of Science and Technology, Beijing, People's Republic of China

ABSTRACT: This paper discusses clearly the difference in meaning between rock material failure and rock project failure while losing its stability. Based on the idea, it is incorrect to use the same criterion to predict the rock failure and underground project failure in numerical calculation. With the aid of system energy principle and catastrophe theory, the criterion of energy catastrophe for predicting rock project system failure is put forward and used in the calculation of the finite element method (FEM). Proved by the real project, the idea and criterion are correct and of certain value, the results of calculation coincide with the state of the mining project.

1 ROCK FAILURE AND ROCK PROJECT FAILURE

In underground rock project, the rock around the excavation space undergoes the process from minim deformation to failure. In general rock failure is caused by that the capacity of rock material for resisting deformation or stress reaches certain limitation which is defined by some strength criteria or deformation failure criteria. However, the failure of rock project structure within certain area is because of failed, the project structure was still in function.

So, it is incorrect to use rock material failure criterion to judge rock project system stability with the finite element method (FEM) or analytic math method.

What more, the failure criterion of rock material is set up according o the test result of stress-strain of rock sample in lab. The result is far away from the real state of rock project. Therefore, when we study the stability of underground rock project, we should put emphasis on rock project system failure, and set up a criterion to judge the failure of rock project system. Based on the method of system energy analysis, this paper put forward the strain energy catastrophic failure (SECAF) criterion for predicting the stability of underground rock project.

2 CRITERION OF ENERGY CATASTROPHE FOR ROCK PROJECT SYSTEM FAILURE

2.1 *Failure criteria of rock and rock project system*

The rock failure can be determined by stress-strain curve as shown in figure 1. The loading process can be divided into two phases. In phase OA, rock is in steady deformation state, saying strictly, the rock is undergoing microscope failure and inner crack number in it is gradually increased. In phase AB, the rock is in unsteady deformation state and presents macroscopic failure behavior. Point A stands for the critical state, the stress is called as strength limit. Generally, it is thought that rock falls when reaches point A.

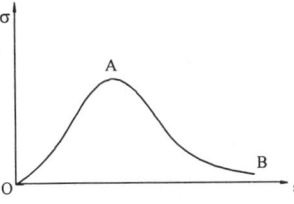

Fig.1 rock sample stress — strain curve

Mohr-column criterion and Griffith criteria are commonly used to judge the failure of rock material, which can be expressed as the formula:
$f(\sigma_1, \sigma_2, \sigma_3) = 0$.

In fact, in-situ rock mass, there are a lot of cracks, joints and many kinds of soft structure planes in field rock mass which weaken the strength and stability of rock mass or rock project. But, it is difficult and indistinct for us to describe the inner structure of rock mass. According to the system theory, this is a "black box" problem, and we can not use a definite mathematical or mechanical model or other failure criterion to analyze the stability state of rock project system.

2.2 Setting up of energy catastrophe criterion

The excavation of rock project is a process compared with both storage and dissipation of system energy. The whole process is from steady state to critical state before system failure. The whole deformation process of rock mass is considered as a approximate static process, and the whole influenced area of the rock project is a system, we take the unit (for FEM) strain energy of rock mass to be the status variable and the stability of underground project is examined by the status of strain energy in the whole system.

In the process of FEM analysis, assuming the strain energy of the Gauss integration point of the unit P (for FEM) as following:

$$E_p(k) = \int_{v_p} \sigma_{ij} \varepsilon_{ij} dv_p \quad (1)$$

where:
- σ_{ij} : stress of the unit;
- ε_{ij} : strain of the unit;
- v_p : volume of the unit P;
- k : load increment number;

In order to determine the rough range of failure area of the rock system, we add up the strain energy of all the units which meet the yielding condition. i.e. $F = (\sigma_i, H_i) = 0$. (here, σ_i stands for the unit primary stress, H_i is a general yield parameter). Then total strain energy of the plastic area in the rock system is gotten as follows:

$$E = \sum_{i=1}^{N} \int_{v_p} \sigma_{ij} \varepsilon_{ij} dv_p \quad (2)$$

In the formula (2), N means the total number of units which are in yield state. Since the strain energy E is changed with the process of loading and unloading induced by excavation, it can be described by a consecutive energy function $E=f(t)$, (here, t means the time' scale of loading), $f(t)$ is expressed in the form of Taylor series:

$$E = f(t) = f(0) + \frac{\partial f}{\partial t}\bigg|_{t=0} t + \frac{\partial^2 f}{\partial t^2}\bigg|_{t=0} t^2 + \frac{\partial^3 f}{\partial t^3}\bigg|_{t=0} t^3$$
$$+ \cdots \cdots \frac{\partial^n f}{\partial t^n}\bigg|_{t=0} t^n \quad (3)$$

We select the most prominent terms and the above formula is simplified as following:

$$E = \sum_{i=1}^{4} a_i t^i \quad (4)$$

$$a_i = \frac{\partial^i f}{\partial t^i}\bigg|_{t=0}$$

take Tshirnhams Transformation (4) $t \to x - A$,
$A = \frac{a_3}{4a_4}$,

formula (4) is transformed into:
$$\overline{E} = b_4 x^4 + b_2 x^2 + b_1 x + b_0 \quad (5)$$

(5) can be farther changed into:
$$\overline{e} = \frac{x^4}{4} + \frac{x^2}{2} u + xv \quad (6)$$

where: $\overline{e} = \frac{\overline{E}}{4b_4}$, $u = \frac{b_2}{2b_4}$, $v = \frac{b_1}{4b_4}$;

$$\{b\} = [s]\{a\} \quad (7)$$

[s] stands for transformation matrix (Stuart.G.W 1988).

formula (6) is CUSP model of catastrophe theory (F.H.Ling 1987), (L.H.Jia 1995). According to CUSP diverge theory, we can get the catastrophic condition of rock mass, as follows:
Given: $\Delta = 4u^3 + 27v^2 \quad (8)$
Δ is called energy catastrophe threshold.
   if $\Delta \leqslant 0$ the system falls;
   if $\Delta > 0$ the system no falls;
Formula (6)~(8) are the energy catastrophe criterion for rock project system used in FEM analysis. Using the multinominal fitting method, the strain energy function series, i.e. $\{E\} = \{E(1), E(2,) \cdots\cdots E(m)\}$ can be transformed into the form: $E(t) = \sum_{i=1}^{n} a_i t^i$;

where: a_i is constant determined by regression (L.H.Jia 1995). Then, $E(t)$ can be transformed into the style that CUSP model needs and may be convenient for computer program design.

2.3 Application of the energy catastrophic criterion in FEM analysis

In order to apply the energy catastrophe criterion to study the stability problem of rock excavation system, we made up a FEM program whole schematics is shown in Fig.2

The program has been used to analyses the stability of stopes with cutting and filling method in Jinchuan Nickel Mine, China.

We have designed two models:
(1) mining area is filled with cement and no pillar left ; Model-1
(2) mining area is filled with cement and some pillars left ; Model-2

The results of calculation of FEM as following:

In Model-1, the energy catastrophe threshold of the rock mass system (including boundary ore, country rock and filling body) $\Delta_1 > 0$ it means the system is stabile.

In Model-2, the energy catastrophe threshold of pillars $\Delta_2 \leq 0$, but the energy catastrophe of the rock mass system (including boundary ore, country rock, pillars and filling body) $\Delta_2 > 0$. It means the pillars may be fall mean while the system is stable.

The results coincides correctly with the real situation in the mine. This example proves the correctness and feasibility of the energy catastrophe criterion proposed in this paper.

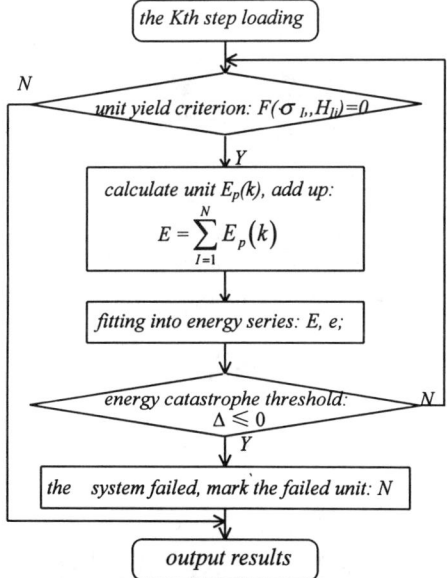

Fig.2 the schematic of FEM program of the system failure criterion

3 CONCLUSION

From the view of system energy catastrophe, the paper put forward the energy catastrophe criterion for predicting failure (losing stability) of rock project system. It overcomes the shortcomings of traditional method used in FEM analysis in which the failure criteria for rock materials were applied to predict the failure of rock project system.

REFERENCES

J. N.Tao.1989. Base for analyze rock project stability *Symposium of New development in rock mechanics:*129-141. Shengyang, China.

D.R.Zhu.1994. Fall criterion for rock project. *Journal of China coal society, vol.19.No.1,Feb,1994*:15-19. Beijing, China.

F.H.Ling.1987.The catastrophe the theory and its application. *Press of Shanghai Jiaotong University*, Shanghai, China.

L.H.Jia.1995. Simulating plastic softening and underground excavation with elastic-plastic large deformation finite element method. *Journal of Northeastern university, vol.16.Sep,1995*:104-109.Shengyang, China.

X.M.Zhou.1993. An FEM analysis of the stability of large scale backfill-rockmass integrated bodies in the mine area No.2 of Jingchuan.co. *CJRME, vol.12.No.2*:95-99.China.

Stuart.G.W.1988. Introduction of matrix calculation. *Press of Shanghai science and technology.* Shanghai, China.

A study of the frictional-slide behaviour in the post-failure region of soft rock

Xiao Li & Sijing Wang
Institute of Geology, Chinese Academy of Sciences, Beijing, People's Republic of China

ABSTRACT: By means of a servocontrolled testing machine a series of triaxial compression tests on mudstone specimens were carried out under the conditions of continuous loading and cyclic loading respectively. The experimental results show that the post-failure behaviour of soft rock is controlled by the so-called main fracture planes. The predominant deformation mechanism in the post-failure region is the frictional-slide effect between the cracked blocks on the main fracture planes. A shear sliding model was put forward, and the prediction of the model agreed well with the results obtained from the tests.

1 INTRODUCTION

Knowledge of the behaviour of fractured rock is of considerable practical interest. For instance, in soft or squeezing rock roadway in coal measure strata, or at great depths of mines, fracturing of rock around excavation is usually unavoidable because the stress concentration caused by the excavation is often beyond the peak strength of surrounding rock mass. Therefore, a broken zone will develop around the excavation, which can cause the roadway to deform or its floor to heave. The stability of roadway, in fact, is related closely to the mechanical behaviour of fractured rock, and the supporting object of soft rock roadway is just the fractured rock (Dong 1994). Moreover, the behaviour of fractured rock is also of the great engineering importance in the case of room-and-pillar stoping with panels because the characteristics of yield pillars under in-situ conditions are very similar to the post-failure behaviour of the relevant rock elements (Bieniawski 1970).

The determination of the complete stress-strain curves, especially the post-peak curve, of rock specimen is one of the most useful approaches to studies on the mechanical behaviour of fractured rock (Farmer 1983). Many authors have paid more attention to the strain-softening behaviour in the post-peak region of rocks (Frantziskonis 1987). However, there have been disputes on the strain-softening mechanism of rock so far (Chen 1996). For example, the strain-softening, in essence, belongs to whether the constitutive properties of material or the structure characteristics. What kind of mechanical tools is more appropriate to describe the strain-softening behaviour. Moreover, for the soft rock in coal measure strata, which uniaxial compressive strength σ_c is below 20MPa, the test data themselves are very poor because of the difficulty for collecting and processing specimens. Therefore, the post-failure behaviour for the weaker mudstone (σ_c=12~18MPa) was studied in laboratory and a great many of test data were obtained in this paper. The experimental procedures and major results are introduced hereunder.

2 THE DEFORMATION BEHAVIOUR OF MUDSTONE

The rock for testing was collected from Maoming colliery in Guangdong province of China, and the rock is carbonous mudstone which belongs to Tertiary of Cainozoic group. Its mineral components are mainly kaolinite, illite, montmorillonite and quartz, and its porosity is 19.7%, volume weight 2.1g/cm^3. The specimen appears ash-white and homogeneous. When the rock encounters water, it will swell and disintegrate in half an hour.

All specimens were cylinders 50mm in diameter by 120mm in high. The specimens were cored from the same direction of two 0.4m cubic blocks which were obtained from the floor and roof of a main roadway in Maoming colliery respectively. The specimens were prepared so that all dimensional tolerance meet the ISRM standard.

A electro-hydraulic servocontrolled testing machine (MTS815) made in USA was used for this study. The closed-loop system was controlled by the axial displacement of specimen. A group of testing data were auto-collected by computer every quarter second during the whole testing process. The six confining pressure values to be chosen were 0, 5, 10, 15, 30 and 40MPa respectively. More than twenty specimens were tested and the testing repeatability was better. The average coefficient of discrepance was smaller than 9%.

Based on the load-displacement data collected by computer in the tests, the axial complete stress-strain curves of floor mudstone under the continuous loading condition are given in Fig.1.

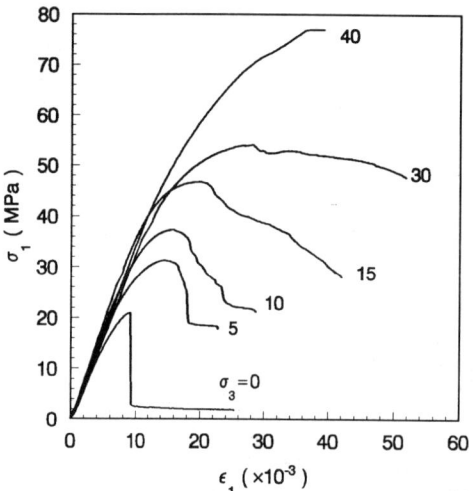

Figure 1. The stress-strain curves of floor mudstone under the continuous loading condition

As shown in Fig.1, the confining pressure has great influence on both the peak strength and the post-failure deformation behaviour of mudstone. With the increasing of confining pressure, the residual strength increases, and the stress-strain curve in the post-failure region becomes gradually flatter. When the confining pressure reaches 30MPa, approximate ideal plastic flow occurs for the specimen. Increasing the confining pressure continually to 40MPa (about double times larger than the uniaxial compressive strength), the phenomenon of strain-hardening occurs. The complete stress-strain curves obtained from the tests are very typical. The confining pressure required from the brittleness to ductility for the mudstone is about 30MPa.

3 THE FRICTIONAL-SLIDE BEHAVIOUR OF FRACTURED ROCK

In order to reveal the strain-softening mechanism of rock, cyclic loading tests in triaxial compression were also conducted. In the tests specimens were first loaded in a hydrostatic pressure condition to the value of the predetermined confining pressure, then the confining pressure was held constant and the specimen was loaded and unloaded in the axial direction until the test is over. The axial stress-strain curve under the cyclic loading on the roof mudstone are illustrated in Fig.2.

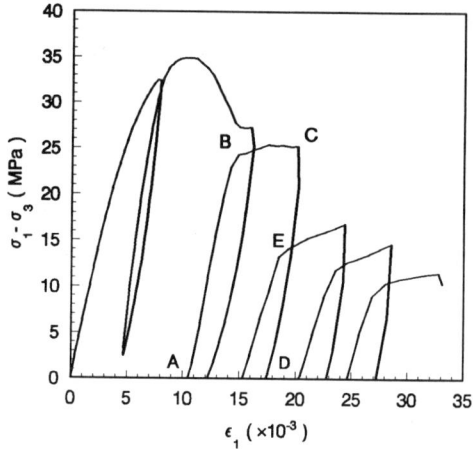

Figure 2. The stress-strain curve of roof mudstone under the cyclic loading condition (σ_3=10MPa)

The experiment results show that the procedures of cyclic loading and unloading can not change the mechanical properties of rock. As shown in Fig.2, the envelope curve of all loading and unloading points is just the complete stress-strain curve under monotonic loading, and this shows that the pre-peak and the post-peak behaviour of rock can be studied by cyclic loading tests.

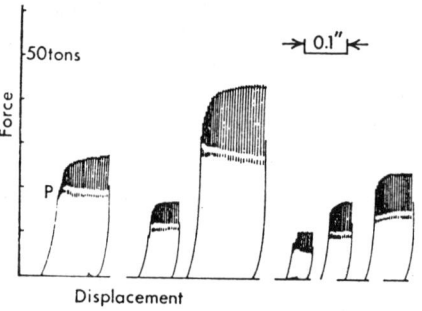

Figure 3. The typical stick-slide curves of rock (after Jaeger 1976)

An important phenomenon is found in the cyclic loading tests that distinct frictional-slide deformation occurs in the post-failure region of rock. If a loading-unloading cycle ABCD in Fig.2 is compared with the typical stick-slide curve of rock (as shown in Fig.3), it is seen that both are very similar. The points on post-failure stress-strain curve correspond to the starting points of slide after loading again (as shown in Fig.2 point B). With the increasing of deformation, the stress of the starting slide will decrease gradually (as shown in Fig.2 point E). Namely, the force necessary to initiate sliding will decrease continually. Therefore, the strain-softening phenomenon for the whole specimen occurs in a average sense.

From the relation curve between axial strain ε_1 and lateral strain ε_3 as shown in Fig.4, we can see that the curve can be divided into two stages, the pre-peak stage and the post-peak stage. The curve in the pre-peak stage appears concave and the strain value is very small. The deformation in this stage dominantly stems from the rock material itself, and the propagation and coalesence of microcracks and the material damage are the principal deformation mechanism. However, it is found in the post-peak stage that ε_1 and ε_3 show good linear relations, and the generalized Poisson's ratio $v = \varepsilon_3/\varepsilon_1$ keeps a constant value which is usually more than 0.5, or even more than 1.0. It is difficult to explain and describe this behaviour by means of the theories of continuum mechanics.

mention. Rock specimen in the post-failure region has been essentially parted to form several blocks by the main fracture planes. So the predominant deformation mechanism in the post-failure region of soft rock is frictional-slide effect between the parted blocks along the main fracture planes. As long as the cracked blocks can keep structural stability under the restraint of confining pressure, the sliding deformation of rock specimen will become great more, which is often companied with great lateral dilatancy ($v > 0.5$).

In order to demonstrate the sliding deformation characteristics after failure of rock, a shear sliding model, based on the fracture pattern of tested specimens, is put forward as shown in Fig.5.

From the geometric relation of Fig.5, we get

$$\Delta_1 = \Delta \sin\theta \qquad (1)$$
$$\Delta_2 = \Delta \cos\theta \qquad (2)$$
$$\varepsilon_1 = \Delta_2/L = (\Delta/L)\cos\theta \qquad (3)$$
$$\varepsilon_3 = \Delta_1/D = (\Delta/D)\sin\theta \qquad (4)$$
$$v = \varepsilon_3/\varepsilon_1 = (L/D)\tan\theta \qquad (5)$$

The symbols of above equations are shown in Fig.5.

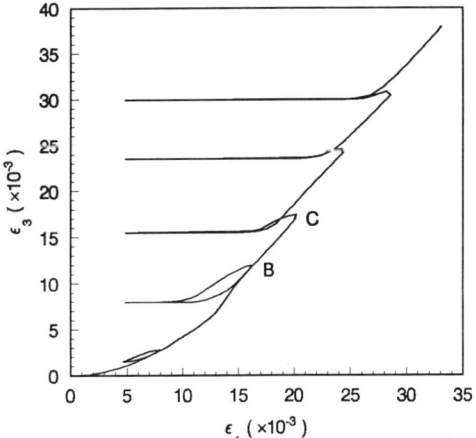

Figure 4. the $\varepsilon_1 \sim \varepsilon_3$ curve of roof mudstone under the cyclic loading condition ($\sigma_3 = 10$MPa)

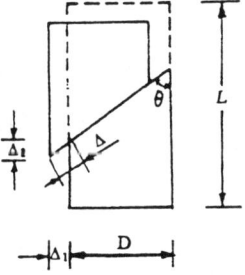

Fig.5 The shear sliding model

Based on above experimental results, we can find that one or two visible macrocracks at the specimen surfaces have initiated when the peak strength is reached. These macrocracks usually run through between two ends surfaces or end and lateral surfaces of rock specimen. We call these macrocracks as main fracture planes. As soon as the main fracture planes appear, the mechanical behaviour of rock specimen is mainly controlled by the planes and the influence of microcracks and micropores on the deformation properties of specimen becomes too insignificant to

If the fractured angle θ of specimen is obtained from the test, the generalized Poisson's ratio v can be calculated from equation (5). On the other hand, the value of v can be also directly measured from Fig.4. The values obtained by the two methods are shown in Table 1. The difference between them is less than 10%. It is thus clear that such a simple sliding model can also agree well with the test results. It can be seen from equation (5) that v is related to both the length/diameter ratio and the fractured angle of specimen, and it is a constant. This also verifies the structure effect sliding along the main fracture planes during the post-peak deformation of rock specimen.

4 CONCLUSIONS

1. The post-peak strain-softening behaviour is

Table 1 Comparison of the predicted and tested generalized Poisson's ratio v

Code No. of specimen	Confining pressure (MPa)	Ratio of height to diameter	Fracturing angle (°)	Predicted value of v	Tested value of v	Error (%)
FL-1	5	2.40	76	0.60	0.66	9
FL-2	10	2.44	57	1.67	1.58	5
FL-3	15	2.04	53	1.43	1.48	3
RO-1	5	1.91	63	0.97	1.07	9
RO-2	10	2.26	57	1.47	1.43	3

actually a kind of structure effect of fractured rock, and the friction caused by the mutual sliding between the cracked blocks is the predominant deformation mechanism in post-failure region of rock.

2. The generalized Poisson's ratio v of the fractured rock is a constant which may be more than 0.5. Therefore, deformation of fractured rock is often characterized by great lateral dilatancy.

3. The shear sliding model put forward in this paper has given a better description to the geometric relations of the post-failure rock, and the model agreed well with the results obtained from the test.

ACKNOWLEDGMENTS

This Project 49602039 is supported by National Natural Science Foundation of China (NSFC). All tests were carried out in the Centre-Laboratory of Rock Mechanics & Ground Control, China University of Mining and Technology, Xuzhou, Jiangsu, China.

REFERENCES

Bieniawski, Z.T. 1970. Time-dependent behaviour of fractured rock. *Rock Mechanics*. Vol.2, 123-137.

Chen, Q.M. 1996. The mechanical behaviour of broken rock under low confining pressure and its application. *Mining Science and Technology*, Guo & Golosinski (eds), Balkema, 739-742.

Dong, F.T. etc. 1994. Roadway support theory based on broken rock zone. *J. of China Coal Society*. Vol.19, No.1, 21-32

Farmer, I.W. 1983. *Engineering behaviour of rocks*, second edition. Chapman and Hall, London. 77-80, 135-139.

Frantziskonis, G. & Desai, C.S. 1987. Constitutive model with strain softening. *Int. J. Solids Structures*, Vol.23, No.6, 733-768.

Yaeger, J.C. & Cook, N.G.W. 1983. *Foundations of rock mechanics*, Chapman and Hall, London. 57-63.

The effect of the number of joints on the failure pattern of discontinuous rock mass around circular hole

Eui-Kwon Cho & Young-Seok Kim
Chonbuk National University, Chonju, Korea

ABSTRACT : In order to investigate the failure pattern around the circular hole in discontinuous rock mass, uniaxial and biaxial compressive tests were carried out for limestone, FEM analysis was performed. Test conditions are presented as follow. Number of joints are changed of 2 steps(n = 1,3) and inclination of joint changed of 4 steps(α = 0°, 30°, 60°, 90°).

1. INTRODUCTION

Underground structures such as tunnels, shafts, storages and appurtenences are structures completely encased and housed into existing rock mass. So In relation with stability of structure, it is very important factor to understand and grasp the behavior of rock masses. By the way, the in-suit rock is not intact rock, but contain natural defects such as joints, fractures, faults and as if well known.

And then current studies tend to carried out compression and shear test using relatively small specimen approached the condition of in-suit rock. And when circular hole was excavated in jointed non-rock material such as plaster and cement mortar many studies for mechanical behavior of material around the hole were performed.

In this study, to search failure patterns of the discontinuous rock around circular hole which it was excavated in the jointed rock masses, we studied follows.

1. Under the uniaxial and biaxial compressive test, the effect of the variation of the number of joints and the inclination of joint on the failure pattern of rock mass around circular hole.
2. By FEM analysis, the effects of the variation of the number of joints and the inclination of joint upon the stress distribution.

2. SPECIMENS

Specimens used in this study is Cheon-ho Limestone (located Yusan-myon Iksan city Chonbuk Korea) and the size of specimen is 200 × 100 × 45mm(length × width × depth).

The artificial joint plane of specimen was prepared by rock cutting saw and grade by surface grinding machine with diamond wheel. And number of joints(n) is one and three, and inclinations of joint(α) is 0°, 30°, 60° and 90° for longitudinal axis.

The circular hole was placed on the center of specimen. In case of single joint, the joint plane(central joint in case of three joints) was corresponded with the center line of hole

The interval distance between joints and diameter of hole are 32mm. the joint plane was adhered by chloroethyrene emulsion adhesives and was dried in nature.

Completed shapes and Material properties of Cheon-ho limestone are showed in and Fig 1 and Table 1 respectively.

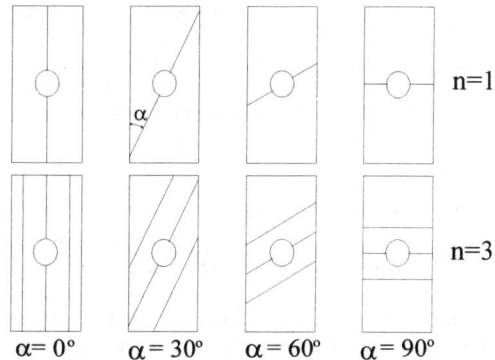

Fig. 1. Schematic diagram of specimens used in test.

Table 1. Material properties obtained in test.

Specimen	Limestone	Remark
Young's Modulus	6.12×10^5	(kg/cm^2)
Poisson's Ratio	0.275	
Density	2.69	(g/cm^3)
Uniaxial compressive Strength (kg/cm^2)	772.4	Intact
	456.7	Intact(hole)
Biaxial compressive Strength (kg/cm^2)	835.2	Intact
	712.1	Intact(hole)

3 TESTING METHODS

3.1 Testing appratus

Testing appratus are divided into the universal testing machine and the biaxial loading equipment, the universal testing machine as automatic pressure control system, is largely divided into loading unit and control/measurement unit, and the maximum loading capacity is 100 ton.

In the biaxial compressive equipment, the lateral load(constraint load) is generated by injucting oil into ram (Samsung, sr-103, Korea) which it is capacity of 10 ton

3.2 Testing method

3.2.1 Uniaxial and biaxial compressive test

To observe failure patterns of jointed specimen, the uniaxial compressive test, on constantly maintaining displacement velocity of ram at 5mm per minute, was performed on specimen with various condition

In the biaxial test, after arranging the center of ram on the same level with the center of hole, lateral load was applied up to 5 ton and constantly maintained it. Axial loading employed the same condition with uniaxial test.

And to hold uniformly up lateral load, the lateral displacement permitted because when axial load is applied on specimens, lateral load increased.

4. RESULTS OF EXPERIMENT

4.1 Failure pattern under the uniaxial compression

The failure pattern of specimen with one joint is shown Fig. 1, the thick line in the figure is the joint, and numbers are the generation order of cracks.

In case of specimen with only a hole and specimen with α = 0° and 90°, the first tensile crack occurred in the crown and invert of hole (in case of α = 0°, separation of joint plane).

The second shear crack occurred toward the maximum shear stress direction on right and left of hole, together with the expansion and propagation of the first failure plane, and the third additional shear failure formed in the circumference of hole.

In case of the specimen with α = 30°, the failure was only the sliding failure along the joint plane and it is thought that sliding occur because the joint plane is nearly placed in the acute angle of the shear failure plane.

In case of α = 60°, sliding appeared along the joint plane, but it was very small. The second tensile crack was created in the crown and invert. And together with the propagation and expansion of second cracks, the third additional shear failure occurred on the joint plane, and an the ultimate failure is the second tensile failure plane.

Failure patterns of specimen with three joint planes are shown in Fig. 3, in case of specimen with α = 0° and 90°, the tensile cracks(in case of α = 0°, separation of center joint) was created in the crown and invert.

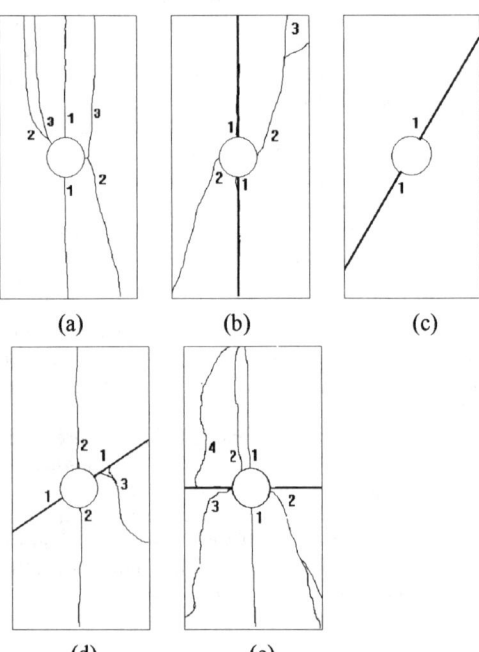

Fig. 2 Failure pattern of specimens with one joint under uniaxial compressive test.
(a) Intact rock with a hole, (b) α=0°
(c) α=30°, (d) α=60°, (e) α=90°

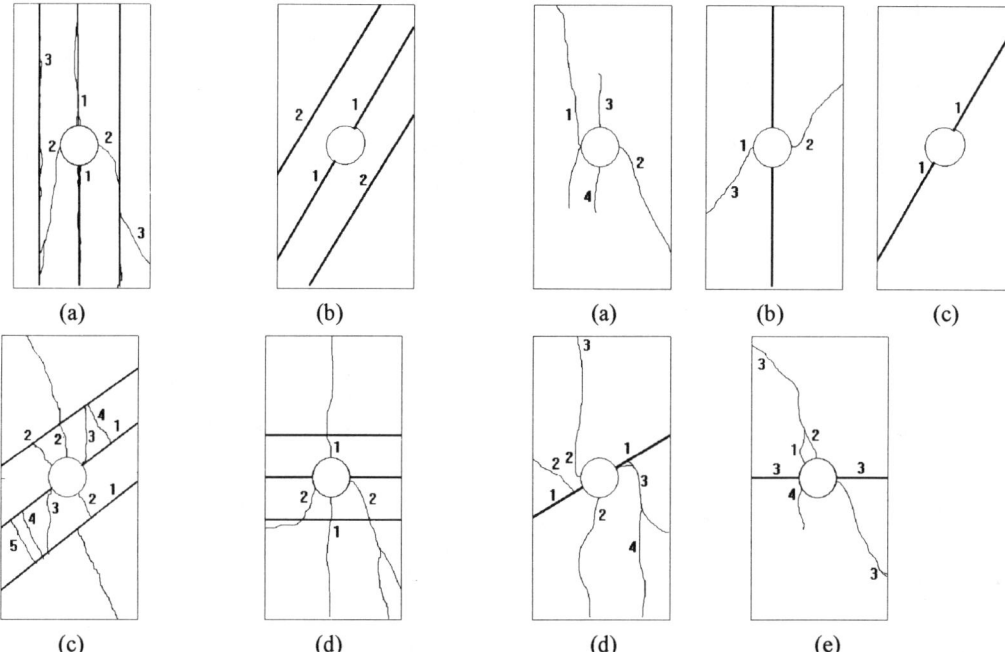

Fig. 3 Failure pattern of specimens with three joints under uniaxial compressive test.
a) α=0°, (b) α=30°, (c) α=60°, (d) α=90°

Fig. 4 Failure pattern of specimens with one joint under biaxial compressive test.
(a) Intact rock with a hole, (b) α = 0°
(c) α = 30, (d) α = 60°, (e) α = 90°

The second conjugate shear cracks which the shear crack occurred in the springing is closed with each joints was created together with the propagation and expansion of the first cracks. And the third additional cracks followed and the ultimate failure plane is first tension one.

In case of α = 30°, the sliding failure was brought about along each joints. In case of α = 60°, the sliding failure was observed along each joint plane, but was very small. And the second and third compressive shear crack occurred in material between upper and lower joints, and cracks and joints was closed. And the failure of inner material followed additionally.

4.2 Failure pattern under the biaxial compression

Failure patterns of the specimen with one joint are shown in Fig. 4, in case of specimen with only a hole and specimens with α= 0° and 90° respectively, the shear crack was created around the hole, was expanded and propagated and this crack became main failure plane.

On the hand, the very small crack occurred in the crown and invert, but it not affected on the failure of specimen.

In case of α = 30°, only sliding failure along the joint was brought about, and it is the same failure pattern in the uniaxial test.

In case of α = 60°, the sliding failure was appeared along the joint plane but it was very small. The expansion shear crack created in the springing was propagated toward the shear direction. Together with expansion of the second failure plane, the third shear crack was occurred from the hole, but the third crack is not affected on the main failure factor.

Fig. 5 is shown failure patterns of specimens with three joints, the shear crack occurred toward about 45° in case of specimens with α = 0° and 90°. The second shear cracks together with the expansion of the first cracks was propagated through the joint plane, and then specimen failed.

In case of α = 30°, the only sliding occurred along each joints. In case of α = 60°, firstly, the very small sliding failure was formed along the joints, the shear failure was created in perpendicular to the joint plane. And the additional cracks occurred in material between upper and lower joint planes.

597

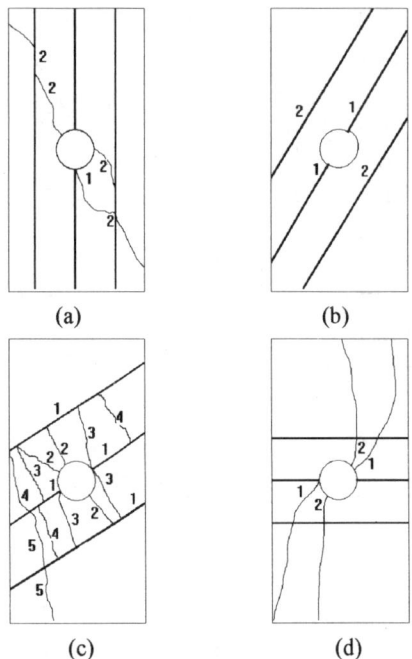

Fig. 5 Failure pattern of specimens with three joints under biaxial compressive test.
a) α=0°, (b) α=30°, (c) α=60°, (d) α=90°

Table 2 Material properties used in FEM analysis

	Block	Joint
Young's Modulus(kg/cm^2)	6.12×10^5	2.8×10^5
Poisson's Ratio	0.275	0.369
Density(g/cm^3)	2.69	1.85

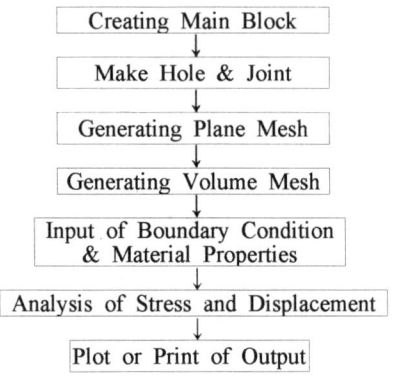

Fig. 6. Flow chart.

4.3 FEM analysis and results

To compare above test results with stress distributions by FEM analysis on jointed rock masses around the circular hole, FEM analysis was carried out under the same conditions with above test. The model used in this analysis assumed the plane stress condition and FEM program is MACTRAN[1] which use in only macintosh personal computer.

To investigate stress distributions of jointed rock with various conditions which number of joints(n) is one and three and inclination of joint is 0°, 30°, 60° and 90°, material properties in this FEM analysis use values obtained in test. Material properties is in shown Table 2, the procedure of analysis is shown in Fig. 6.

4.3.1 Stress distribution around circular hole

Fig. 7 is shown the maximum principle stress contour of models with three joint planes and inclination angle α = 0° and 60° and Fig. 8 is

---
1) The programer is Jae-Young Lee as professor, agricultural college, Chonbuk Nat. Univ. This Program can carry out mechanical behavior of 3-dimension structure (stress, deformation and heat conduction analysis etc.)

shown shear stress contour of model with three joint planes and inclination α = 30° and 60°. In case of α = 0°, the peak tensile stress is distributed in the central joint plane(or crown and invert) and the compressive stress in the springing, so that, it is anticipated that the failure of model occur by the tension stress.

As shown in Fig. 7(b) and Fig. 8(b), the compressive stress of the model in case of α = 60° is fully distributed, shear stress is distributed on section at about 45° to principle stress direction and correspond with failure plane of specimens in test, so that, it is anticipated that model failed by shear stress.

In case of α = 30°, the lowest shear stress is distributed on the central joint plane, so that, it is forecasted that sliding of the model occur along the joint plane by the shear stress.

In the biaxial condition, the stresses are showed symmetric distribution in case of specimens with α = 0° and 90° and is mirror symmetric distribution along the joint(or center one) in case of α = 30° and 60°. all of models are distributed compressive stress, stresses is discontinuously distributed along the joint planes.

The Maximum principle stress contour of the model with three joints and α = 60° and 90° under the biaxial condition is shown in Fig. 9.

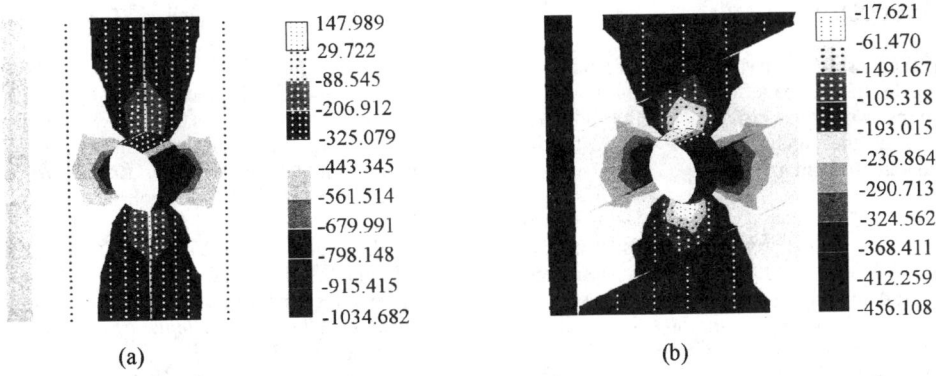

Fig. 7 Max. principle Stress contour of specimen with three joints in uniaxial compressive condition. (a) α = 0°, (b) α = 60°

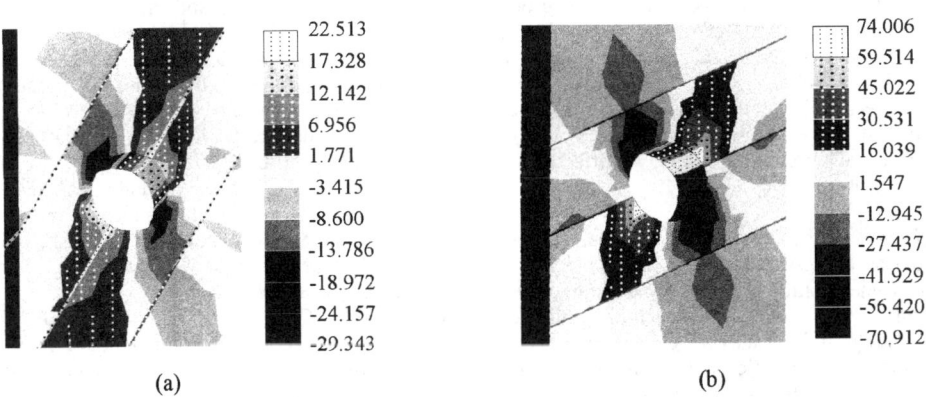

Fig. 8 Shear stress contour for specimen with three joints in uniaxial compressive condition. (a) α=30°, (b) α=60°

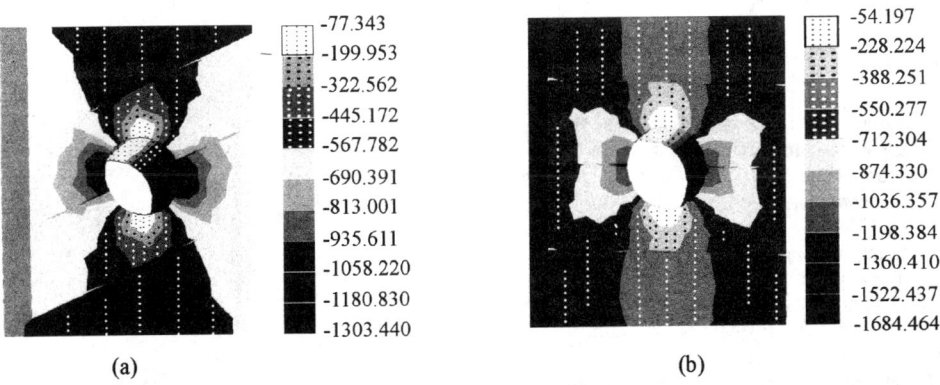

Fig. 9 Max. principle stress contour for specimen with three joints in biaxial compressive condition. (a) α = 60°, (b) α = 90°

5. CONCLUSIONS

From the result of tests and FEM analysis to investigate failure patterns of discontinuous rock specimens with the variation of the number of joints and the inclination of joint, following conclusions are obtained.

1. Under the uniaxial compressive test, failure pattern are divided into 3 groups, namely 1) sliding failure($\alpha = 30°$), 2) expansion tension failure and 3) sliding and tensile failure.
Especially, specimen with three joints and $\alpha = 60°$ is compressive shear failure.
The effect of the number of joints scarcely appear at $\alpha = 90°$ and the effect of the inclination of the joint conspicuously appear in $\alpha=30°$

2. Under the biaxial compressive test, the failure patterns are divided into 2 groups regardless the number of joints, namely, 1) sliding failure($\alpha = 30°$) and 2) shear failure ($\alpha = 0°$, $60°$ and $90°$)
The effect of number of joints distinctly appeared at $\alpha = 60°$ and influence of the inclination of joint is largely effected in specimen with $\alpha = 30°$.

3. From results of FEM analysis, stresses distributions are divided into 2 groups along the joint(or center joint), namely, 1) symmetric($\alpha = 0°$ and $90°$) and 2) mirror symmetric distribution ($\alpha = 30°$ and $60°$).
The effects of tensile and compressive stress appear in the uniaxial condition, but in the biaxial condition appear only the effects by the compressive stress.
And stresses are discontinuously distributed by joint plane, so that, this is predicted that stresses are released in joint plane.

REFERENCES

1. B. Amadei, W. Z. Savage, 'Analysis of borehole expansion and gallery tests in anisotropic rock masses', Int. J. Rock Mech. Min. Sci. & Geomech. Abstr., vol. 28, No. 5, 1991, p.383-396.
2. B. Amadei, 'Strength of a regularly jointed rock mass under biaxial and axisymmetric loading conditions', Int. J. Rock Mech. Min. Sci. & Geomech. Abstr., vol. 25, No. 1, 1988, p. 3-13.
3. B. C. Haimson, I. Song, 'Laboratory study of borehole breakouts in Cordova Cream: a case of shear failure mechanism', Int. J. Rock Mech. Min. Sci. & Geomech. Abstr., vol. 30, No. 7, 1993, p. 1047-1056.
4. E. Z. Lajtai, V. N. Lajtai, ' The collapse of cavities', Int. J. Rock Mech. Min. Sci. & Geomech. Abstr., vol. 12, 1975, p. 81-96,
5. H. S. Yang, H. K. Lee, 'Scale model test and numerical analysis for rock behaviors around the excavation opening with discontinuous model materials', J. Korean Institute Mineral & Mining Eng., vol. 27, 1990, p. 390-411.
6. H. Y. Chung, H. S. Yang, H. K. Lee, 'A study on the deformation behavior of soft rock around the underground opening', J. Korean Institute Mineral & Mining Eng., vol. 25, 1988, p. 332-339.
7. J. W. Kim, H. K. Lee, 'A study on the deformation behaviors around mining roadways in layered rocks', J. Korean Institute Mineral & Mining Eng., vol. 25, 1988, p. 320-331.
8. W. Zhu, P. Wang, 'Finite element analysis of jointed rock masses and engineering application', Int. J. Rock Mech. Min. Sci. & Geomech. Abstr., vol.30, No.6, 1993, p.537-544.
9. R. T. Ewy, 'Yield and closure of directional and holizontal wells.', Int. J. Rock Mech. Min. Sci. & Geomech. Abstr., vol 30, No. 7, 1993, p. 1061-1067.
10. S. P. Lee, C. I. Lee, 'A numerical analysis of stress and deformation behavior around a cavern in a discontinuous rock mass', J. Korean Institute Mineral & Mining Eng., vol. 27 1990, p.268-282.
11. T. Aoki, C. P. Tan, W. E. Bamford, 'Effects of deformation and borehole failures in sat-saturated shales', Int. J. Rock Mech. Min. Sci. & Geomech. Abstr., vol. 30, No. 7, 1993, p. 1031-1034.

Fractal characteristic of width of branching cracks in coal mass

Jin Shen
Institute of Rock and Soil Mechanics, Chinese Academy of Sciences, Wuhan, People's Republic of China

Yangsheng Zhao & Kanglian Duan
Mining Technology Institute, Shanxi Mining University, People's Republic of China

ABSTRACT: Based upon the concept of the fractal law of diameter exponent followed by the branching phenomena of tree trunks and of blood vessels of biological or floral system in nature[1], the authors have performed lab measurement of the widths of the main cracks and micro-ones branching from the former on a great of number of coal specimens with various metamorphic degrees sampled from coal fields of six mining areas. The results show: (a) after branching of the cracks in coalmass, there exists a relationship between main crack and those branching from it which also, in statistic sense, follow the above stated diameter exponent law, $d^\Delta = d_1^\Delta + d_2^\Delta$; (b) all the values of Δ are less than 2, having an average value of $\Delta \approx 1.16(1.06 \sim 1.23)$, and (c) the branching angle and the diameter exponent are varing in inverse proportional manner. What is described in this paper is a very significant exploration and study, which is very valued for its reference to many aspects in engineering.

1 INTRODUCTION

The distributing rule of rockmass cracks has been being a formidable problem hindering the development of rock mechanics. Researchers have, for many years, guided rock engineering constructions in some way only through the statistical results of in situ observation or investigation on a concrete geological body. In recent years, fractal geometry, founded by Mandelbrot, has been widely applied to the research on rock mechanics, and especially, a great deal of explorative work about the distributing rule of rockmass cracks has been carried out. Seeking and determining the fractal rule of rockmass cracks based upon measuring data for guiding engineering construction has become a research topic on which many researchers focus their attention.

As concerns the study of the distributing rule of rockmass cracks, it can be generally divided into the following main types, i.e., study on the fractal effect of the roughness of a single crack[1]; studying on the distributing form of a crack system[2,4]; quantitative study on the relationship between the number and the size of cracks' distribution in a rockmass[3] and on the fractal effect of branch angle of branching cracks[2]. Up to now, however, no publication has been reported concerning the study on the rule of the widths of rockmass cracks before and after their branch. Mandelbrot et al.[1] has carried out through study on the branching phenomena of the biological circle in nature, such as those of tree trunk, of artery and bronchi of human being, and of river branch etc.. The studying results have shown that the biological and floral system in nature, for example, diameters of a tree trunk and its the exponent law as follows: $d^\Delta = d_1^\Delta + d_2^\Delta$, where d d_1 d_2 are the diameter of the trunk before brancing and that of the branch after branching respectively, Δ is diameter exponent. What interests us is if the crack branch of rockmass does follow aforementioned law which the system of living things on the earth follows. It well known that once cracks initiate in rockmass, they will, in general, extend to some lengths and then branch to a certain extent and finally taper out. Such criss-cross cracks and fissures, like blood vessels of a human being, comprise the passage way of underground water, natural gas and other liquids, depending on which the liquids can be stored or drained. In addition, the branching crack system plays a role of potential mechanism to adjust the interaction of solids and liquids, in this sense, it can be reasonably considered that there must exist a harmonic inherent rule of crack distribution in rockmass.

For the above reason, the authors have carried out, since 1991, systematically experimental study about the branching effect of cracks on a set of specimens of coal strata and the results show that there exists an ideal fractal rule about the widths of the crack and its branches before and after branching.

2 FRACTAL ANALYSING APPROACH TO CRACK BRANCHING

There are various crack branching phenomena in rockmass, but they have a common character, viz., irregularity of branching. The formation of crack branching system is assumed like that the new crack branch group at branching step $(i+1)$ is such a sub–branching group which initiates continually on the basis of the crack branch group of the last branching step (i), as a result, a self–similar branching system forms as shown in Fig. 1. Statistically, they can be considered as being of approximately statistical self–similarity, this irregular phenomena can be simulated by fractal geometry. According to fractal geometry, the fractal dimension for crack branching angle (Fig. 1(b)) can be calculated from eq. (1)

$$N = 3 \quad\quad 1/r = 2\cos\pi\beta/2$$
$$d_f = lg3/lg(2\cos\pi\beta/2) \quad\quad (1)$$

in which N—the number of generators,
r—similar ratio,
d_f—fractal dimension.

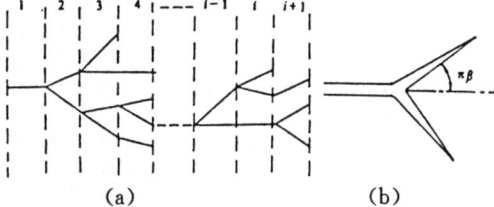

Fig. 1 Fractal model of crack branching

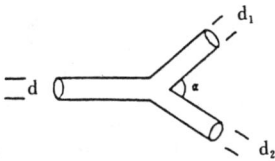

Fig. 2 Tree trunk branching model

Now, Fig. 1(b) is reducible, according to crack width, to that shown as in Fig. 2. According to the rule of branching of biological and floral systems described by Mandelbrot[1], take widths as variables and assume that the widths of a main crack before branching and of its sub–cracks after branching are d, d_1, and d_2 respectively, the width exponent of crack branching, Δ, can be calculated from eq. (2)

$$d^\Delta = d_1^\Delta + d_2^\Delta \quad\quad (2)$$

It should be pointed out that studying the branching effect of rockmass cracks by means of eq. (2) is, in some sense, a statistical rule, which offers a new approach to the study of rockmass crack branching of liquid flowing in rockmass and therefore is after all a valuable attempt.

2.1 Sampling of specimens

Coal specimens for observation were sampled from the main seams of several coal mines, including Kouquan Mine (Datong mine Bureau), Yinying mine (Yangquan mine bureau), Ximing mine (Xishan mine bureau), Gu Shuyuan mine (Jincheng mine bureau), and mine 5 of Hebi mine bureau. The specimens were coalified to different extents, including lowly metamorphosed gas coal and highly metamorphosed poor coal and blind coal. The characters of each coal seam are listed Table 1.

2.2 Study Approach

Firstly, prepare the lamp sections of various coal seams from in situ sampled in the same way as that used for coal petrography research, and then investigate or observe these lamp sections using a set of 10 × 10 microscope to find the real branching crack carefully (there exist a great number of intersections of inner weak planes in a coal–petrological mass which fairly look like branching cracks in outward). Afterwards, observe and measure the cracks widths before and after branching and the branching angles. In measuring the widths, because of irregularity of the cracks, the selected points at branching for measuring should be quite representative, and in general 3 to 5 widths are measured at these points. Finally, weighted average calculation is performed on the measured widths according to the crack length characterized by them. This weighted average width is taken as that for study.

3 FRACTAL RULES OF BRANCHING CRACKS

In the above described method, 84 sets of cracks in all were observed through a microscope to

Table 1 Characters of Specimens of Coal Seams

Sampling Place	Geological Time	No. of Seam	Code of Experiment	$R°_{Max}$ (%)	V	I	E	Mineral	Quality
Mine No. 1 (Yang quan)	Shanxi Group (Permo carboniferous system)	3	YQ	1.8180	91.05	5.09	0	3.86	black jack
Mine No. 5 (He bi)	Shanxi group (Triasic carboniferous system)	2	HB	1.9060	85.25	9.06	0	5.09	poor coal
Ximing Mine	Taiyuan Group (Permo carboniferous system)	8	XM	1.9816	78.85	13.42	0	7.73	poor coal
Yinying Mine	Taiyuan Group (Permo carboniferous system)	15	YY	3.2009	90.9	5.49	0	4.42	fat coking coal
Gushuyuan Mine	Taiyuan Group (Permo carboniferous system)	3	JC	41570	89.10	7.51	0	3.39	anthracite
Kouquan Mine	Taiyuan Group (Jurassic period)	5	L	0.8140	72.50	20.3	4.8	2.4	Weak coking coat

obtain more than 500 data. All these measured data were analysed and studied in detail and among them, those which had lower reliability were forsaken. The final results are listed in Table 2.

Based upon the analyses performed on the coal specimens of different quality sampled from 6 mines, the following conclusions have been reached:

1. The widths of coal—rock mass before and after branching fairly, in statistcal sense, follow the exponent law, $d^\Delta = d_1^\Delta + d_2^\Delta$. The diameter exponent with an average value of 1.66(1.06~1.23) are all less than 2 and the average variational coefficient is 22.3%(0.159~0.29), which shows the above stated rule is basically tenable.

2. The average branching angle after branching is 58.9°(56°~62°) and herefrom the calculated average fractal dimension of branching cracks is 1.67(1.66~1.68). The variational coefficients are very low (less than 3%), which shows that the rule under consideration is strictly tenable.

3. It can be seen from Table 2, that there exists a certain correlation between the diameter exponent Δ and the branching angle α of cracks. For almost all specimens except those from Yangqun Mining, the branching angle decreases with the increment of the diameter exponent and similarly, the branching angle decreases with increment of the ratio of the sum of the branching crack widths to the width of the main crack. According to Xie Heping[2], the fracture toughness of material crack development, k/k_0, will increase with the increment of the branching angle during crack development, which shows that the greater the branching angle is, the less probability for crack to develop. Provided the surface—energy for the same rock material at the crack tip is constant, the width of a crack with a smaller branching angle is necessarily greater than that of another crack with a greater branching angle. The reason for this lies in that the wider the crack is, the more energy it needs. The diameter exponent of crack branching more reasonably explains, from another angle, the branching rule of crack development.

Table 2 Date Describing Crack Branching Characters of Coal Specimens

coal specimen	diameter exponent(Δ)	variational coefficient(η)	$\frac{d_1+d_2}{d}$	branching angle	fractal dimension(d_f)
XM	1.06	0.159	1.00740	62°	1.68
JC	1.14	0.23	1.0654	58.7°	1.67
YY	1.16	0.224	107957	58.7°	1.67
L	1.17	0.20	1.0532	57.3°	1.67
HB	1.2	0.29	1.12713	56.0°	1.66
YQ	1.23	0.237	1.0759	60.8°	1.674
average	1.16	0.223		58.9°	1.67

4 APPLICATION OF DIAMETER EXPONENT RULE OF BRANCHING CRACK TO ROCK SEEPAGE

The main passage ways for the seepage in a rockmass comprise, like the blood vessels, cracks with different characters from each other, for instance, different lengths and different widths. In a human being's body, the energy is transmitted first through main vessels then stepwise to smaller and smaller blood vessels and finally through capillaries to cell organs, whereby the organs live and multiply. Similarly, when men inject water into rock strata, the water is injected first through boreholes and then transmitted to secondary cracks and finally reaches small fissures, voids and pores. As the result, the water is stored in small cracks or pore and damp the rockmass to achieve some engineering goal. On the contrary, man's engineering activity of developing underground water or natural energy resources (e. g. natural gas and petrolium) is the inverse course of the above water—injection, viz. liquids stored in micro—cracks and pores flow, through branching cracks, into main cracks and then into large size cracks and finally they are drained out from underground through boreholes. It is the branching crack that links up microcracks, pores and fissures. Consequently, it is of more significance to study the seepage character of branching cracks or the interaction of liquids and solids.

Now analyse the following geometrical—mechanical model of branching cracks (Fig. 3).

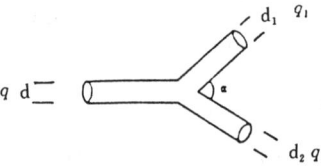

Fig. 3　Geometrical-mechanical model of branching cracks

According to the presumption of equally wide crack model, the movement rule of water flowing in cracks is expressed in eq. (3)

$$q = K_f J_f \qquad (3)$$

where q is the average flowing velocity, K_f the percolation coefficient of cracks and J_f the inner hydraulic gradient

$$K_f = gd^2/12v \qquad (4)$$

in which g is acceleration of gravity, d the crack width and v viscous factor for liquid flowing.

In substitution of the branching crack width into eq. (3) and eq. (4) respectively, one can get

$$q = (gd^2/12v)J_f \quad q_1 = (gd_1^2/12v)J_{f_1}$$
$$q_2 = (gd_2^2/12v)J_{f_2}$$

Assuming that the rock media around branching cracks are very poor in percolation, one can obtain the following equilibrium of flow capacity:

$$q = q_1 + q_2$$

namely

$$(gd^2/12v)J_f = (gd_1^2/12v)J_{f_1} + (gd_2^2/12v)J_{f_2}$$

It can be reducible to

$$d^2 J_f = d_1^2 J_{f_1} + d_2^2 J_{f_2} \qquad (5)$$

According to eq. (2)

$$d^\Delta = d_1^\Delta + d_2^\Delta$$
$$d = (d_1^\Delta + d_2^\Delta)^{1/\Delta}$$

Substituting it into eq. (5), then

$$(d_1^\Delta + d_2^\Delta)^{2/\Delta} J_f = d_1^2 J_{f_1} + d_2^2 J_{f_2} \qquad (6)$$

Provided diameter exponent $\Delta = 2$ and $J_{f_1} = J_{f_2} = J_{f_{12}}$, then one can get

$$(d_1^2 + d_2^2)J_f = (d_1^2 + d_2^2)J_{f_{12}}$$

This equation indicates that when the diameter exponent $\Delta = 2$, it is unnecessary to increase external energy for either main—micro branching crack or micro—main crack liquid flowing, in other words, for $\Delta = 2$, in either case (main—micro or micro—main) the liquid flowing in branching cracks is favourable.

If diameter exponent Δ is less than 2, for example $\Delta = 1$, in substitution of it into eq. (6) then

$$(d_1 + d_2)^2 J_f = d_1^2 J_{f_1} + d_2^2 J_{f_2}$$

Obviously, the above equation is tenable only when $J_f < J_{f_{12}}$, which shows that while extracting fluids such as petrolium, natural gas and water etc., owing to artificial-made negative pressure and pressure gradient of pore or fissure fluids, the pressure gradient of the fluids in branching cracks is liable to be higher than that of main cracks. This fact indicates that the branching crack system with a diameter of $\Delta < 2$ is more favouable for draining or extracting fluids from rockmass. On the contrary, this system makes injection of water into rockmass difficult, the reason for this is that the condition of $J_t > J_{f_{12}}$ should be satisfied for injection of fluids, therefore injection can be realized only when the flowing in and flowing out volume at branching cracks are not in equilibrium, this is obviously not an ideal situation.

If diameter exponent $\Delta > 2$, it is more conveninet to assume $d_1 = d_2$, $J_{f_1} = J_{f_2} = J_{f_{12}}$, then eq. (6) can be reducible to

$$(d_1^\Delta + d_2^\Delta)^{2/\Delta} J_f = (d_1^2 + d_2^2) J_{f_{12}}$$
i.e., $\quad (2d_1^\Delta)^{2/\Delta} J_f = 2d_1^2 J_{f_{12}}$
or $\quad 2^{2/\Delta} d_1^2 J_f = 2d_1^2 J_{f_{12}} \quad (7)$

From which, it can be derived that eq. (7) is tenable to keep an equilibrium state of flow capacity at branching cracks only when J_f is identically greater than $J_{f_{12}}$. This conclusion shows that such behaviour of branching cracks is more suitable to inject fluids into rock mass.

Now analyse the observing results of branching cracks in coal specimens. It has been found from experiments that the diameter exponent of branching cracks in the specimens are all less than two, i.e., $\Delta < 2$, having an average value of $\Delta \approx 1.16$. This character of coalmass is very favourable for draining liquids out from it. Geological study and thermolysis simulation of coal have shown that coal, during its long metamorphic historic period, produced a great deal of hydrocarbon gas, for example, one ton of coal produced about 170m³ hydrocarbon gas during metamorphosing into anthracite, among which only about 20m³ of the gas remain in coal seams. In other words, owing to the above described character of branching cracks, most of hydrocarbon gas escapes away from coalmass through micro-cracks and then through branching cracks, or it can be reasonably explained that the high pressure hydrocarbon gas forces coalmass to possess the above character and therefore smoothly to creat a dynamic equilibrium between fluid and coal mass. This is the mystery of nature.

REFERENCES

B. B. Mandelbrot, The Fractal Geometry of Nature. 1982. New York, W. H. Freeman.

Heping Xie, Fractals in Rock mechanics. A. A. Balkma. 1993.

Zhao Yangsheng. Kang Tianhe. Hu Yaoqing. Permeability Classification of Coal Seams in China. Int. J. Rock mechanics and Mining Science. 1995. Vol. 32. No. 4.

A practical approach for obtaining ground reaction curve for tunnels under squeezing ground conditions

R.K.Goel
CMRI Regional Centre, CBRI Campus, Roorkee, India

J.L.Jethwa
CMRI Regional Centre, Shankar Nagar, Nagpur, India

B.B.Dhar
Central Mining Research Institute, Dhanbad, India

ABSTRACT: Ground reaction curve concept is quite useful for designing the supports specially for tunnels through squeezing ground conditions. An easy to use empirical approach for obtaining the ground reaction curve is proposed in the paper. The approach has been developed using the correlations of estimating support pressures and closures in tunnels proposed by Goel et al. (1995) and Goel (1994) respectively.

1 INTRODUCTION

Support design in tunnels through squeezing ground conditions is a difficult task. Tunnel closure as high as 20 per cent of the tunnel size has been recorded in an Indian tunnel driven through phyllites of lower Himalayas experiencing squeezing ground conditions. This type of ground condition is commonly encountered in tunnels excavated through weak, fractured and jointed rock masses with high insitu stresses.

In order to reduce the closure under squeezing ground condition one can not opt for stiffer supports because it would attract high support load and therefore the support costs would also be high. On the other hand, more flexible support can also not be used as learnt from the ground reaction curve concept. Measured values of support pressure in Indian tunnels have indicated that the support pressure again started rising when the tunnel closure is allowed to exceed 5 per cent of the tunnel size. Therefore, an optimum support system is desired to avoid time and cost over-runs besides safety of the workmen. The problem of optimum support design in squeezing ground condition may be solved by using the ground reaction curve concept.

The analytical approach for obtaining ground reaction curve, though quite popular, requires expertise for selecting the values of input parameters. This approach, therefore, may be difficult for field engineers and geologists.

An easy to use empirical approach for obtaining the ground reaction curve for tunnels through squeezing ground conditions is proposed in the paper. However, initially the ground reaction curve concept is presented in brief.

2 GROUND AND SUPPORT REACTION CURVES CONCEPT

Figure 1 shows ground and support reaction curves between the required support pressure (p) and the normalised tunnel closure (u_a/a). Attempts to restrict any radial closure results in very high pressure requiring a very stiff support system at the tunnel face itself. This would be expansive besides resulting in low construction rate. The support pressure can be brought down by allowing the tunnel to deform. As the tunnel deforms, a broken zone is formed. The advantage of a broken zone can be maintained till the failed rock mass retains some cohesion. If the tunnel is allowed to deform beyond an optimum limit, shown by point 'A' in Fig. 1, the failed rock mass looses its cohesion and the ground arch is destroyed leading to a rise in support pressure. Figure 1 also shows four possible types of support reaction curves. A very stiff support system (represented by curve 1) would not allow desired radial tunnel closure and, therefore, the support pressure would be very high. A very flexible support (Curve 3), on the other hand, would permit excessive deformation leading to

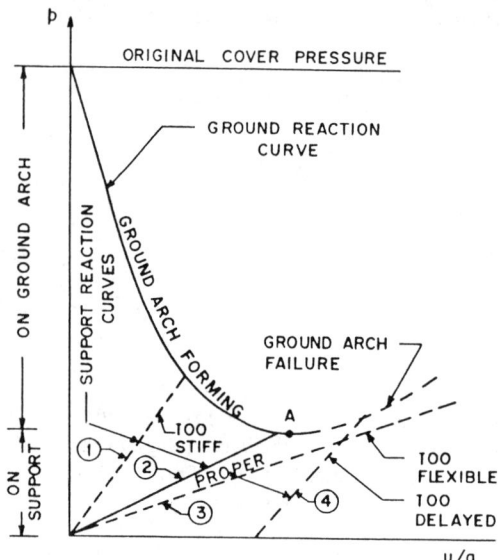

Fig. 1. Schematic presentation of ground and support reaction curves under squeezing conditions

3 THE APPROACH

For developing the empirical approach of ground reaction curve, the correlations suggested by Goel et al. (1995) and Goel (1994) for estimating support pressure and closure respectively have been used. The main input parameters are - (i) Rock Mass Number N (defined as Barton's Q with SRF = 1) and preferred over Q to avoid the uncertainties in selecting the rating of SRF, (ii) tunnel depth H in metres to consider the stress effect in absence of SRF and (iii) tunnel radius a in metres to consider the effect of confinement on insitu crushing strength of the rock mass. The two correlations are :

$$p_{sq} = (f/30) \, [10^{(H^{0.6} \cdot a^{0.1}/50 \cdot N^{0.33})}] \quad (1)$$

$$u_a/a = [(0.095 \cdot H^{0.81})/(N^{0.27} \cdot K^{0.62})] \quad (2)$$

where p_{sq} = ultimate estimated support pressure for tunnels under squeezing ground conditions in MPa; u_a/a = normalised estimated tunnel closure in per cent; f = correction factor for tunnel closure as given in Table 1; H = tunnel depth or overburden in metres; a = tunnel radius in metres; N = rock mass number defined as Barton's Q with SRF = 1; and K = effective support stiffness in MPa.

A rise in the value of correction factor f for

failure of the ground arch. Even if a support of stiffness equivalent to curve is used but its installation is delayed (Curve 4), the ground arch may still be destroyed. A support of proper stiffness shall be installed soon after excavation (Curve 2) in order to obtain the maximum benefits of the ground arch formed due to the broken zone. In other words, the tunnel shall be allowed to deform to an optimum level (say $u_a/a \leq 5$ %) under controlled conditions in order to reduce the support pressure to an optimum level and thereby reducing the support cost.

Table 1. Correction factor f for tunnel closure (after Goel et al., 1995)

S.No.	Degree of squeezing	Normalised tunnel closure, %	Correction factor, f
1.	Very mild squeezing $(270 \, N^{0.33} \, B^{-0.1} < H < 360 \, N^{0.33} \, B^{-0.1})$	1 - 2	1.5
2.	Mild squeezing $(360 \, N^{0.33} \, B^{-0.1} < H < 450 \, N^{0.33} \, B^{-0.1})$	2 - 3	1.2
3.	Mild to moderate squeezing $(450 \, N^{0.33} \, B^{-0.1} < H < 540 \, N^{0.33} \, B^{-0.1})$	3 - 4	1.0
4.	Moderate squeezing $(540 \, N^{0.33} \, B^{-0.1} < H < 630 \, N^{0.33} \, B^{-0.1})$	4 - 5	0.8
5.	High squeezing $(630 \, N^{0.33} \, B^{-0.1} < H < 800 \, N^{0.33} \, B^{-0.1})$	5 - 7	1.1
6.	Very high squeezing $(800 \, N^{0.33} \, B^{-0.1} < H)$	> 7	1.7

NOTE: Normalised tunnel closure is defined as radial tunnel closure expressed in per cent of tunnel radius tunnel closures beyond 5 per cent is due to the increase in the loosening pressure which is reflected in a rising 'ground reaction curve' (Fig. 1). Tunnel closures should normally not be allowed to exceed 5 per cent of the tunnel size. However, in case of weak rock masses with depths exceeding 500m, it may be necessary to permit higher closures to bring down temporary support requirements to manageable levels for ease of supporting at face which is necessary for faster drivage. Such higher tunnel closures will be associated with larger plastic zones which will be mobilising relatively higher ultimate support pressures requiring higher support capacities. In other words, attempts to reduce support requirements closer to tunnel face will be associated with a thicker tunnel lining.

The correction factor f in Table 1 does not take into account the method of excavation. For larger tunnels (dia. > 9m) under high squeezing ground conditions, full face excavation will not be possible and therefore, heading and benching method of excavation would have to be adopted. In heading and benching method, the tunnel wall closures would be excessive and the value of f would also be very high (Singh et al., 1997).

It may be noted here that the correlations for predicting support pressures (Eq. 1) and tunnel closures (Eq. 2) are backed by the measured values of support pressures and closures in Himalayan and other Indian tunnels experiencing squeezing ground conditions. Detailed data sheet can be referred in Goel (1994).

The parameters of Eqs. 1 and 2 may be collected without much difficulty at the design stage itself. The ground condition, non- squeezing or squeezing, can be find out by using the approach of Goel et al. (1995).

Using Eqs. 1 & 2, the ground reaction curves (GRC) are obtained. The methodology is described in the following paragraphs in the form of worked example. For the example, the tunnel depth H and the rock mass number N have been assumed as 500m and 1 respectively and the tunnel radius as 5m.

3.1 GRC Using Eq. 1

In Equation 1, as described earlier, f is the correction factor for tunnel closure. For different values of permitted normalised tunnel closure (u_a/a), different values of f are proposed in Table 1. Using Table 1 and Eq.1, the support pressure (p_{sq}) values have been estimated for the assumed boundary conditions and for various values of u_a/a (column 1) as shown in Table 2. Subsequently, using the p_{sq} (column 3) and u_a/a (column 1) from Table 2, GRC has been plotted for u_a/a up to 5 per cent (Fig. 2).

3.2 GRC Using Eq.2

For obtaining GRC from Eq. 2, the following equation of support stiffness would also be used.

$$K = [\,p\,/\,(u_a/a)\,] \qquad (3)$$

It is important to mention that u_a/a values for estimating K from Eq. 3 should be a dimensionless quantity and not in per cent. It means that instead of 1 per cent, the u_a/a value would be 0.01.

Using the values of u_a/a (dimensionless corresponding to percent value) and p_{sq} from columns 1 and 3 respectively of Table 2 in Eq. 3, K values (column. 4, Table 2) have been obtained.

Table 2. Showing calculations for constructing GRC using Eqs. 1 and 2

Assumed u_a/a (%)	Correction factor (f)	p_{sq} from Eq. 1 (MPa)	K from Eq. 3 using col. 1 & 3 (MPa)	u_a/a from Eq.2 for K at col. 3 (%)	f for u_a/a at Col. 5	p from Eq. 1 (MPa)	p from Eq. 3 using col. 4 & 5 (MPa)
(1)	(2)	(3)	(4)	(5)	(6)	(7)	(8)
0.5	2.7	0.86	172	0.59	2.6	0.82	1.03
1	2.2	0.7	70	1.04	2.2	0.69	0.73
2	1.5	0.475	23.75	2.05	1.4	0.44	0.48
3	1.2	0.38	12.66	3.02	1.15	0.36	0.38
4	1.0	0.317	7.9	4.02	1	0.31	0.32
5	0.8	0.25	5.06	5.37	0.85	0.27	0.27

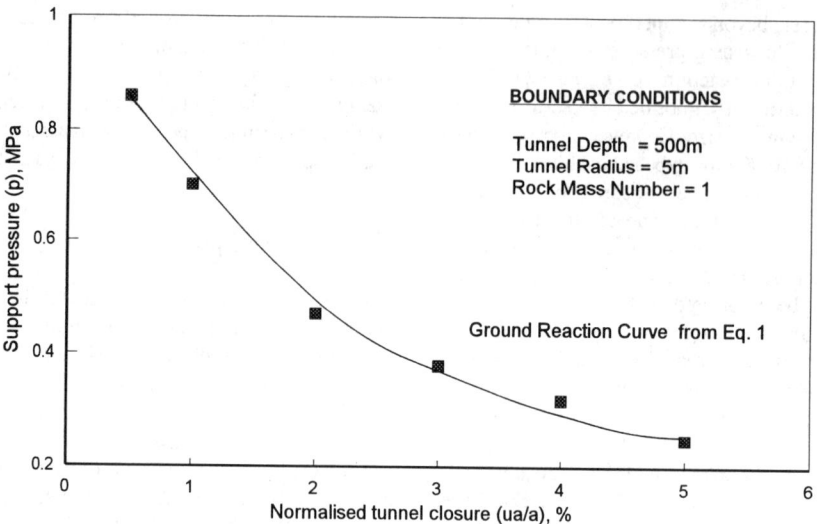

Fig. 2. Ground reaction curve obtained from Eq. 1

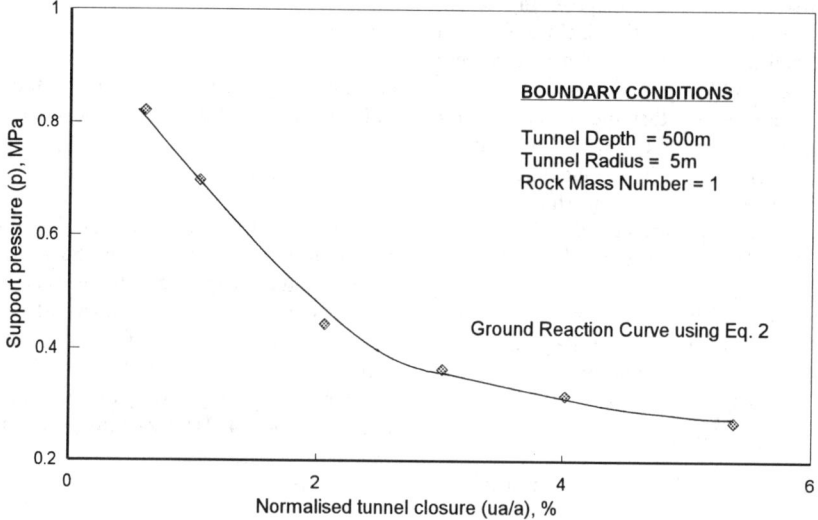

Fig. 3. Ground reaction curve obtained from Eq. 2

Using this K value in Eq. 2, normalised tunnel closure (u_a/a) is calculated for the given boundary conditions (H = 500m and N = 1) and tabulated in column 5, Table 2. This value of normalised tunnel closure, subsequently, is used to obtain support pressure from Eq. 1 (Column 7, Table 2) or from Eq. 3 (Column 8, Table 2). Three sets of values of support pressure and normalised closures are available for plotting three GRC. One set of data is given in Columns 1 and 3 (Fig. 2), second set is from columns 5 and 7 (Fig. 3), whereas third set is represented by columns 5 and 8.

It is interesting to see that though the two equations (Eqs 1 and 2) have been developed using different data, the ground reaction curves obtained from these two equations (Figs. 2 and 3) are practically identical. This has generated confidence on the applicability of the suggested approach.

It may be highlighted here that the approach is simple, reliable and users' friendly because he values of the parameters can be easily obtained in the field.

4 CONCLUSIONS

The approach for obtaining the ground reaction curve, discussed in the paper, is simple and reliable. The designers can use this approach for tunnel support designs in the jointed and weak rock masses experiencing squeezing ground conditions.

The worked example is for normalised tunnel closures up to 5 per cent. However, the approach can be utilised for higher values of closure by taking higher values of correction factor f, as mentioned in the paper.

REFERENCES

Barton, N., Lien, R. and Lunde, J. (1974). Analysis of rock mass quality and support practice in tunnelling and a guide for estimating support requirements, report by the Norwegian Geotechnical Institute.

Goel, R. K. (1994). Correlations for predicting support pressures and closures in tunnels, Ph.D. Thesis, Nagpur University, India.

Goel, R.K., Jethwa, J.L. and Paithankar, A. G. (1995). Indian experiences with Q and RMR systems, Tunnelling and Underground Space Technology, Vol. 10, No. 1, pp. 97-109.

Hoek, E. and Brown, E.T. (1980). Underground excavations in rock, Institution of Mining and Metallurgy, London.

Singh, Bhawani, Goel, R.K., Jethwa, J.L. and Dube, A.K. (1997). Support pressure assessment in arched underground openings through poor rock masses, Paper accepted for publication in Engineering Geology, Amsterdam.

AE characteristics on progressing fracture within solid media in multiple-stage rock shear tests

T. Shiotani, K. Matsumoto, M. Tsutsui & H. Chikahisa
Technological Research Institute, Tobishima Corporation, Chiba, Japan

ABSTRACT: In-situ rock shear tests are effective techniques for determining shear strength of rock masses. The tests are adopted to important structures such as dams and nuclear plants for estimation of the foundation strength. In the conventional tests, over 3 specimens under similar geological conditions are essential to determine the strength. A multiple-stage rock shear test, if the following problem is solved, will be a promising method which enabled us to determine the rock shear strength reasonably. It is a how to control the shear loading in each stage. In other words, the establishment of successful premonitory phenomena to the final fracture is essential in the multiple tests. Acoustic Emission (AE) is a well known premonitory phenomenon generated with the release of strain energy which is stored within the materials, and it leads us to useful information of the process of fracture. In the paper, the AE technique is applied to the multiple-stage rock shear tests, and the validity of AE, as a stage control index with estimation of the fracture in each stage, is discussed through the laboratory tests of mortar. Results confirm that AE activity affords us accurate controls of the tests and effective estimation of the progressing fracture.

1 INTRODUCTION

Current in-situ rock shear tests are performed for estimation of the foundation strength. Because the tests entail enormous costs and labor, its civil engineering applications are limited only to important constructions, such as dams and nuclear power plants and so on. On the other hand, a multiple-stage rock shear test is a useful technique because the test enabled us to obtain more promising results by testing only one specimen. One specimen test brings us some advantages: it takes labor and costs less than the conventional shear tests. It has some difficulties, however, the control of shear loading in stages is the most formidable aspect. In other words, the establishment of successful premonitory phenomena to the final fracture is essential in the multiple tests.

Concerning the method for evaluation of the fracture in multiple-stage tests, ISRM (Kovari et al 1983) suggested that "The axial load is... increased... until the corresponding peak strength is observed" in multiple failure states of triaxial tests. However, in stress control tests in brittle materials, the load reaching the peak strength means attaining the eventual final failure that occurs unexpectedly. (Funato et al 1991). Therefore, successful results may not be acquired without the establishment of effective indexes as to evaluate the fracture processes sensitively in every multiple-stage.

Upward movement of block with the shear-dilatancy is one of the keys to solve this problem. Barton (1973) and Ledanyi et al. (1970) studied in detail relations between the dilatancy and shear processes in rock materials. In the case of applying the movements by the dilatancy to an index of controlling the shear loads in multiple-stage tests, however, it must be assured that the relations are the universal behavior in every kind of material. Tanaka et al (1977) examined these relations due to the geological conditions: joint characteristics and rock properties, and they concluded that the fracture processes and displacement patterns are strongly related to the geological conditions and they are tightly dependent on the joint conditions/ a variety of rock themselves. Yoshinaka et al. (1974) discussed the joint shear strength of granite, and they concluded that the joint strength of granite is strongly influenced by the magnitude of normal stress rather than an influence of weathering states in rock. Barton (1971) mentioned that there are certainly relationships between the joint roughness and the joint shear strength. Hence, it is summarized that the upward movements of block with dilatancy in rocks are different in geological conditions and influenced by their joints, and therefore, it is difficult for shear-loading control to be an index on the beginnings of the fracture.

Generally, light, heat and sound are physical phenomena premonitory to the failure. AE (Acoustic Emission) is a well known active index belonging to the sound, and it is closely related to the states of fracture in the materials. AE measurements of the rock have been performed by seismologists. Mogi (1962a, 1962b), Utsu (1965) and Sholz (1968) reported that the AE occurring mechanisms are very similar to those of earthquakes, and the fracture process could evaluate by b-value/ m-value defined by the AE peak-amplitude distributions. Moreover the b-value was refined recently as being a precursor of slope failure by Shiotani et al. 1994.

In the rock shear process, a number of AE studies (e.g. Koerner et al. 1981) were performed. Ishida et al (1986), for example, applied AE to the in-situ direct shear test, and they mentioned that the AE occurrence rate is closely associated with the fracture of rock, and compared AE with displacement in fracture process, and emphasized the superiority of AE to the displacement in the sensitivity of the fracture process. Funato et al (1991) also indicated this advantage of AE as a premonitory phenomenon, and they concluded that there are certain relationships between the AE activity and the beginnings of the dilation. However, because the shear loading control at each stage is difficult in the multiple stage shear tests, successful works applying AE techniques have not been reported as far as we know.

In a paper, basic issues which should be considered in AE applied multiple-stage shear tests are studied. Because, in cyclic tests applying AE techniques, there is a relation between loads and AE activity that is called the Kaiser effect: in the loading process, there are no emissions at all until the previous maximum stress. Firstly, the establishment of the Kaiser effect is examined in the tests. Secondly, the possibility of AE as a stage control index is discussed in every vertical loading level. Finally, cohesion and angles of internal friction of Coulomb's failure criterion are determined from every fracture state evaluated by AE, and the both coefficients are compared with those from the conventional shear tests.

2 ACOUSTIC EMISSION

Acoustic Emission (AE) is a stress wave produced by sudden movement within stressed materials. The sources of AE are defect-related deformation processes such as crack growth and plastic deformation, and therefore, a seismic wave is the large-scale AE.

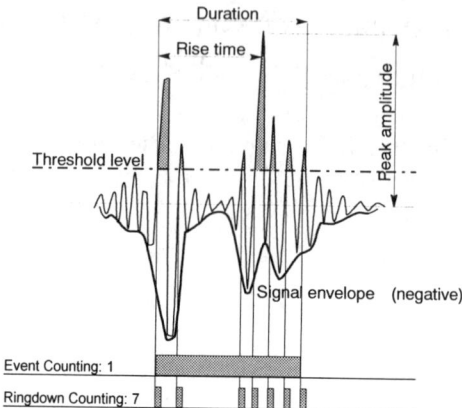

Figure 1. Illustration of conventional AE parameters in the AE envelope monitoring.

2.1 AE envelope monitoring

In the modern instrument approach, each AE signal is measured in detail as shown in Figure 1. The figure illustrated conventional AE parameters in the AE envelope monitoring. In AE monitoring, the sensitivity is governed by the threshold levels set by the operator, and only signals exceeded the threshold are acquired as AE. When we count the AE signals, there are two counting methods: event counting and ring-down counting as shown in the figure. Because the ring-down counts is defined by the number of times the AE signals crosses the threshold, it reflects the magnitude of emission source, while the event counts, that is defined the number of waves, don't reflect any information of waves. Accordingly, the ringdown counting is well used counting method and one of the effective AE parameters for evaluating fracture progressing. There are an other prameters in the AE wave as described below;

Peak amplitude; the peak voltage attained by the AE waveforms. This is a simple measurement of knowing the signal size. Amplitude is conventionally expressed in decibels relative to 1 microvolt at the sensor.

Energy; means the measured area under the rectified signal envelope in our AE equipment, the definition however, is dependent on each manufacturer.

Duration; the time from first to the last threshold crossing, and closely related to the ring-down count.

Rise time; the time from first threshold crossing to the signal peak, used for signal filtering aimed at discrimination of AE signals from electromagnetic noise.

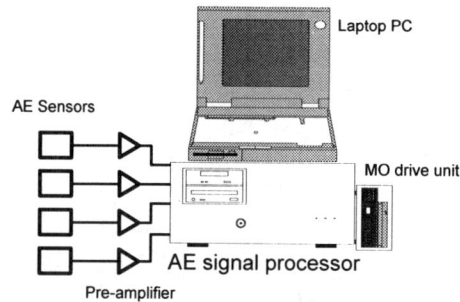

Figure 2. AE monitoring instrumentation.

2.2 AE monitoring system

Figure 2 shows AE monitoring instrumentation. AE is detected over threshold of 40dB by 150 kHz resonant AE sensors (Pac R15), and converted to an electric signal, amplified 40dB at a pre-amplifier, and processed some parameters in a signal processor (Pac Spartan 2000). Acquired signals as AE are transferred to a lap-top PC, stored in the hard disk, and analyzed in work stations using a magneto-optical disk.

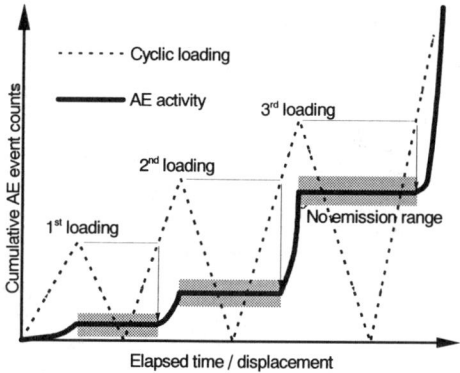

Figure 3. Relationship between AE, time and displacement in cyclic loading tests.

2.3 Kaiser effect

Because AE is produced by stress-induced deformation of the material, it is highly dependent on the stress history of the materials. AE testing is often carried out under conditions of rising load. Figure 3 shows the relationship between AE, time and displacement in cyclic loading tests. The first load application will typically produce much more emission than subsequent loadings. In fact, subsequent loadings should produce no emissions at all until the previous maximum load is exceeded. This behavior was first reported by Kaiser (1953) and has been a leading influence in the development of AE test methodology. Whenever we examine the relations between AE activity and loading levels in cyclic loading tests, maximum pre-loading levels have to always be considered into the interpretations of the test results.

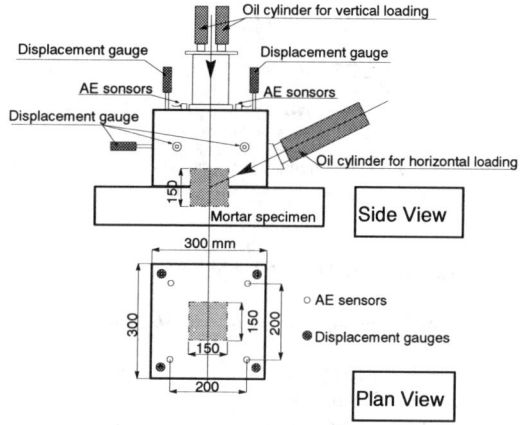

Figure 4. Experimental setup of the multiple-stage shear test.

3 EXPERIMENTAL STUDY

3.1 Test procedure

Figure 4 shows the experimental setup of the multiple-stage shear test. A specimen of mortar (15*15*15cm), covered with high-strength gypsum having compression strength of 2.3kN in wet conditions, was set in an experimental mold. Displacement in three directions are monitored by displacement gauges. Especially in this shear test, to know behavior of the dilation accurately, the displacement gauges having an accuracy of 0.001mm were adopted for vertical displacement. Four AE sensors of 150 kHz resonant type are attached on the upper surface of the covering gypsum with the couplant of high vacuum grease. The tests are performed under the constant strain rate of 1.0mm/ minutes. 1.3MPa (3tf), 2.6MPa (6tf) and 3.9MPa (9tf) of vertical loadings are given for each stage of the multiple tests.

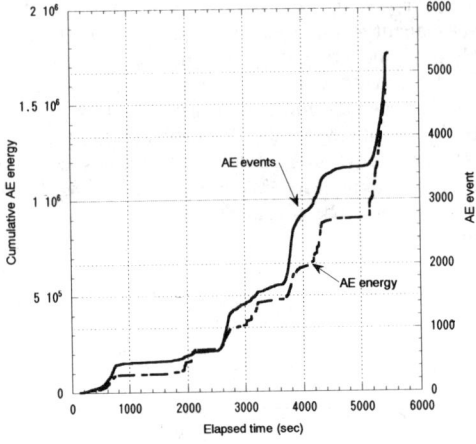

Figure 5. Different tendency between AE events and energy with elapsed time.

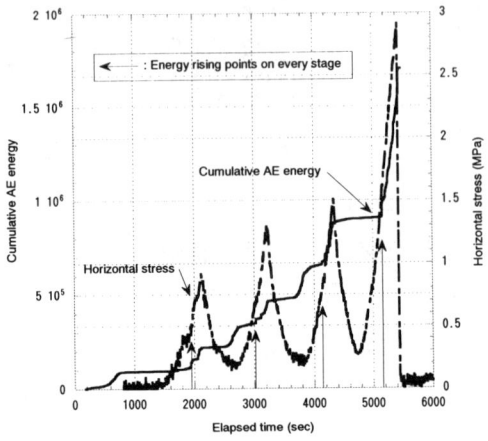

Figure 6. A loading path of horizontal stress and cumulative AE energy with elapsed time.

4 RESULTS AND DISCUSSIONS

4.1 *Evaluation differences due to AE parameters*

Figure 5, for example, shows the different tendency between AE events and energy with elapsed time. In the figure, AE events changes so smoothly that it is difficult to point out the sudden change. While in a AE energy curve, because it increases sharply, it is easy to find out the sudden changing points. Therefore, in this paper, the AE energy is tried to adopt as an index of the progressing fracture.

4.2 *Estimation of fracture stress in every loading stage with non-considering Kaiser effect*

Figure 6 shows a loading path of horizontal stress and cumulative AE energy with elapsed time. The solid lines with arrows indicate suddenly rising points in a cumulative energy curve. Note, the rising points in the figure are determined from a point of view that the suddenly points would exist within a limited range of loading processes. In this paper, "Non-considering Kaiser effect" means "taking no account of maximum pre-loading levels when we determine the rising points". It is summarized from the figure that there certainly exist the suddenly rising points of cumulative AE energy in every loading stage.

4.3 *Estimation of fracture stress in every loading stage with considering Kaiser effect*

Figure 7 shows an enlargement of figure 6, for

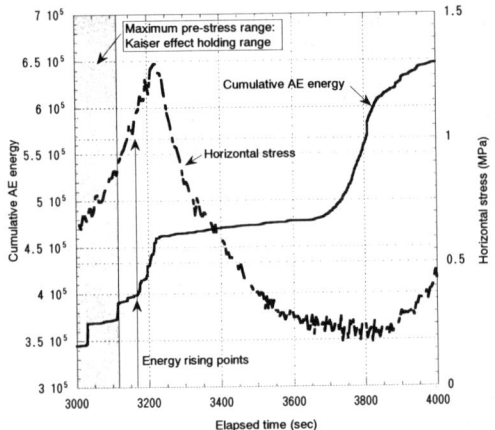

Figure 7. An enlargement of figure 6 from 3000 to 4000 seconds.

example, from 3000 to 4000 seconds. An energy suddenly rising point in the figure is determined from a point of view that the suddenly rising points would exist within limited ranges exceeded maximum pre-loading levels, described as "Considering Kaiser effect" in the paper. It is found from the figure that even if Kaiser effect holds for the next loading, the energy suddenly rising point of 3170 seconds obviously exist. In fact, this tendency could be observed in every loading stage. Therefore, the possibility of the failure estimation by AE was confirmed by this aspect.

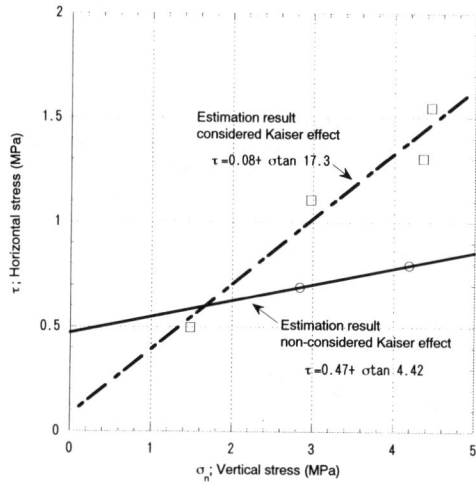

Figure 8. Calculated cohesion and angles of internal friction of Coulomb's criterion fitted it to the AE estimated results.

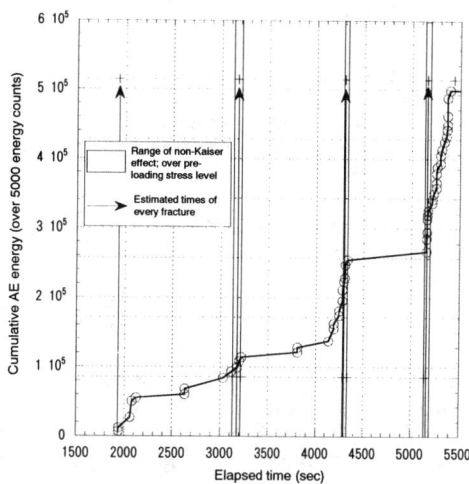

Figure 9. Cumulative AE energy over 5000 with elapsed time.

4.4 Applicability to Coulomb's criterion

Figure 8 shows the calculated cohesion and angles of internal friction of Coulomb's criterion fitted it to the AE estimated results. Cohesion and angles of internal friction from the conventional shear tests were 0.73 MPa and 19 degrees respectively. It is found that both coefficients of non-considering Kaiser effect are different from those of the conventional shear tests. However, in the case considering Kaiser effect, angles of internal friction is excellently identical to the value of the conventional tests, but cohesion is different. It is thought that the difference between both cohesion is attributed to the difference of the methodology for fracture evaluation i.e. the results from multiple-tests are evaluated by premonitory AE, however, those from the conventional tests are measured as the peak-strength being invariably greater than the strength evaluated by AE.

4.5 Possibility of AE energy control for shear loading in Multiple-stage shear tests

In fracture process of each stage, we observed large scale of AE energy as could be distinguished between micro fracture and macro. In this section, the possibility of shear load control by AE energy is discussed. Figure 9 shows the cumulative AE energy over 5000 with elapsed time. Solid lines with arrows express every fracture point estimated by considering Kaiser effect described in section 4.3; the shadowed areas show the range beyond the pre-loading level in each loading stage. Figure 10 shows

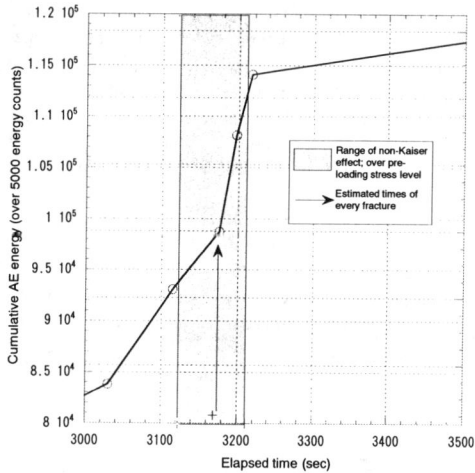

Figure 10. An enlargement of figure 9 from 3000 to 3500 seconds.

an enlargement of figure 9. It is clear from these figures that the suddenly rising points of the energy curve are dramatically conform to those evaluated as fracture in every stage. Consequently, shear load control by AE energy in multiple-stage tests is found out to be possible. In this study, the examination of loading patterns considering Kaiser effect is insufficient. However, an ideal loading pattern in the tests is found i.e. virgin loading range in every stage, as they exceed maximum pre-stress, should take as wide as it can for estimating AE sudden changing points easily.

5 CONCLUSIONS

In this study, AE technique was applied to the multiple-stage shear tests, and the validity of AE as a stage control index with estimation of the fracture in each loading stage was examined through the laboratory tests of mortar, and the following was concluded.

  1) By applying AE technique to the multiple-stage rock shear test, evaluation of the fracture in each stage is possible.

  2) Kaiser effect should be considered in cases where the shear load control by AE are adopted to the multiple tests.

  3) AE energy is able to control the shear loading in the multiple tests efficiently.

  4) When the shear load control by AE is adopted to the multiple tests, virgin loading ranges in each stages should take as wide as it can for estimating AE sudden changing points easily.

REFERENCES

Barton, N.R. 1971. A relationship between joint roughness and joint shear strength. *Proc. Symp.* ISRM, Nancy: 1-8.

Barton, N.R. 1973. Review of a new shear strength criterion for rock joints. *Eng. Geology* 7(4).

Funato, A., T. Yokoyama & H. Wada 1991. In-situ multistage direct shear test. JSCE. *Proc.23rd Symp. Rock Mech.*: 192-196. (in Japanese)

Goodman, R.E. 1963. Subaudible noise during compression of rocks. *Geol. Soc. Am. Bull.* 74: 487-490.

Ishida, T, T. Kanagawa, S. Sasaki & Y. Urasawa 1986. AE monitoring during the in-situ direct shear test applied to an underground cavern. JSCE. *Jour. Geotech. Eng.* 376: 141-149. (in Japanese)

Kaiser, J. 1953. Erkentnisse und folgerungen aus der messung von geräuschen bei zugbeanspruchung von metallischen werkstoffen. *Arch. Einsenhüttenwesen.* 24: 43-45. (in German)

Koerner, R.M., W. M. McCable & A. E. Lord Jr. 1981. Overview of acoustic emission monitoring of rock structures. *Rock Mechanics* 14: 27-35

Kovari, K, A. Tisa, H. H. Einstein & J. A. Franklin 1983. Suggested methods for determining the strength of rock materials in triaxial compression.
Revised ver., *Int. J. Rock Mech. Min. Sci. & Geomech. Abst.* 20(6): 283-290.

Mogi, K. 1962a. Study of elastic shocks caused by the fracture of heterogeneous materials and its relations to earthquake phenomena. *Bull. Earthq. Res. Inst.*, Univ. Tokyo 40: 125-173

Mogi, K. 1962b. Magnitude-frequency relation for elastic shocks accompanying fractures of various materials and some related problems in earthquakes (2nd paper). *Bull. Earthq. Res. Inst.*, Univ. Tokyo 40: 831-853.

Shiotani, T, K. Fujii, T. Aoki & K. Amou 1994. Evaluation of progressive failure using AE sources and improved b-value on slope model tests. JSNDI. *Progress in Acoustic Emission VII*: 529-534.

Sholz, C. H. 1968. The frequency-magnitude relation of microfracturing in rock and its relation to earthquakes. *Bull. Seismol. Soc. Am.* 58: 399-415.

Tanaka, T. & M. Furuta 1977. On the rock mass shear strength by in-situ block shear test. *Eng. Geology* 18-1·2: 1-12. (in Japanese)

Utsu, T. 1965. A method for determining the value of b in a formula $logn=a-bM$ showing the magnitude-frequency relation for earthquakes. *Geophys. Bull.*, Hokkaido Univ.,13: 99-103

Yoshinaka, R & M. Furuta 1974. Joint shear strength of granite. *Eng. Geology.* 15-2: 12-22. (in Japanese)

Evaluation of in situ hydromechanical properties of rock fractures at Laxemar in Sweden

Jonny Rutqvist & Chin-Fu Tsang
Earth Sciences Division, Lawrence Berkeley National Laboratory, Calif., USA

Daniel Ekman & Ove Stephansson
Division of Engineering Geology, Royal Institute of Technology, Sweden

ABSTRACT: Hydraulic injection tests were carried out on hydraulic conductive zones in the sub-vertical borehole KLX02 at Laxemar area in Sweden. The purpose of the testing was to determine storativity, transmissivity, and the stress dependency of the fracture transmissivity. Seven open fractures in three highly conductive borehole sections were tested at a depth of 270, 315 and 340 meters. Pulse, constant head injection and step-pressure tests were conducted with moderate to high injection pressures. The injection tests were analysed using a numerical model that accounts for coupled fluid flow and mechanical deformation in both the joint and the surrounding rock. The results show that the hydromechanical properties are very sensitive to the *in situ* conditions such as past shear displacement and mineral fillings. The transmissivities of the most hydraulic conductive fractures are relatively insensitive to normal stress due to a large initial hydraulic aperture caused by past shear dilation.

1 INTRODUCTION

The coupling of geohydrological and geomechanical processes in hard fractured rocks is an issue of concern in the performance assessment of nuclear waste repositories. Fractures and fracture zones are the dominating features for flow and contaminant transport in the bedrock. Laboratory results have shown that the fluid flow through natural fractures is strongly affected by the stress normal to the fracture. In the field, hydraulic conductivity generally decreases with depth partly because increased stress compresses the open fractures to a smaller aperture. However, despite high stresses, one can often find highly conductive fractures deep in the bedrock. Such fractures may be kept open by previous shear displacement which creates flow channels between the unmated fracture surfaces. A drill core of a borehole intersecting such a fracture may can be tested in the laboratory for stress versus transmissivity. However, the drill core represents only a point in the fracture plane which may not be representative of the *in situ* fracture hydromechanical behavior.

In this study, *in situ* hydromechanical properties are determined on fractures intersecting the 1700-meter deep borehole KLX02 at the Laxemar area near Äspö Hard Rock Laboratory in Sweden. The fractures are located at three highly conductive zones at depths of 270, 315 and 340 metres. The hydromechanical properties are back-calculated by coupled numerical modeling of single borehole multiple pressure injections tests according to Rutqvist (1995).

2 FIELD TESTS AT LAXEMAR

The KLX02 borehole is cased to 200 metres depth and has a diameter of 76 mm for depth greater than 200 metres. The bedrock consists mainly of granite and diorite and the fracture frequency is in general low, especially down to 700 meters and between 1100-1500 meters. The three test zones could be identified from flow, temperature and electrical resistivity logging as the most hydraulic conductive zones in the upper 700 metres of the borehole (Ekman, 1997). The zones at 270 and 340 meters coincide with the intersection of fracture zones a few meters wide, which also were identified by borehole radar. The rock in these zones is reported to be crushed and fractured granite with a frequency of 10 to 20 fractures per meter. Pulse injection tests together with images from a high resolution TV system showed that the transmissivity in each zone was dominated by flow through a few open fractures. The most hydraulic conducting fractures appear to have unmated rough surfaces with open channels between contact points and parts filled or coated with calcite or chlorite. The orientation of the fractures at 270 and 340 meters are oblique to the fracture zone and may be strike-slip shear fractures. The hydraulic conductive zone at 315 meters consists of a major and a minor fracture and there is no increased fracture frequency in the vicinity.

The tests were carried out using hydraulic fracturing instrumentation. The downhole equipment consists of a double packer with 0.65 m packer separation, a pressure transducer and a valve for instantaneous change of fluid pressure (Ekman, 1997).

Three types of injection tests were conducted in a sequence as follows (Figure 1a):
1) Pulse test
2) Constant head injection test
3) Multiple pressure test (hydraulic jacking)
4) Pulse test

The pulse tests are conducted to determine the transmissivity and the flow dimension of the fracture plane close to the borehole intersection within a radius of less than 1 meter (Figure 1b). The constant head injection test is conducted to characterize the hydraulic properties farther away from the borehole and of the network of intersecting secondary fractures. The test gives information on the transmissivity, flow dimension and can also be used to define the location and type of outer hydraulic boundaries.

The hydraulic jacking test, is conduced by a stepwise increase of the fluid pressure. At each step, the well pressure is kept constant for a few minutes until the flow is steady (Figure 1a). The flow rate at each pressure step is strongly dependent on the fracture aperture and normal stiffness of the fracture in the vicinity of the borehole (Rutqvist, 1995). Thus, it is essentially a near field test for determination of the near field fracture hydromechanical properties.

3 EVALUATION OF THE FIELD TESTS

The hydraulic tests at each fracture are interpreted by simulating the tests using the finite element model ROCMAS (Noorishad et al, 1992).

3.1 Numerical model

The ROCMAS code is designed to model the coupled hydromechanical behavior of both the host rock and the fracture intersecting the borehole. The field test environment is discretized into an axisymmetric finite element model with its center coinciding with the borehole (Figure 2).

The actual finite element mesh is by symmetry discretized in the right upper corner of the model in Figure 2 and consists of 380 elements with detailed refinement near the borehole and the fracture. At the initial stage, an initial stress and fluid pressure are prescribed in the entire model. The boundaries of the finite element mesh are placed sufficiently far from the fracture so that they do not affect the results during the time of injection.

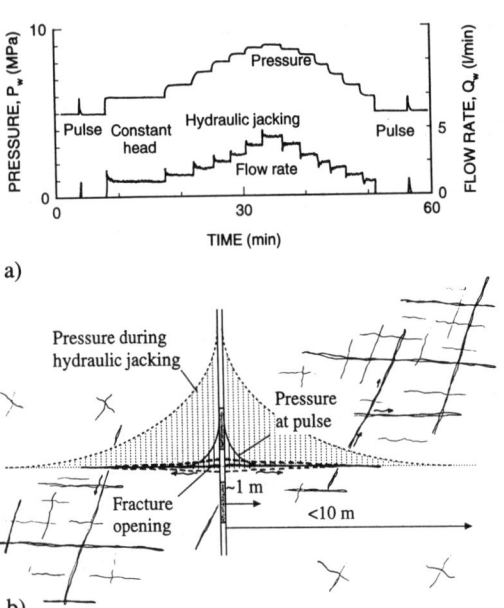

Figure 1. Hydraulic injection tests for coupled hydromechanical evaluation. a) Well pressure and well flow as a function of time. b) Radius of influence for various tests.

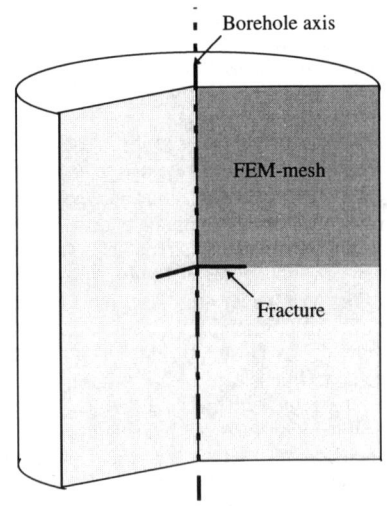

Figure 2. Finite element model of the field test.

3.2 Hydromechanical coupling by Goodman's model

The simplest non-linear joint normal closure model is the Goodman's original model (Goodman, 1974). It can be written as a simple expression by reformulating the original equation into terms of effective normal stress, σ'_n and the current maximum closure. It is here defined as a mechanical aperture, b_m giving a stress versus aperture relation as (Figure 3):

$$\sigma'_n = \frac{A_i}{b_m} \qquad (1)$$

where A_i is a constant defined as:

$$A_i = \sigma'_{ni} \cdot b_{mi} \qquad (2)$$

where σ'_{ni} and b_{mi} is the effective normal stress and mechanical aperture, respectively, at some initial or reference state. This implies that the relation between the stress and mechanical aperture is completely defined by only one parameter, A_i. The normal stiffness, k_n of the fracture is:

$$k_n = \frac{\sigma'_n}{b_m} \qquad (3)$$

The transmissivity, T of a fracture depends on the size of interconnected voids and is related to a hydraulic fracture aperture (b_h) which can be defined according to Witherspoon et al. (1980) as:

$$T = \frac{b_h^3 \rho_f g}{12 \mu_f} \qquad (4)$$

where ρ_f and μ_f are the density and dynamic viscosity of the fluid, and g is the gravitational acceleration.

The hydraulic aperture at a given normal stress is assumed to be:

$$b_h = b_{hr} + f \cdot b_m \qquad (5)$$

where b_{hr} is the residual hydraulic aperture when the fracture is mechanically closed and f is a factor that compensates for the deviation of flow in a natural rough fracture from the ideal case of parallel plate type of fracture surfaces (Witherspoon et al, 1980).

Equation (1), (4) and (5) completes a relation between the fracture transmissivity and effective normal stress:

$$T = \frac{\rho g}{12 \mu_w} \left[b_{hr} + \frac{A_i \cdot f}{\sigma'_n} \right]^3 \qquad (6)$$

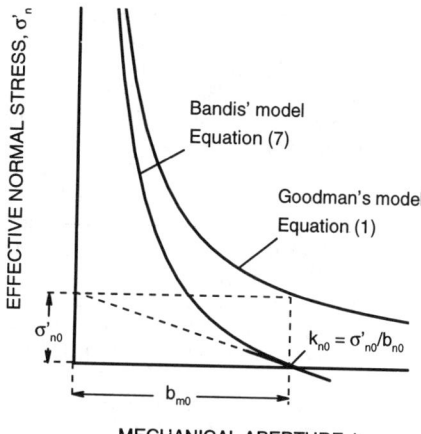

Figure 3. Mechanical aperture as a function of effective normal stress for Goodman's and Bandis' models.

3.3 Hydromechanical coupling by Bandis' model

Bandis' joint model (Bandis et al, 1983) is defined by two constants and provides a greater flexibility of the stress versus normal displacement function. It can be shown that Bandis' joint model is similar to Goodman's but with the curve displaced along the stress axis. Therefore, Bandis' joint model can be written in an analogous way to equation (1) as:

$$\sigma'_n = \frac{A_i}{b_m} - k_{no} b_{m0} \qquad (\sigma'_n > 0) \qquad (7)$$

were k_{no} and b_{m0} are normal stiffness and mechanical aperture, respectively, at zero stress (Figure 3).

Equation (4), (5) and (7) completes a relation between the fracture transmissivity and effective normal stress with bandis' model:

$$T = \frac{\rho g}{12 \mu_w} \left[b_{hr} + \frac{A_i \cdot f}{\sigma'_n + k_{no} b_{m0}} \right]^3 \qquad (8)$$

3.4 Modeling of field tests

Coupled numerical modeling of a pulse test shows that the fracture storativity is controlled by the deformability of the rock matrix surrounding the fracture (Rutqvist, 1996). The numerical modeling of field tests also shows that the storativity is in general close $3.5 \cdot 10^{-8}$ for granite with the above mentioned test equipment. The results of the evaluation also show that the flow near the well in most fractures is approximately radial but with an irregular response (Figure 4). In some cases, before jacking, the flow was non-radial indicating inhomogeneous hydraulic properties (Figure 4).

At all fractures, a steady flow was obtained after some time of constant head injection (Figure 5). This can be interpreted as that the injected fluid has reached a constant pressure boundary which may be an intersecting joint with a much larger hydraulic aperture. Alternatively, the apparent steady flow may be due to the fact that the flow becomes spherical in a well-connected network of fractures in the rock mass. The constant head injection tests indicate an apparent constant pressure boundary at a radial distance of less than 10 meters away from the borehole.

Rutqvist (1995) showed that the pressure flow response in a high pressure test is strongly dependent on the effective stress versus transmissivity relationship for the fracture plane immediately around the borehole. This relationship is defined in equations (6) and depends on three parameters:
1) Normal stiffness parameter $A_i \cdot f$
2) Residual hydraulic aperture, b_{hr}
3) Initial effective normal stress, σ'_{ni}

The strong dependency on these parameters implies that they can be back-calculated from the pressure flow response without much influence of other parameters. This also means that the radial distance to the constant pressure boundary as obtained from the constant head injection test is not a critical parameter.

Figure 6 shows field test results and the numerical modeling of the hydraulic jacking test at 267 meters depth. At the first cycle of step-wise increasing pressures, the flow rate increases as a non-linear function of pressure. A temporal peak-pressure is obtained at a flow rate of 1.3 liters/minute before the pressure begins to decrease with an increasing flow rate. This can be interpreted that the fluid pressure is high enough to break apart healed regions of the fracture plane. The opening may also be accompanied by a shear displacement if the fracture is inclined to direction of the *in situ* principal stresses. The subsequent step-pressure cycle to takes a different path because of the change in hydromechanical properties due to shearing and fracturing. These changes in hydraulic properties are also observed in the pulse response after the hydraulic jacking (Figure 4).

The multiple pressure injection tests at 267 meters were modeled with Goodman's joint model before fracturing and Bandis' model after fracturing (Figure 6).

3.5 Results of evaluation

The results of the evaluation are presented in Table 1 and Figure 7. In most cases, the Goodman's joint

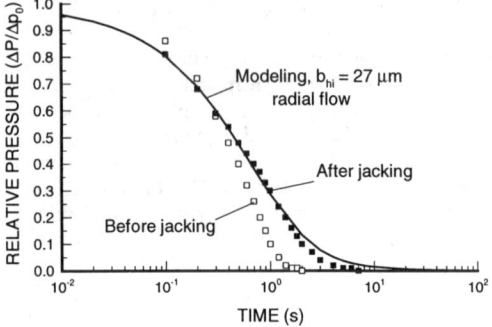

Figure 4. Modeling and field response of pulse injection tests at 267 meters.

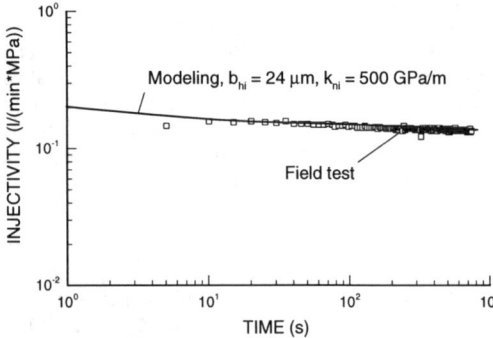

Figure 5. Modeling and field test response of a constant head injection test at 267 meters. $P_w = 5.8$.

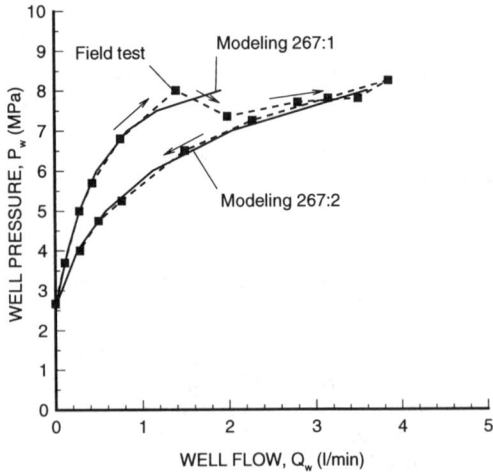

Figure 6. Modeling and field response of a multiple pressure test at 267 meters.

model gave a satisfactory match to the field data or the quality of the field data was not good enough to perform a matching to get the additional parameters for the Bandis' joint model. At 267 and 316 meters depths two sets of test results are presented in Table 1 for data before and after fracturing.

Figure 7 presents the stress versus transmissivity relationship according to Equation (6) for the undisturbed conditions before fracturing. The stresses at the test intervals are more or less constant with an initial effective normal stress of about 4 MPa. Identification of this stress level in Figure 7 shows that the transmissivity varies two orders of magnitude depending on the fractures that are tested. This scattering naturally depends on different characteristics such as roughness, past shearing and mineral filling. The fractures at 316 and 336 show a major decrease in transmissivity with increasing normal stress. These fractures have the lowest normal stiffness and at the same time a small residual aperture. The fracture at 316 meters shows a negative residual hydraulic aperture. This can be interpreted that the fracture is closed for water flow because of clogging of soft mineral filling at a high normal stress. As a consequence, a fracture with the same characteristics as the fracture at 316 meters depth would be practically closed if it was located at 500 meters depth. On the other hand, fractures with the equivalent characteristics as the fracture at, for instance, 338 meters would be open and have a large hydraulic transmissivity, even at high stress and great depth.

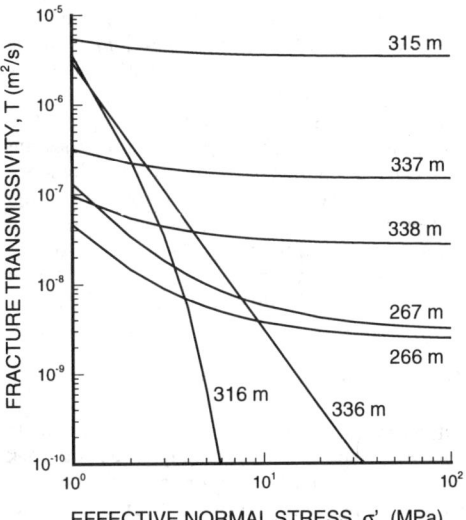

Figure 7. Transmissivity versus effective normal stress for fractures at KLX02 before fracturing.

4 DISCUSSION

Comparing our results to small scale laboratory tests on natural tensile fractures we can say that these *in situ* shear fractures have a much higher initial or residual aperture while the stiffness is about the same order. This implies that the change in hydraulic aperture due to a change in normal stress is small compared to the initial hydraulic aperture, and therefore, the transmissivity is relatively insensitive to *in situ* stresses. These results are in agreement with earlier field studies by Carlsson and Olsson (1977) who concluded that the hydraulic conductivity of fractures zones in general decreases slower with depth than for the more competent rocks.

One consequence of our findings is that the hydraulic conductivity is more likely to depend on normal stress and depth in the upper few hundred meters of the bedrock. In this region both the mated joints in the more competent rock and fractures

Table 1. Results of evaluation of multiple pressure tests at Laxemar using coupled numerical modeling.

Depth	Residual hydraulic aperture, b_{hr}	Joint parameter, $A_i \cdot f$	Joint parameter, $k_{n0} \cdot b_{m0}$	Initial effective normal stress, σ'_{ni}	Initial hydraulic aperture, b_{hi}	Initial "hydraulic" normal stiffness, k_{ni}/f
(m)	(μm)		(MPa)	(MPa)	(μm)	(GPa/m)
266	14	24		4.9	19	1000
267:1	15	19		4.4	24	500
267:2	-41	694	5.8	4.4	27	150
315	148	30		3.9	164	1000
316:1	-29	190		3.9	20	80
316:2	3	138		3.9	38	110
336	0	153		4.1	37	110
337	56	17		4.1	60	1000
338	32	17		4.1	36	1000

zones contribute to the hydraulic conductivity. However, at greater depths, the mated joints will be closed and the conductivity will be dominated by relativity stress-insensitive shear fractures.

These studies suggest that it is very important to test fractures *in situ*. The high pressure injection method can be improved further by simultaneous measurement of the mechanical joint opening at the borehole wall. This would give further information about the factor f and the mechanical joint stiffness, and at the same time reduce uncertainty due to components such as the flushing of joint filling.

5 CONCLUSIONS

This study shows that it is possible to back-calculate the *in situ* hydromechanical properties of rock fractures through a combination of high pressure injection testing and coupled hydromechanical modeling. Shear fractures in tectonic zones could be tested although the core samples from those sections were completely crushed. The flow in these strongly hydraulic conducting zones is dominated by a few open shear fractures. The hydromechanical evaluation shows that these fractures are relatively insensitive to changes in normal stress due to a large initial hydraulic aperture caused by past shear dilation. The hydromechanical properties of natural fractures are strongly dependent on past shearing and cementation or dissolution of mineral fillings and should therefore be tested *in situ*.

ACKNOWLEDGEMENTS

This study was supported by the Swedish Nuclear Fuel and Waste Management Company (SKB) and by a post-doctoral fellowship to the first author from the Wenner-Gren Center Foundation in Sweden. The field tests were conducted in cooperation with the Vattenfall Hydropower Company in Sweden. Work is partially supported by the Department of Energy, under contract No. DE-AC03-76SF00098.

REFERENCES

Bandis, S., A.C. Lunsden & N. R. Barton 1983. Fundamentals of Rock Joint Deformation. *Int. J. Rock. Mech. Min. Sci. & geomech. Abstr.* 29:249-268.

Carlsson A. & Ohlsson T. 1977. Hydraulic properties of Swedish crystalline rocks. Hydraulic conductivity and its relation to depth. Bull. Geol. Inst. Upps., 71-84. Uppsala University, Sweden.

Ekman, D. 1997. Rock stress, Hydraulic Conductivity and Stiffness of Fractures Zones in the Laxemar Borehole, Småland, Sweden. Masters Thesis. Division of Engineering Geology, Royal Institute of Technology, Sweden.

Goodman, R. E. 1974. The mechanical properties of joints. Proc. 3rd Congr. ISRM. Denver, 1A:127-140.

Noorishad, J., C.-F. Tsang, & P.A. Witherspoon 1992. Theoretical and field studies of coupled behaviour of fractured rocks - 1. Development and verification of a numerical simulator. *Int. J. Rock. Mech. Min. Sci. & geomech. Abstr.* 29:401-409.

Rutqvist, J. 1995. Determination of hydraulic normal stiffness of fractures in hard rock from hydraulic well testing. *Int. J. Rock Mech. Min. Sci. & Geomech. Abstr* 32:513-523.

Rutqvist, J. 1996. Hydraulic pulse testing in single fractures in porous and deformable hard rocks. *Q. J. Eng. Geol.* 29:181-192.

Witherspoon, P. A., J.S.Y. Wang, K. Iwai & J. E. Gale 1980. Validity of the cubic law for fluid flow in a deformable fracture. *Water Resources Res.* 16:1016-1024.

Physical simulation of energy exchange aimed at detection of dangerous forms of rock mass failure

E.V. Lodus
The State Research Institute of Mining Geomechanics and Mine Surveying, VNIMI, St. Petersburg, Russia

ABSTRACT: The dangerous dynamic forms of rock mass failure occurring in the stages of construction and exploitation of underground structures can be studied through physical simulation of energy exchange on the rock specimens in laboratory tests. The equipment and measuring techniques have been developed to simulate the energy exchange during deformation and failure of rock mass as close as possible to field conditions. The simulation involves complicated loading conditions, rheology, rock movements, gradients of mechanical stresses, thermal gradients, structural non-uniformity of rock mass, types of stress state, duration of deformation, and degree of stiffness of loading systems.

1 INTRODUCTION

The penetration into the underground space inevitably entails the failure of rocks. The rock mass failure is caused by two energy sources, one of which, W_1, is an energy of rock-crushing tools and the other, W_2, is an energy of elastic recovery of rocks around the failed rock. The energy (W_1+W_2) is expended for deformation and disintegration of the rocks (W_3), for relative displacements of surrounding rocks (W_4), and for movements of failed mass (W_5). This may be written as

$$W_1+W_2+W_3+W_4+W_5 \qquad (1)$$

If $W_5 > 0$, the rock failure may manifest itself in dangerous forms such as rock bursts, sudden outbursts, earthquakes etc. The prediction of a dangerous pattern of rock mass failure involves evaluation of the components of Eq. (1) for specific conditions of energy exchange.

It is impossible to determine the energy-exchange components in situ because of the complexity of energy-exchange process by itself and the inaccessibility of rock mass for direct observations. In our opinion, the best method to solve this problem is the physical simulation of energy exchange in laboratory tests of rock specimens. The specimens under test are ideally suited for simulation of rock mass failure and movement. The energy of rock mass can be realized in form and magnitude by a specifically designed loading press. Various mechanisms for mechanical treatment of specimens act as rock-crushing tools. The more exactly and fully the laboratory tests reproduce actual energy exchange conditions, the more reliable will be the prediction of a dynamic pattern of failure.

The simulation of energy exchange requires that the special features of rock mass behaviour should be taken into account, in particular, complicated stress-strain relationship, rheology including creep phenomena and stress relaxation, relative displacements of rock mass parts under load, presence of gradients of mechanical stresses and thermal gradients, and structural non-uniformity of rock mass. Each group of these parameters is to be simulated by the specially developed procedure on the specially designed equipment.

2 SIMULATION OF LOADING CONDITIONS IN ROCK MASS

The complicated stress conditions are typical for any area of rock mass in the vicinity of mine workings. Any variation in loading conditions gives rise to a change in the energy level and hence affects the nature of energy exchange. To simulate these conditions, the REZHIM 1 testing unit is used. A range of load variations that can be

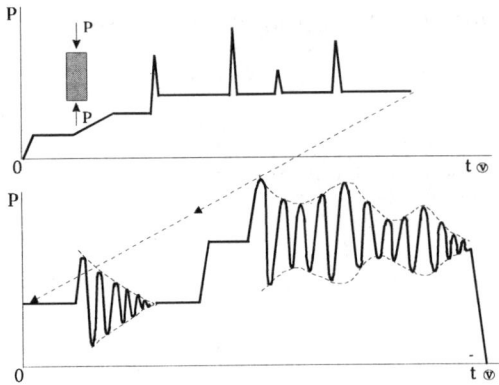

Figure 1. Variation of load P with time t.

realized on this testing unit is shown in Figure 1. The test unit provides for a smoothly increasing or decreasing load on a specimen, a long-term constant load, a stepwise increasing impact load, an impact load at varying pulse multitude, one-cycle and two-cycle damped fluctuations of loading, and a conditionally-momentary unloading. The duration of loading in every above-mentioned case, loading level, and the order of alternating various loading modes are chosen based on knowledge of actual stress conditions of the rock in question. On the REZHIM 1 testing machine we have a possibility to carry out a full programme of tests without having to remove the load from the specimens, as is the case in the in situ conditions.

3 STUDY OF RHEOLOGICAL PROCESSES

The rheological processes, i.e. creep phenomena and relaxation have an impact on the energy-exchange process due to changes in the behaviour of rock materials in the pre-failure stage of deformation. Which of processes will dominate - creep or stress relaxation - is determined by a degree of stiffness of the loading system. To initiate one or the other rheological process the testing machines with a soft or stiff loading system are required. For the creep to be induced the REZHIM 1 testing equipment with a soft loading system is used, creating a longterm constant load. When the load value and rock properties are such that the creep leads to failure of the specimen under test, the failure will be of dynamic form. This may be attributed to a high store of energy W_2 on the soft loading systems and consequently to a high magnitude of energy W_5.

To initiate the stress relaxation the stiff loading systems are used in particular IZGIB testing unit whose principle of operation is as follows: a specimen is being pushed into the stiff matrix with a cam slot whose curvature radius progressively decreases. As the specimen moves the bending load increases and the bending deflection f becomes greater (Figure 2). At the instant t_1 the specimen is brought to rest, the bending deflection remains fixed and the stress relaxation in the specimen material begins.

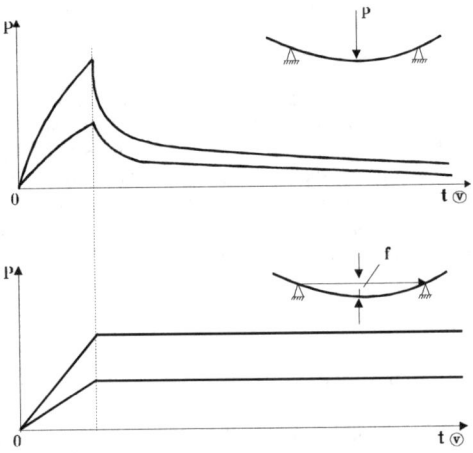

Figure 2. Rheological process (stress relaxation)

4 MODELLING THE RELATIVE DISPLACEMENTS OF ROCKS

Energy W_4 being expended for relative displacements of elements of rock mass is dependent on such factors as properties of rocks on contact planes, contact conditions, and time-dependent variation of forces P_1 and P_2 (the elements being held in place by the force P_1 and shifted by the force P_2). (Figure 3).

The displacement may be initiated both through increasing the force P_2 within a period of time between t_2 and t_3 (Figure 3b) and decreasing the force P_1 within a period of time between t_4 and t_5 (Figure 3a), much as the displacements of rocks are induced in situ. The PODVIZHKA loading system provides for modelling the relative displacements of rocks, with the diagrams plotted from which the energy expenditure for rock displacements under various conditions on contact planes can be determined.

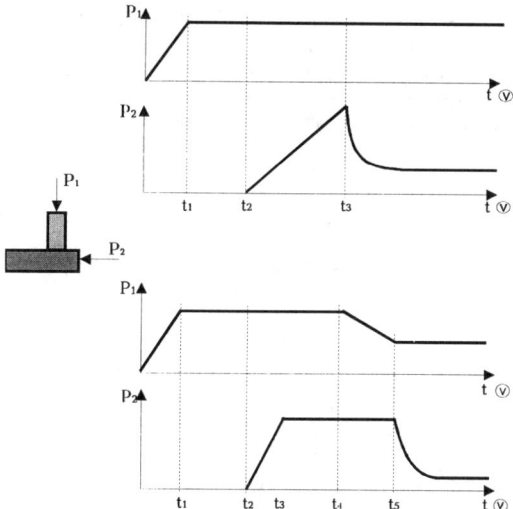

Figure 3. Variation of forces P_1 and P_2 with time in simulation of rock displacements

5 SIMULATION OF GRADIENTS OF MECHANICAL STRESSES

The simulation of gradients of mechanical stresses has a twofold purpose - to reveal the role of the mechanical gradients in the rock mass deformation and energy-exchange and to reveal the role of energy W_1 expended by the rock-crushing tools. In order for the mechanical gradients to be created, the specimen under load is mechanically treated (Figure 4).

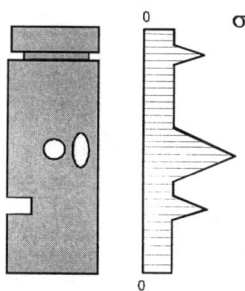

Figure 4. Simulation of gradients of mechanical stresses

As the specimen is loaded by a stiff or a soft loading system some holes and slots of various size and shape are made in it, which are arbitrarily oriented relative to each other and to loading direction. The order of machining the specimen and the rate of drilling the holes and slots in it represent, as a scaled model, the rock mass behaviour during drivage of mine openings.

6 SIMULATION OF THERMAL GRADIENTS

In order that the thermal gradients be created, the specimen under load is unevenly warmed up by using the heat sources that come in contact with the surface of the specimen at some points or over some areas. The sources of elevated temperatures may be placed in the slotted holes which have been drilled to initiate the gradients of mechanical stresses. In this case one may study the influence of a number of factors, i.e. the effect of thermal and mechanical gradients and of rock-crushing mechanisms. (Figure 5).

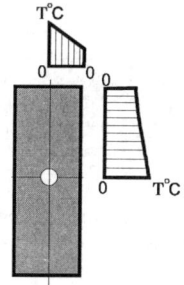

Figure 5. Thermal gradients

7 SIMULATION OF STRUCTURAL NON-UNIFORMITY OF ROCK MASS

The influence of the structural features of rocks such as composition, grain size, lamination, and density both on the stress behaviour and the conditions of energy exchange can not be studied on the rock specimens. The importance of each structural factor individually can not be assessed at all because in actual practice these factors undergo changes simultaneously and to a variable extent.

The most suitable material to study the important part that the structural factors play in the development of rock deformations and in the energy exchange is artificial polycrystals. Artificial polycrystals are used to simulate the structural features of salt rocks. By mechanical mold pressing of a weighed amount of salt crystals an artificial polycrystal with specified structural parameters is produced. The uniaxial compression strength σ, density ρ and porosity $P\%$ of the polycrystal are dependent, to a certain limit, on the force of pressing of the weighed amount of crystals (Figure 6).

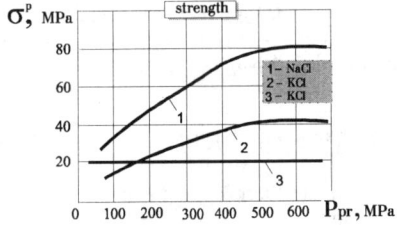

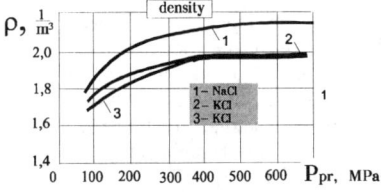

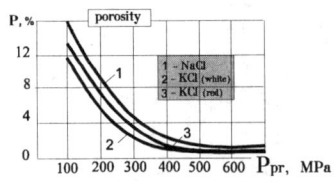

Figure 6. Strength (a), density (b) and porosity (c) dependent on pressing force for artificial polycrystals

With increasing density of rock salt polycrystals from 1.73 t/m^3 to 1.97 t/m^3 their strength, plasticity and energy indices increase many times (Figure 7).

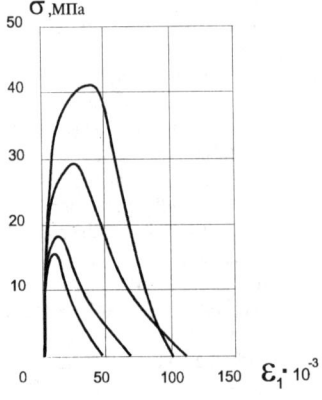

Figure 7. Influence of density on properties of salt rocks

As illustrated (in Figure 8) the properties of salt rock depend to a large extent on its composition. The strength diagrams show a regular variation of stress-strain values according to a relationship between rock salt and potast salt in the polycrystal.

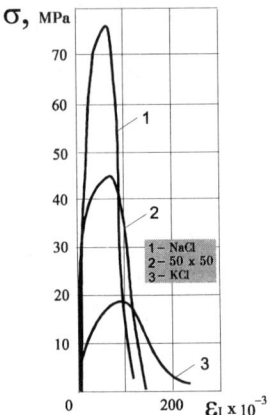

Figure 8. Influence of composition on properties of salt rocks

The increase or decrease of an average grain diameter also affects the rock properties (Figure 9).

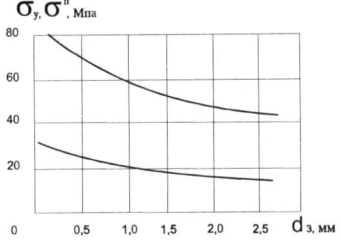

Figure 9. Influence of average grain size on properties of salt rocks

The artificial polycrystals are put through laboratory tests in much the same manner as the specimens of natural rocks and enable one to assess the role of the structural factors in the energy-exchange process.

A type of stress-state and a loading rate are also of decisive importance in the energy-exchange process as well as in the development of strains, which may be illustrated by an example of artificial polycrystals of salt rocks. Figure 10 show the curves of strains for rock salt polycrystals, obtained in compression tests at varying side pressure. As Figure 11 show, the rate of loading affects the fracture energy, W_3, and plastic properties rather than the strength. Line 1, line 2 and line 3 correspond to the duration of loading 24 hours, 1 hour and 1 second, respectively.

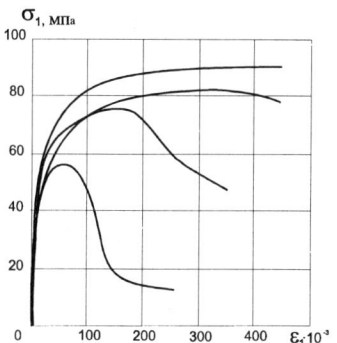

Figure 10. Influence of the stress state type on deformation pattern of artificial polycrystals

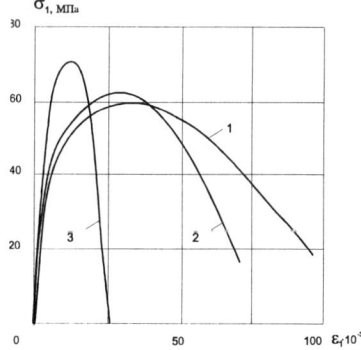

Figure 11. Influence of loading duration on deformation pattern of artificial polycrystals

8 ANALYSIS OF SIMULATION RESULTS

The results of simulation represented in graphical form are best suited to analysis. For this purpose, two resultant diagrams are required, one of which describes the energy expended for deformation and destruction and movement of rocks, and other describes the energy stored on a loading system as well as energy of rock-disintergating machine. From the graphical representations we can detect the presence and assess the magnitude of an excess energy that is responsible for dynamic pattern of failure. An example of graphical analysis can be seen in Figure 12. The curve 1 represents the energy that is required for the rock failure and movement, and the curve 2 the energy stored on a loading system and rock-crushing tool. W_5 is an excess energy that causes dynamic failure.

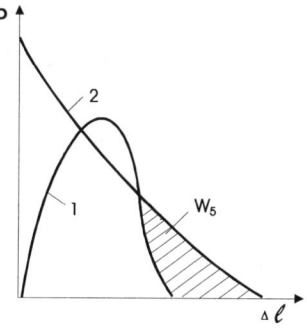

Figure 12. Graphical analysis

Here, the graphical analysis is given in a very simplified version. It is self-evident that the graphical analysis should be extended to include rheology, structural features, type of stress state and other conditions of actual energy exchange.

9 CONCLUSIONS

The dangerous dynamic forms of rock mass failure are due to the energy exchange between a loading system and the failed rocks. The rock mass failure mode can be predicted by physical simulation of the energy-exchange process on the rock specimens in laboratory tests with reproducing peculiar features of energy exchange in rock mass as exact as possible.

Evaluation of JRC from image processing of borehole wall

Jaedong Kim
Kangwon National University, Chunchon, Korea

O. Stephansson
Royal Institute of Technology, Stockholm, Sweden

ABSTRACT: A new approach to investigate the characteristics of rock joints was done with borehole wall images. Images captured by borehole scanner on the KAS03 borehole excavated at Aspo in Sweden were used for this study. An image processing procedure for recognizing fracture boundary revealed on a borehole wall and a JRC evaluation method from the processed and measured data of a certain joint or joint set were developed using computer codes.

1 INTRODUCTION

During several decades until now, rock core recovered from borehole has been one of the most strong tools to see the inside of a rock mass and it really showed direct and clear results for this purpose. But in these days it becomes possible to get more useful geological informations not only from rock cores but from borehole walls using image processing techniques.

Image processing technique has been developed in the field of computer engineering and so many other related fields accept this to achieve their specific purposes. Rock engineering field also shows this sort of research trends in investigation of rock mass through boreholes. Recently developed electronic and optical equipments, for example, borehole camera or borehole scanner enable to obtain images of borehole walls down to deep rocks.

Through these images, more geological informations can be extracted, especially for the intervals that core was lost. And borehole wall can still provide an useful information in addition to the one from the core.

In this research one type of borehole wall scanner called BIPS(Borehole Image Processing System) was used to get borehole wall images. A borehole named KAS03 was selected for this research, which was excavated to survey the geological conditions of Aspo Hard Rock Laboratory in Sweden.

In this paper, a sequence of steps for image processing of borehole wall using macro commands built in image analyzer and a computer code to evaluate a JRC (Joint Roughness Coefficient) value for a certain fracture or fracture set is discussed.

2 SITE AND STEREOGRAPHIC PROJECTION OF FRACTURES

Borehole wall images of KAS03 borehole excavated at Aspo area in Sweden(Figure 1) were captured by using borehole scanner. The depth of borehole is just in excess of 1000m and the average plunge is about 82o. The most common rock is granite down to the depth of approximately 565m.

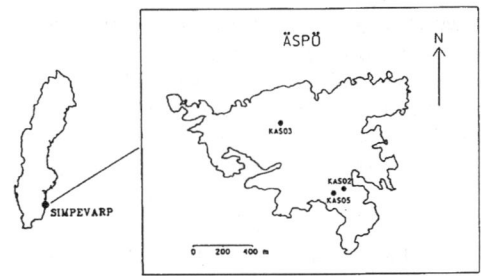

Figure 1. Geological location of the Simpevarp area in Sweden and the location of borehole KAS03 at Aspo.

As the Aspo HRL(Hard Rock Laboratory) is located at the depth of 450m, research was focused around this depth.

BIPS itself has some functions of analysis, those are visualization of borehole wall image in different ways, hole inclination calibration, orientation and aperture analysis of fracture encountered with a borehole, generation of geological database including rock type and several statistical data treatments.

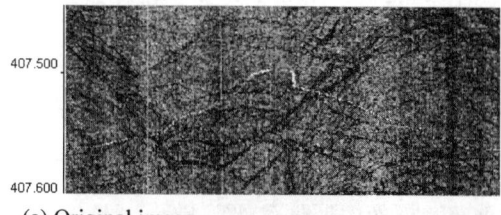
(a) Original image.

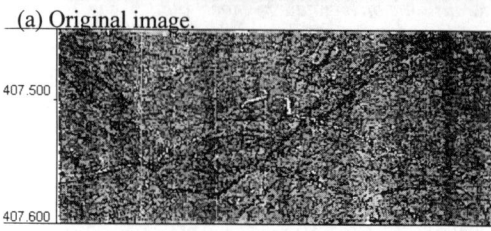
(b) DELIN process.

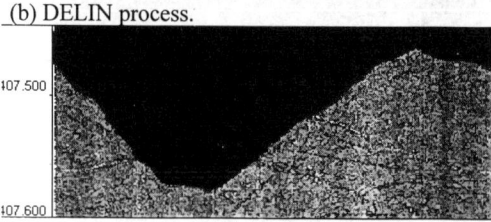
(c) Conversion of the gray value of the upper-part of boundary to black.

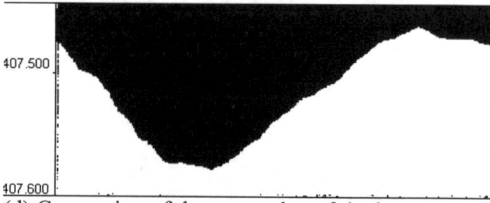
(d) Conversion of the gray value of the lower-part of boundary to white.

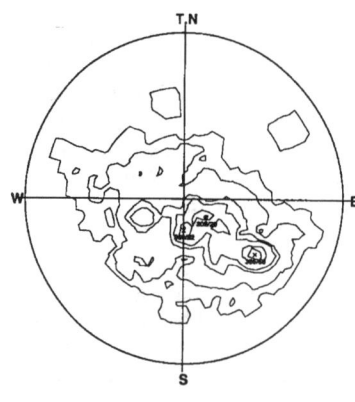

Number of Data : 309
Joint Set 1 : 306/59   7% or more
Joint Set 2 : 308/20   6-7%
Joint Set 3 : 000/22   6-7%
Sampling Depth : 370.3 - 426.3

Figure 2. Wulff stereographic projection of 309 fractures analyzed from KAS03 borehole.

Figure 3. A series of images showing the image processing sequence.

Total 309 fractures were recognized from the borehole wall images captured by BIPS and stereographic analysis was executed. From the Wulff stereonet diagram as shown in Figure 2, three distinct joint sets were analyzed. To evaluate a representative JRC value for each joint set, every ten fractures positioned at the center of the pole concentration of each set were chosen for image processing. These joints were assumed to have typical characteristics of each set.

3 IMAGE PROCESSING OF BOREHOLE WALL

Several steps of image processing technique were applied to the fracture images revealed on a borehole wall. They are delineation, manual fracture boundary tracing, transformation of gray values and measurement of the locations of pixels consisting fracture boundary.

Delineation(DELIN) process was for making fracture boundary to be clear according to the gray value differences and on this processed image manual fracture boundary was traced by digitizer. The image then can be separated into two parts by the boundary. After transforming the gray values of upper and lower parts into complete black and white, the fracture boundary could be measured automatically according to the extreme difference of gray values adjacent to the boundary. Figure 3 shows a series of processed images which were obtained at each step.

4 JRC EVALUATION

To evaluate a JRC value of a certain fracture, some geometrical projection procedure on the measured fracture profile is necessary. Figure 4 shows a profile of a fracture revealed on the borehole wall and a basic sinusoidal curve based on its dip direction and angle.

As JRC means the degree of roughness along a certain line, a roughness profile can be the difference between the magnitudes of theses two curves.

Figure 5 shows the roughness profile and its projected one on the direction perpendicular to the fracture surface.

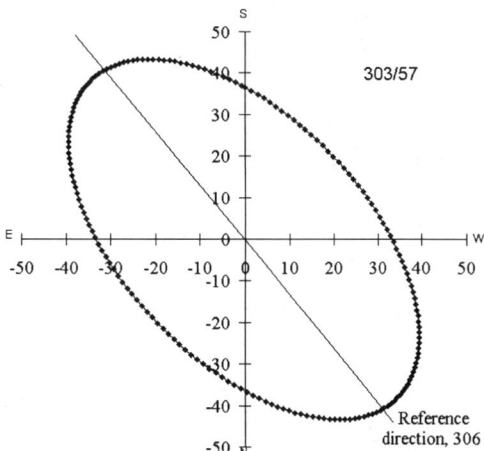

Figure 6. The elliptical periphery that the roughness profile then can be mated.

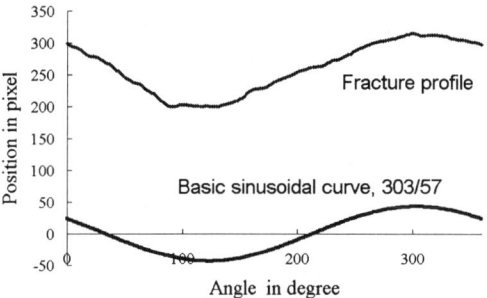

Figure 4. A fracture profile measured from the processed image of and its basic sinusoidal curve.

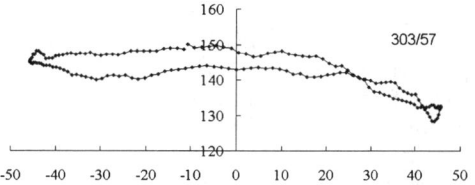

Figure 7. The roughness profile projected on the reference direction.

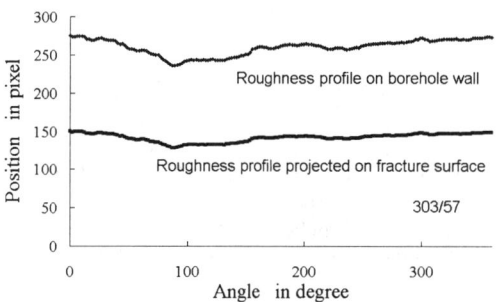

Figure 5. The roughness profile and its projection on the direction perpendicular to the fracture surface.

The roughness profile then can be mated with the elliptical periphery, which reveals a real area(Figure 6) intersected by borehole wall and fracture. Figure 7 is the roughness profile projected on the reference direction line which was set as in Figure 6. Reference direction can be chosen arbitrarily according to the one that the shear deformation is expected to occur.

Tse and Cruden(1979) present an interesting research result that the root mean square, Z_2 of the first order derivative of roughness profile (Equation (1)) has a strong correlation with JRC (Equation (2)). They made data sets for their analysis by digitizing the ten typical roughness profiles originally illustrated by Barton and Choubey(1977).

$$Z_2 = \sqrt{\frac{\sum_{i=0}^{M}(y_{i+1} - y_i)}{M(\Delta x)^2}} \quad (1)$$

where Z_2: root mean square, Δx: constant small interval, M: number of data.

$$JRC \approx 32.2 + 32.47 \log_{10} Z_2 \quad (2)$$

This idea of analysis for the calculation of JRC has recently been examined by several authors, McWilliams et al(1990), Roberts et al(1990), Yu and Vaysade(1990). These last authors found that JRC value is dependent on the sampling interval along the profile and modified Equation.(2) to (3).

$$JRC \approx AZ_2 - B$$

Δx(mm)	A	B	
0.25	60.32	4.51	(3)
0.5	61.79	3.47	
1.0	64.22	2.31	

Considering the lowest resolution of borehole wall image adopted in this study was about 1.0mm/pixel, the data interval was set to 1.0mm of constant value. Hence the roughness profile data was rearranged using a numerical interpolation technique.

There needs an another step of calibration to compensate a round-off error for evaluation of JRC. This error was caused in measuring the locations of fracture boundary from the image. As the boundary consisted of a sequence of neighbor pixels and the roughness profile could be measured with unit of pixel, digits under decimal point were rounded off. So with a precise view, roughness profile shows a stepwise variation in its magnitude.

Figure 8 shows a possible amount of round-off error that can be included in the cases that real profile data was rounded off under decimal point and under 10^{-1}. Compared with the case rounded off under 10^{-1}, this represents that it needs about ten times as high as the current resolution to eliminate the round-off error. This calibration curve was used for calibrating the calculated JRC value from Equation (3).

The JRC values obtained after calibration are shown in Figure 9. The JRC values are 16.7 ± 2.6, 12.3 ± 3.0 and 9.5 ± 3.7 for joint set 1, 2 and 3, respectively.

5 CONCLUSION

In this paper, images of borehole wall KAS03 at Aspo captured by borehole scanner were used to evaluate JRC value of a certain joint or joint set. An image processing procedure for recognizing fracture boundary and evaluation method of JRC for a certain joint or joint set were developed.

Image processing and fracture profile measurement procedure were written using macro commands which are popular in the field of image analysis. JRC evaluation procedure was written in FORTRAN code.

The result of JRC evaluated in this research needs to be compared with the one measured on the same real fracture surface to confirm its validity.

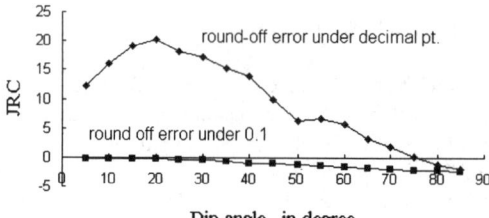

Figure 8. Round-off error variation according to the dip angle.

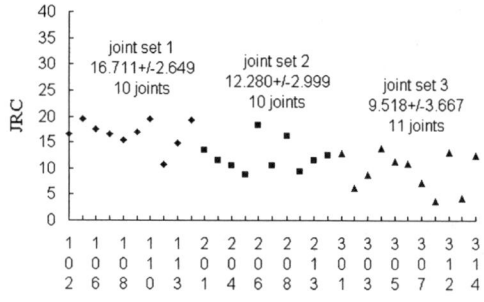

Figure 9. Result of JRC evaluation after calibrating the round off error.

ACKNOWLEDGEMENTS

The financial support by the Korean Research Foundation is acknowledged.

REFERENCES

Tse, R. & D.M.Cruden 1979. Estimating joint roughness coefficients. *Int.J.of Rock Mech. & Mining Sciences* 16:303-307.

Barton, N. & V.Choubey 1977. The shear strength of rock joints in theory and practice. *Rock Mechanics* 10:1-54

McWilliams, P.C., J.C.Kerkering & S.M.Miller 1990. Fractal characterization of rock fracture roughness for estimating shear strength. *Proc. of Int.Conf.on Mechanics of Jointed and Faulted Rock.* 331-336. Vienna: Balkema.

Roberds, W.J., M.Iwano & H.H.Einstein 1990. Probabilistic mapping of rock joint surfaces. *Proc. of Int.Symp.on Rock Joint.* 681-691. Loen: Balkema.

Yu, X. & B.Vayssade 1990. Joint profiles and their roughness parameters. *Proc. of Int.Symp.on Rock Joint.* 781-785. Loen: Balkema.

Fractal geometry and mechanical behavior of rock fracture surfaces

J.A. Wang
Institute of Fractal Mechanics, Beijing Graduate School, China University of Mining and Technology, People's Republic of China

Marek A. Kwaśniewski
Rock Mechanics Laboratory, Faculty of Mining and Geology, The Technical University of Silesia, Gliwice, Poland

ABSTRACT: In order to confirm the physical significance of fractal parameters for quantifying the surface roughness of rock joints, a theoretical model of a rock joint has been proposed on the basis of contact mechanics. The representative shape of asperity at contact has been specified by fractal dimension (D) and the y-intercept (A) in this model. The model considers different contact mechanisms, including elastic deformation, shear sliding and fracture of asperity. Theoretical investigation based on the fractal-contact model of a rock joint reveals several important relationships between the surface roughness and the mechanical behavior of the joint: (i) High values of fractal dimension give rise to higher normal stiffness and shear stiffness. On the other hand, a low value of the intercept results in a high normal stiffness; (ii) The ratio of normal/tangential force has less effect on normal stiffness, but results in different normal deformation depending on the magnitude of the normal force applied; (iii) The shear stiffness appears to be normal force dependent. The higher the normal force, the higher the shear stiffness. Numerical simulation of shearing of rock joints under direct compression has shown that the constitutive model derived in the study agrees well with the experimental results.

1 INTRODUCTION

The roughness of joint surfaces significantly affects the mechanical properties of joints and behavior of rock masses. In a conventional approach, difficulties are encountered when trying to describe effectively the roughness due to the scale effect and, therefore, to quantitatively relate roughness to mechanical response. The present research has been designed to gain a better understanding of the geometric surface structure of rock joints and the resulting mechanical behavior under loading conditions (Wang 1994).

The investigation started with a characterization of joint surface roughness. By employing a non-contact scanning instrument - an automated 3D laser profilometer, various types of rock fractures have been measured with high accuracy. These joint surfaces included extension, shear and hybrid fractures induced in rock samples of Coal Measure sandstones of various grain size and mechanical properties.

The scanning results have been analyzed according to the theory of fractal geometry, which provides for the scale invariant property of irregular objects. By comparison with the prescribed geometric profiles, two structural aspects have been distinguished: the roughness and waviness of a real rock fracture surface. The geostatistics based fractal analysis employs two basic parameters to quantify the surface roughness - fractal dimension (D) and the y-intercept (A) on the log-log plot of variance of the rough surface vs lag. Fractal dimension characterizes the irregularity of the rough surface while the y-intercept is statistically analogous to the slope of asperity of the rough surface.

Extensive investigation on fractal dimension and the intercept has been carried out within the present study (Kwaśniewski & Wang 1993, 1995). It has been found that natural fracture surfaces behave like self-affine fractals. Fractal dimension is spatially distributed over the surface and significantly different along the direction of crack propagation and orthogonal to it. The specific values of fractal dimension and the y-intercept suffer from scale effect and therefore these quantities are the statistical parameters as applied to describe joint roughness. These scale effects arise from the sampling size, sampling interval and the resolution of the scanning instrument. The study shows that the increasing sampling size and decreasing sampling interval may, to some extent, provide constant estimates of fractal dimension and the intercept.

To reveal the intrinsic relationship between the roughness, properties and mechanical behavior of rock joints, experimental investigations have been carried out in the laboratory by means of triaxial compression and shear testing devices. Shear tests under uniaxial compression were conducted at three different angles of shearing: 35°, 40° and 45°. Such an arrangement allowed the joint behavior to be investigated under different normal forces. It has been found that roughness parameters of joint surfaces influence their mechanical properties and behavior in a complex manner. In general, a high fractal dimension and a low value of the intercept give rise to a high normal and tangential stiffness.

Experimental results demonstrate that the shear strength of a rock joint primarily depends on the normal force. The influence of roughness on shear strength is of secondary importance and, in most cases, is not a single parameter dependent. Statistical analysis shows that the cross effects of fractal dimension and the intercept can be well expressed in terms of variance. The tensile strength of the rock material also plays a role in the fracture of the joint surfaces.

The process of surface damage of rock joints regarding loading history has also been investigated within the present study. It has been found, *i.a.*, that the evolution of surface roughness relates primarily to the plastic shear work in the loading process. It is also dependent on the original surface roughness and tensile strength of the rock material. On the basis of the experimental data, the empirical evolution law of surface damage has been developed in reference to the above parameters (Kwaśniewski & Wang 1997).

Many constitutive models of rock joints have been developed so far. These models can be roughly divided into two groups: one based on the global joint response (see Bandis 1993; Saeb & Amadei 1993 for a recent review), the other originates from the local contact mechanism (Swan 1983; Sun 1985; Brown & Scholz 1985, 1986; Plesha 1985; Zubelewicz et al. 1987; Plesha & Haimson 1988; Cook 1992; Qiu et al. 1993). Since it has been ascertained that the global mechanical behavior of rock joints results from the contact mechanisms at local asperities, it is of interest to the present study to consider the constitutive relationships of rock joints accounting for the local surface roughness.

The analysis of rough surfaces in contact has, over the years, progressed from the simple asperity-on-asperity approach of Archard (1957), through more elaborated solutions by Greenwood & Williamson (1966), Nayak (1973) and O'Callaghan & Cameron (1976) to the complex random process models of Bush et al. (1975, 1979) and a numerical technique of subsurface stress analysis developed by Bailey & Sayles (1991).

In a standard approach, the deviation of a rough surface from its mean plane is assumed to be of a random character described by statistical parameters such as the standard deviations of the surface height, slope and curvature (Nayak 1971). It has already been well established that these topographical parameters are highly scale-dependent. In consequence, predictions of the contact models based on such parameters may not be unique to a given pair of interacting surfaces. This is why employing scale invariant parameters in both quantification of surface roughness and constitutive modelling of joints would be a welcome endeavor. The major asset of fractal characterization of surface roughness is scale-independence; the fractal geometry provides information on surface structure at all the length scales that exhibit the fractal behavior. Using the fractal approach it is possible to consider the multiscale nature of rough surfaces in the contact process.

Majumdar & Bhushan (1990, 1991) developed a fractal elasto-plastic contact model. It considers a flat plane in contact with a statistically isotropic rough surface at tip of asperities which is described by fractal dimension D and a characteristic length scale G. Based on Hertzian contact theory, the model shows that for elastic deformation, the normal load P and the real area of contact A_r are related as $P \sim A_r^{(3-D)/2}$; for plastic deformation, the load and contact area are linearly related. The model predicts that the number of contact spots contributing to a certain fraction of the real area of contact remains independent of load although the spot size increases with load. Moreover, the load-area relation and the fraction of the real area of contact in elastic and plastic deformation are quite sensitive to the fractal parameters of surface roughness. Following this approach, Zhou, Leu & Blackmore (1993) have shown that fractal geometry can also be used to analyze wear processes. A fractal model for wear prediction was developed which connects fractal geometry, contact mechanics and wear theory. The model predicts the wear rate in terms of two fractal parameters: the fractal dimension and the topothesy.

The two fractal contact models mentioned above account for normal contact behavior in the absence of tangential traction across the interface. For rock joints, the interlock of asperities is a common phenomenon which determines the shear behavior under loading. The purpose of the present paper is to present a constitutive model of rock joint which has been derived on the basis of fractal geometry and contact mechanics. Some important contact mechanisms at asperities, such as elastic, frictional plastic deformation, and surface damage, have been considered and introduced into the model to show the influence of fractal parameters on the mechanical behavior of rock joints subjected to both normal and tangential loads.

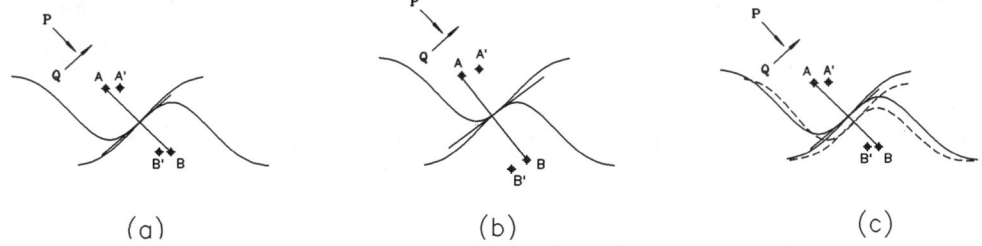

Fig. 1. Analytic modes of contact at asperities: (a) elastic, (b) plastic sliding, (c) fracture.

2 CONSTITUTIVE MODEL OF ROCK JOINT

Experimental study shows that the mechanical behavior of rock joints is related to the geometric surface structure. The global response of a joint under loading is controlled by the local contact mechanisms. Analytically, the modes of contact at asperities can be specified by:

1. *Elastic contact* - There is no evident slide between two surfaces of a joint. Deformations in normal and tangential direction are recoverable upon the removal of load (Fig. 1a);

2. *Plastic sliding* - One asperity slides in respect to another when basic frictional resistance is overcome ($Q \geq \mu P$). Surface damage might be associated with the sliding and the deformation is unrecoverable (Fig. 1b);

3. *Fracture of asperity* - In absence of a significant tangential force ($Q < \mu P$), when the resultant stress due to normal force P or the combination of normal and tangential forces exceeds the strength of the holding material, fracture of asperity normally occurs. This results in an extra permanent deformation and lower order of contact (Fig. 1c).

Contact mechanics applies to stress and deformation analysis with the assumption of a smooth, continuous shape of contact. For a rock joint, the surface structure is randomly extended and can be characterized by fractal parameters, *i.e.* fractal dimension D and the intercept A.

The variogram suggests that surface roughness is composed of numerously superimposing asperities. Providing that a single asperity has a sinusoidal form, it can be expressed by a periodic function

$$z(x) = \sqrt{A}\, r^{(2-D)} \sin\frac{\pi x}{r} \quad (0 \leq x \leq r) \quad (1)$$

where r is defined as the wavelength of the representative asperity (Fig. 2).

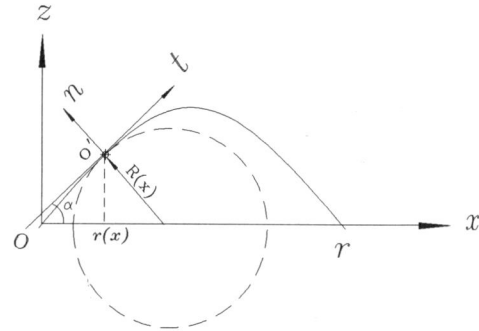

Fig. 2. Schematic asperity.

At any instant contacting point of the asperity, the radius of curvature is given by

$$R(x) = \frac{\left[1 + \left(\frac{dz}{dx}\right)^2\right]^{3/2}}{\left|\frac{d^2z}{dx^2}\right|} = \frac{\left[1 + \pi^2 A\, r^{2(1-D)} \cos^2\frac{\pi x}{r}\right]^{3/2}}{\pi^2 \sqrt{A}\, r^{-D} \sin\frac{\pi x}{r}} \quad (2)$$

Based on contact mechanics (Johnson 1985) the constitutive relationship at local contact asperities has been derived in an incremental form:

$$dP = \left\{\frac{3}{4}\frac{E^2 P}{(1-\nu^2)^2}\left[1 + \pi^2 A r^{2(1-D)}\cos^2\frac{\pi x}{r}\right]^{3/2}\right\}^{\frac{1}{3}} d\delta_n$$

$$= k_n\, d\delta_n \quad (3)$$

$$dQ = \frac{2(1-\nu)}{2-\nu} k_n \left(1 - \frac{Q}{\mu P}\right)^{\frac{1}{3}} d\delta_t = k_t d\delta_t \quad (4)$$

where P and Q are normal and tangential force acting across the asperity, δ_n and δ_t are normal and shear displacements, k_n and k_t are coefficients of normal and tangential stiffness, μ is a coefficient of kinetic friction, and E and ν are Young's modulus and Poisson's ratio, respectively. It should be noted that the coefficient of tangential stiffness (k_t) is dependent not only on surface roughness but also on the ratio of tangential and normal forces applied.

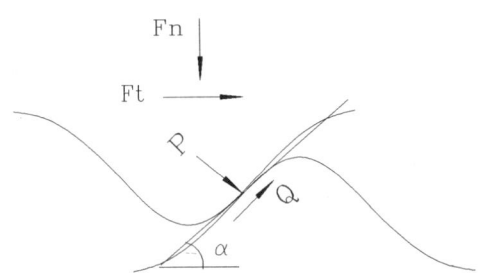

Fig. 3. Transformation from local to global coordinates.

Combining Eqn. (3) and (4) into a matrix form, the constitutive relationship between forces and displacements at asperities can be given as

$$\begin{Bmatrix} dP \\ dQ \end{Bmatrix} = \begin{bmatrix} k_n & 0 \\ 0 & k_t \end{bmatrix} \begin{Bmatrix} d\delta_n \\ d\delta_t \end{Bmatrix} \quad (5)$$

The off-diagonal components in the stiffness matrix are zero, with regard to the symmetrical deformation for both positive and negative $d\delta$ at an instantaneous contact point. It is clear that when the tangential force exceeds the friction resistance at the contact point, sliding takes place along the asperity and a new contact configuration is reached. The equilibrium state of contact requires assessment of force and displacement in reference to the current contact condition in successive incremental steps of load or displacement.

Under sufficiently high normal force, or when the friction within the interface is large enough to exclude net motion of the two surfaces ($Q < \mu P$), brittle fracture might occur to the contact point. According to contact mechanics, stresses induced by normal and tangential forces can be calculated from the distribution of forces within the contact area

$$\sigma_t = \sigma_t^p + \sigma_t^q \quad (6)$$

where σ_t is the tensile stress caused by the normal and tangential stress components σ_t^p and σ_t^q.

Under the global system, the constitutive relationship is given by

$$\begin{Bmatrix} dF_n \\ dF_t \end{Bmatrix} = \begin{bmatrix} K_{nn} & K_{nt} \\ K_{tn} & K_{tt} \end{bmatrix} \begin{Bmatrix} du_n \\ du_t \end{Bmatrix} \quad (7)$$

where F_n and F_t are the normal and shear forces applied to the joint (Fig. 3), and u_n and u_t are the global displacements in normal and tangential directions, respectively. The explicit expressions of contents in the stiffness matrix in Eqn. 7 are given by

$$K_{nn} = k_n \cos^2\alpha + k_t \sin^2\alpha \quad (8)$$

$$K_{nt} = K_{tn} = (k_n - k_t) \cos\alpha \sin\alpha \quad (9)$$

$$K_{tt} = k_n \sin^2\alpha + k_t \cos^2\alpha \quad (10)$$

where k_n and k_t are coefficients of normal and tangential stiffness at local contact asperities (cf. Eqn. 3 and 4). The angle of the transformation from local to global system can be calculated from

$$\alpha = \tan^{-1}\left(\frac{dz}{dx}\right) = \tan^{-1}\left[\pi \sqrt{A} \, r^{(1-D)} \cos\frac{\pi x}{r}\right] \quad (11)$$

The present constitutive model has been derived on the basis of local contact mechanics, and therefore, can comprise different mechanisms of contact at asperities in a uniformed program. When slip or fracture occurs to asperity, the empirical surface damage law which has been developed on the basis of experimental investigations (Wang 1994; Kwaśniewski & Wang 1997) is extended into the model to up-date and modify the contact position as well as the shape of asperity.

3 NUMERICAL IMPLEMENTATION

A number of examples will be presented here to show the performance of the constitutive model and to investigate the effect of surface roughness on the mechanical behavior of rock joints. Shear tests under direct compression were numerically simulated, where the prescribed displacement $u(\theta)$ was applied obliquely to the joint plane. The basic parameters used in the modelling of rock joint were as follows: $E = 30106$ MPa, $\nu = 0.147$, $\sigma_T = 8.0$ MPa, $\mu = 0.5$, $\theta = 40°$, and $\Delta u(\theta) = 0.05$ mm.

In the modelling procedure, the contact condition - elastic, plastic sliding or fracturing at contact of asperities - is automatically assessed through the following criteria:
1. elastic contact: $Q < \mu P$ and $\sigma_t < \sigma_T$;
2. sliding contact: $Q > \mu P$;
3. fracture at contact: $\sigma_t > \sigma_T$.

The contact configuration is up-dated according to the current state of contact. When sliding takes place, the plastic work is calculated in case the tensile stress in a vicinity of the contact area exceeds the tensile strength of the material. The shape of asperity is then modified by recalculating fractal dimension D_s and the y-intercept A_s according to the empirical evolution laws of surface damage (see Kwaśniewski & Wang 1997).

Numerical simulation reveals several significant aspects of fractal parameters influencing the mechanical behavior of rock joints:

Effect of fractal dimension (D). At a constant value of y-intercept ($A = 1.5 \times 10^{-3}$), different values of fractal dimension were used ($D = 1.1$, 1.3, 1.5, 1.7) to study the mechanical response of a joint under shearing. As a result it has been established that high fractal dimension gives rise to high normal and shear stiffnesses (Fig. 4). This finding has a high correspondence with experimental observations (Wang 1994; Kwaśniewski & Wang 1997).

Effect of the y-intercept (A). At a constant value of fractal dimension ($D = 1.5$), four values of the y-intercept were used ($A = 1.0 \times 10^{-2}$, 1.0×10^{-3}, 1.0×10^{-4}, and 1.0×10^{-5}) in the calculations. As it has been shown elsewhere (Kwaśniewski & Wang 1995), the higher values of the y-intercept indicate steeper slopes of asperities. Numerical simulation shows that smaller y-intercepts (smooth surface) produce higher normal stiffness (Fig. 5). The results are consistent with the physical sense and agree well with experimental measurements. As shear stiffness is proportional to normal stiffness, a higher value of the intercept produces a lower shear stiffness. Although this effect is convincingly supported by experimental evidence, its mechanical background has not yet been fully established. Perhaps, in more complex contact situations, the fracture of steeper asperities is responsible for the behavior.

Effect of shearing angle (θ). In shearing tests under direct compression, different shearing angles give different ratios of normal to shear force. The bigger the angle θ, the higher the normal force.

Within the present numerical study it was interesting to investigate the stiffness of a joint with fractal dimension $D = 1.5$ and the y-intercept $A = 1.5 \times 10^{-3}$ under different angles of shearing ($\theta = 35°$, 40° and 45°). As it is shown in Figure 6, the variation of shear angles has a minor influence on normal stiffness, only higher ultimate normal displacements were obtained for higher shearing angles. However, the effect of angles of shearing on shear stiffness is much pronounced, i.e. high angle of shearing produces a relatively high shear stiffness.

A number of actual laboratory shear tests have also been simulated to explore further the performance of the model. Comparison of the predictions with the measurements is presented in Figure 7. As can be seen, at low level of load, the predicted normal deformation satisfactorily agrees with the measurements. At a high load level, however, the predicted normal deformation gradually deviates from the measurements. Since the present constitutive model has been derived based on the contact geometry of a single pair of asperities, the interaction of neighboring asperities has been ignored. In fact, as increasing the load, some small asperities emerge to form a large asperity and some new contact spot is created. This problem has been investigated by Majumdar & Bhushan (1990, 1991). The present model could be improved by considering the contact size-distribution and the number of contact spots with respect to the load. Future work will take this factor into account.

In the pre-failure part of shearing, the shear deformations predicted by the model correspond highly with experimental results. After peak shear strength, there occurs a competitive friction mechanism between the joint surface and the boundaries of the shear testing device. The post-failure behavior of shear is, therefore, partially controlled by the boundary conditions. The numerical simulation of the post-failure behavior of rock joint might be more instructive if compared with results of direct shear experiments in which normal force could be specified.

4 CONCLUDING REMARKS

The aim of the study was to give a better understanding of the influence of roughness on the mechanical behavior of rock joints. Based on contact mechanics, a constitutive model of rock joint has been derived. The assumption was adopted that smooth sinusoidal asperities in the representative scale are brought into contact under transformed normal force P with the presence of tangential traction Q. In the model, the representative shape of

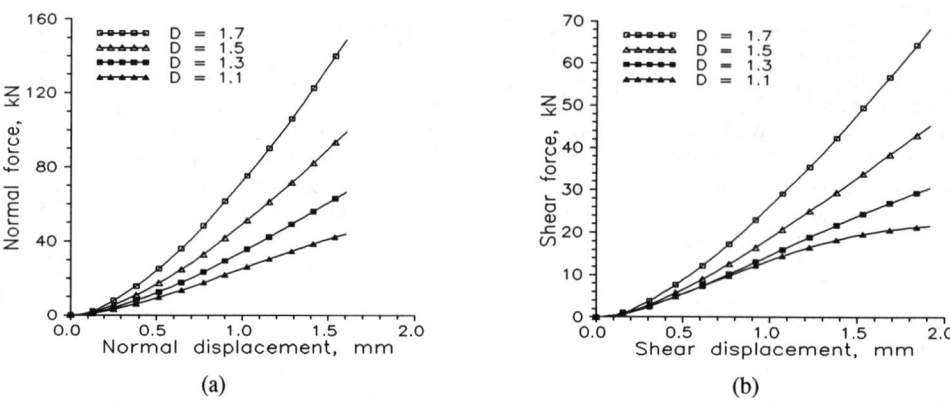

Fig. 4. Effects of fractal dimension D on (a) normal and (b) shear deformation of joint.

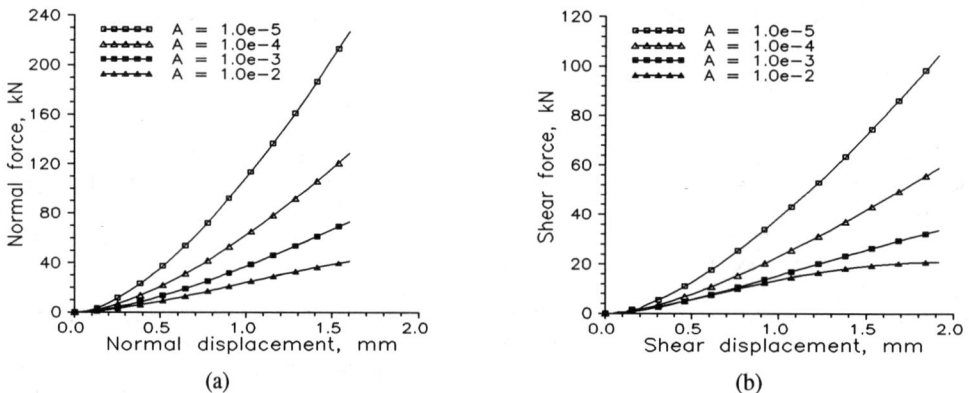

Fig. 5. Effects of intercept A on (a) normal and (b) shear deformation.

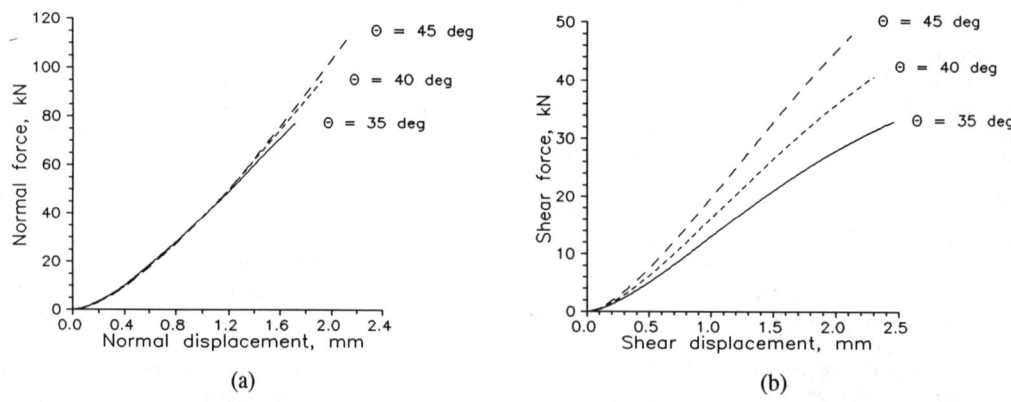

Fig. 6. Effects of shear angle θ on (a) normal and (b) shear deformation.

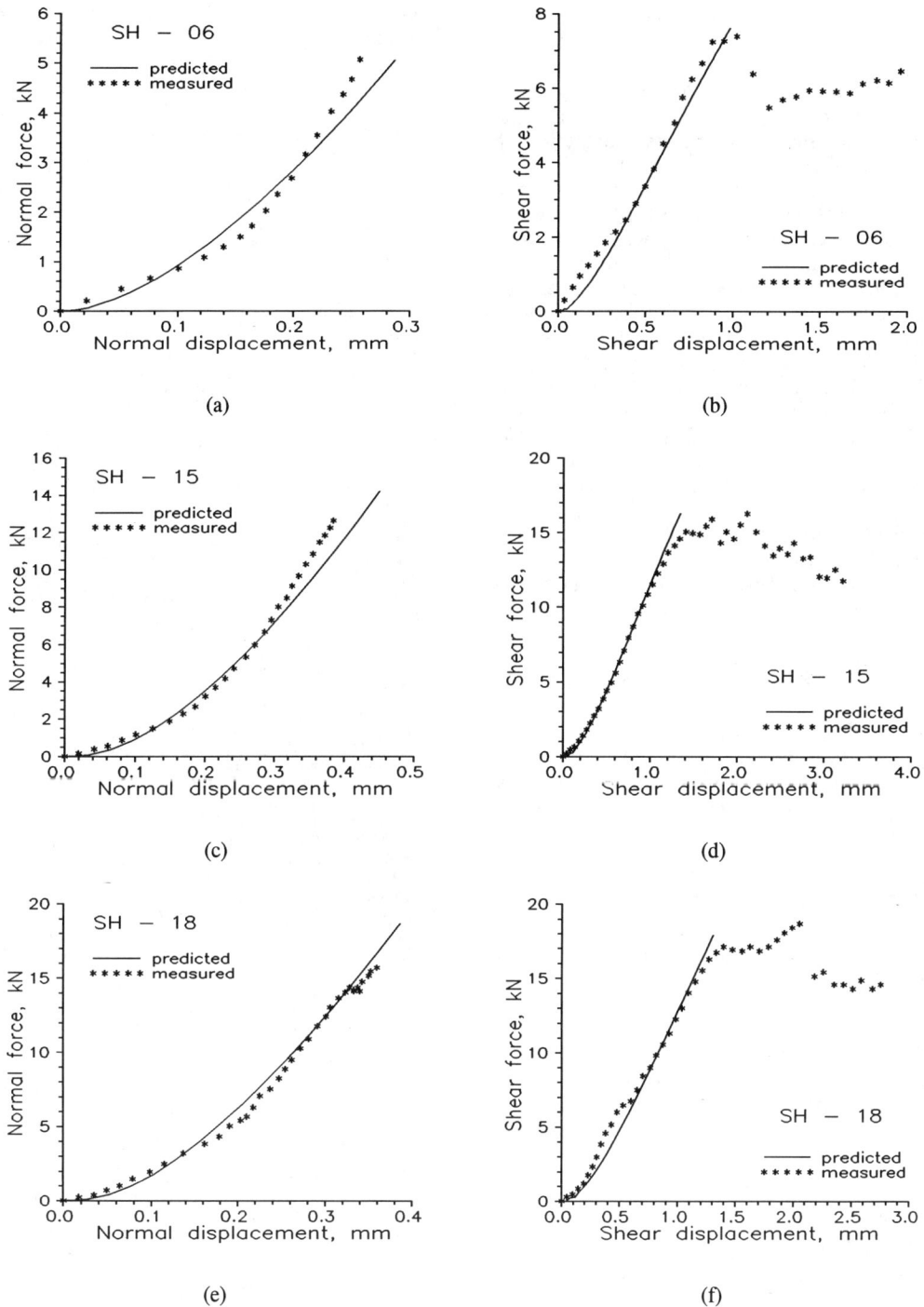

Fig. 7. Modelling of modified shear tests (SH-06, SH-15, SH-18); (a), (c), (e) - normal deformation, (b), (d), (f) - shear deformation.

asperity at contact is specified by geostatistics/fractal geometrical parameters - fractal dimension D and the y-intercept A.

Numerical investigation based on the constitutive model of rock joint reveals several significant features of surface roughness influencing the mechanical behavior of a joint. High values of fractal dimension give rise to higher normal stiffness and shear stiffness. On the other hand, the low value of the intercept produces a high normal stiffness. These results are conclusively in agreement with the experimental measurements. However, the effect of high shear stiffness under lower intercept has not been fully interpreted from the physical point of view. The fracture of contact asperity is perhaps responsible for this behavior which, in fact, requires further investigation.

The ratio of normal/tangential force has less effect on normal stiffness, while it results in different normal deformation depending on the magnitude of the normal force. The shear stiffness appears to be normal force dependent also. The higher the normal force, the higher the shear stiffness.

Numerical simulation of actual laboratory shear tests under direct compression was carried out using the proposed constitutive law. The simulation results show that the model predicts both normal deformation at low normal load and shear deformation prior to the peak shear strength which correspond highly to observations. Further work is needed however to take into account the effect of the increase in contact spots with load, and the increase in the total contact area due to the emergence of neighboring asperities.

ACKNOWLEDGEMENT

The investigation, which was carried out at Rock Mechanics Laboratory, Faculty of Mining and Geology, The Technical University of Silesia, was supported jointly by State Committee for Scientific Research and Staszic Colliery in Katowice, Poland through research grant No. 231/CS6-9/92. The support is gratefully acknowledged.

REFERENCES

Archard, J.F. 1957. Elastic deformation and the laws of friction. *Proc. R. Soc. London, Ser. A* 243:190-205.

Bailey, D.M. & R.S.Sayles 1991. Effect of roughness and sliding friction on contact stresses. *ASME J. Trib.* 113:729-738.

Bandis, S.C. 1993. Engineering properties and characterization of rock discontinuities. In J.A.Hudson & E.T.Brown (eds), *Comprehensive Rock Engineering*, Vol.1 - *Fundamentals*: 155-183. Oxford: Pergamon Press.

Brown, S.R. & C.H.Scholz 1985. The closure of random elastic surfaces in contact. *J. Geophys. Res.* 90:5531-5545.

Brown, S.R. & C.H.Scholz 1986. Closure of rock joints. *J. Geophys. Res.* 91:4939-4948.

Bush, A.W., R.D.Gibson & T.R.Thomas 1975. The elastic contact of a rough surface. *Wear* 35:87-111.

Bush, A.W., R.D.Gibson & G.P.Keogh 1979. Strongly anisotropic rough surfaces. *ASME J. Lubr. Technol.* 101:15-20.

Cook, N.G.W. 1992. Natural joints in rock: Mechanical, hydraulic and seismic behaviour and properties under normal stress. *Int. J. Rock Mech. Min. Sci. & Geomech. Abstr.* 29(3): 198-223.

Greenwood, J.A. & J.B.P.Williamson 1966. Contact of nominally flat surfaces. *Proc. R. Soc. London, Ser. A* 295:300-319.

Johnson, K.L. 1985. *Contact Mechanics*. Cambridge: Cambridge University Press.

Kwaśniewski, M.A. & J.-A.Wang 1993. Application of laser profilometry and fractal analysis to measurement and characterization of morphological features of rock fracture surfaces. In J.P. Piguet et F.Homand (eds), *Géotechnique et Environnement*: 163-176. Vandœuvre-lès-Nancy: Sciences de la Terre.

Kwaśniewski, M.A. & J.-A.Wang 1995. On the fractal character of rock fracture surfaces. 18th Winter School of Rock Mechanics, Szklarska Poręba, March 1995, p. 71. (in Polish)

Kwaśniewski, M.A. & J.-A.Wang 1997. Surface roughness evolution and mechanical behavior of rock joints under shear. *Int. J. Rock Mech. & Min. Sci.* 34:3-4, Paper No. 157.

Majumdar, A. & B.Bhushan 1990. Role of fractal geometry in roughness characterization and contact mechanics of surfaces. *ASME J. Trib.* 112:205-216.

Majumdar, A. & B.Bhushan 1991. Fractal model of elastic-plastic contact between rough surfaces. *ASME J. Trib.* 113:1-11.

Nayak, P.R. 1971. Random process model of rough surfaces. *ASME J. Lubr. Technol.* 93:398-407.

Nayak, P.R. 1973. Random process model of rough surfaces in plastic contact. *Wear* 26:305-333.

O'Callaghan, M. & M.A.Cameron 1976. Static contact under load between nominally flat surfaces in which deformation is purely elastic. *Wear* 36:79-97.

Plesha, M.E. 1985. Constitutive modeling of rock joints with dilation. In E.Ashworth (ed.), *Research and Engineering Applications in Rock Masses*, Vol.1:387-394. Rotterdam: Balkema.

Plesha, M.E. & B.C.Haimson 1988. An advanced model for rock joint behavior: Analytical, experimental and implementational considerations. In P.A.Cundall, R.L.Sterling & A.M.Starfield (eds), *Key Questions in Rock Mechanics*: 119-126. Rotterdam: Balkema.

Qiu, X., M.E.Plesha, X.Huang & B.C.Haimson 1993. An investigation of the mechanics of rock joints - Part II. Analytical investigation. *Int. J. Rock Mech. Min. Sci. & Geomech. Abstr.* 30(3):271-287.

Saeb, S. & B.Amadei 1992. Modelling rock joints under shear and normal loading. *Int. J. Rock Mech. Min. Sci. & Geomech. Abstr.* 29(3):267-278.

Sun, Z. Asperity models for closure and shear. In O.Stephansson (ed.), *Fundamentals of Rock Joints*: 173-183. Luleå: CENTEK Publishers.

Wang, J.-A. 1994. *Morphology and Mechanical Behavior of Rock Joints*. Ph. D. Thesis, p. 236. Faculty of Mining and Geology, Silesian Technical University, Gliwice, Poland.

Zhou, G.Y., M.C.Leu & D.Blackmore 1993. Fractal geometry model for wear prediction. *Wear* 170:1-14.

Zubelewicz, A., K.O'Connor, C.H.Dowding, T.Belytschko & M.Plesha 1987. A constitutive model for the cyclic behavior of dilatant rock joints. In C.S.Desai, E.Krempl, P.D.Kiousis & T.Kundu (eds), *Constitutive Laws for Engineering Materials - Theory and Applications*, Vol.II:1137-1144. New York: Elsevier.

Tests on mechanical properties of model fracture zones

Ömer Aydan
Tokai University, Shimizu, Japan

Yasuhiro Shimizu
Meijo University, Nagoya, Japan

Tomoyuki Akagi
Toyota National College of Technology, Japan

Toshikazu Kawamoto
Aichi Institute of Technology, Toyota, Japan

ABSTRACT: Fracture zones are commonly found in the Earth's crust. Therefore, it is almost impossible not to encounter these zones during large rock excavations. These zones are always problematic causing heavy water in-flow, instability, squeezing etc. during excavation. Therefore, the assessment of mechanical characteristics of these zones is quite important. Nevertheless very few studies undertaken so far on this aspect of fracture zones. In this paper, the results of tests on the mechanical properties of model fracture zones are presented. Tests involve uniaxial compression tests, direct shear tests and elastic wave velocity measurements. Furthermore, several techniques for assessing mechanical characteristics of fracture zones is briefly outlines and the assessed mechanical properties by these techniques are compared with experimental results on model fracture zones.

1 INTRODUCTION

Fracture zones are commonly found in the Earth's crust and it is almost impossible not to encounter such zones during large rock excavations. These zones cause very severe problems such as heavy water in-flow, squeezing, instability etc. during excavation of rock engineering structures. Therefore, the assessment of the strength and deformability of these zones is of great concern for rock engineers. As well known, fracture zones are consisted of intact rock blocks bounded by joints with or without gouges.

The simple method for assessing mechanical characteristics of fracture zones is based on the use of mechanical characteristics of the weakest material, namely, gouge in the fracture zone. Although this method is on the safest side, it underestimates their mechanical characteristics. For a more reasonable and economical characterization, their actual structure must be taken into account. However, this requires enormous efforts involved in characterization as well as mechanical and numerical modelling. Furthermore, very few studies are undertaken so far on this problem.

Experimental studies on the fracture zones are mainly carried by two disciplines, namely, geophysics and rock mechanics. The geophysicians are mainly interested on the thermo-hydro-mechanical characteristics of gouges present in fracture zones and faults in relation to the earthquake prediction projects. On the other hand, rock mechanicians and engineers are concerned with the stability of underground and surface structures passing through fracture zones. Tests by rock mechanicians and engineers are mainly carried out by using the direct shear test technique. During laboratory tests, the type, granulometry and thickness of filling materials are generally varied. In-situ direct shear tests are carried out by cutting out samples along the fracture zones which are generally too expensive to perform.

In this paper, the results of an experimental program undertaken on model fracture zones are presented. The experimental program involves laboratory uniaxial and direct shear tests. In addition, elastic wave velocity measurements are carried out as an index for in-direct characterization of these zones. The experimental results such as elastic moduli, uniaxial compressive strength, shear strength etc. are compared with those obtained from several averaging techniques which utilise the mechanical and geometrical characteristics of constituents of these zones.

2 TEST SET-UP AND EXPERIMENTS

In this study, both uniaxial tests and direct shear tests on model fracture zones are carried out.

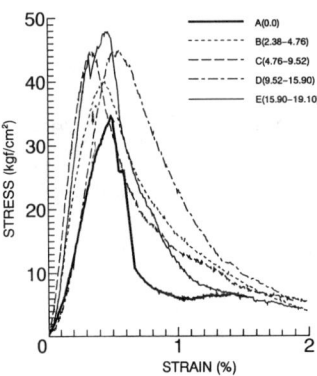

Fig. 1 Typical stress-strain curves of samples with different grain size

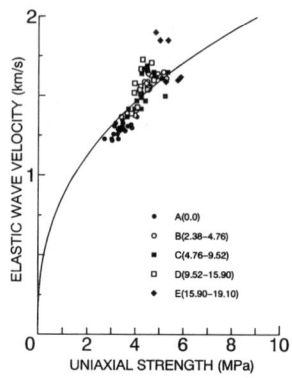

Fig. 2 Experimental uniaxial strength - elastic wave velocity relation

2.1 Uniaxial Tests

In uniaxial compression tests, specimens composed of materials representing intact blocks and gouge were prepared by varying the grain size of intact material while keeping its volume fraction between 21-28 %. Figure 1 shows shows typical stress-strain relation obtained in uniaxial compression tests on samples with different grain sizes. The stress-strain relation of intact material is not shown in the same figure as its uniaxial strength is 15 times the gouge model material. Model A corresponds to gouge material. Before each test, a series of measurements on elastic wave velocity of samples was performed. Figure 2 shows the relation between the elastic modulus and uniaxial strength of model fracture zones.

2.2 Direct Shear Tests

Direct shear tests were carried out on samples which were 150 mm long, 75 mm wide and 75 mm high. The main purposes of tests were to investigate the effects of fracture zone thickness and of grain sizes of intact particles on the shear response and strength characteristics of model fracture zones. The normal stress applied to samples was 10, 15 and 20 kgf/cm^2. The maximum normal stress was kept below the uniaxial strength of model fracture zone in order to prevent out-of plane failure in samples. The thicknesses of fracture zone were 5, 10, 15, 20 mm. Figure 3 shows the direct shear response of model fracture zones A and C. When the thickness of fracture zone is 75 mm, it corresponds to the complete fracture zone material. As seen from these figures, the effect of fracture zone thickness has no significant effect on the peak and residual strength of model fracture zones. On the other hand, the increase in thickness result in lower shear stiffness of these zones.

The grain size of intact particles in fracture zones was varied. 5 set of samples having different grain sizes were prepared. Figure 4 shows the shear responses of 4 set of samples having a fracture zone thickness of 20 mm. As seen from the figures, the increase in grain size generally results in higher shear strength and shear stiffness.

3 ASSESSMENT TECHNIQUES

There are presently three different assessment techniques, namely, 1) Empirical techniques, 2) Equivalent material techniques, and 3) Numerical techniques. These techniques are briefly outlined in the following sub-sections.

3.1 Empirical Techniques

There are presently three main approaches to assess the mechanical properties of fracture zones, namely, (a) Jointing index methods, (b) Rock mass classifications and (c) Elastic wave velocity methods.

(a) Jointing Index Methods - JIM: These methods are based on an index defined as the ratio of length of sample to discontinuity spacing or number of block contained within the sample. The earliest formula was proposed by Protodyakonov (1964) for coal seams and other formulas were proposed by Goldstein et al (1966) and Vardar (1977).

(b) Rock Classification Methods - RCM: Although there are several rock classifications used in many countries, it seems that RMR and Q-system are the most widely known rock classifications (Bieniawski 1974, Barton et al. 1974). The Q-value and RMR are generally less than 0.1 and 25 for fracture zones, respectively.

(c) Elastic Wave Velocity Methods - EWVM: In one of these methods (EWVM-1), a reduction factor R defined as

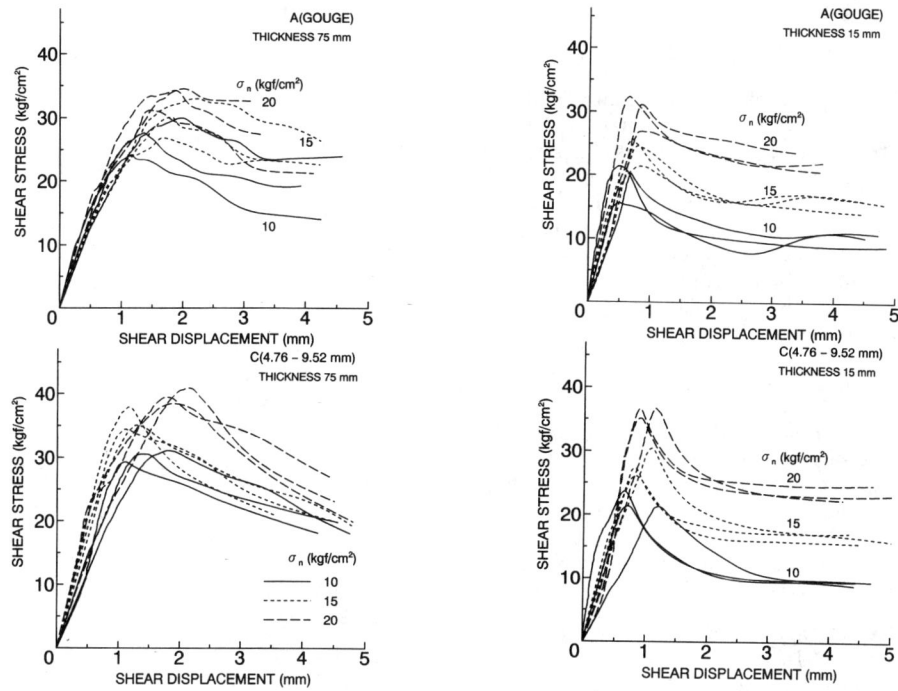

Fig. 3 Typical shear stress-displacement curves of samples with different fracture zone thicknesses

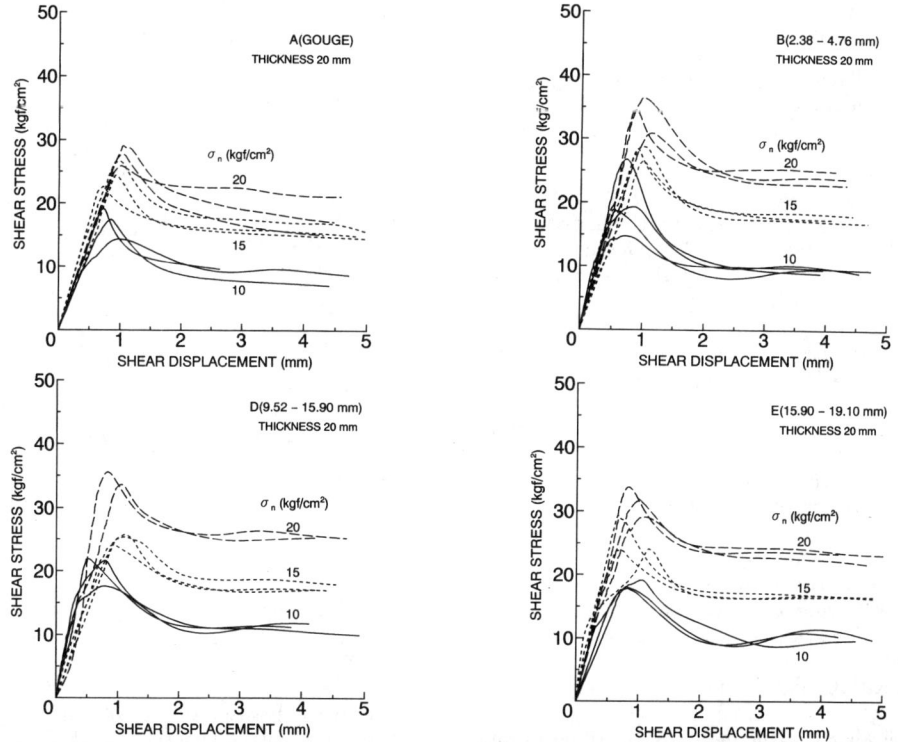

Fig. 4 Typical shear stress-displacement curves of samples with different grain size for a fracture zone thickness of 20 mm

$$R = \left(\frac{V_p^f}{V_p^i}\right)^2 \quad (1)$$

is used for obtaining macroscopic properties of fracture zones from the properties of intact material is known. This method is commonly used to estimate the mechanical properties of fracture zones as well as those of rock masses in tunnelling in Japan.

Another approach (EWVM-2) could be the direct use of the elastic wave velocity of the mass and empirical relations developed by Aydan et al. (1993). The elastic wave velocity of these zones will be generally less than 1.6-1.7.

3.2 Equivalent Material Techniques

Equivalent continuum modelling starts with the first attempts by Voigt (1910) and Reuss (1929) for multiphase materials. The approach of Reuss now constitutes the basis of *mixture theory* for multiphase materials. Since then many models are proposed and the models applicable to fracture zones in rock mechanics are:

1-) Mixture theory based either on Reuss or Voigt models,
2-) Micro-structure models (Aydan et al. 1992, 1996b), and
3-) Homogenization technique (Kawamoto and Kyoya 1993).

The main characteristics of these models are described in detail and compared by Aydan et al. (1995).

3.3 Numerical Techniques

One can find several numerical techniques such as Finite Element Method (FEM) (Goodman et al. 1968, Ghaboussi et al. 1973) Discrete Finite Element Method (DFEM) (Aydan et al. 1996a), Distinct Element Method (DEM) (Cundall 1971) and Discontinuities Deformation Analysis (DDA) (Shi 1988) which are developed for fractured and jointed media. These methods can also be used to assess the mechanical characteristics of fracture zones. Such an application was made by Kawamoto et al. (1990) with the use of finite element method. They presented several applications of this technique to model mechanical behaviour of fracture zones under uniaxial and direct shear loading conditions. With some improvements, this method has a high application potential for assessing the mechanical behaviour of fracture zones.

4 COMPARISONS AND DISCUSSIONS

The applicability of the methods described in the previous section to the uniaxial compression and direct shear tests on model fracture zones is given and discussed in this section.

4.1 Uniaxial Compression Tests

Figures 5 and 6 compare experimental results for elastic modulus and uniaxial compression strength with those predicted by different techniques described in the previous section. Among equivalent methods, the predictions given by the microstructure theory (GPLSM, GSLPM) are generally consistent with experimental results. It is interesting to note that all experimental results are bounded by predictions by Reuss and Voigt models which may be interpreted as upper and lower bounds. The predictions by elastic wave velocity methods are somewhat different form experimental results. Nevertheless, the differences between predictions and experimental results are not

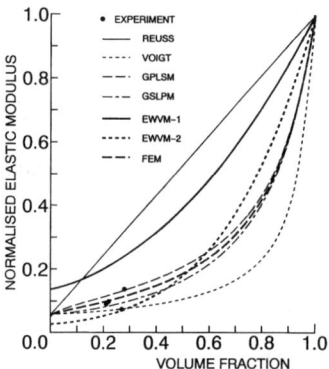

Fig. 5 Comparison of predictions by different methods with experimental results

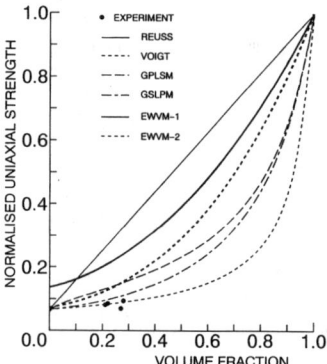

Fig. 6 Comparison of predictions by different methods with experimental results

so pronounced. This implies that they may be usefull during feasibility studies. The predictions by jointing index methods is not plotted in the same figure as there is a number of possible paths for block size and volume fraction relation. However, separate computations indicate that the formula given by Protodyakonov (1964) is also applicable to fracture zones, although it overestimates their mechanical characteristics.

4.2 Direct Shear Tests

The effects of grain size of intact particles and thickness of model fracture zones on their peak and residual shear strengths are presented and discussed in this sub-section. Various assessment techniques described in Section 3 are almost non - applicable to this aspect of fracture zones as it seems that the geometry of interface between fracture zone and intact rock plays an important role on the strength properties. Figure 7 shows the effect of grain size of intact particles on the peak and residual strengths of model fracture zones for a fracture zone thickness of 20mm. As seen from the figure, it seems that the increase in grain size results in higher strength characteristics.

Figure 8 shows the effect of fracture zone thickness on sample set C of model fracture zones. Although the experimental results are scattered, it seems that the strengths of samples are slightly less than those of composite fracture zone material itself. This is probably due to the existence of interfaces between the fracture zone and adjacent intact blocks and their geometry. Although utmost care was taken to prevent shearing along the interfaces, experimental results seem to have been infuenced by the interface failure. Nevertheless, their shear strength is generally equal or greater than that of gouge material (see for example Figure 3).

5 CONCLUSIONS

A series of uniaxial compression and direct shear tests was carried out on model fracture zones in laboratory and test results are presented in this paper. Then the available methods for assessing the mechanical characteristics of fracture zones are briefly described and their validity are checked by uniaxial compression and direct shear tests on model fracture zones. Although actual fracture zones are more complicated, tests on model fracture zones

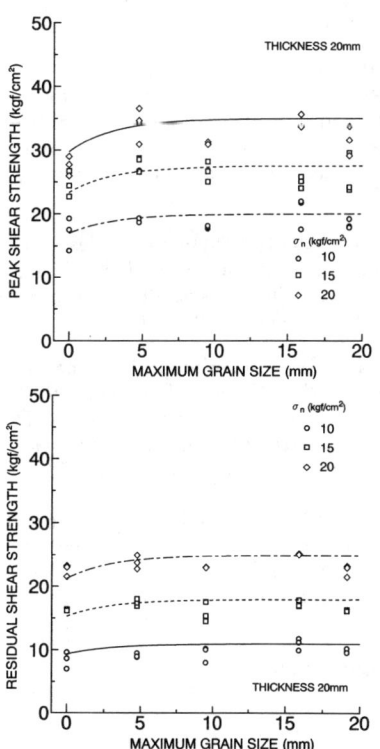

Fig. 7 The effect of grain size on peak and residual shear strength

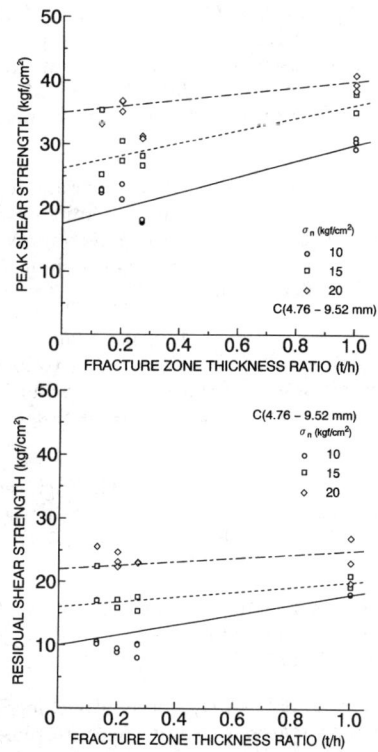

Fig. 8 The effect of fracture zone thickness on peak and residual shear strength

yield some information on their mechanical behaviour and some insights how to assess their mechanical characteristics. As far as uniaxial compression tests are concerned, the best estimation is given by the finite element method. The equivalent models may be also effective in this respect. Methods based on rock classifications are almost inapplicable to fracture zones while elastic wave velocity methods may be usefull for assessing the mechanical characteristics of fracture zones. However, it should be noted that they may overestimate the mechanical characteristics. Direct shear test results showed that the existence of intact blocks within fracture zones results in higher shear strength and stiffness. Furthermore, the strength of fracture zones consisting intact rock blocks is greater than that of gouge material (i.e. clay). Therefore, the actual geometry and constituents of fracture zones must be taken into account when their mechanical characteristics are assessed.

REFERENCES

Aydan, Ö., N. Tokashiki, T. Seiki and F. Ito 1992: Deformability and strength of discontinuous rock masses. Int. Conf. on Fractured and Jointed Rock Masses, Lake Tahoe, 256-263.

Aydan, Ö., T. Seiki, G.C. Jeong, T. Akagi (1995). A comparative study on various approaches to model discontinuous rock mass as equivalent continuum. *2nd Int. Conf. on Mechanics of Jointed and Fractured Rock*, Vienna, 569-574.

Aydan, Ö., Mamaghani, I.H.P., Kawamoto, T. 1996a: Application of discrete finite element method (DFEM) to rock engineering. North American Rock Mechanics Symposium. Montreal, 2, 2039-2046.

Aydan, Ö., N. Tokashiki, T. Seiki 1996b: Microstructure models for porous rocks to jointed rock masses. The third Asia - Pacific Conference on Computational Mechanics, Seoul, 3, 2235-2242.

Barton, N., Lien, R., Lunde, I. 1974. Engineering classification of rock masses for the design of tunnel supports. Rock Mechanics, 6(4), 189-239.

Bieniawski, Z.T. 1974. Engineering Rock Mass Classifications. John Wiley & Sons, New York, 251 pp.

Cundall, P. A. 1971. A Computer Model for Simulating Progressive, Large-Scale Movements in Blocky Rock Systems. Proc. Int. Symp. on Rock Fracture, II-8, Nancy, France.

Ghaboussi, J., E.L Wilson and J. Isenberg 1973. Finite element for rock joints and interfaces. *J. Soil Mechs. and Found. Eng. Div., ASCE*, SM10, 99, 833-848.

Goldstein, M., Goosev, B., Pyrogovsky, N., Tulinov, R., Turovskaya, A. 1966. Investigation of mechanical properties of cracked rock. 1st ISRM Congress, Lisbon, 1, 521-524.

Goodman, R.E., R. Taylor and T.L. Brekke 1968. A model for the mechanics of jointed rock. *J. Soil Mechs. and Found. Eng. Div., ASCE*, SM3, 94,637-659.

Hoek, E., Brown, E.T. 1980. Emprical strength criterion for rock masses. ASCE J. Geotech. Engng. Div., 106, GT9, 1013-1035.

Kawamoto, T., Kyoya, T., Ichikawa, Y. 1990. Evaluating material properties of faulted zones by image analysis and numerical methods. Proc. Int. Symp. on Rock Joints, Loen, 59-66.

Kawamoto, T., T. Kyoya 1993: Some applications of homogenization method in rock mechanics. Seminar on Impacts of Computational Mechanics to Engineering Problems, Sydney.

Protodyakonov, M.M., Koifman, M.I. 1964. Uber den Masstabseffect bei Untersuchung von Gestein und Kohle. 5. Landertreffen des Internationalen Buros für Gebirgsmechanik, Deutsche Akademie der Wissenschaften, Berlin, 3, 97-108.

Reuss, A. 1929. Berechnung der Fliessgrenze von Mischkristallen auf Grund der Plastizitätsbedingung für Feinkristalle. Z. Agnew. Math. u. Mech., Vol. 9, 3965-3985.

Shi, G.H. 1988. Discontinuous Deformation Analysis: A New Numerical Model for the Statics and Dynamics of Block System. *Ph.D. Thesis, Department of Civil Engineering, University of California, Berkeley*, 378p.

Vardar, M. 1977. Zeiteinfluss auf des Bruchverhalten des Gebriges in der Umgebung von Tunbeln. Veröff. D. inst. F. Bodenmech., Univ. of Karlsruhe, Heft 72.

Voigt, W. 1910. Lehrbuch der Kristallphysik, Leipzig, Teubner.

Physical model and numerical model analyses of jointed rocks

B. Wang
Mining Laboratories, CANMET, Natural Resources Canada

S. Kwon & H. D. S. Miller
Rock Mechanics and Explosives Research Center, University of Missouri-Rolla, Mo., USA

D. H. Lee & H. K. Lee
Department of Mineral and Petroleum Engineering, Seoul National University, Korea

ABSTRACT: This paper presents physical and numerical model studies of the behavior of excavations in blocky rocks. Physical model tests were carried out using a base friction test machine. A Block-Spring Model was used for the numerical model studies. Qualitative studies were conducted to investigate the deformation behavior of the rocks in relation to different geometrical conditions of the excavations and the rock joints. The results from the physical models and the numerical models were compared. Good agreement has been observed between the two types of models.

1. INTRODUCTION

The deformational and failure behavior of rock blocks around an excavation located in highly jointed rock mass are very complicated. In order to investigate the rock response after excavation, Trollope (1969) developed clastic mechanics. Based on the clastic mechanics, Ling (1985) tried to determine the stress distribution around an excavation in the country rock. However, clastic mechanics has limitations on its application to actual situations, since the joints are assumed uniformly distributed. For irregular joint systems, it is necessary to use computer simulations and/or physical model tests to investigate the deformation and failure mechanism around an excavation. The Distinct Element Method (Cundall and Lemos, 1985) has been used to simulate the jointed rocks by several researchers (Kwon et al. 1995).

Recently a Block-Spring Model (BSM) has been developed for analyzing stress and deformation of jointed rock masses (Wang and Garga, 1993). The model simulates the rock masses using an assemblage of blocks consisting of a grid of differently shaped polygons. It is especially suitable for analyzing blocky rocks. The model has been applied to various mines to study the behavior of the underground rock excavations and rock slopes (Wang et al. 1995).

The research work presented in this paper consists of qualitative studies of the rock response to varying geometrical conditions of the excavations and the joints. Several base friction models were constricted which include horizontally bedded rock models with different opening sizes and models with different dip angles of the rock joints. Identical computer models using BSM were also constructed with similar geometrical conditions. The results of the two types of model could therefore be compared.

2. BASE FRICTION MODEL TEST

2.1 *Introduction*

There are at least three different ways to simulate gravity load: (a) tilting; (b) centrifuge; and (c) base friction. The base friction test has clear advantages for generating the effect of gravity compared with tilting experiments, which are difficult to control, and centrifuge models, which are expensive and limited to relatively small dimensions (Bray and Goodman 1981). Because of that, base friction model testers were widely used to demonstrate qualitative modes of deformation and failure and to examine interactions among geological factors in slope stability as well as to investigate underground excavation problems. In the base friction model test, the gravity effect is simulated by the friction force generated between the model material and a moving belt. The similarity between the friction force in the model test and the actual stress due to gravity has been demonstrated by Bray and Goodman (1981).

In the model test as shown in Figure 1, the increment of the drag force (dT) with depth is,

$$dT = \mu \cdot \gamma \cdot dV = \mu \cdot \gamma (t \cdot dx \cdot dz) \qquad (2.1)$$

where, μ is the coefficient of the friction between the model and the belt; γ is the unit weight of the model material; and t is the thickness of the model. From the above equation, the normal stress in the model in the direction parallel to the belt movement can be determined as,

$$\sigma_z = \mu \cdot \gamma \cdot z \qquad (2.2)$$

This shows that the friction force in the base friction model test increases linearly with depth similar to the gravity force.

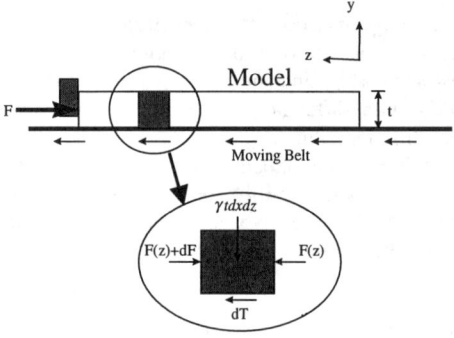

Figure 1. The base friction model.

2.2 Base friction model tester

The base friction model frame developed for this study can simulate the slope stability and underground excavation situations with a maximum model size of 1 m high and 0.7 m wide. The test machine consists of: (a) the continuous conveyer belt on which a sandpaper of carborundum 60 mesh was attached in order to increase the friction force; (b) a motor with which the moving speed of the belt is controlled; (c) acryl plate with 1 mm grid for measuring the displacements of the blocks around an opening; and (d) a camera to take pictures for determining the deformation of the blocks. The displacements of the blocks around the excavation were measured by enlarging photographs taken with a short time interval. Figure 2 shows a typical example of an actual observation from a base friction test.

Tiles were chosen as the suitable model material for the tests, because of the reduced friction between each tile compared with the friction between the belt and model. Table 1 lists the important physical properties of the tiles used in this study.

Table 1. Properties of the physical model materials.

Size (cm×cm×cm)	Weight (g)	Density (g/cm^3)	Friction Angle Face-Face	Friction Angle Base-Belt
2.1×4.6×0.6	11.15	1.92	20.3	42
4.6×4.6×0.6	26.12	2.05	20.6	52

2.3 Test procedure and results

Each test was carried out using the following procedure:

1. Install the tiles with a specific angle on the belt.
2. Run the belt for a while to tighten the model.
3. Remove the blocks to make an excavation as shown in Figure 3.
4. Move the belt with a constant speed of 9 cm/min.
5. Take pictures at pre-determined times.
6. Measure the X and Y displacements of each block after enlarging the pictures.

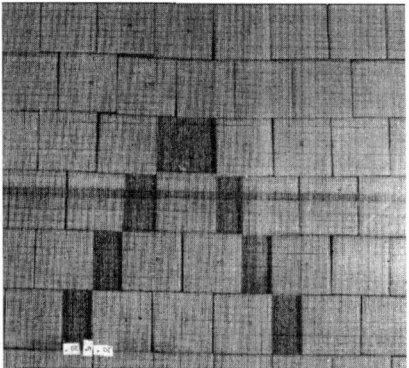

Figure 2. Roof collapse with horizontal joints.

The influence of several parameters such as volume of the blocks, opening width, and dip angle of the major discontinuity could be investigated by means of these tests. Figure 4 shows one of the test results. When the major joint angle is as high as 60°, the roof blocks slide along the major joint planes. Figures 5 and 6 show the influence of the major joint angle on the roof block displacements. The overall displacements of the blocks, A, B, C, D, E, F, and G, in the roof and rib were measured at time 10 and 20 seconds after excavation and plotted in Figures 5 and 6.

When the major joint angle was 60°, the deformation of the blocks was much higher than when it was 30° and 45°. In contrast, there was no significant difference in the displacements of the rib blocks, E, F, and G, between the cases of 30° and 45°.

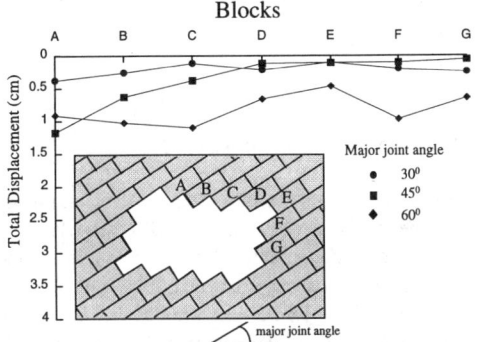

Figure 5. Total displacements of unstable blocks at 10 seconds after excavation.

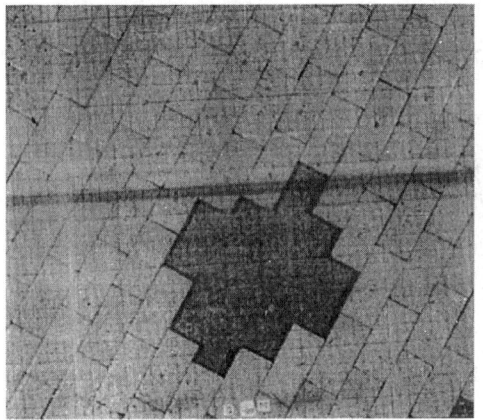

Figure 3. Initial state of a model with major joint angle of 60°.

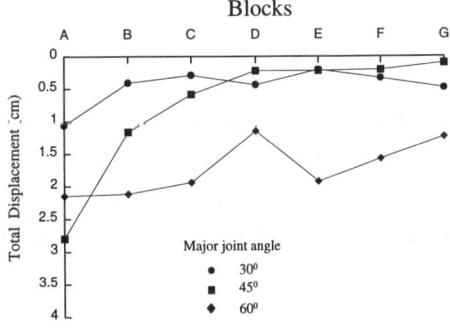

Figure 6. Total displacements of unstable blocks at 20 seconds after excavation.

3. COMPUTER SIMULATION USING BSM

3.1 Description of the BSM

BSM is a numerical method for modeling the behavior of jointed rocks (Wang and Garga 1993). It is an established static approach based on the equilibrium condition of the blocks. The model assumes that the blocks are subject to gravity, initial ground stress, contact forces between the blocks and other forces, e.g., groundwater pressure

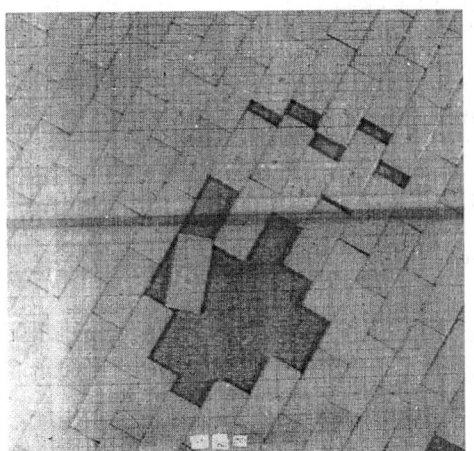

Figure 4. A post failure phenomena of the physical model with major joint angle of 60°.

and the forces from the ground support system. The blocks are assumed to be in an equilibrium condition under these forces. A stiffness matrix is constructed based on the equilibrium equations of the blocks. The displacements of the blocks are obtained when the equilibrium equations are solved. The contact forces between the blocks are then calculated from the relative displacements between the blocks. An iterative procedure is used to simulate the progressive failure/movement of the blocks.

3.2 The BSM models and results

In order to compare the numerical results with those from the physical models, six BSM models were constructed: three with horizontal joints and different block/opening sizes and three with varying dip angles of the joints. Figures 7 and 8 are two typical BSM models. The geometry of these models corresponds to that of the physical models discussed earlier.

In the BSM model, the vertical stresses are calculated as follows:

$$\sigma_z = \gamma' z \tag{3.1}$$

Where γ' is the unit weight of the material. In the case of the physical model, it is the friction imposed on the model by the belt; z is the depth of the point of interest.

By comparing eq. (2.2) and eq. (3.1), it is noted that the unit weight of the material used in the BSM model can be derived as follows:

$$\gamma' = \mu \gamma \tag{3.2}$$

or

$$\gamma' = tan(\phi) \gamma \tag{3.3}$$

where ϕ is the friction angle between the base of the physical model and the belt as listed in Table 1; γ is the unit weight of the physical model material, also given in Table 1.

It should be noted that this is a qualitative analysis for studying block deformation and failure behavior related to the size of the opening and to the geometrical conditions of the rock joints. The parameters used in the numerical models were material properties which were used in all the models under varying conditions.

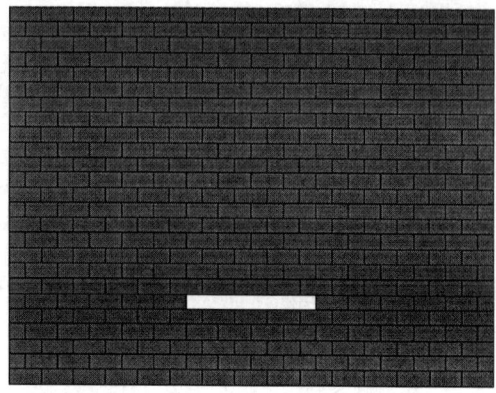

Figure 7. A BSM model with horizontal joints (block size: 2.1 cm × 4.6 cm)

Figure 8. A BSM model with joints dip 45° (block size: 2.1 cm × 4.6 cm)

For comparison purposes and corresponding to the physical model, the block displacements at a fixed iteration step are obtained.

The BSM results (Figure 9) indicate that, with all other parameters fixed, increasing the span of the opening may cause an increase in the displacements, which is obviously true. Figure 10 shows the displacements of the blocks with varying dip angle of the joints. It is noted that the blocks tend to be more vulnerable to failure when the dip angle is higher. The BSM results are in agreement with the physical model results discussed earlier. A comparison between the BSM results and the physical model results are plotted in Figures 11 and 12. It is noted that the physical model and the BSM model indicate similar displacement patterns.

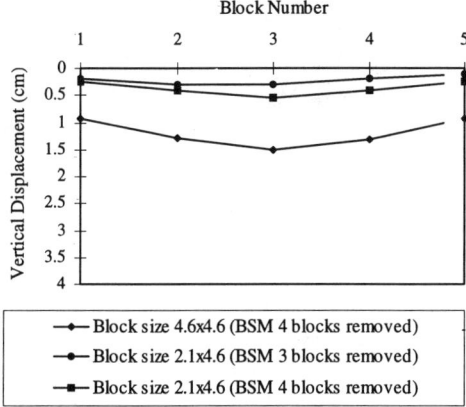

Figure 9. BSM results: a comparison of the displacements of the blocks for different spans of the opening (horizontal joints)

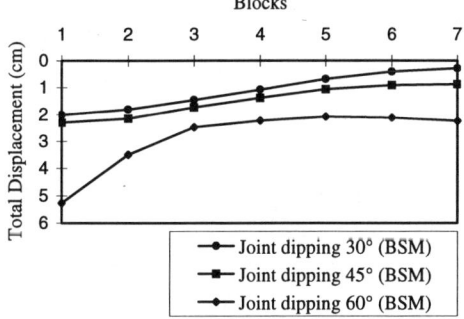

Figure 10. BSM results: a comparison of the displacements of the blocks for different dip angles of the joints

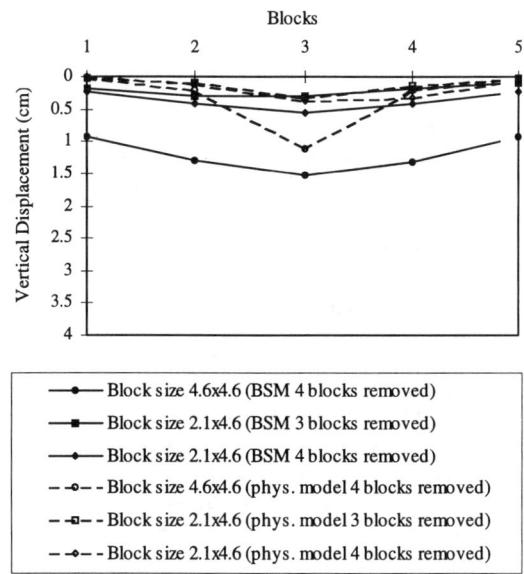

Figure 11. Comparison between BSM results and the physical model results (horizontal joints)

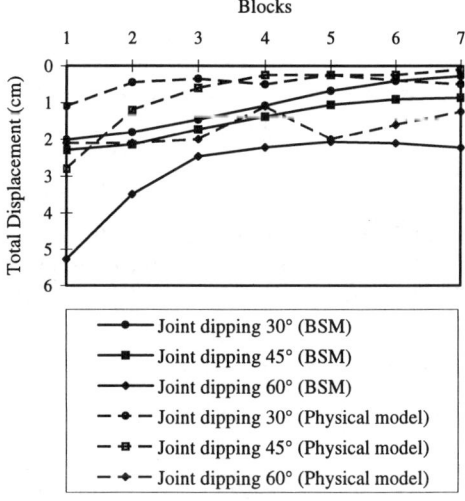

Figure 12. Comparison between BSM results and the physical model results (joints with different dip angles)

4. CONCLUSIONS

From the physical model tests and computer simulations using BSM, the following conclusions were drawn. (1) When the major joint angle is low, the width of the excavation is the most critical factor for the rock stability. In the horizontal joint models, the displacements of the roof blocks increase with an increase in the width of the excavation. (2) The major joint angle is particularly important for assessing the stability of the excavation in a jointed rock mass. There is a significant increase in displacement with an increase in joint dip angle. (3) The deformation behavior of the blocks around an excavation predicted by BSM is similar to that observed from the physical model tests. BSM can be a useful method for predicting the deformation behavior of excavations containing highly jointed rocks.

5. REFERENCES

Bray, J.W. & R.E. Goodman 1981. The theory of base friction models, *Int. J. Rock Mech. Min. Sci. & Geomech. Abs.* 18: 453-468.

Cundall, P.A. & J.V. Lemos 1985. A generalized distinct element program for modeling jointed rock mass. *Proc. of the Int. Symp. on Fundamentals of Rock Joints*: 335-343.

Kwon S, D.H. Lee & H.K. Lee 1995. Failure and deformation behavior of rocks around an opening excavated in blocky-jointed rock mass, *Mechanics of Jointed and Faulted Rock, Vienna,* pp.789-793.

Ling, L.Y. 1985. Calculation of country rock stress and loose pressure of the excavation in the clastic rock mass of alternate joints. *Numerical Methods in Geomechanics in Nagoya*: 1101-1109.

Wang, B. & Garga, V.K., (1993); A numerical method for modelling large displacements of jointed rocks — Part I: Fundamentals; *Canadian Geotechnical Journal*, Vol.30, No.1, pp.96-108.

Wang, B., Y.S. Yu and T. Aston (1995); Stability assessment of an inactive mine using the numerical model. *The 3rd Canadian Conference on Computer Applications in the Mineral Industry*, Montréal. pp.390-399.

Wang, B., Y.S. Yu and S. Vongpaisal, (1995); A case study of sub-level retreat mining at Detour Lake Mine using BSM models. *The 2nd International Conference on Mechanics of Jointed and Faulted Rocks, Vienna, Austria,* pp.927-932.

Trollope, D.H. 1969. The mechanics of discontinua or clastic mechanics in rock problems. *Rock Mechanics Engineering Practice*: 275-320.

Effects of the thermal cracks on the physical and mechanical properties of the Pocheon granite

Yong-Kyun Yoon
Department of Mineral and Energy Resources Engineering, Semyung University, Jecheon, Korea

ABSTRACT: A study of the effects of temperature on the physical and mechanical properties of the Pocheon granite was carried out. Pocheon granite samples were thermally cracked at themperatures up to 600°C. Author has measured the effective porosity, uniaxial compressive strength, Young's modulus, Poisson's ratio, axial & lateral strain, elastic wave velocity, and permeability as a function of maximum temperature attained during a thermal cycle. Permebility experiments using a transient pulse method have been performed at confining pressures of 20-40(MPa) and pore pressures of 8MPa and 10MPa. Young's modulus, uniaxial compressive strength, and Poisson's ratio decreased with increasing temperature and changed more rapidly between 500°C and 600°C. Permeability increased with increasing temperature but decreased with confining pressure. Permeability is found to be most sensitive indicators to the thermal cracking.

1 INTRODUCTION

High level radioactive waste disposal studies and geothermal energy recovery all require a better understanding of the mechanical and transport properties of rocks at a elevated temperature. If a rock is heated, thermal stresses are developed because of differential thermal expansion caused by the thermal expansion differences of neighboring mineral grains and the thermal expansion anisotropy in grains(Bauer & Johnson 1979). Differential thermal expansion can extend and widen pre-existing cracks, and create new boundary and intragranular cracks(Baur & Johnson 1979). Since such thermally induced cracks can increase the crack porosity of a rock, and it can influence substantially the mechanical and transport properties of rocks, a fundamental understanding of thermal cracking mechanism is required(Bauer & handin 1979; Page & Heard 1981; Alm 1982; Heard & Page 1982; Heuze 1983; Homand-Etienne & Poupert 1989; Lau et al. 1991; Yoon & Lee 1996).

Measurements of the physical and mechanical properties of a rock perturbed by heat are best performed in the laboratory where some set of mechanical - thermal - hydrological conditions of lithostatic pressure, stress, temperature, and pore pressure may be carefully controlled(Heard 1980).

In this paper results from uniaxial compression tests and permeability tests performed on Pocheon granite samples after thermal cycling are presented.

2 STARTING MATERIAL

Pocheon granite consists of 43.1% quartz, 30.4% K-feldspar, 17.3% plagioclase, 4.3% sericite, and 3.5% biotite. Grain size ranges 0.5mm to 3mm with the dimension of quartz being 2-3mm. Optical microscopy observations for unheated samples show that the intercrystalline boundaries are entirely sealed off and the intracrystalline cracks are not nearly present. Initial interconnected porosity was determined at 0.62% by means of measuring the weight difference between water-saturated and vacuum-dried samples. The longitudinal and shear-wave velocities of fully-dried samples were determined from ultrasonic pulse method and their values are 2990m/s and 1890m/s, respectively. Uniaxial failure of Pocheon granite occurs at 189MPa.

Right cylindrical samples(41mm to 42mm dia.) were cored from the large block and their ends were ground flat and parallel to 0.05mm. The length to diameter ratio of samples for performing uniaxial compressive tests is 2 and lengths of samples for measuring permeability range 50mm to 55mm.

3 EXPERIMENTAL METHODS

In order to avoid provoking thermal shock, different cylindrical samples were slowly heated($\leq 3°C$/min) to predetermined temperatures(200, 300, 400, 500, and 600°C) and then cooled slowly in a muffle furnace(Lee & Lee 1995). Since all thermal cracking occurs during the first heating cycle, samples were heated only once(Richter & Simmons 1974; Bauer & Johnson 1979). The maximum heating temperature was maintained for 2hr.

Optical microscope examinations for heated samples thermally cycled to 600°C reveal that heating to 200°C results in partial opening at intercrystalline boundaries. At 400°C, most intercrystalline boundaries are cracked and cracks which had existed have widened. Additionally, a few intracrystalline cracks are observed. At 600°C, the boundaries of all crystals open and the intracrystalline cracks are pronounced.

All measurements of uniaxial compressive strength, tangent Young's modulus, Poisson's ratio, axial & lateral strain, elastic wave velocity, effective porosity, and permeability were measured at room temperature.

3.1 *Uniaxial compression tests*

The testing machine used in a uniaxial compressive testing is a closed-loop servo-controlled testing machine linked to a microprocessor-based data acquisition & display system. A servo-controlled hydraulic ram can apply forces of 4.5MN. Measurements of axial load are made using a strain gage type load cell of 500kN. Axial displacements are measured with LVDT attached to the driving apparatus. Measurements of axial strain are determined using a axial extensometer and ones of lateral strain are carried out using circumferential extensometer located around the circumference of the sample at mid-height. Axial displacement was used as a feedback signal. The displacement rate applied was varied in the range 1.76 to 1.91μm/s to attain 20×10^{-6} constant strain rate with different samples. Six tests were carried out for each maximum temperature.

3.2 *Permeability tests*

The permeability of thermally treated samples was measured using a transient pulse method developed by Brace et al. (1968), which is to measure the decay of a small change of pore pressure imposed at one end of a sample with time. In this method, two water reservoirs of known volume are connected to either side of a sample, and pressure equilibrium is established in the reservoir-sample-reservoir system. Then a differential pore pressure is suddenly applied in one of both reservoirs containing water under pressure. The subsequent pressure-time histories of two reservoirs are monitored, and an exponential curve fit is imposed. Hence, the natural logarithm of pressure decay relates linearly to time, and the slope of the resulting line is a function of the permeability of the sample(Yamada & Jones 1980).

Permeability for each sample thermally cycled to different maximum temperature was measured at each confining pressure and at each pore pressure, in which the confining pressure was increased in steps of 5MPa from 20MPa to 40MPa, and pore pressure applied at each confining pressure step was 8MPa and 10MPa.

4 RESULTS AND DISCUSSION

The elastic wave velocity is plotted against maximum thermal cycle temperature in Figure 1. The velocity is found to decrease with temperature. Since the elastic wave velocity is sensitive to microcracking in rocks, the decreases are ascribed to thermal cracking due to differential thermal expansion of the mineral grains. Aside from the longitudinal velocity having the higher value than the shear velocity, both exhibit almost identical functional forms with temperature. Between room temperature and 500°C, the velocity decreases in an approximately linear relationship with temperature. Between 500°C and 600°C, the velocity decreases

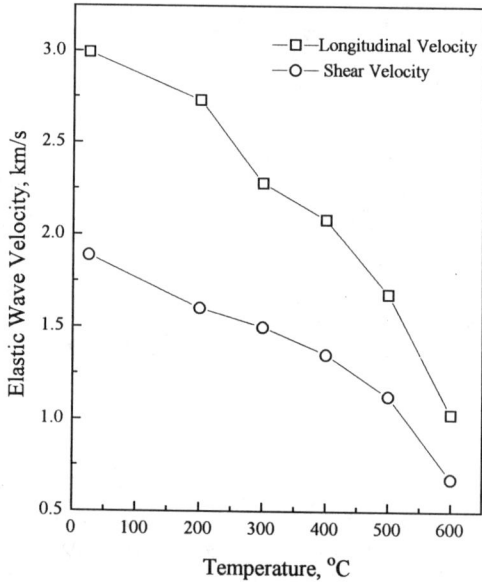

Figure 1. Variation of elastic wave velocity as a function of maximum thermal cycle temperature.

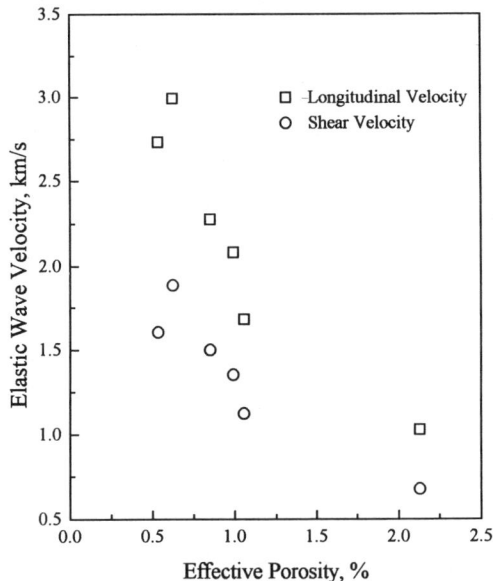

Figure 2. Elastic wave velocity vs effective porosity.

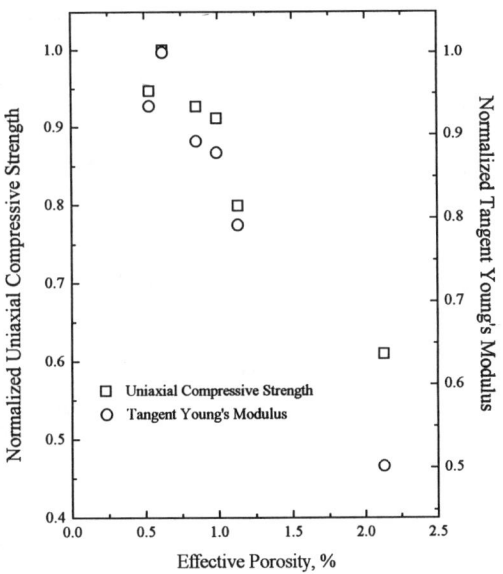

Figure 4. Normalized uniaxial compressive strength and tangent Young's modulus vs effective porosity.

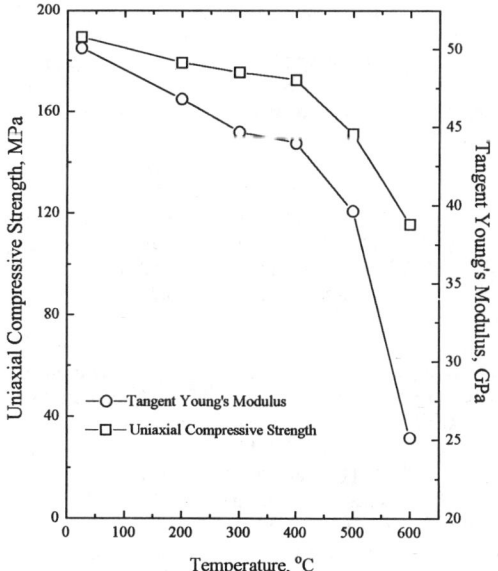

Figure 3. Uniaxial compressive strength and tangent Young's modulus as a function of maximum thermal cycle temperature.

more rapidly. Reductions of the longitudinal velocity are 44% and 66% at 500°C and 600°C, respectively. Figure 2 shows an approximately bi-linear relationship between the elastic wave velocity and the effective porosity of samples heated to progressively higher temperatures. From room temperature to 500°C, the velocity decreases rapidly with the effective porosity. From 500°C upwards, the velocity decreases at a reduced rate. This represents that a change of the effective porosity caused by thermal cracking entails the variation of the elastic wave velocity.

Uniaxial compressive strength and tangent Young's modulus are plotted against temperature in Figure 3, where tangent Young's moduli were determined at 50% level of failure stresses of different samples. The variations in uniaxial compressive strengths and tangent Young's moduli are weak until 400°C. From 400°C upwards, both diminish at a pronounced increased rate. At 600°C, the tangent Young's modulus decreases more rapidly than the uniaxial compressive strength. Reductions of the uniaxial compressive strength and tangent Young's modulus are 40% and 50% at 600°C. Figure 4 shows the variations of the normalized uniaxial compressive strength(S_c/S_{c0}) and tangent Young's modulus(E/E_0) with the increase of effective porosity of samples heated to maximum thermal temperatures, where S_c & E are the temperature-dependent values, and S_{c0} & E_0 are the reference values of non-treated

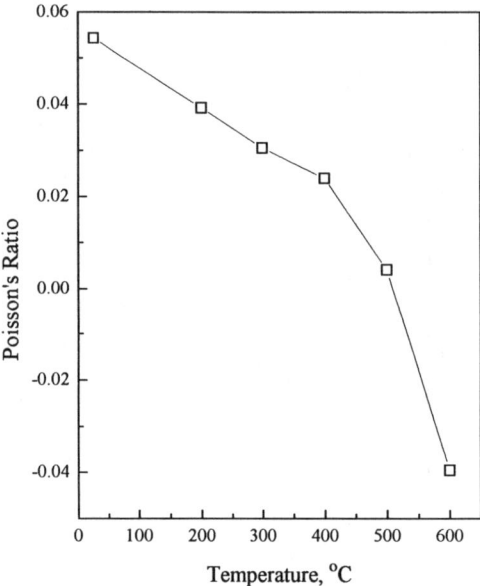

Figure 5. Poisson's ratio as a function of maximum thermal cycle temperature.

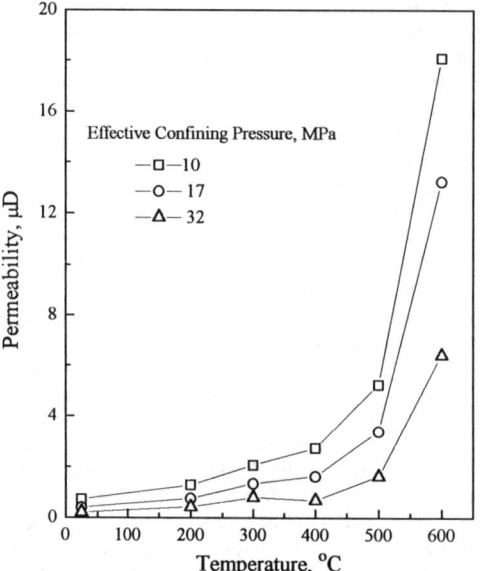

Figure 6. Permeability as a function of maximum thermal cycle temperature.

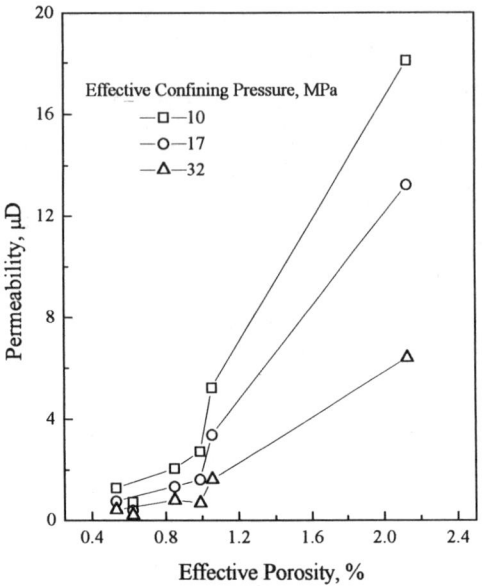

Figure 7. Permeability vs effective porosity.

samples. The decreases of S_e/S_{∞} & E/E_0, and the increase of the effective porosity are sufficiently related to each other.

Figure 5 illustrates the relationship between Poisson's ratio and maximum thermal cycle temperature. Poisson's ratio was determined at 50% level of failure stress like tangent Young's modulus. Poisson's ratio shows a steady decrease with temperature. Negative Poisson's ratio at 600°C is induced by negative lateral strain. At all temperatures above 300°C, and at the beginning of loading, negative lateral strains are measured. Stress levels, which a transition from negative lateral strain to positive lateral strain occurs, appear to be 26, 29, 49, and 88% at 300, 400, 500, and 600°C on a failure stresses basis. This results are similar to the finding by Homand-Etienne & Poupert(1989).

Figure 6 shows the measured relation between the effective confining pressure(confining pressure minus pore pressure) and permeability with temperature. Permeability increases with temperature without relation to the applied effective confining pressure. Above 500°C, permeability increases at the greatest rate. The increases of permeability are induced by thermal cracking which have increased interconnectivity between pre-existing and newly formed thermal cracks, and widened crack aperture. At all temperatures, effective confining pressure have an effect on permeability. The reduction of permeability with increasing effective confining pressure mirrors reduction of the aperture of microcracks in rocks. Figure 7 shows permeability versus effective porosity at each confining pressure. Relationship between permeability and effective porosity represents an approximately bi-linear fashion. Below 400°C, interconnection of cracks has influence on the increase of permeability, whereas widening of cracks is more important at the higher temperatures.

5 CONCLUSIONS

Physical and mechanical properties data from these high temperature tests are worthy in studying high level radioactive waste disposal and geothermal energy recovery.

The study for physical and mechanical properties of thermally cycled Pocheon granite exhibit that thermal cracks increase with the intensity of thermal treatment. Elastic wave velocity, tangent Young's modulus, uniaxial compressive strength, and Poisson's ratio decrease with maximum thermal temperature, whereas the effective porosity and permeability increase with temperature. From 500°C upwards, the variations of properties appear to be very great.

Negative lateral strains appear above 300°C. These are expected to be induced by the development of boundary cracks and their increase with temperature.

This study indicates that permeability is more sensitive to thermal cracking than are other properties.

REFERENCES

Elm, O 1982. The effect of water on the mechanical properties and microstructures of granitic rocks at high pressure and high temperatures. *Proc. 23rd U.S. Symp. Rock Mechanics*: 261-269.

Bauer, S.J. & B. Johnson 1979. Effects of slow uniform heating on the physical properties of the Westerly and Charcoal granites. *Proc. 20th U.S. Symp. Rock Mechanics*: 7-18.

Brace, W.F., J.B. Walsh, & W.t. Frangos 1968. Permeability of granite under high pressure. *J. Geophys. Res.* 73(6): 2225-2236.

Heard, H.C. 1980. Thermal expansion and inferred permeability of Climax quartz monzonite to 300°C and 27.6 MPa. *Int. J. Rock Mech. Min. Sci. & Geomech. Abstr.* 17: 289-296

Heard, H.C. & L. Page 1982. Elastic moduli, thermal expansion, and inferred permeability of two granites to 350°C and 55 Megapascals. *J. Geophys. Res.* 87(B11): 9340-9348.

Heuze, F.E. 1983. High-temperature mechanical, physical and thermal properties of granitic rocks-A Review. *Int. J. Rock Mech. Min. Sci. & Geomech. Abstr.* 20(1): 3-10.

Homand-Etienne, F. & R. Poupert 1989. Thermally induced microcracking in granites:characterization and analysis. *Int. J. Rock Mech. Min. Sci. & Geomech. Abstr.* 26(2): 125-134.

Lau, J.S.O., R. Jackson, & B. Gorski 1991. The effects of temperature and pressure on the mechanical properties of Lac du Bonnet grey granite. *Rock Mechanics as a Multidisciplinary Science*, Roegiers(ed): 313-323.

Lee, H.W. & C.I. Lee 1995. A study on thermal shock, thermal expansion and thermal cracking of rocks under high temperature. *J. Korean Society Rock Mech.* 5(1): 22-40.

Page, L. & H.C. Heard 1981. Elastic moduli, thermal expansion, and inferred permeability of Climax quartz monzonite and Sudbury gabbro to 500°C and 55Mpa. *Proc. 22nd U.S. Symp. Rock Mechanics*: 103-110.

Richter, D. & G. Simmons 1974. Thermal expansion behavior of igneous rocks. *Int. J. Rock Mech. Min. Sci. & Geomech. Abstr.* 11: 403-411.

Yamada, S.E. & A.H. Jones 1980. A review of a pulse technique for permeability measurements. *Soc. Pet. Eng. J.* 20: 357-358.

Yoon, Y.K. & H.K. Lee 1996. The effect of the thermal stress on the mechanical behavior and permeability of rocks - I. mechanical behavior. *J. Korean Society Rock Mech.* 6(1): 1-9.

Assessment of parameters for rock weathering classification: A case study on Malanjkhand Copper Project, India

K. Seshagiri Rao & Anand S. Gupta
Department of Civil Engineering, Indian Institute of Technology, Delhi, India

ABSTRACT: This paper deals with the problems related to the existing classification of weathered rock mass and outlines a new method for the quantitative assessment of weathering in rocks. The classification system is based on detailed laboratory and field studies conducted on three crystalline rocks of India. However, in the present paper the method is discussed for India's largest base metal mine - Malanjkhand Copper Project. Emphasis is placed on quantitative expression of the weathering, so that prediction and performance of the rock can adequately be judged in the field.

1 INTRODUCTION

For any rock engineering project, rock mass characterization is a difficult task, particularly at the site where weathering has considerably effected the rock mass. In common practice, assessment of weathering extent of rock is carried out with the help of standard guidelines of a classification system.

Number of classifications have been recommended for the assessment of weathering in rock material and mass. Most of them are descriptive and use the ambiguous terms which do not adequately express the variable nature of the zones of weathering.

Present study is the part of detailed geotechnical study on weathered rock material and rock mass carried out over few crystalline rocks including granite of Malanjkhand Copper Mine. Several profile studies were under taken at different depths of quarry slopes (from surface to 100 m below) of the open pit with an intention that the study will assist in the modification of mine design in expansion plan. Based on detailed field and laboratory study, a method is developed for determining the rock mass quality.

Malanjkhand granite (middle Proterozoic age) is medium to coarse grained (porphyritic type) and varies in colour. The bedrock is studied at the foot wall of the quarry. The granitic rocks are intruded by dikes and quartzitic veins. Mineralization is associated with these veins of quartz and quartz-monzonite. Apart from several minor criss-cross joints, four major sets of joints were observed in the area. The location of studied area is shown in Fig.1.

2 LABORATORY AND FIELD STUDY

In field, rock material was studied by simple field instrument like Schmidt hammer test and also by visual identification. The rock mass was examined through the procedures given by ISRM (1981), IAEG (1981) and Geomechanics classification, (Bieniawski, 1974). Altogether, 13 profiles were studied but here, only one will be discussed. The rock materials are extensively studied in laboratory by point load (σ_{tp}), Brazilian (σ_{tb}) and uniaxial compressive strength (σ_c) tests. In addition, some indices, like quick absorption index (Hamrol, 1961) and slake durability index were also evaluated. Moreover, intrinsic properties such as composition, texture, and microfracturing were studied through petrological microscope and XRD.

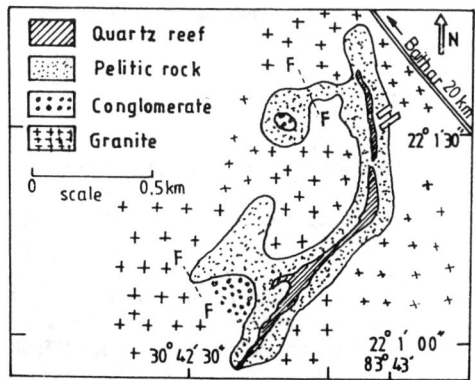

Fig. 1 Geological map of the Malanjkhand area.

3 RESULTS OF LABORATORY TESTS

Laboratory test results, as presented in Table 1, show that weathering processes appreciably changed the mechanical and other properties of Malanjkhand granite. Apart from this general conclusion it has been noticed that weathering seems to be influencing the tensile strength more drastically than compressive strength, particularly in the initial stages of weathering. In an attempt to categorize the weathered granite on the basis of index properties, it has been found that no single strength index is appropriate to distinguish the weathering extent of granite rock (Gupta and Rao, 1996).

4 ROCK WEATHERING CLASSIFICATION

The proposition of this classification is an attempt to classify the extent of weathering based on simple field tests and profile study. This is also performed with an intention to provide quantitative estimation of the extent of weathering.

The classification involves detailed assessment of weathering by visual identification and engineering characterization in both material and mass scale. Each possible parameter of rock mass influenced by weathering is identified and studied carefully in the classification system. For this, step-wise procedure is given in following:

1. Identification and characterization of weathered rock at material scale.

2. Delineation of weathering zones by observing the marked changes in spatial distribution of weathered material and assessment of representative weathered material predominant in the respective weathering zone.

3. Identification and description of rock mass parameters influenced by weathering.

4. Assigning the rating for different weathering zones and classification as per given range of total rating for classification.

4.1 Identification of material: Weathered rock material classification.

4.1.1 Visual identification and grading

The parameters which change gradually with progressive weathering have been visually identified under five points. They are, viz. (a) discolouration, (b) textural changes, (c) disintegration, (d) decomposition and (e) relative strength. The important observation with description and identified grades for granite are mentioned in Table 2. The terms suggested for the weathering extent has kept the same as advised by IAEG (1981) and ISRM (1981).

Table 1. Results of the index properties of granite.

Index Properties	Weathering Grades			
	W_0	W_1	W_2	W_4
QAI %	0.014	0.51	1.16	9.03
US Vel. m/s	5983	3691	1849	178.6
σ_{tp} MPa	10.42	7.90	2.87	0.16
σ_{tb} MPa	16.13	14.47	1.91	0.97
σ_c MPa	132.3	101.7	48.4	2.64
E_{t50} GPa	36.88	19.46	12.99	0.36

4.1.2 Quantitative approach to the material classification

The direct use of index properties such as unconfined compressive strength (σ_c) and point load index (σ_{tp}) has been found futile in characterization of weathering extent (Gupta and Rao, 1996). A proposed new index (R_s), strength ratio [($\sigma_{cWeathered}$ / σ_{cFresh}) x 100] seems to be promising and very useful in quantifying the weathering state particularly with respect to degradation in material strength due to weathering. The study shows that other indices such as coefficient of weathering (k) (Iliev, 1966) slake durability (Sd) (Franklin and Chandra 1972) and quick absorption index (QAI) (Hamrol 1961) can also be used as good indicators of weathering. Shown in Fig. 2 is the correlation observed for R_s with K for several rocks and with Sd for the three crystalline rocks tested.

The proposed classified ranges of each weathering index for crystalline rocks are presented in Table 3 against corresponding weathering grade. Along with index range values, selected ratings have also been suggested for each weathering stage.

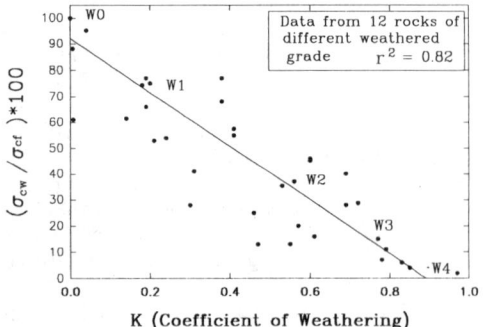

Fig.2 Relationship between R_s and K.

4.2 Delineation of weathering zones in profile

In general, profiles developed in granitic rocks show heterogeneous pattern of weathering. They were recognized through the visual changes in spatial distribution of weathering grades. The problem of demarcation of such weathering zones is resolved by dividing the exposure in small units (sub-zones) and classifying each separately. The spatial distribution of weathered material is assessed as volumetric percentage (to the nearest of 5%) of different weathered material and average of proportion decides the predominant representative grade for the rating.

4.3 Assessment of rock mass parameters

In the assessment of state of weathering of rock mass, joints have been observed as a most important factor in controlling the distribution of weathering effects within the rock. In the present study, the weathering zones were identified through three important mass parameters, viz. state of joint weathering, number of joints, and joint width and filling. Roughness of joints and joint wall strength of joint wall which alter significantly by weathering, control the overall strength and deformational behaviour of rock mass. Though, roughness of wall rock depends on type of failure, weathering at joint also changes the pre-existing roughness at the surface. Hencher and Richards (1989) observed that in practice weathered joints can exhibit higher shear strength than their less weathered counterpart. But in case of highly weathered walls where weathered material will act as gauge material, the strength along the wall will be reduced significantly even in absence of any filled up materials. Schmidt hammer test is used to assess the degree of weathering at joint wall. Although advance of weathering is observed to be closely related with joint spacing in any rock mass, however irregular distribution of weathering shows virtually no relationship with jointing. In the present study consideration of filling material is also taken under this parameter with an assumption that joint filling is generally composed of the weathered product of same wall rock material. The description of a typical weathering profile is given in Table 4 along with classified terms depending on the final rating (R_w) value and RMR values.

4.4 Assignment of ratings

All the important elements of rock mass, affected by weathering, have been considered by using a rating system. Not all the parameters are of equal importance in the assessment of rock mass strength thus a numerical weightage is assigned to each parameter according to influence of weathering on it. The final rating for the rock mass is made by summing up the weighted values determined for the individual parameters in each zone. Higher value of final rating (R_w) reflects less weathering. The recommended rating for each parameter and each class is presented in Table 5. Based on final rating the zones of profile may be classified according to the range of total ratings as suggested in Table 6. In the classification, joint number is given less weightage because physical weathering (jointing) is assessed less influencing than chemical weathering

Table 2. Classification of weathered granite by visual identification.

Grades	Visual Identification Description
Fresh Rock W_0	Grey or pink coloured. No discolouration. Grains are having vitreous lusture. Virtually no major crack present.
Slightly Weathered W_1	No significant staining. Dull lusture of minerals. Grains are tightly bonded. Few felspars are gritty. Hair line crack are visible in small quantity.
Moderately Weathered W_2	Slightly stained. Few grains are gritty in appearance. Altered microcracks are visible, but they are tight. Few felspars (plagioclase) are decomposed. Feldspar can be scratched. Sample can be broken by one firm blow of geological hammer.
Highly Weathered W_3	Discoloured and highly stained into pale brown colour. Most grains are gritty and clayey. Loosely bonded fractured grains of quartz. Microcracks are filled with clays. Few felspars are undecomposed.
Completely Weathered W_4	Completely discoloured. Specks of white clays are present. Loosely bonded grains. Microfractures are open and filled with clay. Sample can be crumbled by fingers.
Residual Soil	Original texture is lost. Samples become granular with virtually no strength.

Table 3. Suggested range of weathering indices.

Grades	W_0	W_1	W_2	W_3	W_4
R_s	80-100	50-80	25-50	10-25	<10
K	0.0	0-0.4	0.4-0.7	0.7-0.8	0.8-1.0
QAI	<0.2	0.2-1.0	1.0-2.0	2.0-4.0	>4.0
Sd	>99	95-99	80-95	50-80	<50

Table 4. Observation on the different zones of the typical profile developed over granite rock.

Zone	DESCRIPTION
1	Structureless layer of soil of variable thickness and different shades of dark to light reddish brown shows different horizons of vegetation, leaching and accumulation. Rock fabric is absent.
2	The zone predominantly contains decomposed rocks (W_4 around 70%). Original rock fabric preserved. Relict joints are visible. Weathering is heterogeneous with little development of corestones.
3	The zone is mostly constituted by Highly Weathered material (W_3 grade more than 70%). The rectangular blocks are separated by thin seams of decomposed and friable materials.
4	This zone is little suffered from decomposition. Friable materials are restricted at narrow seams of joints. Highly and Moderately Weathered material dominantly constitutes the zone.
5	Moderately Weathered material zone represents the exposed part of zone. At few places joints are clearly visible and thinly apart (aperture width= 5-10mm). Joint surfaces are highly weathered in state. The joint no./m is observed ranging from 2-4.
6	The zone is mainly composed of Slightly and Moderately Weathered material grade. Joints shows high roughness and low width of 1-3mm. At few places joints are filled with highly weathered materials. Very thin layer of rim can seen at the margin of few block.
7	The features are same as observed in zone 6 with almost no sign of corestone development.
8	Slightly weathered rock material is more prevalent in this zone. Massive huge blocks are present in this zone with slight staining at surface. Less number of joints and most of them are not continuous.

Table 5. Rock weathering classification based on rating system.

Weathered Material Grade	Symbol	Fresh	Slightly Weathered	Moderately Weathered	Highly Weathered	Completely Weathered
	R_s	80-100	50-80	25-50	10-25	<10
Rating		30	25	15	7	3
State of Joint Weathering	Jwt	Fresh	Slightly Weathered	Moderately Weathered	Highly Weathered	Completely Weathered
Rating		35	28	17	8	3
Number of Joints /m	Jn	<2	2-4	4-8	8-16	>16
Rating		25	20	13	6	3
Joint Width (mm)	Jwd	<1.0	1-2	2-5	5-20	>20
Rating		10	8	5	2	1

that is predominant in the subtropical regions. Rating for joint number and joint width is combinely given 1/3 rd weightage out of the total rating and the rest 2/3 rd weightage is given to chemical weathering and also microfractures that are combined results of chemical and physical weathering. The maximum rating assigned for each parameter is further graduated by intermediate values.

In view of the non-availability of the field data and any relationship between mass parameters and effect of weathering in mass, a simplistic approach is adopted. Each parameter's rating is classified based on the division made for R_s. Final rating, R_w also graduated in similar fashion. This can be justified in a way that sometimes, a sample of rock material may represent a small scale model of rock mass, since both have gone through the same geological cycle (Bieniawski, 1984).

Certain classifications like RMR and Q-system are widely used for rock mass strength assessment, the present classification is proposed only for the identification of weathering state at the profile which may also give an approximate expression of strength.

Table 6. Classified range of final rating.

Symbol	Zone	Final Rating
Z0	Fresh Rock	100-81
Z1	Slightly Weathered	80-51
Z2	Moderately Weathered	50-26
Z3	Highly Weathered	25-11
Z4	Completely Weathered	1-10
Z5	Residual Soil	0

4.5 Link to the RMR classification system

The basic structure of the classification is based on Geomechanics classification (Bieniawski, 1974) however, in the present classification, emphasis is given more to the weathering extent and is made for assessment of state of weathering only. Keeping this in view, the profiles were also studied through RMR classification. Since, each zone has some numerical weightage, and expressed in quantitative term as R_w, an attempt has been made to seek possible relationships between assigned rating and other useful properties. This may also help in assessing the engineering behaviour of weathered rock mass to some extent.

Based on the field studies, the three rocks were studied and classified according to the presently developed classification. Along with the suggested R_w, the RMR is also measured for the similar zones. The data obtained for 13 profiles was used to develop a relationship between R_w and RMR, and shown in Fig. 3. The similarities in basic structures and assumptions in finding out the final rating results a very close relationship between R_w and RMR. This is expressed as

$$R_w = 0.9 RMR + 10 \quad (1)$$

and can be generalised as follows:

$$R_w = RMR + 10 \quad (2)$$

where, RMR = final rating in Geomechanics classification; and R_w = final rating in the present classification

4.6 Assessment of strength and deformational properties

Furthermore, certain correlations are also attempted to relate the rating values with material constants of suitable criteria (Hoek and Brown, 1980; and Rao, 1984 and Ramamurthy et al., 1985) in order to predict the strength behaviour of rock mass.

Regarding this, others data available in the limited amount is used for the possible relationships. Figure 4 shows the correlation of B_j/B_i with R_w for two cases, viz. Barcelona granite (Romana, 1983) and Leona rhyolite (McCreath, 1993). The results are in good agreement with exponential relationship. Similarly, for another constant, the relationship between α_j/α_i and R_w is also established. Generalised relationships are expressed as follows:

$$\frac{B_j}{B_i} = \exp\left[\frac{R_w - 100}{30}\right] \quad (3)$$

$$\frac{\alpha_j}{\alpha_i} = \exp\left[\frac{R_w - 115}{140}\right] \quad (4)$$

where, R_w = rating for present Rock Weathering Classification; B_i and B_j = material constants for intact-fresh and weathered jointed rock, respectively, in Rao and Ramamurthy criterion; α_i and α_j = other material constants for intact and jointed rocks in Rao and Ramamurthy criterion.

In all the above cases the R_w is assigned for each class as per the given description. In addition, another relationship between final rating and in-situ deformibility is drawn for the available data set given for Barcelona granite (Romana, 1983). The relationship of $E_{t(in\text{-}situ)}$ (GPa) with R_w is expressed as follows:

$$E_{t(in\text{-}situ)} = \exp\left[\frac{R_w - 27}{16}\right] \quad (5)$$

If the weathered rock masses are studied carefully, these relationships might be helpful in assessment of their engineering behaviour. However, it should be noted that relationships proposed here are based on comparatively very limited field data.

The classification system developed in this study is more applicable in the tropical and subtropical regions where the soil development is marked with thick weathered rock profile. It could specifically be useful during preliminary investigation stage of any rock engineering project.

5 CONCLUSIONS

1. Rock mass properties are inferred in terms of weathering by classifications in material as well as mass scale.

2. Index properties, such as σ_c and E_t are proved much reliable than other indirect index properties.

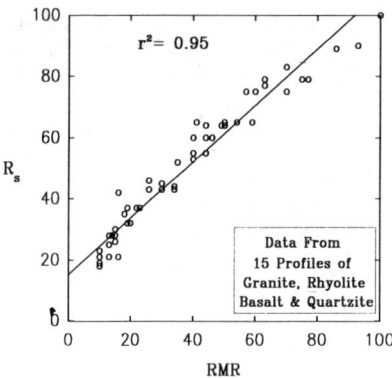

Fig. 3 Relationship between R_s and RMR.

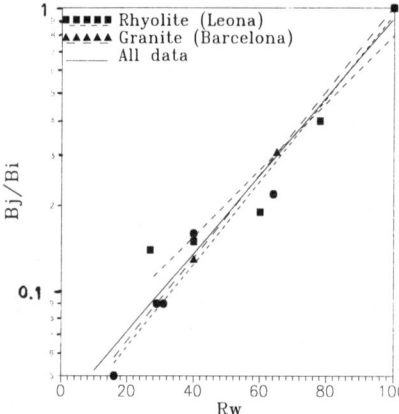

Fig. 4 Variation of the ratio of material constants (B_j/B_i) with R_w.

However, the use of weathering indices seems to be appropriate in classifying the weathered material.

3. In mass scale, where the insitu tests are not feasible, the best possible alternate to characterize the weathering extent is presented. The rock mass elements were studied for granite of Malanjkhand and based on this, the granite mass was classified.

The proposed relationships may be useful in assessment of strength properties of weathered rock mass at any confining state. However, to make the system more reliable and efficient, more data is required.

6 REFERENCES

Bieniawski, Z.T. (1974). Geomechanics classification of rock masses and its application in tunnelling. *Proc. 3rd Cong. ISRM*, 2:A27-32 Denver.

Bieniawski, Z.T. (1984). *Rock mechanics design in mining and tunnelling.* p272. Rotterdam: Balkema.

Franklin, J.A. and Chandra, R. (1972). The slake durability test. Inter. Jl. Rock Mech. & Min. Sci., 9:325-341.

Gupta, A.S. and Rao, K.S. (1996) Engineering behaviour and classification of weathered Rocks. *Ind. Geotechnical Conf.* : 192-195, Madras.

Hamrol, A. (1961). A quantitative classification of weathering and weatheribilty of rocks. *Proc. 5th Int. Conf. Soil Mech. Found. Engg.*, 2:771-774, Paris.

Hencher, S.R. and Richards, L. R. (1989). Laboratory direct shear testing of rock discontinuities. *Ground Engineering*, 22(2):24-31.

Hoek, E. and Brown, E.T. (1980). *Underground excavation in rock.* Instt. Min. Metull., London.

IAEG (Inter. Assoc. Engg. Geol.) (1981). Rock and soil description and classification for geological mapping. *Bull. Int. Assoc. Engg. Jl.* 24:235-274.

Iliev, I.G. (1966). An attempt to measure the degree of weathering of intrusive rock from their physico-mechanical properties. *Proc. 1st Cong. Inter. Soc. Rock Mech.*, 1:109-114, Lisbon.

ISRM (Inter. Soc. Rock Mech.) (1981). Basic geotechnical description of rock masses. *Int. Jl. Rock Mech. Min. Sci.*, 18:85-110.

McCreath, D.R. (1993). Preliminary analysis of quarry slopes in a weathered rock mass profiles. In: *Comprehensive Rock Engineering* (ed.) Hudson, J.A. vol.5, Pergamon, pp 777-806, New York.

Ramamurthy, T., Rao, G.V. and Rao, K.S. (1985). A strength criterion for rocks. Proc. *Ind.Geotech. Conf.* 1:pp.59-64, Roorkee.

Rao, K.S. (1984). *Strength and deformation behaviour of sandstones.* Ph.D. Thesis (Unpublished) submitted to the Deptt. of Civil Engg., Indian Institute of Technology, Delhi, India, p.273.

Romana, M. (1983). In-situ deformibility vs. weathering of granite rock in Barcelona (Spain). *5th Int. Conf. on Rock Mech.*, 1(A):93-99 Melbourne.

Virgin state of stresses and discontinuities of rock masses

Mohamed El Tani
Lombardi Engineering Ltd, Minusio-Locarno, Switzerland

ABSTRACT: The current representation of in situ stress measurements of the earth's crust suggests a functional relationship between stresses. In the light of this observation the static equilibrium of rock masses is explored. The nature of the equilibrium is determined by the ratios of the variations in horizontal stress to that of vertical and shear stresses. Ratios compiled for different parts of the earth's crust show that equilibrium is such as to favour discontinuities.
Practical formulae for calculating rock mass stresses occurring in a consistent rocks are presented. The vertical stress is not necessarily proportional to the thickness of the rock cover measured vertically though this could be the case in deep strata. The differences in altitude of the terrain affect shear stress which is negligible when compared to the vertical stress in deep strata. Two families of line of discontinuity are likely to exist. Each family comprises one or more parallel curves which cannot end within the rock mass.

1 INTRODUCTION

Engineers have long made use of parametric formulae as prediction and design tools. These formulae are usually specific solutions of equations in physics and particularly in elastostatics. Simplifying hypotheses are required to give analytical solutions to these equations. One of the hypotheses formulated to find induced stresses and deformations around a tunnel is that initial stresses are constant and are identified with the vertical and horizontal stresses at the depth of the tunnel. One of the stresses is taken to be equal to the weight of the rock cover and the second to be K times the first. Terzaghi et al. (1952) estimate the radius of action of the induced stresses caused by the excavation of a tunnel to be six times the radius of the tunnel itself, assuming elastic, isotropic rock. Where there is both an elastic and plastic behaviour, this would extend to six times the radius of the elastic-plastic interface. Hence the hypothesis for the state of initial stresses is provided with physical justification in the case of a tunnel being excavated at great depth. The actual depth must be such that relative variations in horizontal and vertical stresses are negligible within the domain which will be subsequently become the domain of the induced stresses. The hypothesis concerning initial stresses which would at the very outside have a local echo has been extended to include rock masses and give an indication of the state of stresses existing within them. Talobre (1957) maintains that nature would endeavour unsuccessfully to achieve an ideal state of stress in deep strata. It is a situation where horizontal and vertical stresses would become principal stresses and equal to the weight of the rock cover. This somewhat singular extension which assumes the ratio K to be equal to unity is known as Heim's Rule. It is constantly modified and adapted to represent in situ measurements of stresses in the upper part of earth's crust.

Developments in the theory of multiphase media and in numerical methods and the introduction of relatively inexpensive computers have brought software within the grasp of just about every engineer enabling him to find numerical solutions to a vast number of the problems he is likely to meet. These numerical codes offer an undoubted advantage since they liberate the engineer from the complications and restrictions of analytical models. He merely has to provide the necessary data and he obtains the desired result. Some data are frequently difficult to define and formulate correctly. This is true of the situation from which the title of the paper takes its name - the state of initial stresses and discontinuities of rocky massifs.

Virgin state of a rock mass is a natural and initial state from which any movement is quantified. It is itself the result of an evolutionary process beginning in a primitive state which is both difficult to imagine and never likely to be determined. The data required to describe it depend on the complexity of the predictive model. Non-linear models necessitate either perfect knowledge of the absolute deformations since an unstressed primitive state or knowledge of the total stresses currently acting on the rocky massif. Unfortunately, it is impossible to ascertain the deformation of a massif whose history and origin are only known in speculative terms. On the other hand, stresses can be measured locally and it is hoped that a deduction can be made - based on a number of these occasional measurements - on the initial stress field of the massif as a hole.

Soundings have been carried out world-wide to determine stresses in the upper part of the earth's crust. We shall begin by looking at some of the results obtained from these soundings and then discuss rock of a consistent nature. We shall then investigate stresses and discontinuities in massifs of consistent rock. An example will demonstrate how the results obtained can be put into practice so as to contribute towards the determination and explanation of the virgin state of stresses and discontinuities of rocky massifs.

2 REPRESENTATION OF IN SITU STRESSES

In situ measurements of the stresses existing in the upper part of the earth's crust are generally grouped by region or by continent. Vertical and horizontal stresses are represented separately as the sum of a residual stress and a stress that varies linearly with depth. This takes the form

$$\sigma_h = r_h + a_h H \quad (2.1)$$

$$\sigma_v = r_v + a_v H \quad (2.2)$$

where σ_h, a_h and r_h are respectively the horizontal stress, its rate of vertical variation and its residue, and σ_v, a_v and r_v are the vertical stress, its rate of vertical variation and its residue. H is the depth. Corresponding rates and residual stresses have been supplied by Hast (1967) for the Fennoscandian region, by Herget (1972) for Canada, by Li (1986) for Northern China, by Haimson (1978) for the USA, by Worotnicki et al. (1976) for Australia and by Gay (1975) for Southern Africa.

The separate presentation of stresses (2.1) and (2.2) assumes that ground level is flat. A connection between stresses and depth has always been sought after and favoured over an actual connection between stresses. Equations (2.1) and (2.2) merge, eliminating depth, and form a single relationship, namely

$$\sigma_h = c + a \sigma_v \quad (2.3)$$

where c and a are determined directly from coefficients r_h, a_h, r_v and a_v and are a function of different region of the world, as follows

$$\sigma_h = -9.4 + 1.8 \sigma_v \quad \text{Fennoscandian} \quad (2.4)$$

$$\sigma_h = -6.0 + 1.6 \sigma_v \quad \text{Canada} \quad (2.5)$$

$$\sigma_h = -3.4 + 1.2 \sigma_v \quad \text{Northern China} \quad (2.6)$$

$$\sigma_h = -7.3 + 0.8 \sigma_v \quad \text{Australia} \quad (2.7)$$

$$\sigma_h = -4.8 + 0.8 \sigma_v \quad \text{USA} \quad (2.8)$$

$$\sigma_h = -6.5 + 0.5 \sigma_v \quad \text{Southern Africa} \quad (2.9)$$

By convention negative stresses are compressive and are expressed in MPa. It appears from relations (2.4) to (2.9) that the asymptotic behaviour of stresses is an horizontal stress that is greater than the vertical stress in the north and lower in the south.

A relationship between stresses would represent a potential phenomenon independent of both depth and edge (EL Tani, 1997). We shall describe rock as being consistent if such a relationship exists and shall express it as follows

$$F(\sigma_h, \sigma_v, \tau) = 0 \quad (2.10)$$

where τ is shear stress. In its simplest form, or undertaking a finite Taylor's series expansion in a suitably chosen area, relationship (2.10) becomes

$$\sigma_h - a\sigma_v - b\tau - c = 0 \quad (2.11)$$

where a, b and c are constant coefficients. They are identified by comparing in turn relationships (2.4) to (2.9) of the stresses found for different regions of the world with relationship (2.11). Coefficient b is found to be zero, a lies between 0.5 and 1.8 and c lies between -9.4 and -3.4 MPa.

Coefficients a and b in (2.11) are also the ratios of the variations in the horizontal stress to that of vertical stress and shear tress. They play an important role, as we shall see later, in the static equilibrium and the discontinuities of consistent rock masses.

3 STRESSES OF CONSISTENT ROCKS

The stress relationship of consistent rocks completes Cauchy's equation of equilibrium

$$\nabla \cdot \Sigma + \Gamma = 0 \quad (3.1)$$

where Σ is the stress tensor, $\nabla.$ is the divergence operator and Γ is the specific weight vector field. The stress tensor is symmetrical and has three independent components in a two-dimensional physical space. A system of rectangular co-ordinates is selected. The horizontal axis oy is directed rightward and the vertical axis oz upward. The stress tensor takes the form

$$\Sigma = \begin{pmatrix} \sigma_h & \tau \\ \tau & \sigma \end{pmatrix} \quad (3.2)$$

where σ represents the vertical stress, σ_h the horizontal stress and τ the shear stress. The specific weight vector field of the rock mass is written as

$$\Gamma = (0, -\gamma) \quad (3.3)$$

where γ is the specific weight scalar field.
By combining (3.1) to (3.3) and (2.11) we eliminate σ_h and obtain a system of two equations with two unknowns

$$\begin{pmatrix} a\dfrac{\partial}{\partial y} & b\dfrac{\partial}{\partial y}+\dfrac{\partial}{\partial z} \\ \dfrac{\partial}{\partial z} & \dfrac{\partial}{\partial y} \end{pmatrix} \begin{pmatrix} \sigma \\ \tau \end{pmatrix} = \begin{pmatrix} 0 \\ \gamma \end{pmatrix} \quad (3.4)$$

The set of first order partial differential equations (3.4) depends explicitly on the ratios a and b Its intrinsic structure that might assume a hyperbolic, a parabolic or an elliptic one depends on the sign of the following discriminant (Zwillinger, 1992)

$$\delta = b^2 + 4a \quad (3.5)$$

Referring to relationships (2.4) to (2.9) it appears that the discriminant (3.5) is positive since coefficient b is zero and coefficient a is positive and lies between 0.5 and 1.8. Hence it can be deduced that the static equilibrium of the different parts of the earth crust assumes a hyperbolic structure. This means that there is a propagation of discontinuities inside the domain of consistent rocks and not a smoothing effect as happens for parabolic or elliptic structures.
The method of characteristics is well adapted to the solving of systems of hyperbolic partial differential equations. Their solution takes a purely algebraic turn when the coefficients are constant. In short, there are two characteristic values which are directly defined based on the coefficients of the hyperbolic equations and are

$$\mu_\pm = \dfrac{-b \pm \sqrt{b^2 + 4a}}{2a} \quad (3.6)$$

Two characteristic straight lines pass through any point P with co-ordinates y_P and z_P. These two lines are

$$(z - z_p) + \mu_\mp (y - y_p) = 0 \quad (3.7)$$

The solution is given along the characteristic straight lines and is

$$(\sigma - \sigma_p) + \mu_\pm (\tau - \tau_p) = \gamma(z - z_p) \quad (3.8)$$

where σ_p and τ_p are the values of the vertical stress and shear stress at point P. The order in which the characteristic values are placed in (3.6), (3.7) and (3.8) must be respected. The characteristic line of slope $-\mu_-$ is given as μ_+ or is the μ_+ part of equation (3.7) and that of slope $-\mu_+$ as μ_-.

4 PRACTICAL COMPUTATION

Consider a point D in the rock mass. Two characteristic lines pass through this point to reach ground level at points L and R (Figure 1).
Let us suppose that the characteristic line passing through L is given by μ_+. Its equation and stresses on this line are given by the μ_+ part of equations (3.7) and (3.8). The characteristic line passing through L is given by μ_-. Its equation and stresses on this line are given by the μ_- part of equations (3.7) and (3.8). Vertical and shear stresses at point D, L and R are written (σ, τ), (σ_L, τ_L) and (σ_R, τ_R).
Vertical and shear stresses at point D can be computed from those of points L and R and their co-ordinates using (3.8) and are

$$\begin{pmatrix} \sigma \\ \tau \end{pmatrix} = \dfrac{1}{\mu_- - \mu_+} \begin{pmatrix} \mu_- & -\mu_+ \\ -1 & 1 \end{pmatrix} \begin{pmatrix} \sigma_R + \mu_+ \tau_R + \gamma(z - z_R) \\ \sigma_L + \mu_- \tau_L + \gamma(z - z_L) \end{pmatrix} \quad (4.1)$$

where z, z_L and z_R are the ordinates of points D, L and R.
Formulae (4.1) can be used to compute shear and vertical stresses for point inside the rock mass from their values on ground level. The ordinates of points L and R can be eliminated from (4.1) in favour of the co-ordinates of point D using (3.7).
To have a better insight inside formulae (4.1) let us suppose that vertical and horizontal stresses are equals and there is no residual stress. In this case coefficients a, b, and c of (2.11) are respectively one, zero and zero. The characteristic values μ_+ and μ_- become respectively 1 and -1. Thus, the characteristic lines are orthogonal and inclined at 45 degrees to the co-ordinate axes.

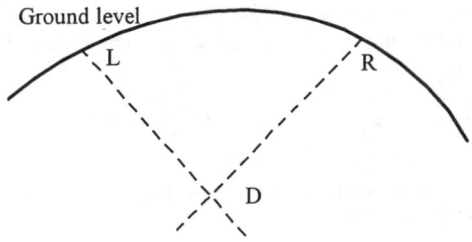

Figure 1 - Intersection of characteristic lines and ground level.

Let us also suppose that at ground level shear and vertical stresses are zero. Formulae (4.1) simplifies and is split in two parts

$$\tau = \gamma \frac{z_L - z_R}{2} \qquad (4.2)$$

$$\sigma = \gamma(z - \frac{z_L + z_R}{2}) \qquad (4.3)$$

Shear stress (4.2) and vertical stress (4.3) inside the rock mass depend only on the ordinates of points D, L and R. Shear stress varies between two extreme values determined by the altitude extreme of ground level. Vertical stress is related to the difference in altitude between D and the mean altitude of L and R. It is not proportional to the thickness of the rock stratum which separates D from the point of ground level with which it would be aligned vertically unless the terrain is always flat from L to R. For deep state where $z_L + z_R$ and $z_L - z_R$ are negligible compared to the depth of D (the origin is taken on ground level), vertical stress will appear to be proportional to the thickness of the rock stratum and the relative value of shear stress in comparison with vertical stress is negligible.

Consider a terrain composed of two plateaux and a descent SO of constant slope (Figure 2). The distance d which separate the plateaux is grater than their difference in altitude h. Shear and vertical stresses in the rock mass are calculated with the same hypothesis which led to formulae (4.2) and (4.3). They have been transcribed directly on to figure 2. Four characteristic lines are drawn, dividing the rock mass into six domains where stresses change form though remain continuous. If the distance between the plateaux diminishes and its value approaches that of the altitude then the domain between the characteristic lines c_3 and c_4 and the descent tends to disappear. The characteristic lines c_3 and c_4 get closer together and join to become one single line. Shear and vertical stresses become discontinuous when crossing through this line but their sum which is equal to γz remains continuous. To finish with this example, a characteristic line can mutate into a line of discontinuity inside the rock mass.

5 DISCONTINUITIES OF CONSISTENT ROCKS

Let us consider a line of discontinuity λ for the stress tensor inside the rock mass. On both sides of

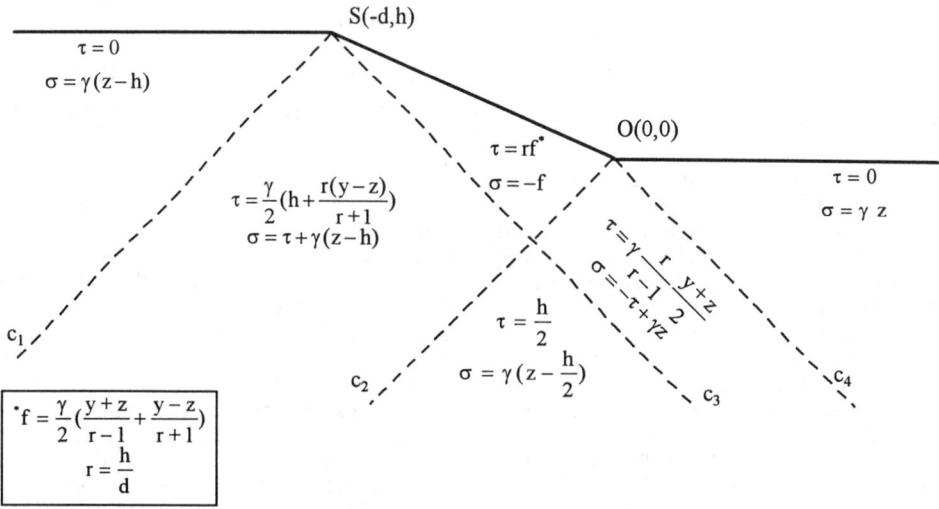

Figure 2 - Shear and vertical stresses in a consistent rock mass where $F(\sigma_h, \sigma, \tau) = \sigma_h - \sigma$. At ground level stresses are zero. c_1, c_2, c_3 and c_4 are four characteristic lines that divide the rock mass into six areas where stresses change form though remain continuous.

the discontinuity where the stress tensor is continuous Cauchy's static equilibrium equation (3.1) must be satisfied. On the discontinuity itself Cauchy's equation is not defined and must be replaced by a new equation dealing with the interaction between the domains situated on either side of the discontinuity. This will be the equilibrium of action and reaction along the discontinuity. The action and reaction are the normal stresses on either side of the discontinuity. Normal stress is the product of the stress tensor and the normal vector to the line of discontinuity. The equation expressing the equilibrium of action and reaction is written

$$[\Sigma]N = 0 \tag{5.1}$$

where N is the normal vector to the line λ and $[\Sigma]$ is the jump in the stress tensor.

The development of equation (5.1) in terms of the components of the stress tensor (3.2) and the components n_y and n_z of the normal vector added to a consideration of relationship (2.11) leads, after some manipulations to

$$\begin{pmatrix} an_y & bn_y + n_z \\ n_z & n_y \end{pmatrix} \begin{pmatrix} [\sigma] \\ [\tau] \end{pmatrix} = \begin{pmatrix} 0 \\ 0 \end{pmatrix} \tag{5.2}$$

where $[\sigma]$ and $[\tau]$ are jumps in vertical stress and shear stress. Since λ is a line of discontinuity the jumps in vertical stress and shear stress cannot be simultaneously equal to zero. The set of equations (5.2) can only be satisfied if its determinant is zero. This gives

$$an_y^2 - bn_z n_y - n_z^2 = 0 \tag{5.3}$$

This equation in turn can only be satisfied if it possesses real roots given by

$$\left(\frac{n_y}{n_z} \right)_{\pm} = \frac{-b \pm \sqrt{b^2 + 4a}}{2a} \tag{5.4}$$

We can see in these roots the characteristic values (3.6) which are real when b^2+4a is positive. They represent the inverse of the slope of the normal vector. Thus the line of discontinuity which has a normal vector of slope $1/\mu_+$ merges into the characteristic line termed μ_- whose slope is $-\mu_+$ and that which has a normal vector of slope $1/\mu_-$ merges with the characteristic line μ_+ with slope $-\mu_-$ (See Section 3).

The result (5.4) is independent of the line of discontinuity chosen. We can deduce from it that at the most two families of lines of discontinuity exist. Each family is formed by one or several parallel lines which merge into characteristic lines.

In the first place let us look at the line of discontinuity whose normal vector has a slope $1/\mu_-$. Taking this root into (5.2) leads to

$$[\sigma] + \mu_-[\tau] = 0 \tag{5.5}$$

The sum $\sigma + \mu_-\tau$ is therefore continuous along the discontinuity, which merges into a characteristic straight line termed μ_+, whereas σ and τ are not continuous. This property will now be used to prove that the line of discontinuity cannot end or be interrupted within the domain of consistency of a rock mass where (2.11) is valid.

Let us assume as a hypothesis that the discontinuity line λ ends in a point P of the rock mass. Let us consider the parallelogram $P_1P_2P_3P_4$ which contains P and whose sides are parallel to characteristic lines (Figure 3). Side P_1P_2 cuts the line of discontinuity in a point Q. Side P_3P_4 does not. Since sides P_1P_4 and P_2P_3 are parallel to a characteristic lines termed μ_+ and have the same slope as the line of discontinuity, we have from (3.8)

$$\sigma_1 + \mu_+\tau_1 = \sigma_4 + \mu_+\tau_4 + \gamma(z_1 - z_4) \tag{5.6}$$

$$\sigma_2 + \mu_+\tau_2 = \sigma_3 + \mu_+\tau_3 + \gamma(z_2 - z_3) \tag{5.7}$$

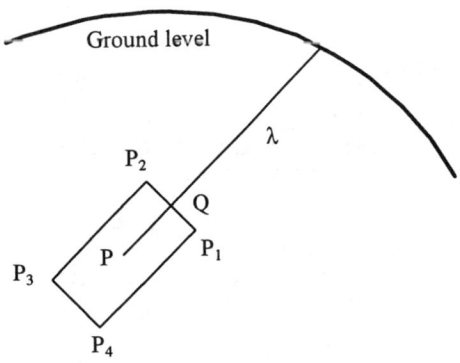

Figure 3 A line of discontinuity originates and ends only on the rim of the domain of consistency of the rock mass.

The subscripted stresses and ordinates refer to points with the same subscript. Let us now bring sides P_1P_4 and P_2P_3 closer together though still keeping them on either side of the line of discontinuity. Since side P_3P_4 does not cross the line of discontinuity σ_3, τ_3 and (z_2-z_3) become equal to σ_4, τ_4 and (z_1-z_4) respectively and it can be deduced from (5.6) and (5.7) that, at point Q

$$[\sigma]+\mu_+[\tau]=0 \qquad (5.8)$$

Equations (5.5) and (5.8) are only compatible if $[\sigma]$ and $[\tau]$ are simultaneously zero. This contradicts the fact that Q belongs to a line of discontinuity and reveals the assumed hypothesis to be absurd. As a result, the line of discontinuity cannot end within the domain of consistency of the rock mass where (2.11) is valid. If a line of discontinuity ends in a point of the rock mass this would mean that it has attained the boundary of the domain of consistency. Results (5.5) to (5.8) and the deductions made are also true if the characteristic values μ_- and μ_+ are interchanged.

Thus, if at a boundary or ground level point $\sigma+\mu_+\tau$ and $\sigma+\mu_-\tau$ are discontinuous two line of discontinuity pass through this point. If $\sigma+\mu_+\tau$ is discontinuous and $\sigma+\mu_-\tau$ is continuous a line of discontinuity of slope $-\mu_-$ originates at this point. If $\sigma+\mu_-\tau$ is discontinuous and $\sigma+\mu_+\tau$ is continuous a line of discontinuity of slope $-\mu_+$ originates at this point.

6 CONCLUSION

Properties of consistent rocks are intimately related to the ratios of variations of horizontal stress to that of vertical stress and shear stress. They have been assumed constants which made easy a bidimensional exploration of consistent rocks.

REFERENCES

El Tani M. 1997. State of stresses and discontinuities of rock masses. *Int. Sym. Eng. Geol. & Environment. Athens*, 6p.

Gay N.C. 1975. In situ Stress Measurements in Southern Africa. *Tectonophysics*, 29, 447-459.

Haimson B.C. 1978. The Hydrofracturing Stress Measuring Method and Recent Field Results. *Int. J. Rock Mech. Min. Sci. & Geomech. Abstr.*, 16, 167-178.

Hast N. 1967. The State of Stresses in the Upper Part of the Earth's Crust. *Eng. Geol.*, 2, 5-17.

Herget G. 1972. Variation of Rock Stresses with Depth at a Canadian Iron Mine. *Int. J. Rock Mech. Min. Sci. & Geomech. Abstr.*, 10, 37-51.

Li F. 1986. In Situ Stress Measurements, Stress State in the Upper Crust and their Application to Rock Engineering. *Proc. Int. Symp. Rock Stress and R. S. Meas.*, 69-78, Stockholm.

Talobre J. 1957. *La Mécanique des Roches Appliquée aux Travaux Publics*. Dunod. Paris.

Terzaghi K. & Richart F.E. 1952. Stresses in rock around cavities. *Geotechnique*, 2, 57-90.

Worotnicki G. & Denham D. 1976. The State of Stress in the Upper Part of the Earth's Crust in Australia... *Investigations of Stress in Rock - Advances in Stress Measurement*, Sydney, Int. Soc. Rock Mech., 71-82.

Zwillinger D. 1992. *Handbook of Differential Equations*, 2nd Edition. Academic Press.

Influence of water saturation on the dynamic response of rock mass surrounding a circular tunnel

Sun-Hoon Kim
Department of Civil Engineering, Youngdong Institute of Technology, Seoul, Korea

Keun-Moo Chang
Nuclear Environment Technology Institute, Korea Electric Power Corporation, Korea

Kwang-Jin Kim
Comtec Research, Stonehaven Court, Clifton, Va., USA

ABSTRACT : A three-dimensional dynamic analysis program was developed to analyze the influence of water saturation on the dynamic response of rock mass. Theoretical formulations incorporated in this computer program are the extension of Biot's two-phase theory to nonlinear region. The program was applied to study dynamic response of the existing tunnel in dry and saturated rocks subjected to a finite cylindrical charge at the center of proposed tunnel. The results show that maximum motions such as velocities and displacements at the springline in saturated rock are 21 % higher than those in dry rock. And Maximum stress ratio defined as the ratio of deviatoric stress to effevtive mean pressure at crown in saturated rock is 100% higher than that in dry rock.

1. INTRODUCTION

Many Practical problems in the field of geomechanics involve quasi-static or dynamic analysis of saturated porous media. And current interest is mainly focused on transient phenomena occurring in earthquakes, explosive loading and consolidation. For all of these the coupling between the deformation of the geomaterials such as soils and rocks and the motion of the pore fluid is of primary importance(Zienkiewicz and Shiomi 1984, Prevost 1986). Biot introduced fundamental analytical work describing the behavior of saturated porous media in a series of papers extending over many years (Biot, 1956). And he established the equation governing the interaction of the solid and fluid media for quasi-static phenomena and then extended them to dynamics. Other investigators have applied Biot's analytic results using techniques which approximate his equations with varying degrees of accuracy and sophistication (Ghaboussi and Wilson 1972). Owing to increasing interest in non-linear applications, a generalized incremental form was derived and large strain and non-linear material behavior were included.

In this study, a three-dimensional multi-phase dynamic analysis program for saturated porous rocks and soils (MPDAP-3D) is developed. Theoretical formulations incorporated in this code are the extension of Biot's two-phase theory to nonlinear region. The numerical analysis using this program is carried out to investigate the influence of water saturation on the dynamic response of underground openings in saturated rock masses.

2. Finite Element Formulation of Non-Linear Two-Phase Medium

2.1 Field Equations

Field equations representing fundamental mechanism of a two-phase medium include principle of effective stress, constitutive equation for skeleton deformation, continuity equation of pore fluid flow, equation of motion for the bulk mixture and equation of motion for pore fluid (Kim 1993).

Principle of effective stress :

$$\sigma_{ij} = \sigma'_{ij} + \delta_{ij} \cdot \pi \tag{1}$$

where σ_{ij} = total stress, σ'_{ij} = effective stress, δ_{ij} =Kronecker's delta, and π = pore water pressure.

Constitutive equation for skeleton deformation :

$$\{d\sigma'\} = [D^{ep}] \cdot \left(\{d\varepsilon\} - \frac{1}{3 \cdot K_g} \cdot \{1\} \cdot d\pi \right) \tag{2}$$

where $[D^{ep}]$= elasto-plastic stress-strain matrix for skeleton, $\{\varepsilon\}$ = skeleton strain, and K_g = bulk modulus of solid grain.

Continuity equation of pore fluid flow :

$$d\pi = \overline{m}_2 \cdot d\varepsilon_v + \overline{m} \cdot n \cdot (d\varepsilon_F - d\varepsilon_v) \quad (3)$$

where $\overline{m} = \dfrac{1}{\left[\dfrac{1}{K_m} - \dfrac{K_s^{ep}}{K_g^2}\right]}$,

$$\overline{m}_2 = \left[1 - \dfrac{K_s^{ep}}{K_g}\right] \cdot \overline{m},$$

where ε_F = volumetric diffusion of pore water,
ε_v = skeleton volumetric strain, n = porosity, K_m = bulk modulus of soil-water mixture with zero effective stress, and K_s^{ep} = elasto-plastic bulk modulus of skeleton.

Equation of motion for the bulk mixture:
$$\sigma_{ij,j} = \rho \cdot \ddot{u}_i + \rho_f \cdot \ddot{w}_i \quad (4)$$

where $\ddot{u}$ = skeleton acceleration, $\ddot{w}$ = apparent water acceleration relative to the solid skeleton, ρ = mass density of mixture, and ρ_f = mass density of pore water.

Equation of motion for the pore fluid:
$$\pi_{,i} = \dfrac{\mu}{\alpha} \cdot \dot{w}_i + \dfrac{\rho_f}{\beta} \cdot \dot{w}_i^2 + \rho_f \cdot \ddot{U}_i \quad (5)$$

where μ = dynamic viscosity of the water, α, β = flow coefficients that are properties of the porous skeleton only, and $\ddot{U}$ = absolute water acceleration.

2.2 Finite Element Formulation

These field equations are described in terms of nodal values and expressed in incremental form. Within each finite element, variables in these field equations can be expressed in terms of element nodal variables using the shape functions.

$$\{\Delta u\} = [N] \cdot \{\Delta u\}_e$$
$$\{\Delta \varepsilon\} = [B] \cdot \{\Delta u\}_e \quad (6)$$
$$\{\Delta w\} = [N] \cdot \{\Delta w\}_e$$
$$\Delta w_{i,i} = \{1\}^T \cdot [N] \cdot \{\Delta u\}_e$$

Stress vector at time step n can be expressed as:
$$\{\sigma_n\} = \{\sigma_{n-1}\} + \{\Delta \sigma\} + \{1\} \cdot \Delta \pi \quad (7)$$

Combining Eqs (1), (2), (3) and (6) yields:

$$\{\Delta \sigma\} = ([D^{ep}] \cdot [B] + \overline{m}_1 \cdot \{1\} \cdot \{1\}^T \cdot [B]) \cdot \{\Delta u\} + \overline{m}_2 \cdot \{1\} \cdot \{1\}^T \cdot [B]) \cdot \{\Delta w\} \quad (8)$$

where $\overline{m}_1 = \left[1 - \dfrac{K_s^{ep}}{K_g}\right]^2 \cdot \overline{m}. \quad (9)$

Global equilibrium equations for the two-phase medium are formulated by principle of virtual work and then linearized to be solved by linear equation solver. Global equilibrium equation at time step n can be expressed by following form:

$$[M] \cdot \{\ddot{d}_n\} + [D] \cdot \{\dot{d}_n\} + [K] \cdot \{\Delta d_n\} = \{P_n\} - \{R_{n-1}\} \quad (10)$$

Introducing a time integration method which incorporates both Newmark's β method and Wilson's θ method (Bathe 1982), we can obtain the following linearized global equilibrium equations which can be solved simultaneously at each time step:

$$[\tilde{K}] \cdot \{\Delta d_n\} = \{\overline{P}_n\} \quad (11)$$

where the generalized stiffness matrix is given by
$$[\tilde{K}] = C_1 \cdot [M] + B_1 \cdot [D] + [K] \quad (12)$$

and the generalized force vector is given by
$$\{\tilde{P}_n\} = \{P_n\} - \{R_{n-1}\}$$
$$- [M] \cdot (C_2 \cdot \{\dot{d}_{n-1}\} + C_3 \cdot \{\ddot{d}_{n-1}\}) \quad (13)$$
$$- [D] \cdot (B_2 \cdot \{\dot{d}_{n-1}\} + B_3 \cdot \{\ddot{d}_{n-1}\})$$

The program MPDAP-3D uses the Generalized Hoek and Brown Model to represent the skeleton constitutive relations of soils or porous materials.

3. Comparison of dynamic response of dry and saturated rock mass.

A parameter study on the structural response of a circular tunnel subjected to a finite cylindrical charge is carried out.

Two 3-D tunnel analyses are performed using the program MPDAP-3D; the first in dry rock and the second in the identical rock, but in a fully saturated condition. The input pressure time histories for these 3-D analyses are obtained from the companion 1-D source calculations. 1-D source calculations include the explosive charge and the free field surrounding rock.

3.1 Model Description

A hypothetical model shown in Figure 1 is considered here to study dynamic response of the

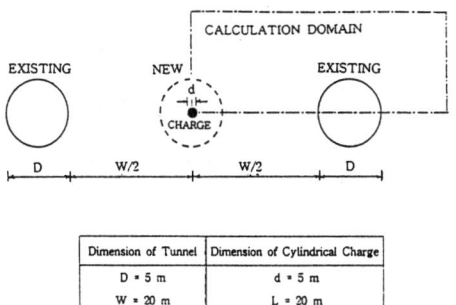

Figure 1. Geometric configuration of existing tunnels and an explosive charge

Dimension of Tunnel	Dimension of Cylindrical Charge
D = 5 m	d = 5 m
W = 20 m	L = 20 m

existing tunnel in dry and saturated rocks subjected to a finite cylindrical charge located at the center of proposed tunnel. For simplicity, gravitational forces are not included in the analysis. Figure 2 shows the schematic view of calculational domain and boundary conditions used for 3-D dynamic analysis. It should be noted that only one eighth of the total domain is modeled due to the symmetry.

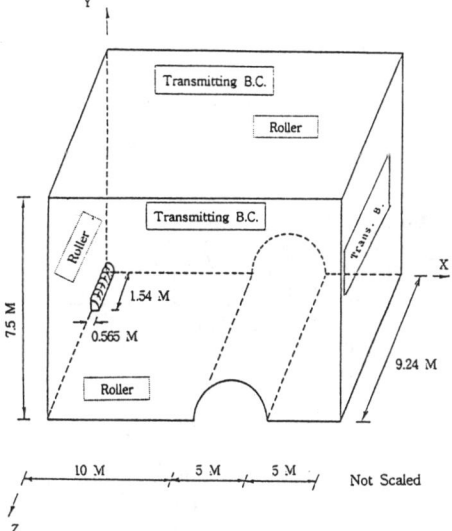

Figure 2. Schematic view of calculation domain and boundary conditions for 3-D dynamic analysis.

The surrounding rock mass is assumed to be the limestone with 13.5% of porosity. The material properties of this limestone are based on the laboratory isotropic and compression test results conducted by Applied Research Associates (Kim et al. 1986). Time histories of stresses and

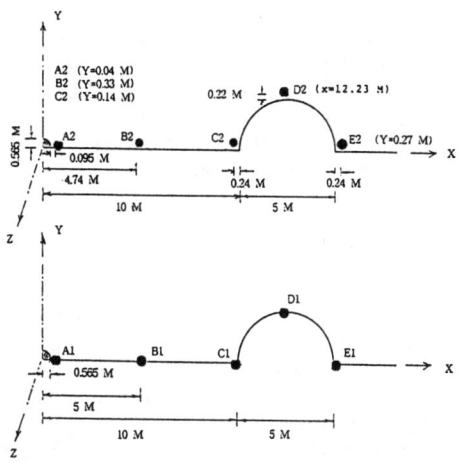

Figure 3. Selected locations for stress and motion time histories
(* Subscriptions 1 & 2 represent location on X-Y plane at Z direction of 0.0 and 0.77 M respectively)

motions are computed at selected key locations shown in Figure 3.

3.2 Three dimensional Dynamic Analysis

Dynamic analyses for both dry and saturated rocks consist of two parts: 1-D source analysis and 3-D tunnel analysis. 1-D source analysis is performed to provide input pressure time histories for 3-D tunnel analysis. This 1-D source analysis includes the explosive charge and the surrounding free field. The explosive charge is modeled by JWL equation. The surrounding rock mass is assumed to be elasto-plastic along the failure envelope and elastic below the failure envelope. Associated flow rule is assumed so that the rock mass is allowed to dilate along the failure envelope.

The 1-D analysis is conducted using a constant time step of 1μ sec and Newmark's γ-damping ($\gamma = 0.7$).

The 3-D analysis is to study the dynamic response of a circular tunnel subjected to a finite cylindrical charge as schematically viewed in Figure 2. The input pressures obtained from 1-D source analysis are specified on the cylindrical surface with radius r=0.565m and length Z=1.54m. The elastic 3-D dynamic analysis is conducted with a constant time step of 40μ sec and Newmark's γ-damping ($\gamma = 1.0$). The use of somewhat higher numerical damping is to suppress the high frequency numerical oscillations.

All the calculational parameters for saturated rock are the same as those used for dry rock except that the void space of the surrounding rock mass is filled with water. No relative flow motions are allowed in the dynamic analysis since the coefficient of permeability for limestone is so low (k = 5.33×15^{-5} cm/s).

3.3 Comparisons of Dry Rock Responses with Saturated Rock Responses

Direct comparisons of dry rock responses with saturated rock responses is to contrast the differences in wave propagation and material responses resulting from saturation of the surrounding rock masses.

Figure 4 shows computed radial stress time histories at a radial distance r=0.56 m along. It is obtained from the 1-D source analysis and is the input pressure time histories used for the 3-D tunnel analyses. At this range, shock wave for dry rock reaches the peak stress of 145 MPa and then rapidly decays to the static stress of 37 MPa. Also shock wave for saturated rock reaches the peak stress of 225 MPa and then rapidly decays to the static stress of 42 MPa. The peak stress in the saturated rock is about 55% higher than that in the dry rock while the static equilibrium stress in saturated rock is only 10% higher than that of dry rock. Such higher peak stress in saturated rock is mainly due to the higher wave speed and less skeleton hysteresis in saturated rock than in dry rock.

Time histories of velocities, displacements and stresses are calculated at selected key locations as outlined in Figure 3.

Figure 5 shows displacements at the tunnel left side wall on springline (Location C_1)

Peak horizontal velocities and percent peak increase due to saturation of rock mass are summarized in Table 1. Increase of peak velocities due to saturation of rock mass ranges from 21% at Location C_1 to 100% at the tunnel right side wall on springline (Location E_1). It should be noted that peak velocities at tunnel crown (Location D_1) occur during the negative phase, indicating that there are strong tunnel rebounding in the horizontal direction at tunnel crown.

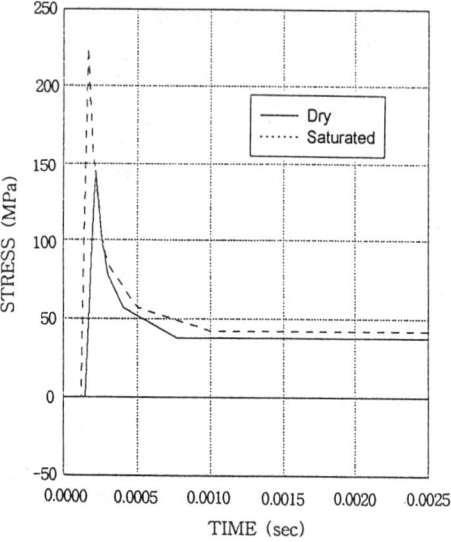

Figure 4. Input pressure time histories used for 3-D dynamic tunnel analysis

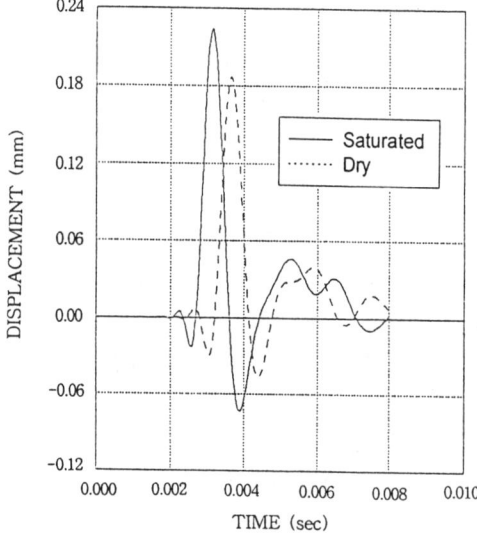

Figure 5. Horizontal displacement time histories at Location C_1

Table 2 summarizes peak values of horizontal and vertical displacements along with percent increase of peak displacements due to saturation of rock mass. Overall the magnitudes of peak displacements around the tunnel are very small. Increase of peak displacements due to saturation of rock mass ranges from 21% at the tunnel left side wall on springlime (Location C_1) to 39% at the tunnel crown (Location D_1). Maximum tunnel displacements occur at the tunnel left side wall on springline : 0.19mm in dry rock and 0.23 mm in saturated rock.

Horizontal stress time histories at locations C_2, and D_2, are shown in Figures 6 and 7 respectively. Both dry and saturated rock calculations show similar stress wave forms at all these selected

Table 1. Comparisons of peak horizontal velocities at selected locations

Velocity \ Location	Peak Velocity (cm/s)		Increase of peak due to saturation (%)
	Dry Rock	Saturated Rock	
Location B_1	77.	96.	25
Location C_1	58.	70.	21
Location D_1	-16.	-23.	63
Location E_1	2.6	5.2	100

Table 2. Comparisons of peak horizontal and vertical displacements

	Dir.	Peak Displacement (mm)		Increase of peak due to saturation (%)
		Dry Rock	Saturated Rock	
Location B_1	H	0.22	0.27	23
Location C_1	H	0.19	0.23	21
Location D_1	H	0.058	0.076	31
	V	0.028	0.039	39
Location E_1	H	0.016	0.02	25

* H : Horizontal Dir. V : Vertical Dir.

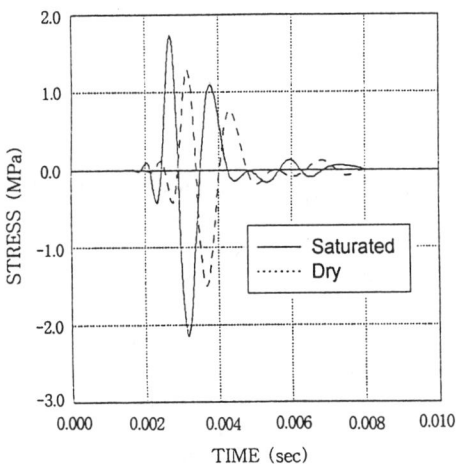

Figure 6. Horizontal stress time histories at Location C_2

locations. Increase of peak horizontal stresses due to the saturation of rock mass ranges from 40% at tunnel crown (Location D_2) to 153% at the vicinity of tunnel right side wall (Location E_2).

Direct comparisons of stress paths between dry and saturated rocks are shown in Figure 8 at Location C_2 and in Figure 9 at Location D_2. At Location C_2 which is near the springline, there are very little differences between the effective stress path in saturated rock and the stress path in dry rock. The total stress path in saturated rock is also plotted in the figure for reference. For saturated rocks, shear strengths are directly related to the effective mean pressure. Thus in assessing safety of potential shear failure in saturated rock, total stress paths should not be used. At Location D_2 which is near crown, the effective stress path

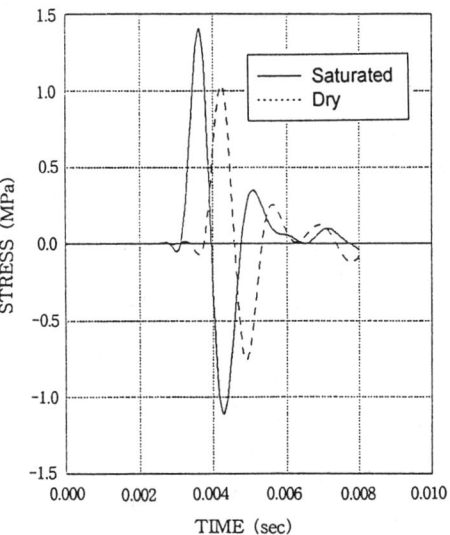

Figure 7. Horizontal stress time histories at Location D_2

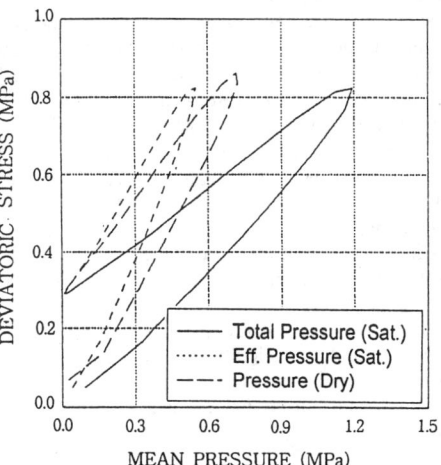

Figure 8. Comparison of stress paths at Locations C_2

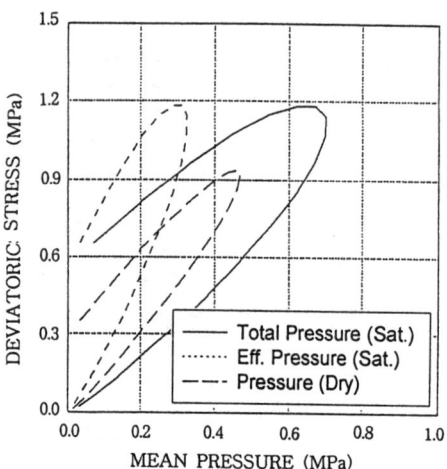

Figure 9. Comparison of stress paths at Location D_2

in saturated rock increases more rapidly toward the failure surface than the stress path in dry rock. At the peak of loading, the stress ratio (defined as the ratio of deviatoric stress to the effective mean pressure) is about 4 for saturated rock and 2 for dry rock so that the saturated rock is more vulnerable to the potential shear damage than the dry rock.

4. CONCLUSIONS

A three-dimensional dynamic analysis program was developed and this program was applied to analyze the influence of water saturation on the dynamic response of rock mass subjected to explosive loadings. Two calculations are compared using an identical explosive charge; the first in dry rock of 13.5% porosity, the second in the identical rock, but in a fully saturated condition. The following key results are as follows.

1) Maximum motions such as velocities and displacements at springline in saturated rock are 21% higher that those in dry rock
2) Maximum stress ratio defined as the ratio of deviatoric stress to effective mean pressure at crown in saturated rock is 100% higher than that in dry rock.
3) The underground structures in saturated porous medium could be significantly more vulnerable to the potential damages associated with high motions and shear failure than those in dry medium.

REFERENCES

Bathe, K.J. 1982. *Finite Element Procedures in Engineering Analysis*. Prentice Hall. New York, USA.

Biot, M.A. 1956. Theory of elastic waves in fluid saturated porous solid, I, II. *J. Acoustical Society of America, 28, 168-191.*

Ghaboussi, J. and Wilson, E.L. 1972. Variation formulation of dynamics of fluid-saturated porous elastic solids. *J. Eng. Mech. Div., ASCE, 98, 947-963.*

Kim, K.J. 1993. Dynamic response of saturated rock masses, Report Vol.1 to Nuclear Environment Management Center, Korea Atomic Energy Research Institute.

Kim, K. J., Blouin, S.E., and Timian, D.A. 1986. Experimental and theoretical response of multiphase porous media to dynamic loads. *Annual Report No.1 to Air Force Office of Scientific Research, Washington D.C.*

Prevost, J.H. 1986. Effective stress analysis of seismic site response. *Int. J. Numer. Anal. Methods Geomech., 10, 653-665.*

Zienkiewicz, O.C. and Shiomi, T. 1984. Dynamic behaviour of saturated porous media the generalized Biot formulation and its numerical solution. *Int. J. Numer. Anal. Methods Geomech., 8, 71-96.*

In situ stress measurement and its application to mining design in Jinchuan Nickel Mine, China

M.Cai
University of Science and Technology Beijing, People's Republic of China

T.Liu & C.Zhou
Jinchuan Non-Ferrous Metal Cooperation, Gansu, People's Republic of China

ABSTRACT: *In situ* stress measurement with borehole overcoring technique was carried out since 1970's in Jinchuan Nickel Mine of China. Through the measurement, 3-D *in situ* stress state and its distribution characteristics in the mine were obtained. Some successful methods for maintaining stability of underground openings and stopping areas based on the measured results of *in situ* stress state are also provided.

1 INTRODUCTION

Jinchuan Nickel Mine is situated in the middle of Hexi Corridor and the edge of Gobi Desert in Gansu Province, north-west of China. It is the second largest nickel deposit in the world with nickel metal production of 40000 tons per annual at the moment.

The formation of the nickel deposit was controlled by the regional main fault F_1 which is 170km long and strikes NW50 ~ 70°, as shown in Figure 1. The nickel ore was borne in ultrabasic rock mass. The main orebody zone in the mine has the same strike as fault F_1 with a dip angle of 50 ~ 70°. It is tens to 500m wide, 6500m long and more than 1000m deep. The NE oriented heterotropic faults separated the main orebody zone into four independent mining areas: No.3, No.1, No.2 and No.4 form west to east. The reserve in No.2 Mining Area covers 75% of the total reserve in the mine and more than 90 % of rich ore is reserved there. At present, the annul ore output is one million tons in No.1 Mining Area and two million tons in No.2 Mining Area with an average grade of 1 ~ 2 percent. The No. 3 and No. 4 Mining Areas have not been excavated yet.

The mine area underwent many tectonic movements and the intrusive actions of magmatic rock. So faults and joints are very developed in the rock mass which caused the rock conditions in the mine very poor and complicated. The mining operation in No. 2 Mining Area has been down to 800m under the ground level. Along with deepening of the mining and development excavations, failure of openings, shafts and stopping areas becomes more and more frequent and serious. To assess and maintain stability of the mine, information on *in situ* stress state is necessary.

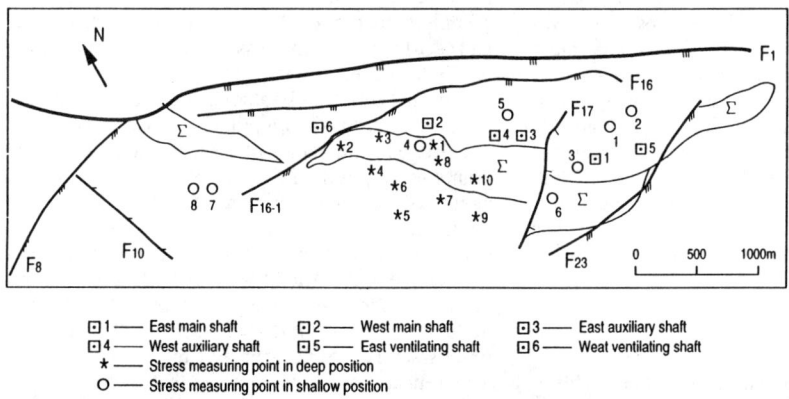

Figure 1. Location of stress measuring points in level projection of the mine

Table 1 Stress measurement results obtained by nickel stress cells in 1970's

Measuring point No.	Depth (m)	σ_1 Value (MPa)	σ_1 Bearing (degree)	σ_1 Dip (degree)	σ_2 Value (MPa)	σ_2 Bearing (degree)	σ_2 Dip (degree)	σ_3 Value (MPa)	σ_3 Bearing (degree)	σ_3 Dip (degree)
1	20	2.4						2.3		
2	44	4.2	20					3.5		
3	375	19.8	3					10.8		
4	480	24.5	335					15.4		
5	460	50.0	347	6	33.4	76	6	28.2	117	81
6	480	32.0	32	6	21.4	137	67	20.6	300	22
7	240	34.4	318	39	21.1	48	0.2	2.6	139	51
8	120	16.8	332	57	12.1	215	16	5.8	117	28

Table 2 Stress measurement results obtained with improved hollow inclusion technique in 1995

Measuring point No.	Depth (m)	σ_1 Value (MPa)	σ_1 Bearing (degree)	σ_1 Dip (degree)	σ_2 Value (MPa)	σ_2 Bearing (degree)	σ_2 Dip (degree)	σ_3 Value (MPa)	σ_3 Bearing (degree)	σ_3 Dip (degree)
1	580	31.18	33.8	6.3	13.74	280.9	74.1	10.88	305.4	-14.5
2	580	24.88	1.9	15.5	13.59	271.3	2.1	12.96	353.6	-74.4
3	580	28.08	35.2	5.0	14.28	88.7	-82.7	11.59	305.8	-6.7
4	580	28.44	36.6	2.2	13.34	299.4	72.9	9.44	307.2	-16.9
5	730	36.95	176.7	-8.8	17.55	2.6	-81.1	13.09	86.8	0.9
6	730	37.86	18.2	1.4	16.79	130.6	86.2	12.22	108.2	-3.5
7	730	34.68	348.0	-5.1	17.34	238.6	-74.9	13.48	259.2	14.2
8	730	31.64	13.2	3.8	18.68	79.9	-80.5	11.59	283.8	-8.7
9	790	40.55	160.6	-1.9	20.55	0.3	-84.3	16.75	70.6	0.7
10	790	37.26	226.0	14.6	18.19	204.2	-74.5	17.66	314.6	-5.6

2 IN SITU STRESS MEASUREMENT RESULTS AND ANALYSES

2.1 Measurement techniques and results

In situ stress measurement at shallow depth (20 ~ 480m) was completed in late 1970's (Liao et al. 1983). 8 measuring points marked as " O " are shown in Figure 1. A nickel stress cell of Hast type with overcoring technique was used at that time. Among 8 points, 3-D stress state was determined at 4 points (No. 5 ~ 8) and only planar stress state was obtained at the other 4 points. To obtain more accurate and more detailed information on stress state in deep position of No. 2 Mining Area, stress measurement at 10 points (marked as "✷" in Figure 1) between depths of 580m ~ 790m was carried out with an improved hollow inclusion technique in 1995. Details of the improved hollow inclusion technique have been described elsewhere (Cai et al. 1995). The measuring results are shown in Table 1 and Table 2.

2.2 Analyses

It took nearly 4 years to obtain the results in Table 1. At each point of No.1 ~ 4 (marked as " O "), overcoring test was conducted only in a vertical borehole, so stress state in a plane perpendicular to the borehole could be determined. Therefore, the results listed in Table 1 for points No.1 ~ 4 are not the real *in situ* maximum and minimum principal stress values. They are two secondary principal stress values in the plane normal to the borehole. To obtain 3-D stress state at points No.5 ~ 8 (marked as " O "), overcoring measurement at three no-parallel boreholes was carried out at each point. In contrast, it spent only 4 months to complete the measurement at 10 points (marked as "✷") with hollow inclusion cells and a 3-D stress state was determined with a single borehole at every point.

From Table 1 and Table 2, It can be seen that the orientation of maximum principal stress in two tables is close, and the values of principal stresses at some points and some levels are also close. However, the values of maximum principal stress in Table 2 are much more uniform than those in Table 1. It is doubtful that the value of maximum principal stress at the depth of 460m (point No.5 in Table 1) could reach 50MPa which was twice the value at the depth of 480m (point No.4). It is reasonable to attribute this to unsatisfactory performance of the nickel stress cell. Theoretical analysis has showed that if the measuring instrument in contact with borehole interferes with deformation of the borehole during overcoring, the stress in the overcore will not be fully relieved. The nickel stress cell is indeed interferes with deformation of the borehole, so it is not very suitable for use with borehole overcoring measurement.

From the measurement results, mainly based on the results obtained in 1995 with the improved hollow inclusion technique, following conclusions on *in situ* stress state in Jinchuan Mine are obtained.

1) There are two principal stresses which are almost horizontal and the third principal stress is almost vertical.

2) The maximum principal stress is horizontal which is about twice the weight of the overburden. This fact indicates that stress state in Jinchuan Mine is dominated by the horizontal tectonic stress field.

3) Orientation of the maximum principal stress is, in average, NNE and approximately vertical to the regional tectonic line.

4) The vertical principal stress is about equal to or little less than the weight of the overburden. The authors have completed *in situ* stress measurement in other metal mines, e.g. Xincheng Gold Mine in east of China, Ekou Iron Mine in central north of China and Meishan Iron Mine in south-east of China. Jinchuan Mine is the only place where the vertical stress is less than the weight of the overburden.

5) The minimum principal stress is also horizontal. It means the difference between the two horizontal principal stresses is large. In average, the maximum horizontal principal stress is about 2.5 times the minimum one. This fact probably is an important reason to cause failure of deep underground excavations.

6) The two horizontal principal stresses ($\sigma_{h.max}$, $\sigma_{h.min}$) and the vertical principal stress (σ_v) approximately increase with depth (H) as shown by following equations:

$\sigma_{h.max} = -0.02 + 0.05103H$ (MPa)
$\sigma_{h.min} = -0.01 + 0.02006H$ (MPa)
$\sigma_v = -0.26 + 0.02559H$ (MPa)

3. APPLICATION OF THE MEASURED RESULTS TO MINING DESIGN

Descending cut-fill stopping method was adopted in the mine on the principle of "rich ore first, poor ore later". Based on the measured results of in situ stress state, following measures were taken to maintain stability of the mine (Yu et al.).

3.1 *Optimum layout of development openings and stopping areas*

It has been found that the stress orientation has much influence on stability of underground openings and stopping areas. If the opening's strike was identical with the major principle stress orientation, the stress surrounding the openings was low and well-distributed. When the angle between the opening's strike and the major principal stress orientation increased, the stress concentration surrounding the opening subsequently increased and uneven stress state occurred. So, orientation of the major principle stress should be considered when to design the layout of development openings. It is better to make strike of the openings as close to the orientation of major principal stress as possible. Of course, production benefit and other geological conditions should also be considered. It is effective for reducing deformation and damage of the openings. The deformation and damage in stopes were also related to orientation of the principal stress. When the stope was put along with strike of the ore body and its axis was vertical to orientation of the major principal stress, the side walls of the stope were easy to deform and became unstable. On the contrary, when axis of the stope was vertical to the orebody's strike, the side walls were quite stable. In addition, the cross-section shape of the opening was also related to the stress state. If the ratio of width to height of the opening was equal to the ratio of horizontal to vertical stresses in the cross section of the opening, the opening would be stable. For example, many transport roadways in poor rock formations were parallel to the strike of the orebody and vertical to the orientation of major principal stress, it caused big difficulty to maintain their stability. In order to solve this problem, the oval-like cross section of the roadways whose width was larger than height and totally-enclosed support with bottom arch were adopted. Thus, the long-term stability of the roadways has been achieved successfully. Based on the same principle, the shape of cross-section of the stope was changed from square of 4m × 4m to hexagon whose hanging and bottom width was 3m, middle width 5m and height 4m. FEM analysis showed that the stress concentration factor in surrounding rock decreased from 6.8 to 3.4, the maximum principal stress value decreased from 91.65Mpa to 46.26MPa.

3.2 *Continuous mining of large area with no pillars*

The original design divided the mining operation into two steps. The first step stopped the panel (mine room) and the second step stopped the pillars between panels. This method would significantly decrease the mining efficiency and output. Furthermore, after the first step, high stress concentration in pillars made them damaged seriously and could not be stopped. So, the mining method was changed to continuous cut-fill method of large area with no pillars. The measured in situ stress results were used to analyze the feasibility of this method by numerical modeling (FEM) and the analyses achieved active conclusions. Using this method, the N0. 2 Mining Area has safely and successfully extarcted a stopping area of 56000m^2 without pillars. This method will be continuously used to stope a mining area of 100000m^2 without pollars.

3.3 Two-step support method

In order to suit the deformation characteristics of surrounding rock under high stress after excavation in poor rock formations, Jinchuan Nickel Mine widely adopted the two-step support method which called "first yield and then resist". The first-step support with shotcrete of thickness less than 5cm and rock bolting should be carried out immediately after excavation. Stress release and creep deformation of surrounding rock were permitted. When the surrounding rock became stable basically, i.e. 30-50 days after the first step support, the second step support was carried out with 10-15cm thick shotcrete and metal net. This method effectively increased the strength and integrity of the support structure and achieved long-term stability of the openings.

3.4 High quality of cement filling

Cement filling can effectively improve the stable state of surrounding rock. Through long-term of study, the supporting mechanisms of fillings have been found as follows:

1) To absorb and transfer stress. When the mined-out area is filled, part of stress is transferred into the filling and redistribution of stress will make the surrounding rock stable.

2) To restrict deformation of the surrounding rock. The filling provides side pressure to the surrounding rock and resists its deformation or displacement so that damage caused by large displacement can be avoided.

3) To separate stress concentration areas. The filling stiffness was small but its deforming capacity was large. So it could play a role to separate stress concentration areas.

According to the stress separating theory, the lower the stiffness of the filling, the greater the effect of stress separating. But, the higher the filling stiffness, the greater the effect of restricting support. So, the coordination of the two kinds of effect should be considered carefully. The elastic modulus of filling currently used accounts for 1 ~ 2% of that of the surrounding rock. Paste filling will be adopted soon, its elastic modulus will be increased to 3 ~ 4% of that of the surrounding rock. It will be hopeful to improve stress distribution in stopes and increase the stability of large scale mining area without pillars.

REFERENCES

Cai M., Qiao L., Li C., Yu J., Yu B. and G. Chen 1995. Application of an improved hollow inclusion technique for *in situ* stress measurement in Xincheng Gold Mine, China. *International Journal of Rock Mechanics and Mining Sciences*, 32: 735-739.

Liao, C. and Z. Shi 1983. *In-situ* stress measurements and their application to engineering design in the Jinchuan Mine, *Proc. Fifth ISRM Congress.*, Melbourne, PP. D. 87-89

Yu, X.,Cai, M. and F. Chi 1985. The philosophy and measures of maintaining stability of underground openings in poor rock formations in Jinchuan Mine District, *Proc. Int. Symp. On Mining Science and Technology*, Xuzhou, China, Paper No. Cd6

Effect of residual tectonic stresses on the TKI-ELI Soma Işıklar Decline stability, Turkey

Y.Özçelik
Department of Mining Engineering, University of Hacettepe, Ankara, Turkey

ABSTRACT: One of the most important parameters of stability of underground openings is the effective strata control. If the strata in the vicinity of the opening are not controlled effectively, it is impossible to perform a safe, efficient and economical operation.

In this paper, some of the reasons and outcome of instability problems observed in the decline opened in clay bearing rocks at Soma Colliery are investigated. Decline under examination have been driven in a recumbent asymmetric syncline, faults and paleoslumpstructure area causing high amounts of residual tectonic stresses. Factors affecting the instability of decline have been studied thoroughly by examining the geology and tectonics of the area, convergence measuring stations to understand in-situ behavior of supports and laboratory tests to find engineering properties of samples taken from nearby convergence measurement stations. As a conclusion, a new strategy and support system are recommended.

1 INTRODUCTION

The amount and characteristics of stress concentrations around an underground opening play the most important role on support stability. Generally, the main factors effecting the stability of underground opening driven in weak rocks are summarized as follows;

- Initial stress condition before excavation,
- Strength & geomechanical properties of formations under excavation,
- Stress area related to excavation geometry & size,
- The positions of tectonic & sedimantological structures,
- Relationship between tunnel axis and principal stress direction,
- Underground water condition,
- Excavation methods & quality,
- Type & stiffness of supports,
- Stress variations results from the excavation near the tunnel (Whittaker and Frith, 1991; Unver and Kargi 1995; Ozcelik and Kulaksiz 1996)

It is aimed in this study to define and analyze the effect of the residual tectonic stress and to recommend a suitable support system to reduce the deformations in TKI-ELI Soma Isiklar Decline. For this purpose, the following studies were performed:

- Determination of structural geology,
- Determination of the engineering properties of samples taken from the decline,
- Determination of the magnitude and characteristics of the deformations.

2 SITE INVESTIGATION AND LABORATORY STUDIES

Soma Isiklar Decline having a cross-sectional area of 17 m^2 has been driven in montmorillonite type clay bearing rocks between +355 and +140 levels with a slope of 12.8° in Isiklar Underground Mine. The height and width of decline were 5.2 and 3.9 meters respectively.

The decline was planned to have a total length of 960 m. It commenced in 1992 and there were no problems until its 300th meter. After this stage, the clay bearing rock with complex tectonic structure has been encountered and large-scale deformations have occurred. The steel sets made of GI-110 profile beam had a spacing of 1 meters until 150th meter and then they were replaced with GI-140 profile as support spacing were reduced down to 25 cm. The thickness of reinforced concrete lining was increased to 1 m where extensive deformations took place.

Strata around the decline had been affected by tectonic movements resulting the formation of folds as anticlines and synclines extensively. There was an recumbent asymmetric syncline in the vicinity of decline forming a complex stress state. However, montmorillonite type clay bearing formations surrounding the decline had high swelling potential affecting stability of the decline inversely. Moreover water inflow has been observed both on the face and on the various locations of decline, a spoil pile of nearby open pit of around 150 m. thick is located right above the decline, while a total number of 5 faults cross the axis of the decline (Figure 1).

Generally, observations indicated that the closures occurring after support setting are mainly due to the factors such as floor-heave (depending on the swelling potential of clay and tectonic structure), reactivating of some structural element and the pressure of the superincumbent strata (by the effect of dead weight of spoil pile of nearby open pit located above). The effects of these factors to the stability of the decline and the picture of some tectonic structure around the decline can be seen clearly from Figures 1 and 2. Figure 2 shows the typical deformation encountered along the decline due to the floor heave, residual tectonic stresses and pressure of superincumbent strata.

The decline has been driven without any detailed study on geological and tectonic structures. Therefore, very serious closure problems reducing cross-sectional area beyond exceptable limits have taken place due to swelling, tectonic structure and dead weight of spoil pile. Six support systems were applied before this research by colliery management but all of them were proved to be inadequate to maintain stability (Ozcelik and Kulaksiz, 1995).

Representative samples were taken around convergence measurement stations and engineering properties of these samples were determined in accordance with ISRM (1981) standards. Summary of test results are presented in Table 1.

3 EFFECT OF PRINCIPAL STRESS DIRECTIONS ON SUPPORT STABILITY

Directions of principal stresses in relation to decline axis is very important for stability and, therefore, support design. Major, intermediate and minor principal stresses are represented by σ_1, σ_2 and σ_3 respectively.

During the formation of folds, in general, side pressure should be of the highest magnitude. In other words, major principal stress must be acting almost parallel to the bedding planes.

It is well known that directions of principal stresses with respect to tunnel or decline axis is very important for stability. Tunnels opened in high horizontal pressure zones are likely to have stability problem at roof and floor; whereas tunnel or decline side wall stability problems are more likely to be faced with in tunnels opened in high vertical pressure zones (Gale, 1991).

Basic tectonic structures determined in the decline and direction of the principal stresses existing in these structures are shown in Figure 3. On the other hand, the complexity of structural geology on the area can be seen clearly from Figures 1, 2 and 3. When Figures 1, 2 and 3 are examined carefully, it is observed that decline stability was adversely affected when decline was driven parallel to recumbent asymmetric syncline axis which is known as the worst position in tunneling, i.e., major principal stress σ_1 acting from sides. Therefore, shear resistance of surrounding rock would be decreased resulting in worst stability conditions.

As a result of these principal stresses the tectonic structures cause extensive deformations on weak zones after excavation.

4 CONVERGENCE MEASUREMENTS RESULTS

Six different locations were selected for in situ measurements of convergence at the decline. Strata movement and deformational characteristics of support around the measurement stations were also determined. The measurement stations that were selected in different geological conditions for better representation of various surrounding rock parameters, are shown on a plan of the decline in Figure 1.

Measurements were taken by using a method developed originally by Whittaker (1976) and modified considering case conditions by Ozcelik (1995). The method of convergence measurement is shown schematically in Figure 4 is capable of defining deformational characteristics of decline in various direction. Measurements were made by using a simple steel tape. Original and deformed profiles are given in Figure 1.

Mid-point of a station was marked by suspending a pendulum from the roof and three steel poles were placed on the floor, one in the middle, and two one meter apart from the mid-point to the left and the right.

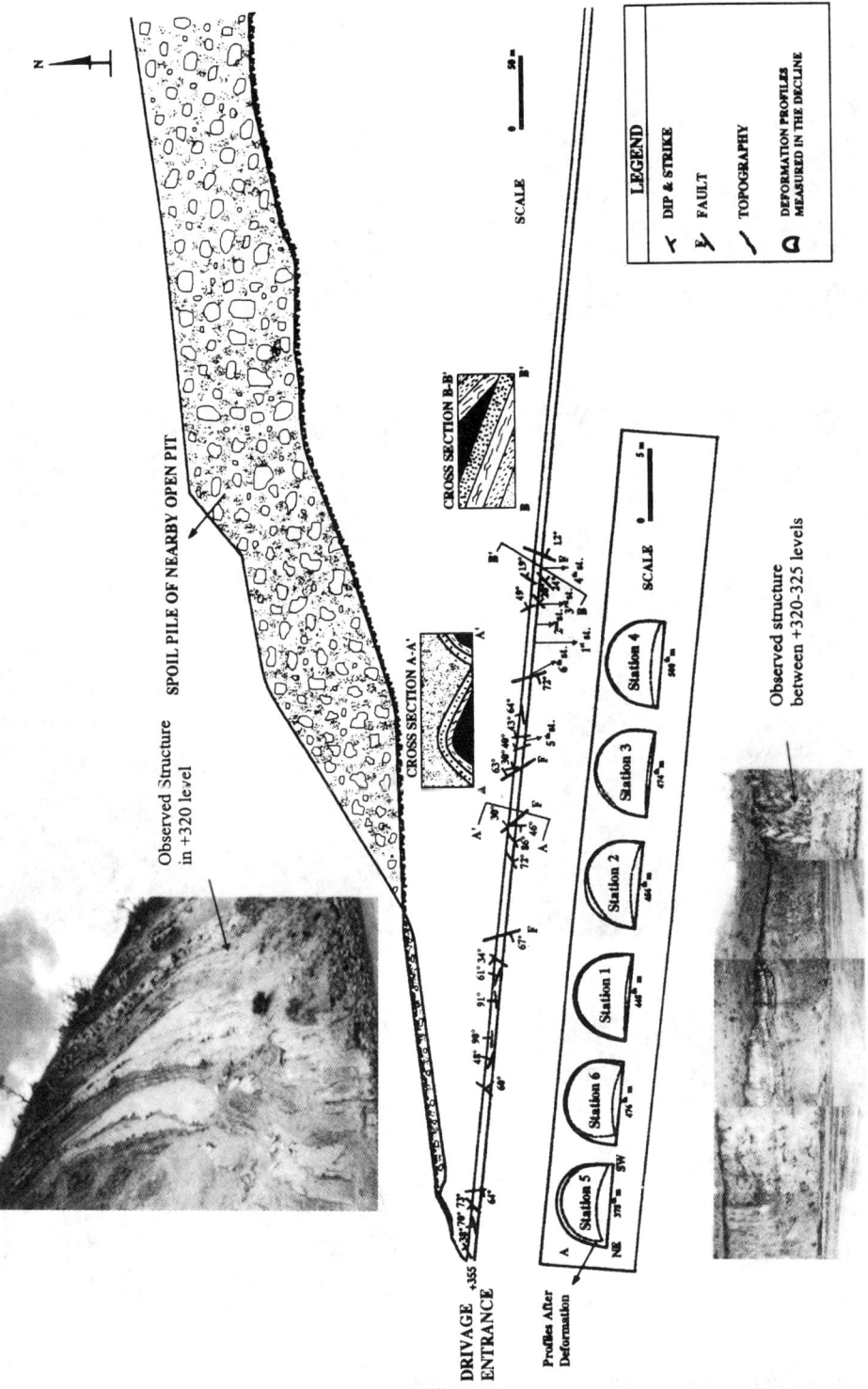

Figure 1. Tectonic structure elements, some cross-sections and profiles of deformation in TKI-ELI Soma Isiklar Decline

Figure 2. Typical deformations encountered along the decline

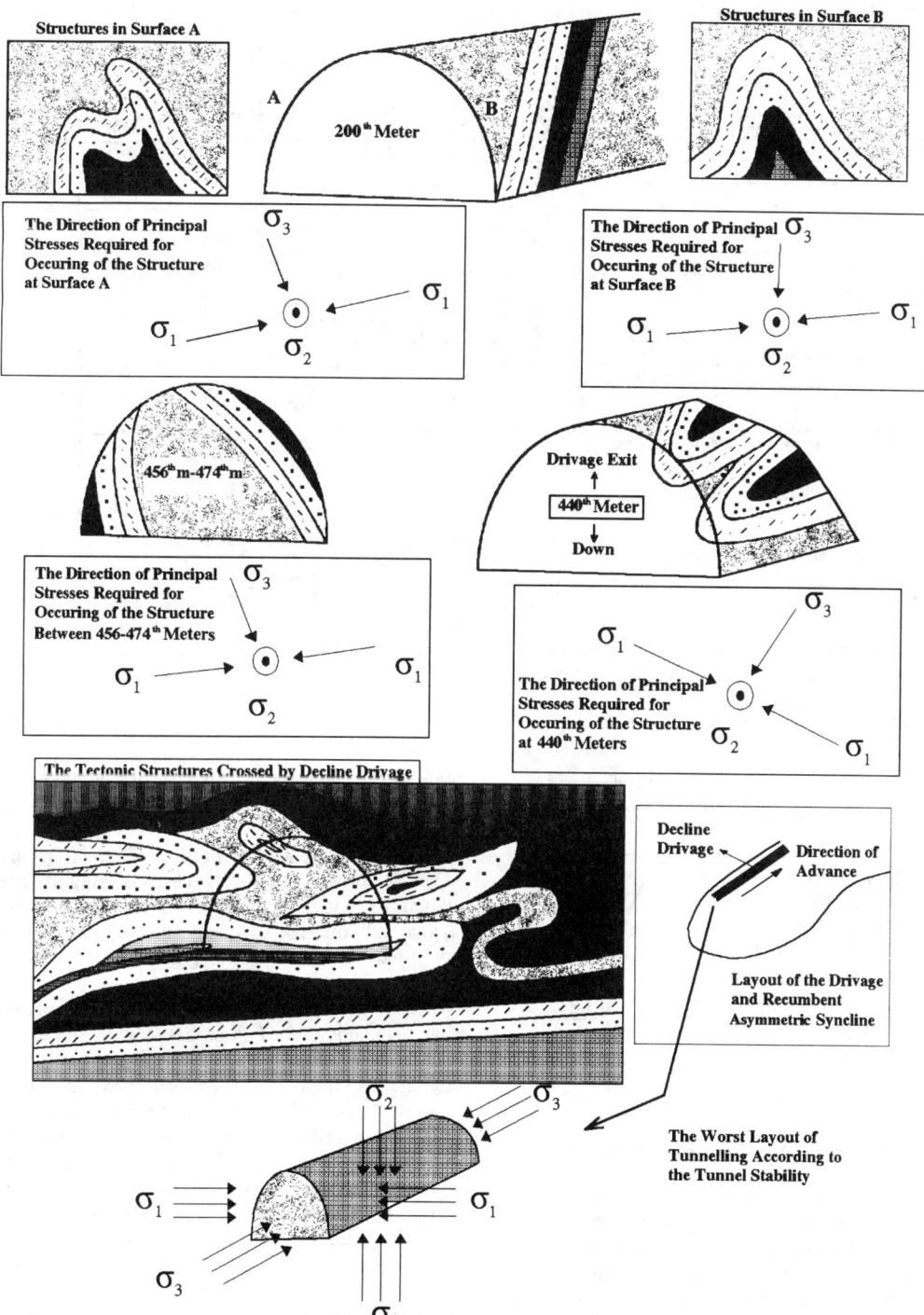

Figure 3. Basic tectonic structures determined in ELI Soma Isiklar Decline and direction of the principal stresses for occurring of this structures

Table 1. Index properties and some geomechanical parameters of the samples taken from the decline

Properties		No1 (448m)	No2 (456m)	No3 (474m)	No4 (500m)	No5 (520m)	No6 (505m)
In Situ Unit Weight (kN/ m^3)		20.2	19.0	21.0	17.5	21.0	17.6
Specific Gravity, Gs		23.8	21.8	24.5	19.6	24.3	20.0
In Situ Water Content, Wn, %		29	27	18.5	37	23	38
Activity, A		2.73	3.51	3.08	4.78	1.67	5.04
Plastic Limit, PL, %		42	38	36	56	33	52
Liquid Limit, LL, %		198	273	113	400	103	420
Plasticity Index, PI, %		156	235	77	344	70	368
Shrinkage Limit, SL, %		10.4	28.5	32.8	19.2	21.5	35.5
Index, Ic**		1.08	1.05	1.23	1.06	1.0	1.03
Grain Size Distribution	Clay (%)	67	77	35	82	70	83
	Silt (%)	27	7	28	5	24	6
	Sand (%)	6	16	37	13	4	11
Soil Classification		CH*	CH	MH**-OH***	CH	CH	CH
Soil Class with Triangular Classification Chart		Clay	Clay	Sandy Clay	Clay	Clay	Clay
Internal Friction Angle ϕ (°)		10-14	9-12	11-15	7-10	11-17	8-10
Cohesion, c (MPa)		0.020-0.030	0.016-0.025	0.030-0.040	0.012-0.018	0.019-0.025	0.017-0.021
Dry Strength (Uniaxial Compression Strength)		Very High	Very High	Medium	Very High	Medium	Very High
Swelling Potential		Very High	Very High	Very High	Very High	Very High	Very High
Swelling Index		0.14	0.10	0.16	0.09	0.22	0.09

* CH : Inorganic clay with high plasticity, swollen clay;
** MH : Inorganic silt, elastic silt,
*** OH : Organic clay with medium or high plasticity.

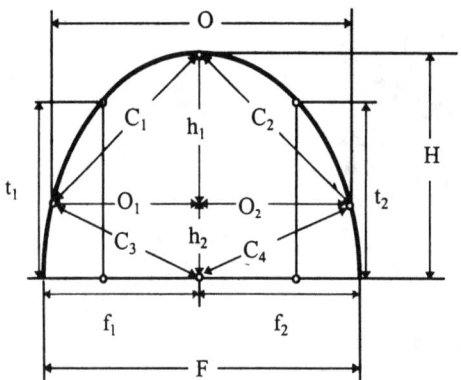

Figure 4. Method of convergence measurement

Side reference points were marked on supports at a height of one meter from floor. Therefore, it was possible not only to measure the vertical and horizontal displacements, but also displacements between the other reference points that made possible to obtain the complete profile of the support after deformation.

Closure measurements between reference points to determine convergence were carried out unregular time a week for a period of different time at all stations.

By examining decline profiles after closure, the most important parameters characterizing closure pattern were found as t_1, t_2 and h_1, h_2. Change of t_1, t_2 and h_1, h_2 distance as an indication of convergence by time is given in Figure 5 for different convergence measurement stations.

Convergence measurement results have shown that high amounts of floor heave with some side closure were observed in all of the stations. While the highest floor heave was observed in station number 4 with 34 cm, the lowest floor heave was observed in station number 3 with 5 cm (h_2). The effect of floor heave on closure at six convergence measurement stations are also presented in Figure 5. However, it was determined that depending to the tectonic structure the closures were mainly in perpendicular the NW-SE direction being in excess of 50 cm (t_1), and the highest closures due to the pressure of the superincumbent strata was observed in station number 1 with 17 cm.

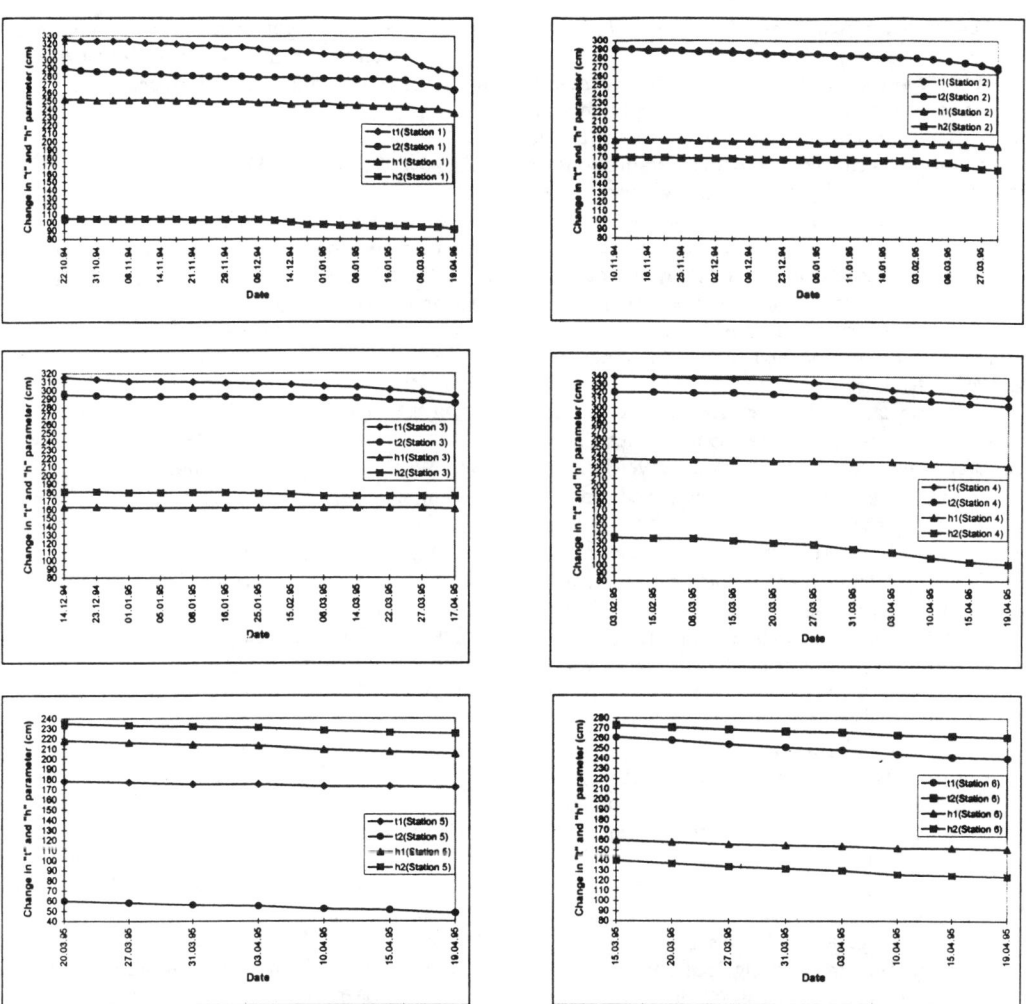

Figure 5. Changes in "t" and "h" parameters as an indication of closures in a period of different time

5 INTERPRETATION OF RESULTS

Samples taken from nearby convergence measurement stations were tested in the laboratory to find various mechanical properties. As the results were assessed, it has been observed that the swelling potential of surrounding rock in decline very high corresponding to a swelling index value of 0.09. Hence, this situation suggested that the swelling was the primary parameter affecting the decline stability. On the other hand, depending on the mechanical properties of rocks such as liquid limit, void ratios, cohesion, internal friction angle values, the amount of deformation was changed. However, many significant relationship between the mechanical properties of rocks and convergence results were obtained.

In general, the amount of convergence increases in formations that have low cohesion, internal friction angle and swelling index. Also, when moisture content, montmorillonite type clay content, swelling potential, liquid limit, activity, void ratios of rocks increases, decline stability decreases.

A detailed survey on geology and tectonics of the area has shown that the decline was driven in a recumbent asymmetric syncline having an inclined axis. Besides swelling, another important factor (second factor) governing decline deformation is residual stresses due to tectonic structures. Nearly 50 cm closure was formed in the direction of the highest residual tectonic stress in the decline.

6 CONCLUSION AND RECOMMENDATIONS

Decline was opened parallel to recumbent asymmetric syncline axis and strike resulting in very poor stability conditions due to the great effect of residual tectonic stresses. Excessive deformations in the decline especially when opened parallel to recumbent asymmetric syncline axis, necessitated floor dinting and side ripping to maintain functions of decline as transportation, ventilation etc. It should be kept in mind that dinting and side ripping would only be a temporary solution. Although repair work was very costly, it promoted further deformation in an accelerated rate.

Recommendations to maintain stability in the light of in situ measurements, laboratory tests and a comprehensive geotechnical study are given below:
- Decline axis should be positioned in the direction perpendicular to recumbent asymmetric syncline axis wherever possible,
- Effective water drainage should be performed,
- In the decline where random loads at various directions were encountered reinforced circular concrete lining must be used as a support system according to the site investigations and laboratory tests. Sharp and perpendicular corners should be avoided in support design.
- Calculations on the deformation of the support requirements by considering the worst conditions have resulted in the use of the reinforced circular concrete lining with thickness of 73 cm in the decline. Moreover, the use of steel-mesh in concrete is suggested in order to stand tensile-stresses. Suggested support system is given in Figure 6. The support design presented here may have to be modified during application in order to get its full advantage. However, rock bolts can be used at places where there are strong rock bands provided that rock bolts extend up to elastic zone around the decline.

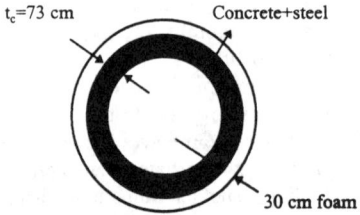

Figure 6. Proposed support system

Finally, this research has indicated that residual stresses due to previous tectonic activity may play the important role together with the swelling in the overall stability of the decline, although, the swelling has a great deal of importance.

7 ACKNOWLEDGMENTS

The author would like to thank to ELI staff for their kind help and cooperation during field studies and Prof. Dr. S. Kulaksiz for his guidance and insight throughout the research and Dr. B. Unver for helping in writing English manuscript of the paper.

8 REFERENCES

Gale W. J., 1991, Strata Control Utilizing Rock Reinforcement Techniques and Stress Control Methods in Australian Coal Mines, *Mining Eng.* Vol. 150, No. 352, January, pp. 247-253.

ISRM, 1981, Basic Geotecnical Description of Rock Masses, ISRM Commission on Classification of Rocks & Rock Masses, *Int. J. Rock Mech. Min. Sci. & Geo. Abstr.* V. 18, pp. 85-110.

Ozcelik Y. 1995, *Geomechanical Properties of Rocks Soma Isiklar Drivage and Support Design*, H.U. MSc Thesis, 174 p. (in Turkish)

Ozcelik Y. and Kulaksiz S., 1995, "The Problems Encountered in ELI Soma Isiklar Decline Drivage, Applied Support Systems and Solution Recommendations" *Bulletin of Chamber of Mining Engineers of Turkey*, Vol. 34, No. 3, pp. 3-14.(in Turkish)

Ozcelik Y. and Kulaksiz S., 1996, Effect of Geomechanical Properties of Weak Rocks in the Soma Isiklar Decline Drivage to the Decline Drivage Stability, *3rd National Symp. in Rock Mech.*, Ankara, pp. 35-45. (in Turkish)

Unver, B., and Kargi, M. A., 1995, Analysis of Instability Problems of Main Roadways at Yeni Celtek, Bolu Underground Mine, *Bulletin of Chamber of Mining Engineers of Turkey*, June, Vol. 34, No. 2, pp. 3-16 (in Turkish)

Whittaker B. N., 1976, Recording Treatment and Interpretation of Roadway Deformation Surveys, *Mining Engineer*, July, pp. 607-617.

Whittaker, B. N., and Frith, R. C., 1990, *Tunneling Design, Stability and Construction*, The Institution of Mining and Metallurgy, England, 460 p.

Application of acoustic emission technique to determination of in situ stress

M. Seto, M. Utagawa & K. Katsuyama
National Institute for Resources and Environment, Tsukuba, Japan

T. Kiyama
Mitsui Construction Co., Nagareyama, Japan

ABSTRACT: In this study the possibility of acoustic emission (AE) technique was investigated to measure in situ rock stress. The rock cores were obtained from one vertically drilled exploratory borehole and two drilled boreholes in underground coal mines. The AE method was suggested to determine in situ rock stress with reasonable accuracy using AE signatures in repeated loadings of a rock core specimen. Based on the results of in situ stress estimation from the AE method, the time interval, up to eight months, did not strongly influence the in situ stress determination using AE method. Cored rock recollected the in situ stress condition reasonably well (within ± 10 %), when compared to the results from over coring and hydraulic fracturing technique. Also, there was significant correlation between overburden pressure and estimated vertical stress from AE method.

1 INTRODUCTION

Reliable evaluation of in situ stress is an important step in the analysis and design of underground excavations, particularly for evaluating the stability of underground structures to prevent failure or collapse of underground openings. Although a number of techniques have been proposed and developed to determine in situ stress, the determination of in situ stress is not an easy task and all suffer from deficiencies and limitations. The main deficiency of established techniques such as over coring method or hydraulic fracturing method is that they are usually expensive and time consuming. Other shortcomings of the techniques are that they are deficient for measuring the in situ stress at depth in remote regions which are hard to access from boreholes or mine workings.

An alternative method for determining the stress state at depth and in remote regions is to take advantage of the Kaiser effect of acoustic emission (AE). This phenomenon, termed the "Kaiser effect", suggests that previously applied maximum stress might be detected by stressing a rock specimen to the point where there is a substantial increase in AE activity. Based on the Kaiser effect, the previous stress can be estimated from the curve of AE activity under monotonically increasing stress. A number of researchers have studied the Kaiser effect in geomaterials since the 1970's, and have discussed the factors that influence stress memory recollection of rocks under uniaxial and triaxial conditions (Kanagawa et al., 1976; Kurita and Fujii, 1979; Houghton and Crawford, 1987; Seto et al., 1989, 1992, 1995; Holocomb, 1993; Utagawa et al., 1995). The technique is functionally workable technology, and is anticipated that the rapid and economical determination of the in situ stress in the rock is possible.

The AE method, however, has some problems in its application to in situ stress measurement. One of the problems is that the Kaiser effect cannot always be clearly observed due to the disturbance by heat, water, and weathering (Yoshikawa and Mogi, 1981). Another problem is that the controversy still exists on that how long the stress memory can be retained. If the Kaiser effect recovered in a short time, it is impossible to apply the AE method to estimation of in situ stress.

In this paper, firstly, using the results of cyclic loading of previously stressed rock specimens, the AE method is proposed to accurately determine the previous stress using AE signatures in the repeated loading (Laboratory tests). In our method, the previous stress can be accurately estimated from AE take-off point in the second and third reloadings, even when the delay time from previous loading is long in the range up to 400 days. Secondly, the applicability of the AE method is discussed based on the estimation results of in situ stresses from cored rock specimens obtained in four different underground sites. The cored rock specimen was repeatedly loaded up to a certain stress level, and AE was measured in each loading to determine the in situ stress (In situ tests).

2 EXPERIMENTAL PROCEDURE

2.1 *Rock specimen in laboratory tests*

In the present study uniaxial compressive stress was repeatedly applied five to six times up to a certain stress level which is nearly twice of the previous stress. The rock specimens tested are Shirahama sandstone, Inada granite, and Tage tuff. The specimen is a rectangular of 30 x 30 mm in square and 60 mm in length.

2.2 *AE measuring system*

In AE measurements, four sensors were put on the confronted sides of the specimen and these sensors formed a parallel oblong plane to the loading axis direction. The AE sensors used in this study were differential typed ones (5.0mm diameter, NF-AE-904DM model) and the resonance frequency was 500kHz. It had high gains between 200kHz and 550kHz. The response frequency band of this system was between 50kHz and 1MHz.

The signals from the AE sensors were amplified 40dB by the pre-amplifiers and sent to the AE measurement system (NF-9600 Local processor) and was amplified further by 40dB. When AE sensors detected AE, these sensors removed the noises which generated at the contact point between the rock specimen and the end piece, hydraulic pressure source etc from AE signals over threshold level (240mV) and detected AE generated only inside the rock specimen. The method to remove the noise was to get rid of the signals whose arrival time difference to these 4 sensors were longer than the prescribed figures. These prescribed figures were determined before the test by the way that we oscillated elastic waves through pressing propelling pencil leads (5mm diameter, 3mm length) on the surface of the specimen, endpiece etc not so as to measure signals from outside the rock specimen.

3 AE METHOD FOR ESTIMATING PREVIOUS STRESS

Figure 1 shows a typical example that indicates the existence of Kaiser effect in a sandstone specimen. Data for a specimen tested 5 min. after the previous 50 repeated loading up to 10 MPa are shown. An arrow indicates the previous maximum stress. The take-off point of AE activity coincides with the maximum previous stress. In all experiments conducted within the short delay time, the existence of Kaiser effect in rock specimens could be clearly observed and the assigned stress from the take-off point of AE signature was within 5 %.

Figure 2 represents the AE signature of sandstone which was previously stressed up to 10 MPa 7 days before the test. An arrow indicates the maximum previous stress, but lots of AEs were produced within the stress level of Kaiser effect. These AEs obscure the Kaiser effect. The longer the delay time is, the more the first reloading produces AEs within the stress level of Kaiser effect (delay time effect). The AEs produced before the

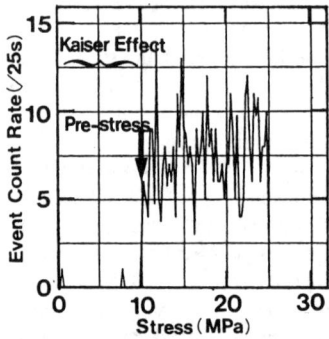

Figure 1 An example of Kaiser effect of sandstone

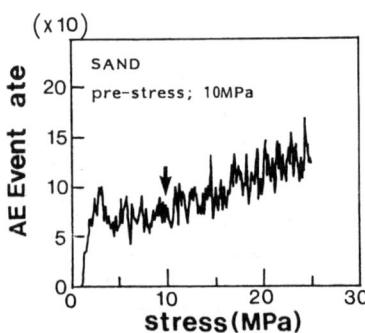

Figure 2 AE behavior of sandstone previously stressed up to 10 MPa 7days before.

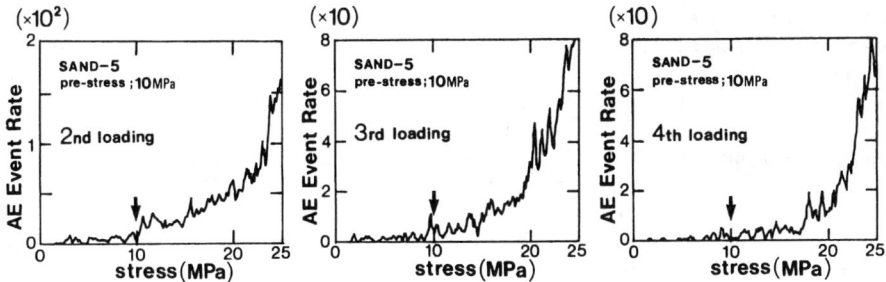

Figure 3 AE behaviors of previously stressed sandstone from the 2nd to 4th reloadings. The AE behavior in the 1st loading is shown in Figure 2.

maximum previous stress level are possibly due to the crack movements (closure, friction, or compaction) inside the specimen (Seto et al., 1995). Figure 3 represents the AE signatures from the second to fourth reloading of sandstone shown in Figure 2. As shown in Figure 3, the AEs below the previous stress level were reduced by the repeated reloadings up to the same stress level, and the AE signature in the second and/or third reloadings gave clear indication of AE increase (AE take-off point). Consequently, based on the results of the laboratory test, even if Kaiser effect was obscured due to delay time effect, the AE signatures in the subsequent reloadings can allow us to determine the maximum previous stress.

4 TESTING PROCEDURE FOR IN SITU STRESS DETERMINATION

The rock cores were obtained from one vertical borehole and boreholes in two different underground coal mines. Rock cores A were obtained from the vertical borehole drilled from the surface up to 180 m depth. The time lag between drilling and test was 5 months. Rock cores B were taken from underground coal mine site (coal mine A), which was located at 685 m depth from the surface. In the site, hydraulic fracturing was conducted using five boreholes to measure the in situ rock stress. The rock core was obtained from one of the horizontally drilled borehole. The time lag of rock core D was 8 months. Rock cores C were obtained from underground coal mine site (coal mine B), which was located at 385 m depth from the surface. In the sites hydraulic fracturing tests were also conducted to measure in situ stresses.

In the present study, the rock core specimen was repeatedly loaded five times up to a certain stress level. The stress level was decided, taking into account of the depth of the site and the uniaxial compressive strength. Rock core specimens were usually stressed up to the stress level of 1.5 to 2 times of overburden pressure estimated from the depth and the density of the rock core. And, AE signature in the second loading was usually utilized to determine the in situ rock stress, because most of rock core specimens did not give a clear AE take-off point in the first loading.

5 IN SITU STRESS MEASUREMENT RESULTS

5.1 *In-situ stress determination from rock cores A from vertically drilled hole*

Figure 4 indicates the depth at which the rock core specimens were obtained. Fourteen rock core specimens were tested to determine the vertical

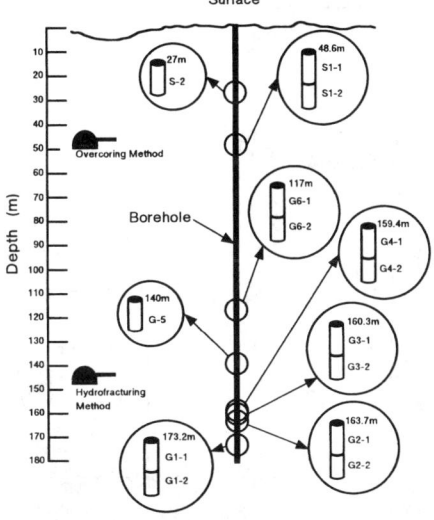

Figure 4 Depths of the tested cores and locations where stress measurements were conducted.

stresses in the site. In the figure, the rock core with "S" is sandy mudstone and the one with "G" is granite. In situ stress measurements were also conducted underground near the borehole by over coring method at 45 m depth and hydraulic fracturing method at 140 m depth.

Figure 5 shows the relation between cumulative ring down counts and stress of granite (G4-1) from the first to fourth loading. The test was conducted 5 months after the drilling. AE take-off point was recognized at 1.96 MPa in the first loading and 1.98 MPa in the second loading. Total number of AEs were much reduced in the second loading when compared to that in the first loading, but note that the AE take-off point was also recognized in the second loading at the almost same stress level as in the first loading. In the second to fourth loadings, another AE take-off point was recognized at about 4.35 MPa.

Figure 6 shows the variation of estimated vertical stresses with depth and the stress measurement results from over coring method and hydraulic fracturing method. The stresses recognized as another take-off point in the AE signatures after the second loading were also plotted in the figure. The estimated vertical stresses from the AE method can be represented by σz (MPa) = $0.0171 \cdot Z$ (m). Main sediments in this field are sandy mudstone and mudstone, but granite is rarely dotted from the depth of 100 m to 180 m. Granite becomes main formation rock below 180 m depth. Average density of the sediments from the surface to 180 m depth is 1.80 (t/m^3). The density agree well with the one estimated from AE method (1.71t/m^3). Consequently, it was concluded that the vertical

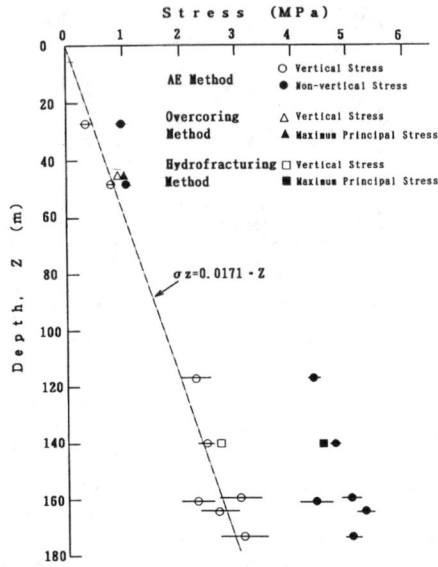

Figure 6 Variation of estimated stressed from AE and measured stresses by other methods with depth.

stress in this site was mostly caused by overburden pressure.

The estimated stresses also agree well with the over coring result at 45 m depth and the hydraulic fracturing result at 140 m depth. And, note that the stresses estimated from another AE take-off point, which was recognized after the second loading, coincides well with maximum principal stress measured by over coring or hydraulic fracturing method.

5.2 In-situ stress determination from rock cores B in coal mine A

Figure 7 shows the underground site in coal mine where rock core samples were obtained from. Rock core specimens used in the AE method were taken from borehole "1" in the figure. Rock type was a sandstone Hydraulic fracturing was also conducted to measure in situ rock stresses in this site.

Figure 8 represents the variation of cumulative AE events with stress from the second to fourth loadings. As shown in the figure, two AE take-off points were recognized in the second and third loadings. While AE generated from the beginning of loading in the first loading and any particular AE take-off points were recognized in the AE signature,

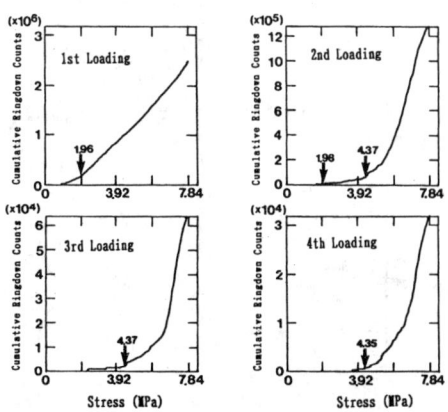

Figure 5 AE behaviors of a rock core specimen (G4-1) from the 1st to 4th loading.

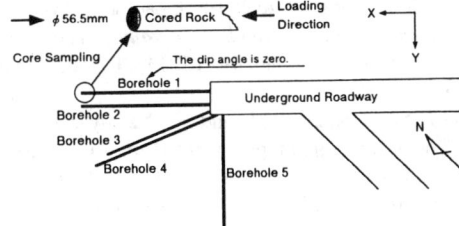

Figure 7 Relative locations of the boreholes in mine A.

AE take-off points can be recognized from the second loading as shown in the figure. From these AE signatures, the in situ rock stress in X direction was estimated as 10.2 MPa. The stress component in X direction calculated from hydraulic fracturing results was 10.5 MPa. Thus, the stress value estimated from AE signatures in the repeated loadings agree well with that from hydraulic fracturing method.

5.3 *In situ stress determination from rock cores C in coal mine B*

Rock cores in the site were shale. The rock core specimen was uniaxilly loaded up to 15 MPa for vertically drilled cores or 25 MPa for horizontally drilled core. Figure 9 indicates an example of AE

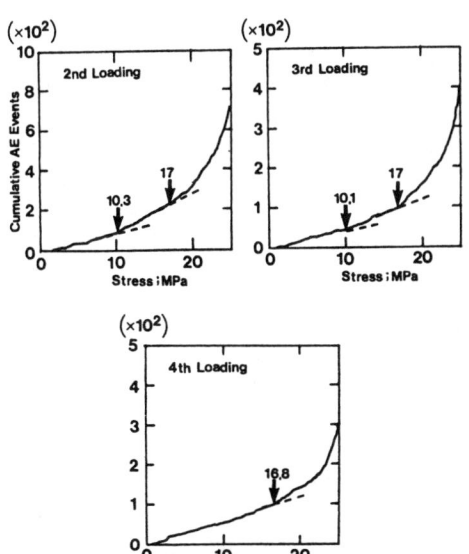

Figure 8 AE behaviors of a rock core specimen from the 2nd to 4th loading.

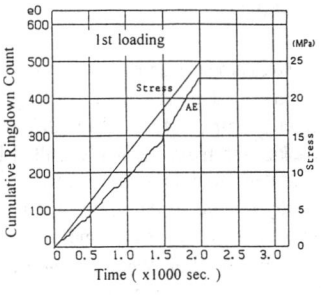

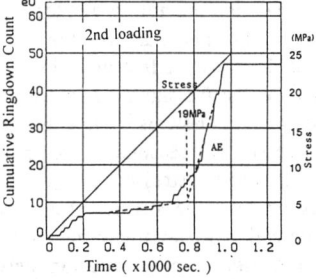

Figure 9 AE behaviors of a rock core specimen in the 1st and 2nd loadings.

behaviour of the rock core specimen from horizontally drilled hole in the first and second loadings. As it can been seen from the figure, AE generates from the beginning and monotonically increases to 25 MPa in the first loading. In the second loading, however, AE take-off point can be recognized at the stress of 19 MPa. In all tests of rock core C AE take-off points were recognized in the first loading, so in situ rock stresses were determined from AE activities after the second loadings.

Table 1 shows the results determined by hydraulic fracturing method and the AE method and an overburden pressure estimated from the depth and the rock density. The determined stresses by the AE method are well consistent with those measured by hydraulic fracturing method. Also, the vertical stress obtained from the AE method agrees well with the overburden pressure.

Table 1 Stress measurement results in Mine B

	Hydraulic Fracturing Method	AE Method	Overberdn Pressure
Vertical Stress (MPa)	8.7	9.8	8.9
Horizontal Stress(strike) (MPa)	18.5	19.0	—

6 CONCLUSIONS

In situ rock stress measurements were performed by using AE signatures in repeated loadings of rock core specimens. In the tests, three different kinds of cored rocks, one from vertically drilled hole and two from drilled hole in underground coal mine, were uniaxially loaded to determine the in situ rock stresses in underground site, and the estimated stresses from AE method were compared with the results from other techniques such as over coring method or hydraulic fracturing method. The main conclusions obtained here are as follows:

(1) Even if the Kaiser effect was obscured due to delay time, the AE signatures in the subsequent loadings could allow us to determine the in situ rock stresses with reasonable accuracy.
(2) The determined stresses by AE were well consistent with the stresses from over coring method and hydraulic fracturing method. There was also significant correlation between the overburden pressure and estimated vertical stress from AE.
(3) The time lag of up to eight months did not deter to evaluate the critical in situ stress condition. Rock core specimen recollected the in situ rock stress reasonably well (within $\pm$ 10%).
(4) The AE method suggested in the paper, in which AE activities after the second loading are utilized to determine the rock stresses, should be applicable to the in situ stress measurement with reasonable accuracy.

REFERENCES

Houghton, D.R. and A.M. Crawford 1987. Kaiser effect gauging: The influence of confining stress on its response, *Proc. 6th ISRM Congress*, Montreal, Canada, Vol.2, pp.981-985

Holocomb, D.J. 1993. Observations of the Kaiser effect under multiaxial stress state: Implications for its use in determining in situ stress, *Geophys. Res. Lett.*, 20, pp.2119-2122

Kanagawa, T., Hayashi, M. and H. Nakasa 1976. Estimation of spatial geostress components in rock samples using the Kaiser effect of acoustic emission, *Rep. No. 375017*, Central Res. Inst. of Electrical Power Industry, Abiko, Japan

Kurita, K. and N. Fujii 1979. Stress memory of crystalline rock in acoustic emission, *Geophys. Res. Lett.*, 6, pp.9-12

Seto, M., Utagawa, M. and K. Katsuyama 1989. Estimation of geostress from AE characteristics in cyclic loading of rock (in Japanese), *Proc. 8th Japan Symp. on Rock Mechanics*, The Japan National Committee for ISRM, Tokyo, Japan, pp.321-326

Seto, M., Utagawa, M. and K. Katsuyama 1992. The estimation of pre-stress from AE in cyclic loading of pre-stressed rock, *Proc. 11th Int. Symp. on Acoustic Emission*, The Jap. Soc. for NDI, Fukuoka, Japan, pp.159-166

Seto, M., Utagawa, M. and K. Katsuyama 1995. The relation between the variation of AE hypocenters and the Kaiser effect of Shirahama sandstone, *Proc. 8th Int. Cong. on Rock Mechanics*, Vol.1, Tokyo, Japan, pp.201-205

Utagawa, M., Seto, M. and K. Katsuyama 1995. Application of acoustic emission technique to detrmination of in situ stresses in mines, *Proc. 26th Int. Conf. Safety in Mines Research Institute Vol.4*, Central Mining Institute, Katowice, Poland, pp.95-109

Yoshikawa, S. and K. Mogi 1981. A new method for estimation of the crustal stress from cored rock samples: Laboratory study in the case of uniaxial compression, *Tectonophysics*, 74, 323-339.

Investigation on the influence of montmorillonite on the stability of the surrounding rock of gallery

S. Zhang, H. Wan, Z. Feng & C. Xu
Department of Resources and Environment Engineering, Wuhan University of Technology, People's Republic of China

X. Xu, J. Yin, B. Wang & T. Cao
Daye Iron Ore Mine, Wuhan Iron & Steel Company, People's Republic of China

ABSTRACT: A hanging gallery was excavated in a diorite rock mass at Daye Iron Ore Mine in 1987. A part of gallery roof collapsed in 1994 and an investigation showed that the collapse was in connection with the expanding pressure caused by the watered montmorillonite filling in the fractures of the rock mass. The content and the expanding pressure of the montmorillonite were determined as well as some other physical-chemical properties. A mechanical calculation indicates that the expanding pressure caused by the watered montmorillonite in a horizontal fracture above the gallery could not lead to the collapse of the gallery. The investigation also makes it clearly that the collapse was closely related to the mining blasting, the seepage of the underground water which made the montmorillonite expanding, mudding and crumbling, and led to formation of caves. A series of measures to improve the stability of the rock mass were put forward. All of their applications on the spot have got good results.

1 ENGINEERING AND GEOLOGY

Since 1993 a serious pressure in the surrounding rock mass of a hanging gallery which was excavated in diorite at -40m sublevel of Jianlinshan Orebody have occured. The collapse of the gallery roof was in connection with the montmorillonite.

Jianlinshan Orebody is a kind of concealed orebody and is situated in a contact zone of diorite with marble. It is a layerlike orebody 650m in length and 40m in average thickness. Occurrence of the orebody is controlled by the contact zone. The rock core ratio is larger than 85%. The rock uniaxial compressive strength is from 16.97MPa to 93.47MPa, the cohesion from 5.74MPa to 31.84MPa, the angle of friction from 39.13° to 44.50°, and the rock mass compressive strength from 8.99MPa to 71.04MPa.

2 INFLUENCE OF THE MONTMORILLON-ITE ON THE STABILITY

2.1 Appearance of the underground pressure

The grouted bolts and the jetting concrete support had been made only in some weak areas because the greater part of the surrounding rock mass maintained stable since the hanging gallery at -40m sublevel was excavated. An underground pressure began to appear in 1993, parts of the rock mass split and peeled off. So the grouted bolt installation and the jetting concrete support of the gallery were carried on all around and the metal nets were installed in parts of the rock mass.

April 1994, we found that there was a cave in the north of the gallery arch outside 1112 sublevel drift. The horizontal section of the cave approximated to an ellipse. Its major axis was 1.2m, its minor axis 0.8m, and its depth approximately 1.2m. Drops of water could be found from the cave and montmorillonite was found at the wall of the cave. Evidently, the cave was formed due to water eroding the montmorillonite filling in a fracture. One meter south to the cave, there was another cave which had a 1.2m major axis and 1m in depth. Drops of water and the montmorillonite were found on the wall of the cave too.

While the mining at -40m sublevel approached toward the gallery, the cracks in the jetting concrete layer of this section became wider and the volume of the seeping water increased gradually. Until December 1994, the gallery had not collapsed during this period the mining of 1112 sublevel drift advanced a 10-m- distance to the caves and the blasting waves directly concussed them. The collapsed place is shown in

Fig. 1. April 1995, we entered the collapsed region to observe, take samples and sketch. Along the gallery, the length of the collapsed region approximated to 3.5m, the widthe to 2.8m, the heigth to 1.5m. Two fracture zones filled with montmorillonite could be found (see Fig. 2). The orientation of one of them is approximately vertical to the gallery axis, its thickness 30 — 40cm. Two caves were justly situated in the zone. Another fracture zone is above the gallery celling about 1m's high, its orientation is approximately horizontal, and its thickness is nearly 10cm. There are two fracture zones filled with montmorillonite a few centimetres in width, and are situated 7m to the west of the collapse place. Their orientation is vertical, and the distance between them is 2m. The walls of the four fractures are smooth. The colour of the montmrillonite is grey-green and is regularly scattered.

With X-ray diffraction analysis, differential thermal analysis, electron microscope scanning, spectrum analysis and chemical analysis, the main clay mineral among the altered soil is identified to be montmorillonite. Its content arranges from 20% to 30%. The testing results may be seen in Table 1 and Table 2.

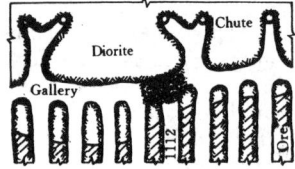

Figure 1. The collapsed place

Figure 2. The collapse of a diorite rock mass filled by montmorillonite

Table 1. Particle sorting of altered minerals

Sample No.	Particle size mm, %						
	>0.25	0.25—0.10	0.10—0.05	0.05—0.01	0.01—0.005	<0.005	0.002
FDB1	42.88	11.96	14.06	12.4	4.8	14.9	9.3
FDB2	13.34	6.32	19.84	11.2	9.2	40.1	30.5
FDB3	37.77	9.81	13.90	9.2	6.8	22.5	16.1
FDB7	24.05	7.60	17.00	9.2	4.8	37.3	30.1
FDB9	8.72	12.10	31.50	24.4	6.0	17.3	10.1

Table 2. Expanding pressure of altered minerals

Sample No.	Expansion rate, %	Dry density g/cm³	Wet density g/cm³	Water content, %	Minimum expanding pressure, MPa	Maximum expanding pressure, MPa
FDB1	26.0	1.87	2.16	15.67	0.125	0.163
	26.0	2.05	2.31	12.96	0.262	0.262
FDB2	43.4	1.94	2.25	15.86	0.200	0.258
	43.4	1.99	2.28	11.71	0.137	0.192
FDB3	51.0	1.99	2.30	15.56	0.296	0.329
FDB7	99.5	1.81	2.33	28.74	0.467	0.582
FDB9	52.5	2.00	2.31	15.56	0.684	0.684

2.2 Hydrologic conditions

The gallery is located in a water insulated diorite rock mass. Its main sections kept dry even in the rainy season of 1994. A few sections appeared moist on the arch ceiling, but the volume of the seeping water on the ceiling of the collapsed place and the nearby place doubling increased as the mining approached toward the gallery. April 1995, just up to the ceiling of the collapsed section, drops of water appeared in more than ten sites. At the same time, the maximum volume of the spring water reached 0.786 ton per day in a deep-drilling hole 1m west to the collapsed section. There also was a spring water 12m west to this hole, the volume of the spring water reached 0.524 ton per day. Drops of water generally appeared in the near sections.

2.3 Analysis of reasons of the collapse

Montmorillonite mineral possesses an active crystal structure to be sensitive to water. It is expanding while absorbing water and contracting while depriving water, in the meantime, they generate a great expanding pressure or a contracting force. Its contracting force makes a lot of tiny cracks and these cracks may further be penetrated. Soaking this dry montmorillonite mineral again, water seeps into the active crystals inside the montmorillonite along the cracks, causes the expanding pressure on the near rock and weaks it. The formation of caves due to washing away the mudding montmorillonite further weaks the near rock.

2.3.1 Influence of air humidity

After the gallery was excavated, the montmorillonite also was exposed in the moist air and absorbed a relative amount of water to expand. Moisture of a drift air arranges from 80% to 90% and seldom changes, so change of the moisture has a little affect on the collapse of the gallery roof.

2.3.2 Influence of underground water

The amount of the seeping water was small and had not a serious affect on the surrounding rock mass at the beginning of the gallery formation because the diorite rock roof was a kind of water insulated mass and the montmorillonite was completely filled inside the fractures.

2.3.3 Influence of the montmorillonite expanding pressure

As the montmorillonite is filled in a few fractures of a hard diorite rock mass, so we may treat the expanding pressure of the montmorillonite by adding it into the stress calculation.

Upon the principle of structure mechanics, we may simplify the rock below the horizontal fracture of the arch ceiling of the gallery to a end-fixed beam with an equal cross-sectional area. The span L, the thickness h, the width b and the cross-sectional area A_0 equal to 4.2m, 1m, 3.5m and 3.5m^2 respectively. There are not enough measured rock stress data *in-situ* to estimate the stress state at Daye Ore Mine, the authors suggest that the average stresses all over the world be taken as the stresses at the mine. Brown and Hoek collected a wide range of stress measurement data (Brown et al. 1978). The rock stresses here should be $\sigma_v = 8.20$MPa and $\sigma_a = 16.88$MPa according to the expression from the linear regress of Brown's data by the author (Zhang 1991). The horizontal stress inside the beam is $N/A_0 = \sigma_a$ (see Fig. 3).

The moment of a certain point x in a end-fixed beam may be defined by

$$M_x = q(-L^2/6 + Lx - x^2)/2 \tag{1}$$

and the maximum stress and the minimum stress in the beam are given by

$$\sigma_{max} = N/A_0 + M/W \tag{2}$$

$$\sigma_{min} = N/A_0 - M/W \tag{3}$$

where W is the section factor of the beam and $W = bh^2/6 = 0.583 m^3$. The solution gives values of $\sigma_{max} = 39.27$ MPa and $\sigma_{min} = 5.68$MPa. So we know that the rock mass is stable when the average compressive strength of the rock mass is 40MPa.

When only considering the rock stresses without the expanding pressure of the montmorillonite, using above method we may get $\sigma_{max} = 37.56$MPa. It is only 4.35% less than the former. We can conclude that the proportion of the expanding pressure to the whole stresses in the rock mass is very small.

2.3.4 Influence of the engineering factors

The influence of the engineering factors on the stability of the gallery may be classified as three obvious phases.

The influence of the gallery excavation.

While excavating the gallery at -40m sublevel, the blasting caused cracks in the surrounding rock mass to a certain extent, made the original cracks splitting and formed an induced stress field. Because the influence scope of the blasting was a few metres generally, the montmorillonite in the fractures kept dry.

The influence of the mining blasting at -30m sublevel. In the first quarter of 1993, the mining at -30m sublevel approached the north of 1112 sublevel drift, just above the hanging gallery at -40m sublevel. The roof thickness of the gallery was only 6.8m. The mining boundary at -30m sublevel is shown by the dotted line at Fig. 4. The mining blasting at -30m sublevel obviously broke the roof. Some cracks penetrated the roof and the passage of the seeping water was formed. The seeping water made the montmorillonite in the fractures Dexpanding and the maximum expanding pressure occurred, which would break the fractured rock nearby. The washing away of the mudding montmorillonite led to the formation and the development of two caves in the gallery roof gradually. This process lasted two years. The roof still kept stable.

The mining blasting at -40m sublevel resulted in the collapse of the gallery roof. The mining at -40m sublevel began in the second quarter of 1993, and had not reached the north of 1112 sublevel drift until the end of 1994. With approaching to the mining blasting, the number of the cracks in the gallery roof increased. The volume of the seeping water double increased. The montmorillonite in the fractures was washed away by the seeping water more rapidly. Finally, the fractured roof collapsed under the direct blasting concussion.

3 THE MEASURES TO IMPROVE THE STABILITY OF THE ROCK MASS

1. At the place where montmorillonite will occur, move the design gallery outside about 3m to 5m, in order to keep a safe distance protecting the design gallery from mining blasting.

2. Smooth wall blasting should be adopted to form a regular arch roof of gallery for the sake of reducing an overexcavated area and blasting concussion.

3. Seal up the surrounding rock of gallery rapidly with jetting concrete to avoid the deliquescence and the weather of the montmorillonite.

4. Preventing jetting concrete from peeling off is a key measure to prevent the montmorillonite in fractures from eroding. For this sake, the following measures may be employed: adopting a reinforced jetting concrete support with anchor bolts widely, improving the strength of concrete properly, ensuring the thickness of the design jetting concrete and increasing the length of anchor bolts to 2.5m or 3m.

5. Drainage in the montmorillonited rock mass with drill holes.

6. Strengthen the surrounding rock of gallery. Whether a mining blasting is at above sublevel or at this sublevel, its affect on the stability of the gallery should be monitored. If some initial jetting concrete breaks occur, the mining blasting should be stopped until the anchoring and the jetting concrete have been renewed or reproduced.

7. Presplitting blasting may be employed in the contact of the orebody with the surrounding rock to prevent the breakage of the gallery from the mining blasting at this sublevel.

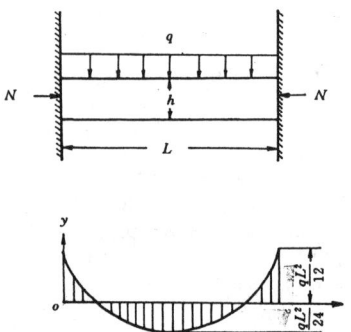

Figure 3. The forces and the bending moments of a end-fixed beam

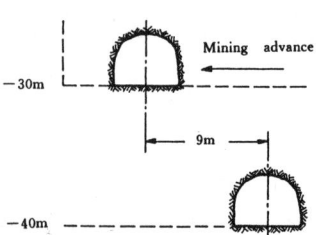

Figure 4. The position of two drifts

8. When caves emerge in a gallery roof due to montmorillonite erosion, fill them up with concrete to prevent them from deterioration.

9. Pour concrete paste to fix the montmorillonited fractures.

10. Increase the sublevel heigth from 10m at present to 15m—20m.

After taking the measures mentioned above, any collapse of the hanging gallery has not appeared.

4 CONCLUSIONS

1. The gallery collapse was in connection with montmorillonite.

2. If only the rock stresses and the montmorillonite maximum expanding pressure had exerted on the gallery roof, the collapse would not have occured.

3. Besides the rock stresses and the montmorillonite expanding pressure, the gallery collapse was in connection with the montmorillonite erosion and the mining blasting closely.

4. Taking comprehensive measures can keep the gallery stable.

REFERENCES

Brown, E. T. & E. Hoek 1978. Trends in relationships between measured *in—situ* stresses and depth. *Int. J. Rock Mech. Min. Sci. & Geomech. Abstr.* 15:211—215.

Zhang, S. 1991. Measurement of rock stresses at Tongkuangyu Mine and estimation of stress state. *Proceedings of the International Symposium on Landslide and Geotechnics*: 209—213. Wuhan: Huazhong Uni. of sci. & Tech. Press.

In situ stress measurement using hydraulic fracturing for shallow tunnel in Korea

Sung-Oong Choi, Hee-Soon Shin & Kwang-Soo Kwon
Korea Institute of Geology, Mining and Materials, Taejon, Korea

ABSTRACT: In situ stress measurements using the hydraulic fracturing technique have been conducted in Korea during the preliminary design site investigation for the Taejon subway tunnel. In situ stresses were measured in 3 shallow boreholes along the axis of a planned tunnel. The magnitude of the maximum horizontal principal stress was in the range of 2.1~2.7MPa between 25 and 50 meters deep below the surface. Its orientation was in the range of 80°~100° clockwise from the true north, and the direction of the proposed tunnel axis was nearly 65°. Rock mass conditions along the proposed tunnel alignment were characterized with closely spaced jointing and slightly to moderately weathered granitic rock. Numerical analysis using the finite difference code was performed considering the measured in situ stress field.

1 INTRODUCTION

As a number of underground projects are planned and constructed, in situ stress measurements have been emphasized in Korea. Especially the engineers who are involved with the design and construction of underground spaces should accept that the geological medium in which an excavation is to be sited is subjected to an unknown pre-existing stress field. Moreover if they want to guarantee the stability of the behavior of an excavation, the unknown pre-existing stresses should be measured in advance to an excavation. The main purpose of measuring the in situ stress is to define the magnitude and direction of the stress field which is an important value in analyzing the stabilities of underground excavations.

There are several methods for stress measurement. Among them, the hydraulic fracturing technique has became a powerful tool. It has the advantage of requiring no prerequested knowledge of the elastic properties of the rock mass because the stresses are directly measured at the test location. And there are no limits in testing depth, and the stress measurement can be conducted before excavations.

The hydraulic fracturing in situ stress measurements have been first conducted at KIGAM (Choi 1994) for the first time in Korea. Since then, there

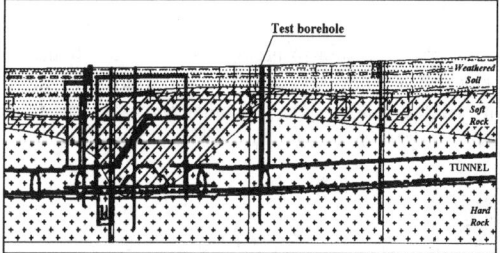

Figure 1. Geological section of the Taejon subway tunnel site including the hydraulic fracturing test borehole.

have been a lot of stress measurements in relation to the constructions of a highway tunnel.

Recently we have conducted a stress measurement for a subway tunnel in Taejon. Figure 1 shows the section of the proposed tunnel site on the Taejon subway including the rock mass conditions and the test boreholes. All of the hydraulic fracturing tests were conducted at depths between 20m and 50m beneath the surface because the subway tunnel will be constructed at depths between 25m and 30m, and it would be meanless to conduct the hydraulic fracturing tests in a weathered soil or weathered rock.

2 PRINCIPLE OF HYDRAULIC FRACTURING

2.1 Conventional hydraulic fracturing

The conventional hydraulic fracturing is based on Kirsch's solution for the stress distribution around a circular hole in a homogeneous, isotropic, elastic material subjected to far-field compressive stresses. It is used in the Hubbert and Willis formular for the critical pressure (breakdown pressure) at the moment of fracture initiation.

$$P_c = 3S_h - S_H + P_{co} - P_o \qquad (2.1)$$

where S_h and S_H are the horizontal far-field stresses, P_{co} is the hydraulic fracturing tensile strength of the rock mass and P_o is the pore pressure in the rock mass. It is assumed that the borehole is vertical, the vertical stress is one of the principal stresses and equals to the overburden stress, the rock is homogeneous, isotropic and initially impermeable and that the induced hydraulic fractures are orientated perpendicular to S_h.

$$S_h = P_s \qquad (2.2)$$

where P_s is the shut-in pressure to keep the fracture open after the pressurizing system in shut-in. Using a linear elastic approach, the other principal stresses can be expressed by;

$$S_H = 3P_s - (P_c - P_{co}) - P_o \qquad (2.3)$$
$$S_v = \rho \cdot g \cdot h$$

which only requires to determine the rock mass density γ, the shut-in pressure P_s and the fracture reopening pressure $P_r = P_c - P_{co}$ at depth h where the fracture is induced.

2.2 Hydraulic fracturing on pre-existing fractures

Most rock masses at depth are characterized by the presence of pre-existing fractures or joints with different orientations with respect to the orientation of the principal stresses. By fluid injection into a sealed-off borehole interval containing such a fracture, it will open as soon as the fluid pressure exceeds the normal stress S_n acting across the arbitrarily oriented fracture plane. In this case the shut-in pressure P_s to keep the fracture open after the pressurizing system is closed equals to the normal stress S_n. Assuming that the stress field varies linearly with depth and the vertical stress S_v is a principal stress, the normal stress S_n acting across the fracture plane of given orientation is related to the far-field stresses by;

$$\begin{aligned}S_n(h_i) = &\rho \cdot g \cdot h_i \cdot \cos^2\alpha_i + 0.5 \cdot \sin^2\alpha_i \{S_{Ho}+S_{ho}\\&+(\delta_H+\delta_h)\cdot h_i+(S_{Ho}-S_{ho})\cdot \cos2(\beta_i-\theta')\\&+(\delta_H-\delta_h)\cdot h_i \cdot \cos2[\beta_i-(\theta'+\theta'')]\}\end{aligned} \qquad (2.4)$$

where α_i and β_i are the dip and dip-direction of the *i*th fracture plane, respectively. θ' is the direction of S_{Ho} clockwise from north and θ'' is the eigenvector of δ_H with respect to S_{Ho}. δ_H and δ_h are the horizontal stress gradients, and S_{Ho} and S_{ho} are the horizontal principal stresses at h=0(at surface).

Eq.(2.4) includes 6 unknowns and the solution requires therefore at least 6 measurements of S_n at various depth on fractures with different orientations. Then the stress field can be estimated by an inversion technique as follows;

$$\begin{aligned}S_H &= 0.5[S_{Ho}+S_{ho}+(\delta_H+\delta_h)\cdot h_i+\Delta]\\S_h &= 0.5[S_{Ho}+S_{ho}+(\delta_H+\delta_h)\cdot h_i-\Delta]\\S_{Hdir} &= 0.5\arcsin[(\delta_H-\delta_h)\cdot h_i \cdot \sin2\theta''/\Delta]+\theta'\end{aligned} \qquad (2.5)$$

where, $\Delta = \{(S_{Ho}-S_{ho})^2 + (\delta_H-\delta_h)^2 \cdot h_i^2$
$+ 2(S_{Ho}-S_{ho})\cdot(\delta_H-\delta_h)\cdot h_i \cdot \cos2\theta''$

This technique is known as HTPF(Hydraulic Tests on Pre-existing Fractures) method (Cornet 1986).

3 TEST EQUIPMENT AND PROCEDURE

3.1 Hydraulic fracturing

Hydraulic fracturing is conducted within a sealed off section of a borehole. The sealing is commonly achieved using a double-straddle packer system consisting of two hydraulically activated rubber packers to pressurize a test interval until a hydraulic fracture is created or a pre-existing fracture is reopened. In general a hydraulic fracture will be created perpendicular to the minimum horizontal principal in situ stress (Hubbert and Willis 1957). The first step in the measurement procedure is the selection of suitable test intervals. The borehole core inspection, borehole camera inspection, and borehole televiewer inspection could be used for that purpose. Ideally, test intervals are selected where no jointing is present. Upon sealing off the selected test interval water is injected into the test interval at a constant rate. Once a hydraulic fracture is created the water injection is stopped and the decay in water pressure is closely monitored to examine the hydraulic fracture close. The next step is the reinjection of water into the test interval to reopen the previously generated hydraulic fracture. After the re-opening of the hydraulic fracture the water flow is again stopped and the pressure decay monitored. This re-opening cycle is repeated several times to confirm a consistent reopening pressure. After a satisfactory pressure-time curve is recorded the test is completed and the double packer system is retreated from the test interval. Figure 2 shows one example of pressure-time, flow rate-time record.

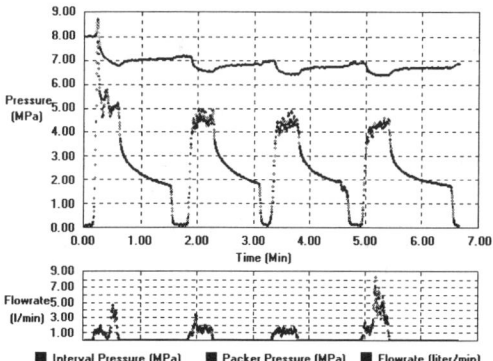

Figure 2. Pressure-time, flow rate-time curve for borehole TB-1 (testing depth 45.0m).

The hydraulic fracturing straddle-packer consisted of two 75cm long inflatable rubber packer elements rigidly connected so as to straddle a 90cm interval. During testing, the packer and test interval pressures and the flow rate were continuously monitored and recorded on a PC via an analog-to-digital converter.

3.2 Hydraulic fracture delineation

Information of the hydraulic fracture orientation is very important for determining the in situ stress field. To confirm that a hydraulic fracture was created and to determine its orientation an impression packer survey or a borehole televiewer survey is conducted. Figure 3 shows the pre-existing crack and the hydraulically induced crack detected by a borehole televiewer.

When the impression packer was lowered into the test hole, the induced hydraulic fracture could be delineated using the wrap kit and the orientation tool.

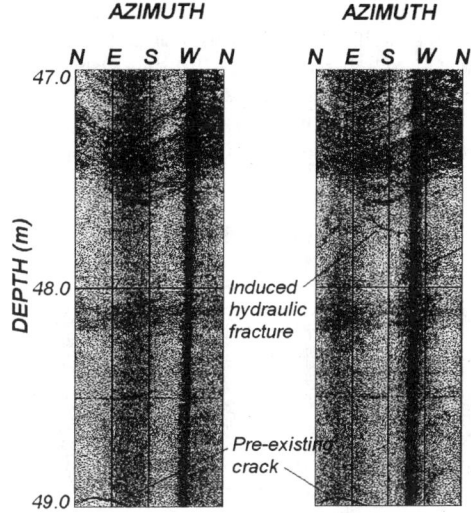

(a) Before test   (b) After test

Figure 3. Results of borehole televiewer inspection before and after the hydraulic fracturing test.

4 RESULTS OF HYDRAULIC FRACTURING

We can conclude the results of hydraulic fracturing tests on 3 boreholes, as listed in Table 1. The minimum and maximum horizontal principal stresses are in the range of 1.2~2.3MPa and 2.1~2.7MPa, respectively. There were some variances in magnitude of in situ stresses, but its orientation showed a good consistency with a range of 80°~100° clockwise from the true north.

5 NUMERICAL MODELLING

A numerical modelling analysis for the proposed subway tunnel was conducted considering the mea-

Table 1. Results of hydraulic fracturing tests.

Borehole	No.1		No.2		No.3	
Depth (m)	27.0	28.5	38.0	41.0	45.0	48.0
P_b (MPa)	2.81	3.83	5.00	5.75	7.77	7.90
P_r (MPa)	1.63	2.33	2.75	3.16	4.28	3.38
S_v (MPa)	0.72	0.76	1.02	1.11	1.21	1.29
S_h (MPa)	1.25	1.55	1.63	1.96	2.28	2.01
S_H (MPa)	2.12	2.32	2.14	2.72	2.56	2.64
T* (MPa)	1.18	1.50	2.25	2.59	3.49	4.52
Orientation	85°± 5°		90°± 5°		95°± 5°	

* T is the in situ tensile strength calculated by initial breakdown pressure and reopening pressure.

sured in situ stress field, and the designed support pattern as well. The 2-dimensional finite difference code, FLAC, was used to analyze the stabilities of tunnel. K_0 was reevaluated and inputted according to the differences between the direction of in situ horizontal stress and the direction of tunnel axis, using the theory of elasticity. Figure 4 shows the generated finite difference meshes considering the excavation steps which are composed by 4 steps totally.

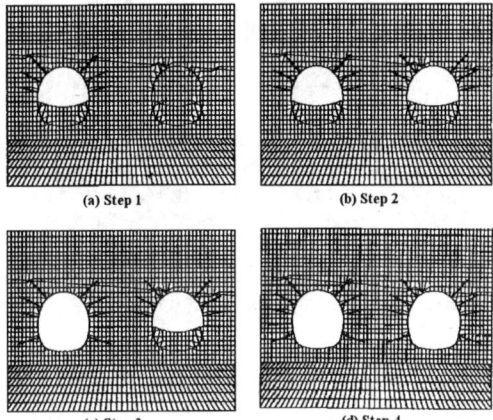

Figure 4. Finite difference meshes for the proposed tunnel considering the excavation step.

As a result of modelling analysis, we can summarize the displacement developed in several concerning points at each step, as listed in Table 2. The maximum vertical displacement in the crown of the right tunnel was 3.52mm which is about two times as big as that of the left tunnel. But it was owing to the different geological condition above the right tunnel.

Table 2. The developed displacement in several concerning points at each excavation step.

Step	Left tunnel		Right tunnel	
	Crown	Side wall	Crown	Side wall
1	1.39	0.65	-	-
2	1.45	0.50	3.38	0.83
3	1.36	1.13	3.52	0.70
4	1.43	0.93	3.22	1.51

6 CONCLUSIONS

A geological survey and a laboratory test on rock specimens are conducted for the evaluation of the in situ rock mass properties which will be used as an input data for a numerical modelling analysis. With these properties in situ stress tensor is also very important value as an input data for the analysis of tunnel stability.

In general, it has been reported that the in situ stress field is various in its magnitude and orientation with respect to a region or a depth. In situ stress field therefore should be measured before excavations at the rock mass in which the underground spaces will be sited, and applied to the stability analysis. Because the relative magnitude of the horizontal principal stress compared to the overburden stress is of major concern for the evaluation of the stability of underground excavations. Hydraulic fracturing is a cost-effective method for the determination of the principal stress components versus other common methods

REFERENCES

Choi, S. O. 1994. Fracture propagation analysis on artificial slot model and in situ stress measurement by hydraulic fracturing. *Ph.D. thesis. Seoul Nat'l University.*

Cornet, F. H. 1986. Stress determination from hydraulic tests on preexisting fractures - the H.T.P.F. Method. *Proc. Int. Symp. Rock Stress and Rock Stress Mea.*:301-311.

Haimson, B. C. 1988. Current hydraulic fracturing interpretation, How correctly does it estimate the maximum horizontal crustal stress? *Trans. Am. Geophys.* 69:1454.

Hubbert, M. K & D. G. Willis 1957. Mechanics of hydraulic fracturing. *Trans. AIME.* 210:153-166.

Ryu, D. W., H. K. Lee & S. O. Choi 1996. Development of integrated hydrofracturing data processing program by statistical approach. *Proc. the Korea-Japan Joint Symp. on Rock Engineering.* 225-230.

Modelling techniques for safety evaluation: Simulation of the coupled behavior

Thermo-hydro-mechanical coupling analysis for underground heat storage: Application of several codes

Hee-Suk Lee, Myoung-Hwan Kim & Hi-Keun Lee
Department of Mineral and Petroleum Engineering, Seoul National University, Korea

ABSTRACT : This paper investigates coupled thermal, mechanical, and hydraulic phenomena in deep rock mass especially for underground heat storage system. In order to understand the basic mechanism of thermal, hydraulic and deformation behavior in rock cavern, disturbed by thermal gradient about 100℃, various numerical experiments were conducted using several codes. The study involves the behavior of fractured rock mass including rock joint. In spite of the limitation of codes modelling fully coupled effects, the heat loss in rock mass, which is a major factor in heat storage, is insignificant in all results.

1. INTRODUCTION

It is one of the major concerns for human being to use and administrate energy, generated by various ways in limited environment efficiently. The problem of energy storage is of great importance for economical use of energy, and many researches have been advanced in aspects of the methodology and technique. In such cases, Much attention has been paid to the heat storage in underground rock mass because of low costs for construction and maintenance compared with long period of storage. Some alternatives of underground heat storage satisfy such requirements. In case of large scale, the storage of thermal water in rock cavern and aquifer can be a solution of these problems.

Generally, the heat stored rock mass is not much exceeding 100℃, thus, thermal source abruptly influence on various aspects of the behavior of rock mass. Mechanical, hydraulic, and thermal characteristics are fully coupled with each other, finally result in another equilibrium state by so complex manner.

These coupled characteristics can be evaluated and predicted through various laboratory tests of models, in-situ heater tests, and artificial numerical simulation. But, field tests are difficult to be performed because of costs, and laboratory tests are limited to simulate the real situation of coupled phenamena, so numerical simulations are powerful tools to evaluate the coupled behavior of these system. This study discusses about the basic process of the underground heat storage system, and was examined the coupling mechanism and influence on rock mass with numerical experiment for various conditions and models. Available code such as FLAC, PLASCON and Geocrack were used for these analysis.

2. BASIC CONCEPTS OF HEAT STORAGE

Supply systems for secondary energy consist of a primary energy source, an energy conversion system, and secondary energy consumers. There will be discrepancies -time-wise and some times local- between energy supply and demand. Overcoming these discrepancies is the basic task of energy storage. If such discrepancies are caused by changes of demand, it is the peak load problem that may be solved, or at least alleviated, by means of energy storage. An energy storage plant may, on the one hand, exhibit lower specific investment costs than a peak load plant, and, on the other hand, it may lower fuel cost despite some storage losses, since excess energy from base load plants with low fuel cost may be used to charge the store. Following definitions are available terms associated with thermal energy storage(Beckmann, 1984).

Thermal Energy Storage(TES) is a physical or chemical process taking place in the store(accumulator) during the charge and discharge operation. The ***store*** consists of the storage vessel(usually thermally insulated), the storage medium, the charging and discharging devices, and the auxiliaries. The storage medium is almost exclusively water/steam. The *storage system* is defined by the manner in which energy for charging the accumulator is extracted from the energy source and in which energy is discharged from the accumulator and is, in many cases, transformed into the required form of energy. By analogy to hydraulic storage systems, the main classification is into flow storage and pumped storage system. The flow storage system gets its charge energy from the heat cycle of the power plant. Pumped storage systems are charged by electrical or mechanical energy. We concerns primarily rock cavern for storage vessel. Underground storage vessels show some advantages in comparison to ground vessels. There seems to be paractically no technical or economical size limaitation, and the cost may be lower than for ground vessels. Also the safety features are excellent. But it has some disadvantages; Special geological conditions must be met for low cost underground vessels, and inspection is more complicated. Storage methods like hot water displacement storage, hot water sliding-pressure (varying pressure), and pressurized air storage(commonly called CAES) are considered for underground heat storage. Figure 1 shows the representative structure of hot water displacement storage from Canadian study. Heat loss is the main problem of heat storage in rock cavern, which is differnt from that of radioactive waste disposal.

3. OVERVIEW OF THERMO-HYDRO-MECHANICAL COUPLINGS

When thermal water/steam, about 100℃, is stored in underground rock mass, the temperature changes the mechanical characteristics of rock, and causes thermal expansion of rock, thus changes the mechanical charateristics of rock mass. Also it changes the viscosity of water, and shape of pore and fractures as a flow path, results in the change of hydraulic flow in rock mass...

The various possible couplings of each systems can be described as below Figure 2. If any changes outbreak at system in states of equilibrium condition, these changes have influence on each other, and varies in dependent manner.

4. DESCRIPTION OF USED CODES

Total three codes-FLAC ver. 3.3(ITASCA, 1995), PLASCON, and Geocrack-were used for modelling the basic TES system. FLAC is the continuum finite difference codes developed by ITASCA, USA, which is an excellent software in geotechnical engineering. This code doesn't solve fully thermo-hydro-mechanical governing equations in same time steps, but solves each governing equations and time steps independently. Thus, some considerations is required for coupling analysis.

Although mechanical model can calculate the coupling behavior simultaneously with other model, thermal and hydraulic model can't calculate that simultaneously. Because of these limitations, this code can't model the fully coupled behavior.

In this study, mechanical steps designated with 100 steps(has no physical meaning) between coupled

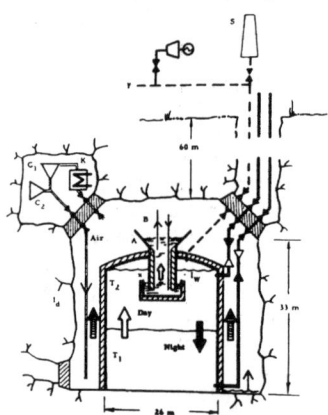

Fig. 1. Hot water displacement accumulator in rock caverns(by Margen).
A narrow sealing passage, B pressure balance tank, C_1 air circulator, C_2 air compressor, I_d dry insulation, I_w wet insulation, K air cooler, S stack, x sealing flow to prevent intrusion of gas, y steam pressurizing supply

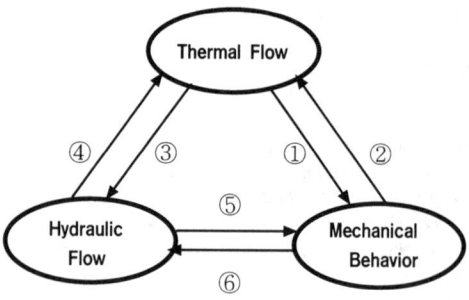

Fig. 2. Thermo-hydro-mechanical couplings diagram.

①-a variations of solid properties with temperature
①-b induced thermal stress
②-a variation of thermal properties with mechanical deformation
②-b heat produced by mechanical dissipation (can be ignored)
③ variations of fluid properties with temperature
④ heat convection by the fluid
⑤ mechanical effect of fluid pressure variation(includes effective stress law)
⑥ variations of porosity and permeability

one thermal and hydraulic time step(has physical meaning). 1 hydraulic and 1 thermal step stands for 1 hour(3600 sec.) Using Fish language, provided in software itself, a thermal and a hydraulic step (including mechanical coupling) proceeds by turns. In this manner approximate coupling effects might be modelled.

PLASCON is the non-isothermal continuum 2D finite element code, initially used for consolidation problem in soils, developed by Lewis., UK. This code uses Biot's consolidation theory, and fully coupled governing equations. In this analysis, The original code was modified for rock mechanics problem, and various options were added such as variable properties with temperature changes. Detailed formulation can be found in the text of Lewis(1987).

Geocrack is originated from Kansas Univ.(Swenson, 1994), and was developed for HDR geothermal reservoir simulation. We used the Win-95 compiled version for the analysis of jointed rock mass. Discrete fluid flow model was used for the modelling of jointed rock mass with finite element method. Finite element model consists of continuum elements(T6, represents rock matrix), interface elements,(represents nonlinear joint stiffness), and fluid elements.

This code handles fully coupled governing equations, but hydraulic coupling is considered only through the flow element. The "Gangi" model(Gangi, 1978) is used to represent the joint opening law. In this model, the relation between joint opening and joint stress is given by,

$$a = a_0 \left[1 - \left(\frac{\sigma}{\sigma_c} \right)^m \right] \quad (1)$$

where a is the joint opening, a_0 is the zero stress joint opening, σ is the joint stress, σ_c is the stress at which the joint is assumed to be closed, and m is a constant. In this model, as the joints close, they become more stiff, as they open they become softer.

5. RESULTS

5.1 Consolidation of an infinite medium with an embedded cylindrical heat source

An analytic solution was provided by Booker and Savvidou for the consolidation of an infinite homogeneous medium, with embedded volumetric heat source of constant heat output.. This problem is applicable to heat storage or underground radioactive waste disposal in case cylindrical heat source exist. Also, this model has been evaluated in similar case by Lewis(1987) and Nguyen(1995).

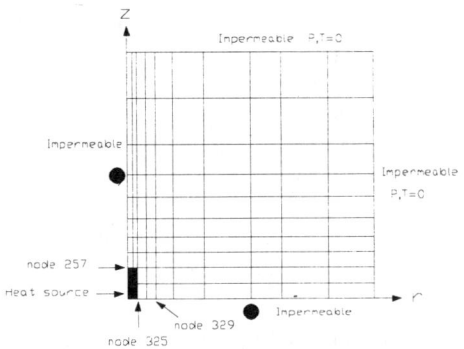

Fig. 3. 2D Consolidation model

Input data used here are as follows: radius of source = 0.3, height of source = 2, q = 1000 E = 6000, ν = 0.4, K = 0.4×10^{-5}, χ = 1.02857, ρC = 40, n = 0.5, $\beta = \beta_s = 0.9 \times 10^{-6}$, $\beta_f = 0.63 \times 10^{-5}$.

Figure 3 shows the finite element mesh and boundary conditions of this model. Assume that initial stress is zero. Figure 4 represents the representitive results in axisymmetric conditions. First, pore pressure change versus time shows the same trend as that of theoretical solution except deviation of 3 points. After pore pressure rises due to thermal expansion of pore fluid, consolidation occurs gradually, thus pore pressure dissipates slowly.

In Figure 4 (b), Temperature rises at each nodes due to conduction, but it is equilibrated at almost simultaneously same time. Node 325 located at cylinder wall, shows the greatest value of pore pressure and temperature.

These results agree well with the analytic solutions by Booker(1985), but model results slightly greater than that. This behavior could be validated from the results of FLAC analysis. Figure 5 shows the examples of FLAC analysis for the same model. Trend is the same, but the shape is different due to the form of time scales

5.2 Rock cavern model

We performed several simulations for various models such as, circular hole, circular shaft, and horse shaped tunnel etc. using PLASCON and FLAC. Properties used in analysis, are refered in the report of Decovalex project(Jing. et al.,1995). Table 1 shows these representative values. In circular hole, the location of rock opening is -100m level, radius is 10 m. Geothermal gradient might be ignored, and initial temperature is 15℃. Pore pressure is assumed to hydrostatic pressure because medium is assumed to be fully saturated from ground. Also in-situ stress is assumed to be K=1, and

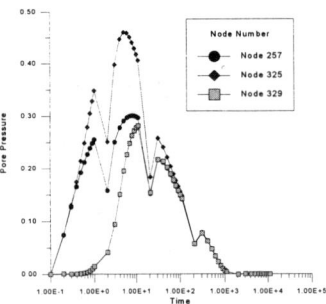

(a) time versus pore pressure

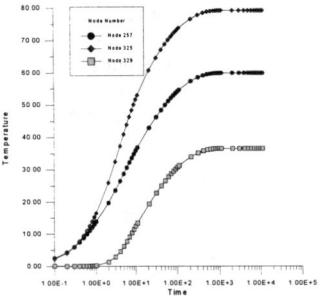

(b) Temperature of several node with time

Fig. 4. Representative results of consolidation model with heat source.

Table 1. Representative properties.

Properties	Input
Rock matrix	
Elastic modulus (E)	60, 12GPa
Poisson's ratio (ν)	0.23
Rock density(ρ_s)	2670kg/m^3
Rock specific heat(C_p)	900J/kg/K
Thermal conductivity(λ)	3W/m/K
linear thermal expansion coefficient(α_T)	9×10^{-6}K^{-1}
Fluid	
Permeability(K)	10^{-10}m/sec
Porosity (n)	0.02
Water density(ρ_w)	1000kg/m^3
Water specific heat(C_w)	4200J/kg/K
Water volumetric compressibility(K_w)	2×10^8
Water expansion coefficint(β_w)	6×10^{-4}K^{-1}

2.67 MPa to facilitate the analysis.

First, Figure 6 shows the representative results of temperature distribution around opening in 810 days after temperature of opening wall was prescribed to 100℃. It can be found that thermal flow extends from the opening wall continuously to outward boundary, but it is very slow. Although long periods(about 3 year) passes, the influenced zone is limited to narrow region near the opening wall due to rock characteristics.

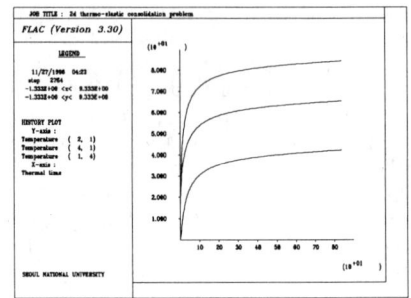

Fig. 5. Temperature change by FLAC for the same model.

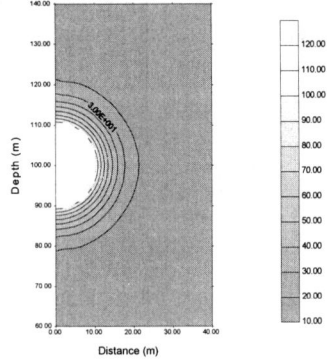

Fig. 6. Temperature distribution at circular opening 810 days after thermal source (by PLASCON).

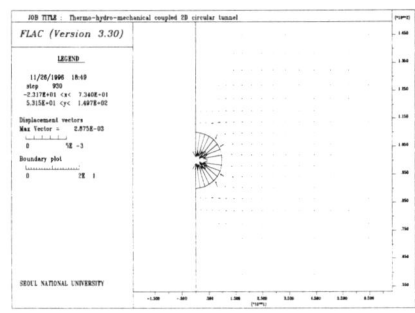

Fig .7 Displacement vector around the opening with heat source.

FLAC results showed same behavior, but due to heat source, stress is comparatively larger than that of PLASCON. Figure 7 shows the representative displacement vector of FLAC results with thermal source. The direction of vector changes at some distances from the opening wall, owing to effects of thermal expansion.

The behavior of thermal and hydraulic flow is consistent. Because of limitation in codes(FLAC),

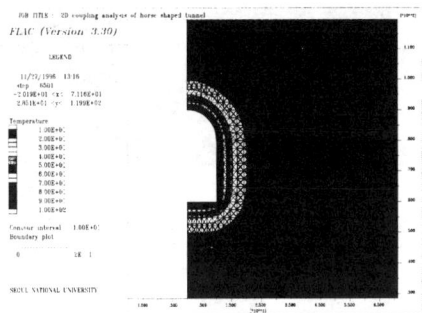

Fig. 8. Pore pressure distribution around the horse shaped tunnel with thermal source.

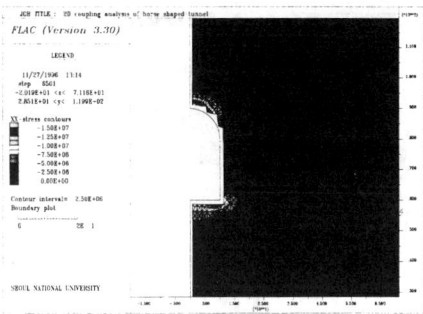

(a) X-direction

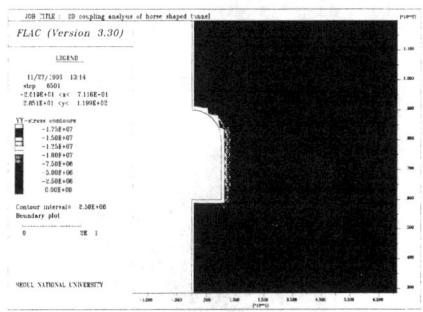

(b) Y-direction

Fig. 9. Representative stress distribution of rock cavern with heat source.

these results might be suspected, so validation with model describing fully coupling effects will be needed in future.

In order to model the hot water displacement cavern, the horse-shaped tunnel model was simulated. It seems that the over behavior showed comparatively similar as circular tunnel, except for the effects of non symmetric shape.

Figure 8 shows the pore pressure distribution around the horse-shaped tunnel. It is noted that zone of large pore pressure exists, probably due to excess-pore pressure which might be generated by the coupled deformation behavior. This distribution involes the flow both to the direction of opening wall and against it from transition zone. It needs further detailed study for understanding complete mechanism.

In all results, heat conduction was so slow that heat loss in rock mass might be insignificant.

Stress distribution around the cavern with heat source can be found in Figure 9. Tangential stress concentration in roof and sidewall is dominant due to thermal coupling effects. These concentration is limited to finite small region, so stability of the cavern might be insignificant.

5.3 Some jointed rock mass models

Some models were constructed in order to investigate thermal flow through discotinuities, and overall coupling behavior. Figure 10 represents example of the circular hole model for jointed rock mass. A quarter symmetricity was achieved, model dimension had 60m x 60m, diameter had 5 m. This problem is known as 'Kirsch problem' in elasticity. A major joint intersected hole from the wall along direction of 45 °. Fluid elements and interface elements was used for modelling of joints. Continuum properties was the same as above example. In-situ stress is assumed to be zero for convenience, initial temperature is 15℃. Joint stiffness was designated to behave like elastic modulus of rock mass. Fluid viscosity had 3×10^{-9} N-day/m^2, initial joint opening was 0.0002 m. Pressure at joint from outer boundary was prescribed to 1 MPa, the other sides was 0.1 MPa.

Fig. 11 represents the temperature distribution at 1000 days after the applying heat source. Symmetric shape along the joint plane is apparent. This shape implies that thermal flow along the joint is slower than that of rock matrix primarily, owing to low conductivity of water and high specific heat. Also, the heat conduction was so trivial that disturbed region is small, compared to long periods.

It suggests that the existence of discontinuity would not influence on the heat loss in deep underground condition. Because joint was compacted in this model, flow rate diminished rapidly at the start point of joint opening.

Stress distribution of jointed rock mass model didn't show significant difference from continuum model. This is mainly due to using the elastic joint model, but slight discontinuity exists in stress distribution. Figure 12 shows these effects. Thermal expansion results in high tangential stress concentration compared with the Kirsch solution. Another models describing multiple jointed rock mass showed similar behavior. Since all models were limited to elastic behavior, these behavior must be checked throughly for other nonlinear models.

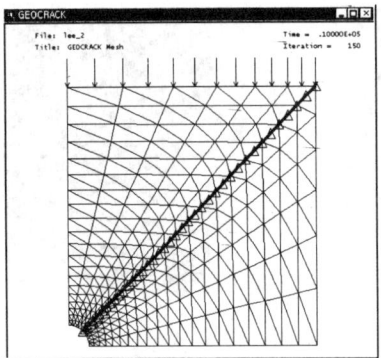

Fig. 10. Example models for discontinuous rock cavern.

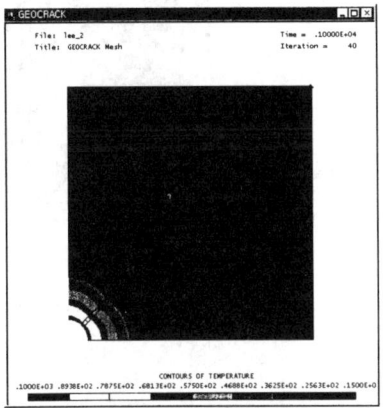

Fig. 11 Representative temperature distribution for jointed cavern.

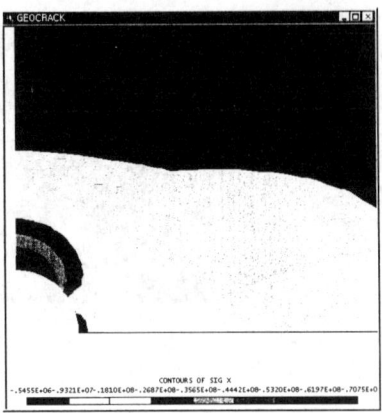

Fig. 12. A stress distribution of x-direction for jointed cavern.

Acknowledgements - This study can be completed with the help of Prof. Swenson(Kansas State Univ.). The author sincerely thanks him for provision of Geocrack.

REFERENCES

Booker, J.R. & Savvidou C. 1985. Consolidation around a point heat source, *Int. J. Num. Anal. Meth. in Geomech.* 9: 173-184.

Gangi, Anthony F. 1978. Variation of whole and fractured porous rock permeability with confining pressure. *Int. J. Rock Mech. Min. Sci. & Geomech Abstr.* 15: 249-257.

Gilli, P. V., Beckmann, G., 1984. *Thermal energy storage:* Springer-Verlag..

Itasca Consulting Group Inc. 1995. *FLAC manual.*

Lewis, R.W. & Schrefler, B.A. 1987. *The finite element method in the deformation and consolidation of porous media*: John Wiley & Sons.

L.Jing, C.F. Tsang & O.Stephansson. 1995 .DECOVALE-X- an international co-operative research project on mathematical methods of coupled THM processes for safety analysis of radioactive waste repositories, *Int. J. Rock Mech. Min. Sci & Geomech. Abstr.* 32(5): 389-398.

Swenson, D.V. DuTeau, R. & Sprecker, T. 1994. Modelling flow in a jointed geothermal reservoir. *submitted to World geothermal conference 1995.*

Swenson, D.V. 1996. *User's manual for Geocrack.*

T.S. Nguyen. & A.P.S. Selvadurai. 1995. Coupled thermal-mechanical-hydrological behavior of sparsely fractured rock: Implications for nuclear fuel waste disposal. *Int. J. Rock Mech. Min. Sci. & Geomech. Abstr.* 32(5): 465-479.

6. CONCLUSION

In this study, basic coupling analysis was performed for various model related to underground TES system. Rock masses disturbed by thermal source, showed large stress concentration, and transition of direction of displacement vectors. Heat loss which was crucial concerns in TES, was of no importance for qualitative aspects. We couldn't realize fully coupled situation due to limitation of codes and elastic behavior, also stress concentration by thermal expansion was seemed to be excessive. Thermal flow in jointed rock mass model showed slower conduction than that of rock matrix, so hydraulic flow in joints had no great influence on heat loss. But, these trends primarily might depend on overall stress conditions.

We has been trying to develop the fully coupled model for describing real rock mass system, thus we will be able to quantify the heat loss of heat storage system.

Experimental and numerical study on the thermo-mechanical behavior of granite

Myoung-Hwan Jang
Korea Mining Promotion Co., Seoul, Korea

Hyung-Sik Yang & Kwang-Ok Park
Chonnam National University, Kwangju, Korea

ABSTRACT : Block heating tests and numerical analyses were carried out to study thermal characteristics of Hwangdeung granites in room-scale block and to evaluate the effects of thermal stress to rock block and the deep cavern. Temperature was distributed radially isothermal from the heat source in the block heating tests. Stresses around a circular cavern showed compressive fields by thermal loading. Magnitude of these stresses were reduced by the boundary pressure. Application with suggested thermomechanical failure criteria indicated that principal stresses around cavity in the high temperature circumstances could be reduced by earth pressure conditions, but the yielded area was enlarged because the rock strength decreased.

1 INTRODUCTION

A potential solution to the problem of isolating radioactive wastes may be burying them in repositories excavated deep below the surface of the earth in rock mass. Several important questions must be studied before the project start. Among them are the effects on a repository of the heat released through the radioactive decay of the wastes. The magnitude of these temperature changes will affect the entire strategy of the use and design of such a repository. The changes of temperature at these waste repositories increase for the first several years after operating, but decrease as time goes by(Kim 1993).

In the 1970s and 1980s, as the needs for in situ tests in the high level nuclear waste repository programs were recognized, several large scale tests were carried out to enhance the understanding of the rock mass behavior of candidate host rocks. Lögster and Voort (1974) conducted a series of block tests on cubical samples of jointed rock to study anisotropic deformation properties of rock mass by using flatjack and hydraulic jack. Pratt et al. (1972) carried out a biaxial block test to study the scale effect on elastic and transport properties of granite samples, but thermal properties were not measured in this test.

In this study, the block heating tests were carried out to study the characteristics of Hwangdeng granite in room-scale block. The granite is supposed to be most possible host rock for radioactive waste repository in Korea.

Numerical analyses were carried out to evaluate the effect of thermal stress for deep circular cavern according to the changes of temperature.

2 ASSUMPTIONS

The mode of heat transfer related with possible temperature distribution. The distribution of temperature within a mechanical component may play an important role. Obviously the material strength and other properties may be a function of temperature especially at high temperatures.

In order to render the heat transfer problem tractable by analytical methods for the field situations, we made the following assumptions: (1) conduction is the only mode of heat transfer, (2) the rock medium can be considered to homogeneously infinite, (3) the heaters and the rock medium have the same constant thermal properties, (4) the heater are contacted with the rock in direct thermal. (5) the rock medium can be considered as infinite one with uniform initial temperature or semi-infinite with the heater drift idealized as an isothermal or adiabatic boundary.

The temperature gradient and boundary condition in elastic body create thermal stress. A generalized Hooke's law can be rewritten as follows:

$$\varepsilon_{ij} = \frac{\sigma_{ij}}{2G} - \delta_{ij}\left\{\frac{\nu}{E}J_1 - \alpha\Delta T\right\} \quad (1)$$

where E is Young's modulus of elasticity, ν is Poisson's ratio, G is the modulus of rigidity, α is the coefficient of linear thermal expansion, T is temperature, and $J_1 = \sigma_x + \sigma_y + \sigma_z$, Kronecker delta δ_{ij}. Equation (1) can be transposed to solve for the stresses, σ_{ij},

$$\begin{aligned}\sigma_{ij} &= 2G\varepsilon_{ij} + 2\delta_{ij}G\left\{\frac{\nu}{E}J_1 - \alpha\Delta T\right\} \\ &= 2G\{\varepsilon_{ij} - \delta_{ij}\alpha T\} + \delta_{ij}\frac{\nu}{1+\nu}J_1\end{aligned} \quad (2)$$

3 EXPERIMENT

3.1 *Apparatus and block samples*

Whangdeung granite was used as rock samples for block heating tests. Granite is supposed to be most possible bedrock for waste repository in Korea. Block size is about $30 \times 30 \times 30$ cm. The physical and mechanical properties of the granite are listed in table 1.

Table 1 Physical and mechanical properties of granite

Properties	Values
unit weight (kN/m^3)	0.0266
absorption (%)	0.27
apparent porosity (%)	0.72
P-wave velocity (m/sec)	3990
S-wave velocity (m/sec)	2100
uniaxial compressive strength (MPa)	152
Brazilian tensile strength (Mpa)	7.5
Young's modulus (Gpa)	35.53
poisson ratio	0.20

Fig. 1 shows the apparatus for block heating test. Thermocouples are inserted in holes of different distances from the heat source at the center of the block. Flatjacks are used to apply the boundary stress.

Heat source is a long cylinder with inserted coil which generates 0.5 kWh of calory. It corresponds to a metal canister which generates about 1 kWh (OWI 1976).

The acquisition system of data was used 12-bit A/D convert, which changed the analog voltage outputs of the thermocouple to digital

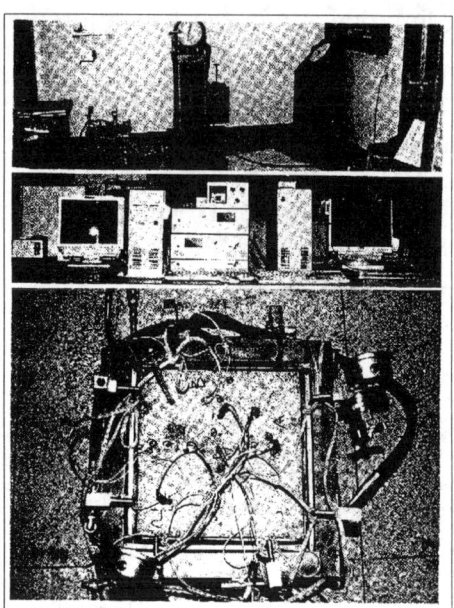

Fig.1 A photograph of block test

numbers usable by the computer. The temperature was measured up tp 8-channels by the thermocouple. The equipment for confining pressure can apply uniform hydraulic pressure up to 70 MPa.

3.2 *Procedure*

The granite block was set up inside flatjack frame. Flatjacks of the same hydraulic line were set up between the block and steel frames. The gaps between Flatjack and block or flatjack and frame were filled adequately with steel plates.

When heat source and the thermocouples were set in the holes preselected, power was supplied on heat source. The heating rate was controled under 2 ℃ per minute for preventing thermal crack from heating (Johnson et al. 1978, Friedmann & Johnson 1978). When heat source was heated up to predetermined temperature, the state was maintained by the control thermocouple unit. Then acquisition system of data was operating, and the pressure of flatjack was maintained to the pressure predetermined by hydraulic device.

4 EXPERIMENTAL RESULTS

In this study, the pressures applied to flatjack were set up to 5 MPa, and temperature of heat source were set at 100, 150 and 200 ℃ respectively. The variation of temperature by the pressure up to 5 MPa was negligible.

Fig.2 shows temperature change measured during 16 hours where heat source is maintained at 200 ℃. Temperature changes fast during first 100 minutes and slows down after 100 minutes. After 200 minutes temperature increases constantly for long time. It took more than 24 hours to reach the steady state even in room scale block test.

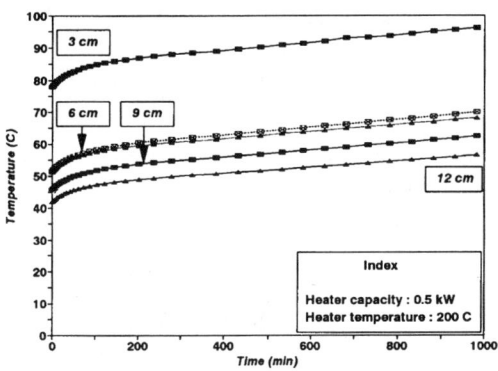

Fig.2 Temperature change as a function of heating time at different points (0.5 kW - 200 ℃ for heat source)

Thermal distribution along radial distance from a heat source is shown in the Fig. 3 for 0, 3, 10, 15 hours of heating. The distribution shows exponentially decreasing along radial distance. Thermal change at beginning of heating was very fast.

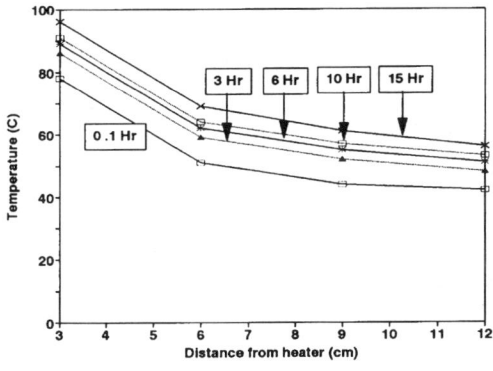

Fig.3 Thermal distribution along radial distance at various heating times

5 NUMERICAL ANALYSIS

Thermal elastoplastic finite element method applied to block and cavern model, based on the algorithm of Owen and Hinton(1980) with Hoek & Brown's empirical failure criterion expended to thermal field.

5.1 Suggested failure criterion

Hoek and Brown presented an empirical criterion which was conceptually based on Griffith's theory. It fits a wide range of rock strength, including jointed or broken rock.

In order to assess failure behavior of structures in rock mass under thermal stress authors adopted empirical failure coefficients as a function of temperature(1996)

$$\sigma_{1T} = \sigma_3 + \sqrt{m_T \sigma_3 \sigma_c + s_T \sigma_c^2} \qquad (3)$$

where,

$$m_T = m(T) = a_0 + a_1 T + a_2 T^2 \cdots a_n T^n \qquad (4)$$
$$s_T = s(T) = b_0 + b_1 T + b_2 T^2 \cdots b_n T^n \qquad (5)$$

and T is the applied temperature in Celsius. They are evaluated by the least-squares curve fitting from the high temperature confined compressive test results.

5.2 Temperature at nodal points

The temperature at nodal points of FEM-mesh was calculated by Dimoshenko's equation.

$$T_r = \frac{(T_i - T_0) \ln\left(\frac{b}{r}\right)}{\ln\left(\frac{b}{a}\right)} + T_0 \qquad (6)$$

Where T_r is the temperature at radius r, a, b are inside and outside radii, T_i, T_0 are the inner and outter wall temperatures.

In the block heating test, the changes of temperature by experimental and analytical methods is shown in Fig. 4. The calculated values by theoretical equation is good agreement with experimental ones.

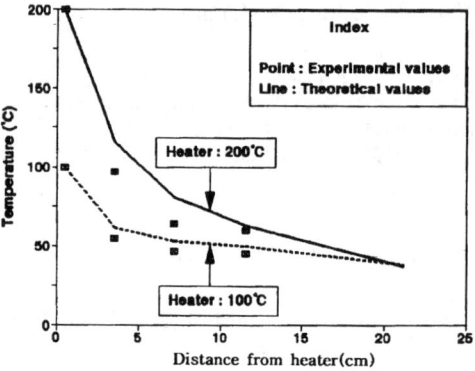

Fig.4 Experimental and calculated temperature distribution

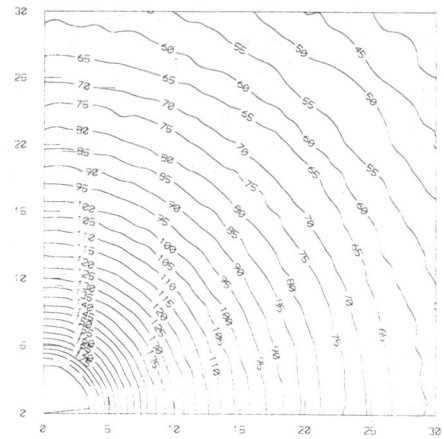

Fig.5 Isothermal contours developend in the model of circular cavern.

5.3 Circular cavern

A circular cavern of 3.5 m radius was modeled. Taking advantage of the symmetry of the problem, only a quarter of the region is analysed. Quadratic element was used in this analysis. Total number of node was 661, which consists of eight node quadrilateral elements. And the total number of element was 200.

Material properties were taken as 41,000 MPa for elastic modulus, 0.18 for poisson's ratio, 180 MPa for compressive strength, and 1×10^{-5} for coefficients of thermal expansion.

Failure coefficients were determined as

$$m_T = 38.97 - 0.081T \tag{7}$$
$$s_T = 1.02 + 0.0011T - 5.07 \times 10^{-6} T^2 \tag{8}$$

Fig. 5 shows the calculated isothermal contours. Boundary conditions of inside and outside was 200℃ and 25℃, respectively. The contours shows concentric circles as the result of block heating test.

Changes of principal stresses caused by thermal loading is shown in Fig. 6 from 100 ℃ up to 500 ℃.

Stresses around heat source shows compressive state. Maximum principal stresses around the cavern were gradually increased up to about a radius' distance and reduced to zero and turned into tensile ones while the minimum principal stresses decreased along radial distance from heat source.

Fig.7 shows the changes of principal stress along radial distance, when the inside temperature of circular cavern was 200℃ and the pressure applied to outside boundary was a

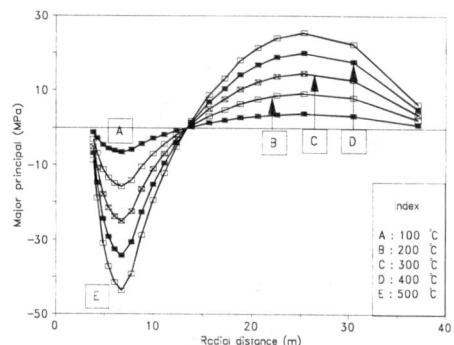
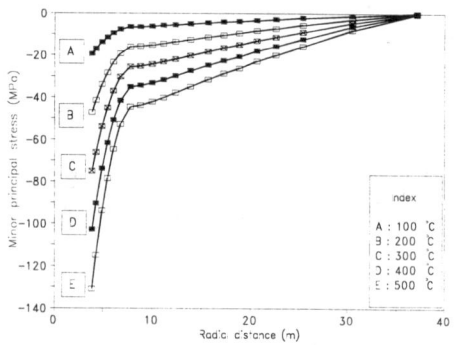

Fig.6 Principal stresses along distance from the heat source, caused by thermal loading

various conditions from 0 up to 20 MPa. In the figure, major principal stress was changed from compressive stresses to tensile ones as outside pressure increased while minor principal stress as

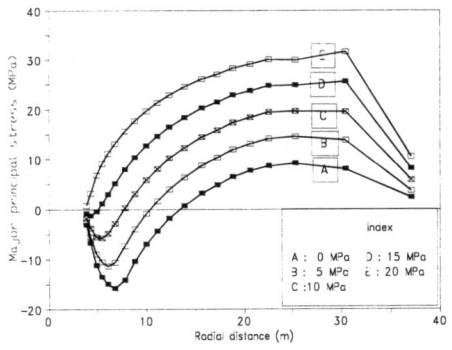

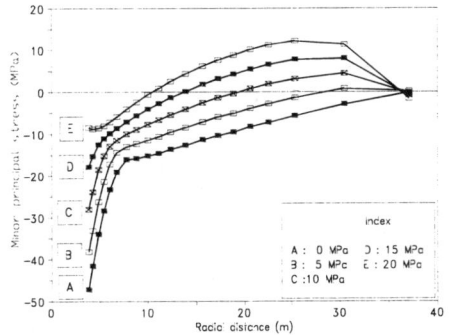

Fig.7 Principal stresses along distance from the heat source, influenced by thermal loading and boundary pressures

compressive ones around the heat source was decreased as outside pressure increased.

Numerical analysis of circular cavern in deep rock shows that compressive stresses around cavern by thermal loading decrease according to boundary conditions of outside, but the relaxation zones should be enlarged by weakening of rock mass according to thermal loading.

CONCLUSIONS

Results of room scale block heating test for Whangdeung granite and numerical analysis for circular cavern model are as follows;

1. Thermal distribution shows exponentially decreasing along radial distance. Thermal change at beginning of heating was very fast.
2. In the block heating tests, the values calculated by theoretical equation was comparatively good agreement with experimental ones.
3. Maximum principal stresses around the cavern were compressive ones and gradually increased up to about a radius' distance then reduced to zero and turned into tensile ones while the minimum principal stresses decreased along radial distance from heat source.
4. Compressive stress by thermal loading was decreased according to boundary condition, but the relaxation zones should be enlarged by weaken of rock's failure strength according to thermal loading

REFERENCES

Kim, K. 1993. Design, execution and analysis of a large-scale in-situ thermomechanical test for siting high level nuclear waste repository. In J.A. Hudson (ed), *Comprehensive rock engineering, principles, practice and projects* 3:881-913

Lögster G. & H. Voort 1974. In situ determination of the deformational behavior of a cubical rock mass samples under triaxial load, *Rock Mech.* 6:65-79

Pratt H.R., A.D. Black, W.S. Brown & W.F. Brace 1972. The effect of specimen size on the mechanical properties of unjointed diorite. *Int. J. Rock. Mech. Min. Sci.* 9:513-529

OWI 1976. *National waste terminal storage program progress report* prepared for the Energy Research and Development Administration

Johnson, B., A.F. Gangi & J. Handin 1978. Thermal cracking of rock subjected to slow uniform temperature changes. *Proc. 19th US Symp. on Rock Mech.* 259-267

Friedman, M. & B. Johnson 1978. Thermal cracks in unconfined Sioux quartzite. *Proc. 19th US Symp. on Rock Mech.* 423-430

Owen D.R.J. & E. Hinton 1980. *Finite element in plasticity : Theory and practice*, Swansea:Pineridge Press Ltd.

Jang M.H. & H.S. Yang 1996. Thermo-mechanical failure criteria of some granites for nuclear waste repository. *Proc. of the Korea-Japan Joint Symp. on Rock Engineering*, 95-99

An analytical solution and application of the coupled flow of fluids around a single well

Zenghe Xu, Xiaohe Xu & Hong Li
Centre for Rockbursts and Induced Seismicity Research, Northeastern University, Shenyang, People's Republic of China

ABSTRACT: Porous solids are taken as deformable body, and the interaction between the deformation of porous solids and the flow of pore fluids is taken account to in present paper. Based on Biot's three-dimensional consolidation theory, the coupled flow of the pore fluids in the porous stratum containing a single round well in the case of the excavation of fix flux is studied. First, a decoulped method is proposed, then under the condition that there is coupled effect between the pore solids and pore pressure, an analytical solutions of fluids pressure distribution and the solutions of displacements, strains and stresses of porous solids are derived. The real case calculation for the porous skeleton being compressible and incompressible are carried out, respectively. The study results of this paper are helpful to the thorough research of the excavation of petroleum and ground water.

1 INTRODUCTION

The research on the fluids flow through porous, deformable solids has provided a proper theoretical foundation for many engineering subject, such as the excavation of oil, gas, ground water and the gas flow in coal strata. The porous media in nature such as soils, rocks will all take place deformation caused by loads and pore pressure, including the deformation of porous skeleton and pore space. The deformation of porous solids will affect the flow of pore fluids, and the pore pressure affect the deformation of porous media in turn. Therefore, in the process of pore fluids flow, there are some interaction between the deformation of porous solids and the flow of pore fluids. In classical fluids dynamics through porous media, porous solids is considered as rigid body being not deformable, or the relation between the deformation of porous solids and the pore pressure is supposed previously (J. Bear.1972). Thus are coupled problem in essence is simplified as a non-coupled ones. The simplification was proven late only to be applied to the case that porous solids take place one-dimensional deformation only and meanwhile the load imposing on the media are kept constant (J.C.Jaeger et al. 1979). But, in general, porous solids deform in all three direction, the manner of the interaction between pore solids deformation and the pore fluids flow is revealed by no means previously. In this case, the flow of the pore fluids in deformable porous solids can't be studied independently as non-coupled problem.

Biot's three-dimensional consolidation theory take into account the coupled effects between porous solids and pore fluids, for the linear elastical and viscoelastical porous solids, Biot's theory is considered as a perfect enough theory (Wang Ren 1984). But, owing mainly to the complexity of the mathematical model of Biot's theory, hitherto, the exact solution are only obtained for very few simple consolidation problems.

The extracation of fluids making use of round well is a principal method in the mining of gas, petroleum and ground water. When the pore pressure on the wall of well is kept constant, a analytical solution of the coupled flow problem has been obtained in note (Xu Zenghe et al. 1992). In this paper, a coupled flow problem with initial-mixed boundary conditions in a plane axial symmetric domain is studied. First, a decoupled method is proposed, then, an analytical solution of pore pressure distribution, and the analytical solutions of the displacements, strains, stresses are derived. Finally, the real cases calculation for the porous skeleton being compressible and incompressible are carried out.

2 THE GOVERNING EQUATION OF THE PROBLEM OF RADIAL COUPLE FLOW

According to Biot's theory, for fluid-saturated porous media, the relation between displacements and strains are the same as elastic media's, and the relations between stresses and strains are

$$\sigma_{ij} = 2G\varepsilon_{ij} + \delta_{ij}(\lambda \varepsilon_v + \alpha p), \quad (i=1,2,3) \quad (1)$$

$$\varepsilon_p = \Delta n = n - n_0 = -\alpha \varepsilon_v + p/Q \quad (2)$$

Where σ_{ij} denotes media's stresses, ε_{ij} strains, $\varepsilon_v = \varepsilon_{11} + \varepsilon_{22} + \varepsilon_{33}$ denotes volumetric strain, λ, G denotes Lame's constants, n_0 denotes the media initial porosity prior to the deformation of porous solids, n after deformation, so ε_p the increment of porosity caused by deforming, p pore pressure, α, Q Biot's constants which also reflect the behavior of the deformation of porelastic solids. The equilibrium equations of porelastic solids taking the displacement form are

$$(\lambda + G)\varepsilon_{v,i} + G\nabla^2 u_i + \alpha p_{,i} = 0, \quad (i=1,2,3) \quad (3)$$

where u_i denotes media's displacement in x_i direction, $\nabla^2 = \dfrac{\partial^2}{\partial x_1^2} + \dfrac{\partial^2}{\partial x_2^2} + \dfrac{\partial^2}{\partial x_3^2}$ Laplace's operator. Biot ignored the effect of the compressiblity of fluids on its mass conservation, i.e. he assumed $\rho = \rho_0 =$ constant, thus the mass conservation equation based on Darcy's law is

$$k(\frac{\partial^2 p}{\partial x_1^2} + \frac{\partial^2 p}{\partial x_2^2} + \frac{\partial^2 p}{\partial x_3^2}) = \frac{1}{Q}\frac{\partial p}{\partial t} - \alpha\frac{\partial \varepsilon_v}{\partial t} \quad (4)$$

where k is seepage coefficient. The equation (3) and (4) are obviously coupled. Because there is the term related to p in equation (3) and the term related to ε_v in equation (4), the solution can't be obtained by solving equation (3) or (4) independently. The fundamental equation (3) and (4) were first established by Biot in 1941 (M. A. Biot 1941). In present paper, we discuss the radial fluid-solid coupled problem with the initial-mixed boundary conditions of fluids and the stresses boundary conditions of solid shown as figure 1. The problem may be studied more conveniently in polar

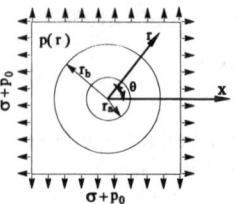

Fig. 1 mechanical model

coordinate, in the case of plane-strains, it has been poven in (Xu Zenghe et. al 1992) that the fundamental equations (3) and (4) may be written as

$$\frac{\partial}{\partial r}[\frac{1}{r}\frac{\partial(ru)}{\partial r}] = -\frac{\alpha}{\lambda + 2G}\frac{\partial p}{\partial r} \quad (5)$$

$$k(\frac{\partial^2 p}{\partial r^2} + \frac{1}{r}\frac{\partial p}{\partial r}) = \frac{1}{Q}\frac{\partial p}{\partial t} - \alpha\frac{\partial \varepsilon_v}{\partial t} \quad (6)$$

the relation of stress-strain and displacement-strain are

$$\begin{cases}\sigma_r = \lambda\varepsilon_v + 2G\varepsilon_r + \alpha p \\ \sigma_\theta = \lambda\varepsilon_v + 2G\varepsilon_\theta + \alpha p \\ \sigma_z = \lambda\varepsilon_v + 2G\varepsilon_z + \alpha p\end{cases} \quad (7.1\text{-}3)$$

$$\varepsilon_r = \frac{\partial u}{\partial r}, \quad \varepsilon_\theta = \frac{u}{r}, \quad \varepsilon_z = 0,$$

$$\varepsilon_v = \frac{\partial u}{\partial r} + \frac{u}{r} = \frac{1}{r}\frac{\partial(ru)}{\partial r} \quad (8.1\text{-}4)$$

Equation (5) and (6) are the governing partial differential equations, they are coupled obviously and can't be solved independently.

3 A DECOUPLED PROCEDURE AND AN ANALYTICAL SOLUTION FOR RADIAL FLUID-SOLID COUPLED PROBLEM

For the sake of convenience, first the following indimensional variable are introduced

$$\bar{r} = r/r_a, \quad \bar{p} = p/p_0, \quad \tau = kp_0/r_a^2 \times t$$
$$\bar{u} = u/r_a, \quad \bar{\sigma}_r = \sigma_r/p_0, \quad \bar{\sigma}_\theta = \sigma_\theta/p_0 \quad (9.1\text{-}6)$$

Substituting the above formula into formula (5), (6), (7.1-2) and (8.1-4) yields

$$\frac{\partial}{\partial \bar{r}}\{\frac{1}{\bar{r}}\frac{\partial(\bar{r}\bar{u})}{\partial \bar{r}}\} = -\frac{\alpha}{\lambda + 2G}\frac{\partial \bar{p}}{\partial \bar{r}} \quad (10)$$

$$\frac{\partial^2 \bar{p}}{\partial \bar{r}^2} + \frac{1}{\bar{r}}\frac{\partial \bar{p}}{\partial \bar{r}} = \frac{p_0}{Q}\frac{\partial \bar{p}}{\partial \tau} - \alpha\frac{\partial \varepsilon_v}{\partial \tau} \quad (11)$$

$$\begin{cases}\bar{\sigma}_r = \dfrac{\lambda}{p_0}\varepsilon_V + \dfrac{2G}{p_0}\varepsilon_r + \alpha\bar{p} \\ \bar{\sigma}_\theta = \dfrac{\lambda}{p_0}\varepsilon_V + \dfrac{2G}{p_0}\varepsilon_\theta + \alpha\bar{p} \\ \bar{\sigma}_z = \dfrac{\lambda}{p_0}\varepsilon_V + \dfrac{2G}{p_0}\varepsilon_z + \alpha\bar{p}\end{cases} \quad (12.1\text{-}3)$$

$$\varepsilon_r = \dfrac{\partial \bar{u}}{\partial \bar{r}},\ \varepsilon_\theta = \dfrac{\bar{u}}{\bar{r}},\ \varepsilon_z = 0,\ \varepsilon_V = \dfrac{1}{\bar{r}}\dfrac{\partial(\bar{r}\bar{u})}{\partial \bar{r}} \quad (13.1\text{-}3)$$

Noting, under the condition of plain-strains, that $\bar{\sigma}_z$ isn't independent of $\bar{\sigma}_r, \bar{\sigma}_\theta$. In the case of the extracation of fluids in fix flux, the flux of fluids on the wall of well $r = r_a$, is known and kept no variation. At the large enough radius $r = r_b$, the pore pressure isn't affected by the fluids extravation and same as the initial pore pressure p_0 before extracation, i.e. on the radius $r = r_b, p(r_b) = p_0$. Because of the same reason, the stresses on the radius $r = r_b$ isn't affected by the extraction of fluids and the stresses concentration near round well. So the initial and boundary conditions are as follows

$$\begin{cases}\bar{p}(\bar{r},0) = 1,\ \bar{p}\left(\dfrac{r_b}{r_a},\tau\right) = 1, \\ \left.\dfrac{\partial \bar{p}}{\partial \bar{r}}\right|_{\bar{r}=1} = \dfrac{V}{2\pi kHp_0};\ \bar{\sigma}_0(1,\tau) = \bar{p}(1) \\ \bar{\sigma}_r\left(\dfrac{r_b}{r},\tau\right) = \bar{\sigma}_0\left(\dfrac{r_b}{r},\tau\right) = \dfrac{\sigma_0}{p_0} + 1\end{cases} \quad (14.1\text{-}5)$$

Integrating formula (10) with respect to $\bar{r}$, leads to

$$\dfrac{1}{\bar{r}}\dfrac{\partial(\bar{r}\bar{u})}{\partial \bar{r}} = -\dfrac{\alpha p_0}{\lambda + 2G}\bar{p} + f(\tau) \quad (15)$$

Superposing (12.1) and (12.2), then substituting the resultant formula into (13.4) produces

$$\varepsilon_V = \dfrac{1}{\bar{r}}\dfrac{\partial(\bar{r}\bar{u})}{\partial \bar{r}} = \dfrac{\bar{\sigma}_r + \bar{\sigma}_\theta - 2\alpha\bar{p}}{2(\lambda + G)}p_0 \quad (16)$$

substituting boundary condition (14.4) and (14.5) into the above formula leads to

$$\left.\dfrac{1}{\bar{r}}\dfrac{\partial(\bar{r}\bar{u})}{\partial \bar{r}}\right|_{\bar{r}=r_b/r_a} = \dfrac{\sigma_\theta - \alpha p_0}{\lambda + G} \quad (17)$$

Substituting the above formula and formula (14.2) into (15) yields

$$f(\tau) = \dfrac{\sigma_\theta + (1-\alpha)p_0}{\lambda + G} + \dfrac{\alpha p_0}{\lambda + 2G} \quad (18)$$

$$\varepsilon_V = \dfrac{1}{\bar{r}}\dfrac{\partial(\bar{r}\bar{u})}{\partial \bar{r}} = -\dfrac{\alpha p_0}{\lambda + 2G}\bar{p} + \dfrac{\sigma_0 + (1-\alpha)p_0}{\lambda + G} + \dfrac{\alpha p_0}{\lambda + 2G} \quad (19)$$

Substituting formula (16) into equation (11) yields

$$\dfrac{\partial^2 \bar{p}}{\partial \bar{r}^2} + \dfrac{1}{\bar{r}}\dfrac{\partial \bar{p}}{\partial \bar{r}} = c\dfrac{\partial \bar{p}}{\partial \tau},\quad c = \dfrac{p_0}{Q} + \dfrac{\alpha^2 p_0}{\lambda + 2G} \quad (20.1\text{-}2)$$

Obviously, according to (20) pore pressure $\bar{p}_i$ is now decoupled with media's displacements $\bar{u}_i$, so the following initial-mixed boundary value problem on $\bar{p}$ may be achieved after incorporating formula (20.1) and initial condition (14.1), and boundary conditions (14.2-3)

$$(I)\begin{cases}\dfrac{\partial^2 \bar{p}}{\partial \bar{r}^2} + \dfrac{1}{\bar{r}}\dfrac{\partial \bar{p}}{\partial \bar{r}} = c\dfrac{\partial \bar{p}}{\partial \tau},\ \bar{r} > 1, \tau > 0 \\ \bar{p}(\bar{r},0) = 1, \bar{r} > 1 \\ \bar{p}(\dfrac{r_b}{r_a},\tau) = 1,\ \left.\dfrac{\partial \bar{p}}{\partial \bar{r}}\right|_{\bar{r}=1} = \dfrac{V}{2\pi kHp_0}, \tau > 0\end{cases}$$

In Problem (I), H denotes the thickness of fluids-bearing formation, and the V fix flux of fluids. problem (I), being the initial-mixed boundary problem of parabolic partial differential, may be taken as the resultant problem after superposing a stable flow problem and a unstable problem.

$$\bar{p}(\bar{r},\tau) = \bar{p}_1(\bar{r}) + \bar{p}_2(\bar{r},\tau) \quad (21)$$

Substituting (21) into the governing equation of problem (I) and letting $\bar{p}_1(\bar{r})$ satisfy non-homogeneous boundary conditions, lead to the following boundary value problem on $\bar{p}_1(\bar{r})$

$$(II)\begin{cases}\dfrac{\partial^2 \bar{p}_1}{\partial \bar{r}^2} + \dfrac{1}{\bar{r}}\dfrac{\partial \bar{p}_1}{\partial \bar{r}} = 0, \bar{r} > 1 \\ \bar{p}_1(\dfrac{r_b}{r_a}) = 1,\ \left.\dfrac{\partial \bar{p}_1}{\partial \bar{r}}\right|_{\bar{r}=1} = \dfrac{V}{2\pi kHp_0}\end{cases}$$

The solution satisfying problem (II) is

$$\bar{p}_1(\bar{r}) = \dfrac{V}{2\pi kHp_0}\ln\dfrac{r_a}{r_b}\bar{r} + 1 \quad (22)$$

The following initial-mixed boundary problem on $\bar{p}_2(\bar{r},\tau)$ may be derived from formula (21), (22) and problem (I)

$$(III)\begin{cases}\dfrac{\partial^2 \bar{p}}{\partial \bar{r}^2} + \dfrac{1}{\bar{r}}\dfrac{\partial \bar{p}}{\partial \bar{r}} = c\dfrac{\partial \bar{p}}{\partial \tau}, \bar{r} > 1, \tau > 0 \\ \bar{p}_2(\bar{r},0) = -\dfrac{V}{2\pi kHp_0}\ln\dfrac{r_a}{r_b}\bar{r} = F(\bar{r}), \bar{r} \geq 1 \\ \bar{p}_2(\dfrac{r_a}{r_b},\tau) = 0,\ \left.\dfrac{\partial \bar{p}}{\partial \bar{r}}\right|_{\bar{r}=1} = 0\end{cases}$$

(III) is the homogeneous boundary value problem of parabolic partial differential equation and its solution is

$$\bar{p}_2(\bar{r},\tau) = \sum_{m=1}^{\infty} c_m e^{-\frac{1}{c}\beta_m^2\tau} R_0(\beta_m\bar{r}) \qquad (23)$$

$$R_0(\beta_m\bar{r}) = Y_0(\beta_m\frac{r_b}{r_a})J_0(\beta_m\bar{r}) - J_0(\beta_m\frac{r_b}{r_a})Y_0(\beta_m\bar{r}) \qquad (24)$$

In formula (23), c_m remains to be determinate, in formula (21) β_m is the positive roots of the following equation

$$Y_0(\beta_m\frac{r_b}{r_a})J'_0(\beta_m) - J_0(\beta_m\frac{r_b}{r_a})Y'_0(\beta_m) = 0 \quad (25)$$

Substituting the initial condition into formula (23) yields

$$F(\bar{r}) = \sum_{m=1}^{\infty} c_m R_0(\beta_m\bar{r}) \qquad (26)$$

Multifying the term of both sides of the above formula by $\bar{r}R_0^2(\beta_m\bar{r})$, then, integrating them in domain $[r_a, r_b]$ produce

$$\int_1^{r_b/r_a} \bar{r}F(\bar{r})R_0^2(\beta_m\bar{r})d\bar{r}$$

$$= \sum_{m=1}^{\infty} c_m \cdot \int_1^{r_b/r_a} \bar{r}R_0^2(\beta_m\bar{r})d\bar{r} \qquad (27)$$

$R_0(\beta_1,\bar{r}), R_0(\beta_2,\bar{r})\cdots R_0(\beta_m,\bar{r})$ are linearly independent and orthogonal each other, so c_m may be derived from formula (27)

$$c_m = \frac{1}{N(\beta_m)} \int_1^{r_b/r_a} \bar{r}F(\bar{r})R_0(\beta_m\bar{r})d\bar{r} \qquad (28)$$

$$N(\beta_m) = \int_1^{r_b/r_a} \bar{r}R_0^2(\beta_m\bar{r})d\bar{r} \qquad (29)$$

According to the integral formula of Bessel function, substituting formula (24) into the above formula produces

$$\frac{1}{N(\beta_m)} = \frac{\pi^2}{2} \frac{\beta_m^2 J'_0(\beta_m)}{J'^2_0(\beta_m) - J_0^2(\beta_m, r_b/r_a)} \qquad (30)$$

In formula (24) and (25), J_0 is the first species, zero order Bessel function, Y_0 the second species, zero order Bessel function. Substituting formula (28) into (23) leads to

$$\bar{p}_2(\bar{r},\tau) = \sum_{m=1}^{\infty} \frac{1}{N(\beta_m)} e^{-\frac{1}{c}\beta_m^2\tau} R_0(\beta_m) \times$$

$$\int_1^{r_b/r_a} \bar{r}' R_0(\beta_m\bar{r}')d\bar{r}' \qquad (31)$$

substituting formula (31) and (22) into (21), the pore pressure distribution function $\bar{p}(\bar{r},\tau)$ in case of the extracting of fix flux and taking into account the coupled effects between fluids and solids may be derived

$$\bar{p}(\bar{r},\tau) = 1 + \frac{V}{2\pi kHp_0} \ln\frac{r_a}{r_b}\bar{r} +$$

$$+ \sum_{m=1}^{\infty} \frac{1}{N(\beta_m)} e^{-\frac{1}{c}\beta_m^2\tau} R_0(\beta_m\bar{r}) \int_1^{r_b/r_a} \bar{r}' \qquad (32)$$

In order to obtain the analytical solutions of stresses, strains and displacements in case of fluid-solid coupled, substituting formula (35) into (19), the resultant formula is first multiplied by $\bar{r}$, then, integrated with respect to $\bar{r}$ in domain $[1, r_b/r_a]$, that lead to the following equation after the arrange of the resultant formula

$$\bar{u} = -\frac{\alpha p_0}{\lambda + 2G}\frac{1}{\bar{r}}\int_1^{\bar{r}} \bar{r}' \bar{p}d\bar{r}' +$$

$$+ \frac{1}{2}(\frac{\sigma_0 - \alpha p_0}{\lambda + G} + \frac{\alpha p_0}{\lambda + 2G})\bar{r} + \frac{a}{\bar{r}} \qquad (33)$$

where a remain to be determinate. By usimg formula (33), strain-displacement relation(13.1) and stress-strain relation (12.1), ε_r and $\bar{\sigma}_r$ may be derived

$$\varepsilon_r = \frac{\partial \bar{u}}{\partial \bar{r}} = \frac{\alpha p_0}{\lambda + 2G}\frac{1}{\bar{r}^2}\int_1^{\bar{r}} \bar{r}'\bar{p}d\bar{r}' - \frac{\alpha p_0}{\lambda + 2G}\bar{p} +$$

$$+ \frac{1}{2}(\frac{\sigma_0 - \alpha p_0}{\lambda + G} + \frac{\alpha p_0}{\lambda + 2G}) - \frac{a}{\bar{r}^2} \qquad (34)$$

$$\bar{\sigma}_r = \frac{\lambda}{p_0}[-\frac{\alpha p_0}{\lambda + 2G}\bar{p} + (\frac{\sigma_0 - \alpha p_0}{\lambda + G} + \frac{\alpha p_0}{\lambda + 2G})]$$

$$+ \frac{2G}{p_0}\{\frac{\alpha p_0}{\lambda + 2G}\frac{1}{\bar{r}^2}\int_1^{\bar{r}} \bar{r}'\bar{p}d\bar{r}' - \frac{\alpha p_0}{\lambda + 2G}\bar{p} + \qquad (35)$$

$$+ \frac{1}{2}(\frac{\sigma_0 - \alpha p_0}{\lambda + G} + \frac{\alpha p_0}{\lambda + 2G}) - \frac{a}{\bar{r}^2}\} + \alpha\bar{p}$$

Substituting the boundary condition on stresses (14.5) into the above equation, the analytical solution of displacement may be derive

$$\bar{u} = -\frac{\alpha p_0}{\lambda + 2G}\frac{1}{\bar{r}}\int_1^{\bar{r}} \bar{r}'\bar{p}d\bar{r}' + (\frac{\sigma_0 - \alpha p_0}{\lambda + G} +$$

$$+ \frac{\alpha p_0}{\lambda + 2G})(\frac{\bar{r}}{2} + \frac{\lambda + G}{G}\frac{1}{\bar{r}}) - \frac{p_0\bar{p}(1)}{2G}\frac{1}{\bar{r}} \qquad (36)$$

according to formula (36), (13.1-2) and (12.1-2) the analytical solution of strains and stresses may be derived

$$\varepsilon_r = -\frac{\alpha p_0}{\lambda + 2G}\frac{1}{\bar{r}^2}\int_1^{\bar{r}} \bar{r}'\bar{p}d\bar{r}' - \frac{\alpha p_0}{\lambda + 2G}\bar{p} +$$

$$+ \frac{1}{2}\Phi(1 - \frac{\lambda + G}{2G}\frac{1}{\bar{r}^2}) - \frac{p_0\bar{p}(1)}{2G}\frac{1}{\bar{r}^2} \qquad (37.1)$$

724

$$\varepsilon_\theta = -\frac{\alpha p_0}{\lambda + 2G} \frac{1}{\bar{r}^2} \int_1^{\bar{r}} \bar{r}' \bar{p} d\bar{r}' +$$
$$+ \frac{1}{2} \Phi (1 + \frac{\lambda + G}{2G} \frac{1}{\bar{r}^2}) - \frac{p_0 \bar{p}(1)}{2G} \frac{1}{\bar{r}^2} \quad (37.2)$$

$$\bar{\sigma}_r = \frac{2G\alpha}{\lambda + 2G} \frac{1}{\bar{r}^2} \int_1^{\bar{r}} \bar{r}' \bar{p} d\bar{r}' +$$
$$+ \frac{\lambda + G}{p_0} \Phi (1 - \frac{1}{\bar{r}^2}) - \frac{\bar{p}(1)}{\bar{r}^2} \quad (37.3)$$

$$\bar{\sigma}_\theta = \frac{2G\alpha}{\lambda + 2G} (\bar{p} - \frac{1}{\bar{r}^2} \int_1^{\bar{r}} \bar{r}' \bar{p} d\bar{r}') +$$
$$+ \frac{\lambda + G}{p_0} \Phi (1 + \frac{1}{\bar{r}^2}) - \frac{\bar{p}(1)}{\bar{r}^2} \quad (37.4)$$

$$\bar{u} = \frac{\alpha p_0}{\lambda + 2G} \frac{1}{\bar{r}} \int_1^{\bar{r}} \bar{r}' \bar{p} d\bar{r}' +$$
$$+ \Phi (\frac{\bar{r}^2}{2} + \frac{\lambda + G}{2G} \frac{1}{\bar{r}}) - \frac{p_0 \bar{p}(1)}{2G} \frac{1}{\bar{r}} \quad (37.5)$$

where $\Phi = \frac{\sigma_0 + (1-\alpha) p_0}{\lambda + G} + \frac{\alpha p_0}{\lambda + 2G}$.

4 AN CALCULATION CASE

If the compression of porous skeleton is so much small that it may be ignored., the compression of porous solids entirely will be only cause by the compression of pore space. Biot's theory has pointed out that, in such case, the following conditions hold

$$\alpha = 1, \quad Q \to \infty \quad (38)$$

Substituting formula (38) into equation (32), yields

$$\bar{p}^*(\bar{r}, \tau) = 1 + \frac{V}{2\pi k p_0 H} \ln \frac{r_a}{r_b} \bar{r} +$$
$$+ \sum_{m=1}^{\infty} \frac{1}{N(\beta_m)} e^{-\frac{1}{c_1} \beta_m^2 \tau} R_0(\beta_m \bar{r}) \times \quad (39)$$
$$\times \int_1^{r_b/r_a} F(\bar{r}') R_0(\beta_m \bar{r}') d\bar{r}'$$

where $\bar{p}^*(\bar{r}, \tau)$ denotes the distribution of pore pressure satisfied condition (38). We make $r_b/r_a = 1000$, $p_0 = 5 MPa$, $\alpha = 0.2$, $Q = 1500 MPa$, $\lambda = 500 Mpa$, $G = 750 MPa$, $H = 4000 cm$, $k = 3 \times 10^{-2} cm/s$, $V = 3000 cm^3/s$, then, for the calcultion of equation (32), mentioned data are made use of, for the equation (39), all the above mentioned data are still made use of except the value of α and Q being now determined by formula (38). By using the results of calculation, the distribution figures of $\bar{p}(\bar{r}, \tau)$ and $\bar{p}^*(\bar{r}, \tau)$ may be drawn and shown as

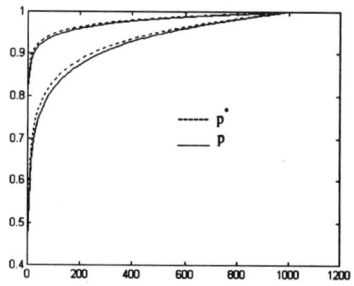

Fig. 2 Distribution of pore pressure of p^* and p

figure 2(a), (b). In figure 2, the dotted lines represent the distribution of $\bar{p}^*(\bar{r}, \tau)$, real line the distribution of $\bar{p}(\bar{r}, \tau)$. It can be easily observed from figure 2 that, in the case of porous skeleton being incompressible, the pore pressure is greater, and the difference between $\bar{P}(\bar{r}, \tau)$ and $\bar{P}^*(\bar{r}, \tau)$ becomes greater with the time.

REFERENCES

J. Bear 1972. *Dynamics of fluids in porous media.* America, Elsevier Publishing Company, INC.

J.C.Jaeger & N.G.W. Cook 1979. *Fundamentals of rock mechanics.* Third Edition, London: Chapman and Hall.

Wang Ren 1984. Mechanical problems for geo-materials (in Chinese). *Mechanics and Practice.* 8(4):1-6.

Xu Zenghe and Zhang Zixia 1992. An analytical solution and application for coupled flow problem under plane axial symmetry condition (in Chinese). *Mechanics and Practice.* 16(6):44-47.

M. A. Biot 1941. General theory of three dimensional consolidation. *J. Appl. Phy.* 12:155-165.

Numerical study on thermo-hydro-mechanical coupling analysis in rock with variable properties induced by temperature

H.J.Ahn
Institute of Technology, Kolon Engineering and Construction Co., Ltd, Korea

H.K.Lee
Department of Mineral and Petroleum Engineering, Seoul National University, Korea

ASTRACT : In this study, thermo-hydro-mechanical coupling analysis considering both general interaction and the variation of rock properties induced by temperature variation was performed for the circular shaft of the diameter of 2m in deep location, which the circular shaft is excavated in the rock mass under hydrostatic stress of 5MPa. In this case, Thermal expansion induced tensile stress vicinity of the circular tunnel. When properties were given as a function of temperature, thermal expansion was increased, but tensile stress zone around wall was little larger.

1. INTRODUCTION

Generally, thermo-hydro-mechanical(THM) interaction in the rock mass is shown in Fig. 1. Interaction separates into the direct interaction that consists effective stress and thermal expansion, and the indirect interaction which consists variations of properties. THM coupling analysis is developed rapidly since 1980's. The direct interaction began from effective stress at the soil mechanics at 1940's, Corapcioglu et al. considered heat effect at 1980s. Governing equation of direct THM interaction in the porous media is systematized by Lewis & Schrefler (1987). Later research, interaction model considering joint effect of rock mass is developed by Witherspoon(1981). Next, about indirect interaction, Aboustit(1985) studied THM analysis considering variation of water viscosity at the Geothermal energy development research. Hart (1986) researched nonlinear THM analysis with theoretical equations. In the most research, indirect interaction is ignored or depend on theoretical equations.

In this study, THM coupling analysis included indirect interaction effects obtained by many experiential researches. Wai's equations(1982) on variety of rock properties by temperature, Wei's equation of permeability(1992) and Bersetto's equation of viscosity(1981) are used for indirection interaction equations and thermo-mechanical aspect is pointed up.

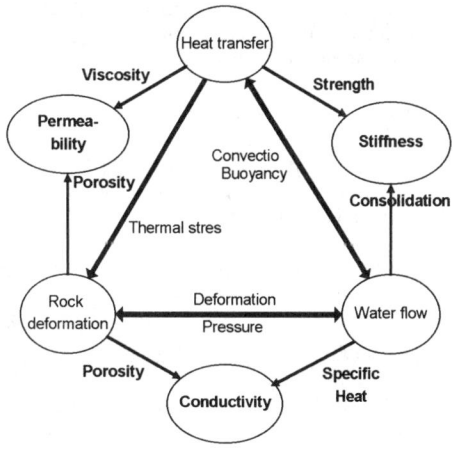

Fig. 1 Relationship of Thermo-Hydro-Mechanical interaction

2. GOVERNING EQUATION AND VARIATION OF ROCK PROPERTIES

2.1 *Governing equation of THM coupling analysis*

Governing equations used in this study are formulated by Lewis & Schefler(1987). The equations are below:

$$K\frac{d\overline{u}}{dt} + L\frac{d\overline{p}}{dt} + TU\frac{d\overline{T}}{dt} = \frac{df}{dt} \quad 1)$$

$$L^T \frac{d\bar{u}}{dt} + H\bar{p} + S\frac{d\bar{p}}{dt} + TP\frac{d\bar{T}}{dt} = \bar{f} \quad 2)$$

$$TL\frac{d\bar{p}}{dt} + TR\bar{T} + TS\frac{d\bar{T}}{dt} = TG \quad 3)$$

At the above equations detailed description is omitted. Governing equations are transformed into matrix form which has four parameters of displacement(x,y), water pressure and temperature.

$$\begin{bmatrix} K & L & TU \\ L^T & S & TP \\ 0 & TL & TS \end{bmatrix} \frac{d}{dt} \begin{Bmatrix} \bar{u} \\ \bar{p} \\ \bar{T} \end{Bmatrix} + \begin{bmatrix} 0 & 0 & 0 \\ 0 & H & 0 \\ 0 & 0 & TR \end{bmatrix} \begin{Bmatrix} \bar{u} \\ \bar{p} \\ \bar{T} \end{Bmatrix}$$

$$= \begin{Bmatrix} \frac{\partial f}{\partial t} + c \\ \bar{f} \\ TG \end{Bmatrix} \quad 4)$$

2.2 Equations of rock properties

Wai(1982) researched variations of rock properties as elastic moduli, possion's ratio, thermal expansion ratio, thermal diffusivity of Ontario granitic Gneiss at 20-200℃ and obtained linear equations. Majors are below;

(1) equation of elastic moduli

For heating : $E_\psi = E_0$ $0 \le \psi \le 80$
$E_\psi = E_0[1 - (\psi - 80) \times 0.0025]$ $80 \le \psi \le 180$
ψ is T(true temperature) - 20

For cooling : $E_\psi = E_0$ $0 \le \psi_{max} \le 80$
$E_\psi = E_0[k_1 + (\psi_{max} - \psi) \times 0.0015]$
$\psi_{max} > 80$, $\psi \le 80$

$E_\psi = E_0[k_1 + (\psi_{max} - \psi) \times 0.0015]$
$80 \le \psi \le \psi_{max}$

about $k_1 = [1 - (\psi_{max} - 80) \times 0.0025]$,
ψ_{max} : maximum temperature ever been.

(2) equation of thermal expansion coefficient
$\alpha_\psi = (6 + \psi/20) \times 10^{-6} / ℃$

2.3 Equation of hydraulic conductivity

(1) equation of permeability
$k_h = k_o \cdot e^{\beta^e \Delta e_v}$
$k_v = k_o \cdot e^{\beta^e \Delta e_h}$

(2) equation of water viscosity
$\mu = \frac{42.7}{(T - 250)}$ $273K \le T \le 333K$
$\mu = \frac{27.9}{(T - 278.8)}$ $333K \le T$

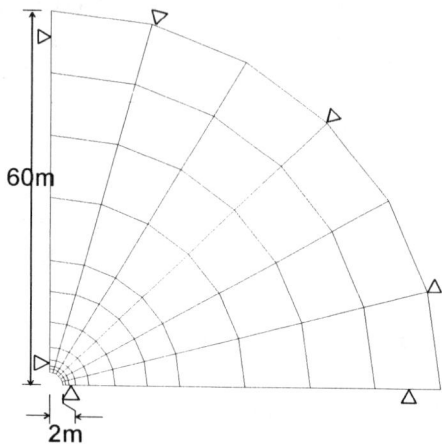

Fig. 2 Finite element mesh of shaft model. Hydraulic pressure boundary is Fixed.

3. RESULT OF THM COUPLING ANALYSIS

3.1 Method and model

In this analysis, TRICOUPL that is modified to consider variation of rock properties from PLASCON of Lewis(1987) is used. This program uses 8 nodes' element and works at elastic media.

The condition of analysis model is circular shaft excavated of diameter of 2m in isometric rock mass subjected hydrostatic stress of 5 Mpa, which was used by Kelsall(1984) and Wei (1992).

Element mesh of this model is shown in Fig. 2 and at this mesh, outer boundary is located out of 60m from the center of shaft.
Properties of rock at 20℃ are shown below.

Properties	E(MPa)	ν
Value	7500	0.25
K_s(/MPa)	β_s(/℃)	β_w(/℃)
0.23×10^{-3}	0.6×10^{-5}	0.21×10^{-3}
β	K(m/day)	K_w(/MPa)
2000	3.11×10^{-5}	0.66×10^{-3}
$(\rho c)_s$(J/m3K)	$(\rho c)_w$(J/m3K)	λ(W/m℃)
2.25×10^6	4.19×10^6	3.2

3.2 Result

THM coupling effects by 200℃ applied to shaft wall was analyzed after the analysis of hydro-mechanical coupling effects by excavation of

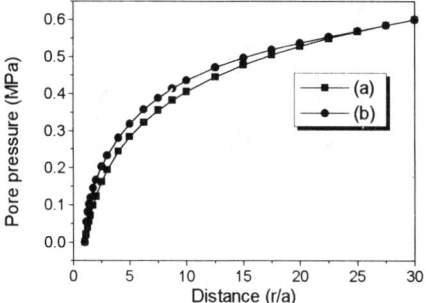

Fig. 3 Pore pressure distribution in H-M coupling analysis; (a) H-M coupling, (b) No coupling.

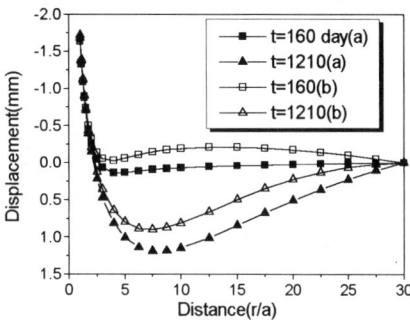

Fig. 5 radial displacement distribution at several time; (a) properties variable, (b) properties const.

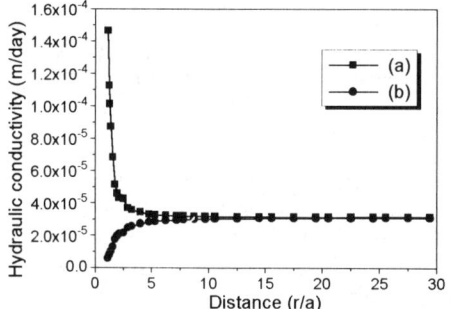

Fig. 4 Hydraulic conductivity distribution in H-M coupling analysis; (a) tangential, (b) radial.

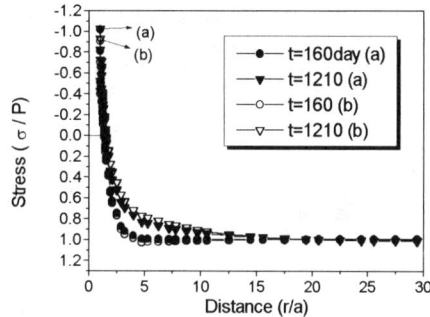

Fig 6. Radial stress distribution at several time; (a) properties variable, (b) properties const.

saturated rock mass. Shown in Fig.3, water pressure distribution changes as the permeability varies with rock deformation. Radial permeability in the vicinity of the tunnel is reduced to about 14% of initial value in Fig. 4. Therefore, water flow is reduced and gradient of water pressure is higher than the case of normal flow analysis including no coupling effect. This result is same as Wei's(1992).

The result of THM coupling analysis when wall of shaft was heated to 200℃ after 110 days from excavation is as shown in Fig. 5 - 8.

Firstly, when the rock properties are constant, deformation at vicinity of shaft is outward expansion in Fig. 5. Therefore tangential stress at the wall of the shaft is reduced from 2 times of initial stress to 1.4 times and radial stress change from a little compressive stress to tensile stress, which is 0.9 times of initial stress in Fig. 6(p is the initial stress of 5 MPa).

Secondly, hydraulic conductivity in the vicinity of shaft is so larger to be about 4 times of initial value in Fig. 7. It is because the permeability increases as displacement increase and water viscosity decrease as temperature increase. This means that water flow is activated and water pressure distribution at the middle point between boundary and shaft become lower as time pass in Fig. 8.

Thirdly, when the rock properties are variable, rock deformation is increased around shaft mainly by elastic modulus reduction in Fig. 5,

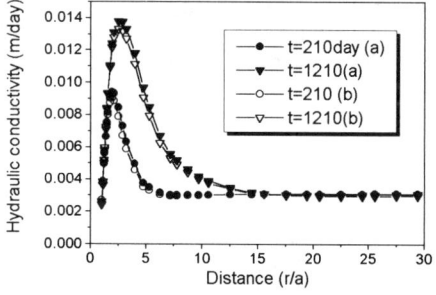

Fig. 7 Radial hydraulic conductivity distribution at several times;(a) properties variable, (b)properties const.

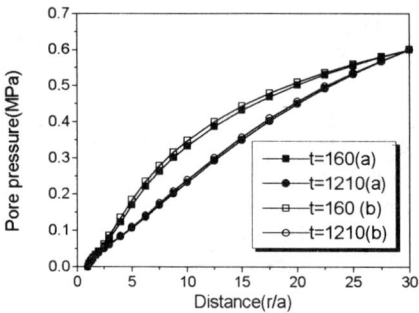

Fig. 8 Pore pressure distribution at several time; (a) properties variable, (b) properties const.

but rock stress change is a little as compared with the displacement in Fig. 6. It's the reason why elastic modulus reduction does not affect stress, but only the displacement, and major properties affecting to stress change is thermal expansion coefficient. Thermal expansion coefficient and stress around shaft increase, as temperature increases.

4. CONCLUSION

The important result of this study is below;
1) Stress redistribution and displacement by shaft excavation change the permeability and affect to water flow. Water pressure distribution increases as the permeability decreases
2) Thermal expansion at the shaft wall induces reduction of compressive stress and produce tensile stress zone around the shaft wall to radial direction.
3) The permeability increases and viscosity decrease as heat transfer progresses. Therefore, hydraulic conductivity increases and pressure distribution decrease.
4) When the properties is variable, displacement by thermal expansion increases as elastic modulus decrease and thermal expansion ratio increase. Stress increases a little because stress distribution is independent on the change of elastic modulus.

REFERENCE

Aboustit B. L., Advani S.H. and Lee J.K.,1985 "Variational principles and finite element sumulations for thermo-elastic consolidation", *Int. J. Num. and Anal. meth. in geomech.* vol. 9, p 49-69.

Borsetto M., Carradori G., and Ribacchi R.,(1981), "Coupled Seepage, Heat Transfer, and Stress Analysis with Application to Geothermal Problems", *Num. Meth. in Heat Transfer*, pp. 233-259.

Hart R.D., John St. C. M., 1986, "A Fully Coupled Thermal-Hydraulic-Mechanical model for non-linear Geologic Systems", *Int. J. Rock. Mech. Min. Sci. & Geomech.* pp. 803-808.

Lewis R.W., Schrefler B.A., 1987, *The Finite Element Method in the Deformation and Consolidation of Porous Media*, John Wiley & Sons.

Sato M., Kamemura K., 1985, "Rock mass behaviour considering water flow and heat transfer", *Fifth Int. Conf. on Num. Meth. in Geolmech.*,Nagoya, pp. 703-710.

Wai R.S.C., Lo K.Y., Rowe R.K., 1981, "Thermal Stress Analysis in Rocks with Nonlinear Properties" , *Int. J. Rock. Mech. Min. Sci. & Geomech. Abstr.* Vol.19, pp.211-220.

Wang J.S.Y., Tsang C.F., Cook N.G.W., and Witherspoon P.A., 1981, "A Study of Regional Temperature and Thermohyrologic Effects of an Underground Repository for Nuclear Wastes in Hard Rock", *J. Geophys. Research.* Vol. 86, pp. 3755-3770.

Wei Lingli, 1992, "Numerical Studies of the Hydro-Mechanical Behaviour of Jointed Rocks", *doctor thesis*, Imperial College.

Analysis of thermo-mechanical behavior of underground cold storage cavern by monitoring and numerical prediction

Yeonjun Park
Department of Civil Engineering, University of Suwon, Korea

Joong-Ho Synn & Chan Park
Korea Institute of Geology, Mining and Materials, Taejon, Korea

Ho-Yeong Kim
Sunkyong Engineering and Construction Ltd, Seoul, Korea

ABSTRACT: The underground, as a cold storage facility, has merits of excellent insulation against external heat/moisture which gives energy saving in long-term stage and easiness of site selection considering efficient circulation. The pilot plant of underground cold food storage is constructed as a R & D scale and technical research has been going on. The mechanical behavior and heat flow around the chilled and cold storage rooms are monitored by installation of multi-point borehole extensometer, stressmeter, thermister string and piezometer. Thermo-mechanical analysis using FLAC is also carried out. Thermal conductivity of rock mass is evaluated and the long-term stabilization of heat flow is predicted by comparison with field monitoring data. This research can contribute to the refined design and operation of commercial sized underground cold storage.

1 INTRODUCTION

In recent years, the underground has been utilized for various purposes. One of the most effective and practical use is the cold storage of food or marine products. As a facility for cold storage, underground has merits of excellent insulation against external heat/moisture which gives energy saving in long-term stage, large heat capacity which is favorable for the control of temperature and moisture, and easiness of site selection considering efficient transportation and circulation of goods.

Underground cold storage facilities have been constructed and operated successfully for nearly 20 years in Norway, Sweden, US and recently Japan. Many researches on underground storage of food have been also conducted in Korea. A pilot plant of underground cold food storage is constructed as a R & D scale in 1995 and thereafter an underground storage terminal for commercial purpose is being constructed in Korea.

During the operation of chilled and cold rooms in the pilot plant, technical researches have been conducted on thermo-mechanical characteristics of rock mass under cooling. The mechanical behavior and heat flow around the chilled and cold storage rooms are monitored by the installed instruments. The thermo-physical properties of the bedrock are evaluated by the laboratory tests. The thermo-mechanical analysis using FLAC is carried out, and the long-term stabilization of the heat flow is predicted by comparison with field monitoring data.

2 GEOLOGY AND LAYOUT OF UNDERGROUND COLD STORAGE CAVERN

The purpose of this project is to achieve the technology and the experience in utilization of underground space, and to setup an underground environmental control standard for various agritural, marine products, and to verify thermal response of the rock mass around cold storage cavern.

The rock type at cold storage site is medium to coarse grained Mesozoic granite. There is one set of major joint along with some random joints. Most of the major joints have steep inclination and run across the tunnel axis. They are estimated not to have big influence on the stability of the cavern. The rock mass is classified as 'Fair' in portal region and 'Good' in deep region by RMR standard. It took 2 months to complete excavation by the conventional drilling and blasting. Main supports are shotcrete and 2.4 m long Swellex bolts. Wire mesh is used where joints are filled with thick clay seam. The dimensions and specifications of the cold storage cavern are presented in Table 1.

The refrigeration and ventilation systems are installed as shown in Fig. 1. Chilled and cold rooms are refrigerated by independent system. The compressor of condensing unit has 10 HP for cold room and 3 HP for chilled room. The condensing is done by air-cooled system. The defrosting is done by heater defrosting system and is controlled by timer or auto-control unit. Ventilation systems are provided for fresh air supply and for cooling of the condensing units. Air supply ducts are installed into the chilled storage room to ventilate the smell from the stored products.

Table 1. Dimensions and specifications of the cold storage cavern

Room	Dimension, m (W x H x L)	Specification
Laboratory	7 x 4 x 12	Room Temp.
Chilled Room	4.5 x 3 x 8	0 °C 60~90 % RH
Cold Room	4.5 x 3 x 10	-25 °C
Mechanical Room	4 x 3 x 5	Room Temp.
Connection Tunnel	3 x 3 x 45	Room Temp.
Total Area	350 m^2	
Total Volume	1038 m^3	

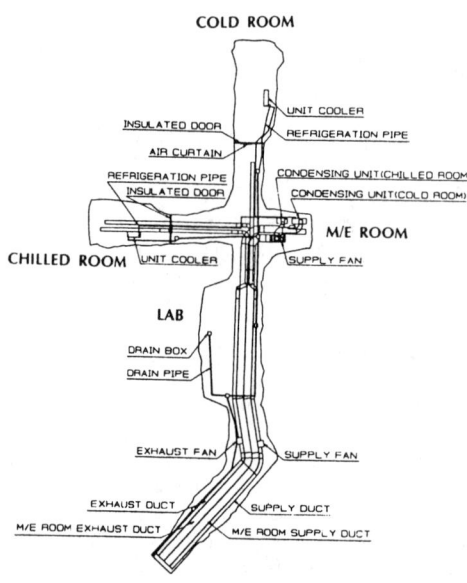

Fig. 1. The refrigeration and ventilation systems

Strip door curtain and air curtain are installed at the chilled and cold room door to prevent heat flow.

3 THERMO-PHYSICAL PROPERTIES OF THE BEDROCK

Thermo-physical properties of granite which is the bedrock are evaluated in laboratory scale. Thermal conductivity is measured using transient hot-wire method in the range from -40 °C to room temperature. Specific heat is measured using DSC(Differential Scanning Calorimeter) in the range from 30 °C to 100 °C, and thermal expansion coefficient is measured using TMA (Thermal Mechanical Analyzer) in the range from 20 °C to 100 °C. Measured thermo-physical properties are presented in Table 2.

Uniaxial compression and Brazilian tension tests are carried out to estimate the temperature dependency of strengths. Temperature range for the test is from room temperature to -60 °C. The result is summarized in Table 3. The uniaxial compressive strength under the temperature below freezing point is increased by about 35 % compared to that under room temperature condition. The tensile strength is reduced a little at freezing point and then increased again as temperature is lowered to -60 °C. This increase of strength seems to be caused by the increase in bonding strength between grains, and it can be said that freezing of rock mass does not cause instability of cold storage cavern.

Table 2. Thermo-physical properties of granite

Temperature range °C	Thermal conductivity W/m.K
- 40	2.69 ± 0.16
- 20	2.71 ± 0.16
- 10	2.71 ± 0.15
- 1	2.63 ± 0.05
27	2.52 ± 0.17
Temperature range °C	Specific heat J/g.°C
35 ~ 100	0.76 ~ 0.85
Temperature range °C	Thermal expansion coefficient, 10^{-6}/K
20 ~ 50	6.65 ± 0.37
50 ~ 80	8.30 ± 0.38
80 ~ 100	10.63 ± 0.46

Table 3. Strength change of granite under varied temperature condition

Temperature condition °C	Uniaxial compressive strength, MPa	Brazilian tensile strength MPa
25	116 ± 5.1	7.7 ± 0.48
0	155 ± 2.7	7.4 ± 0.58
-20	155 ± 8.6	8.0 ± 0.91
-40	151 ± 11.6	8.5 ± 0.19
-60	157 ± 10.3	9.4 ± 1.19

4 FIELD MEASUREMENTS

4.1 Monitoring system

For monitoring of rock behavior and heat flow within the rock mass around the chilled and cold room, multi-point borehole extensometers, strain monitors, piezometers, and thermister strings are installed at several positions as shown in Fig. 2. The maximum measuring depth of extensometer, piezometer and thermister string is 8m, 6m and 8m from the wall, respectively. Extensometers and piezometers have six measuring points, and thermister string has ten measuring points. As well as the installation of these measuring system from inside of the cavern, two strain monitors and three extensometers are installed within vertical boreholes from the surface, which are located above the laboratory, chilled room and cold room, respectively.

4.2 Temperature distribution within rock mass

The operation of the cold storage cavern was started at May 20, 1996 for the cold room and at September 2, 1996 for the chilled room. The target temperature is set as 0 °C for the chilled room and -25 °C for the cold room. According to the monitoring data, the room temperature reached the target temperature in a few weeks for the chilled room and in 4 months for the cold room as shown in Fig. 3. Some small peaks in the graph were caused by unexpected power failure.

Temperature variation measured by thermister string installed at the right side wall and roof of the chilled room is presented in Fig. 4 and Fig. 5, respectively. Each distance written in the graph is the measuring depth from the wall.

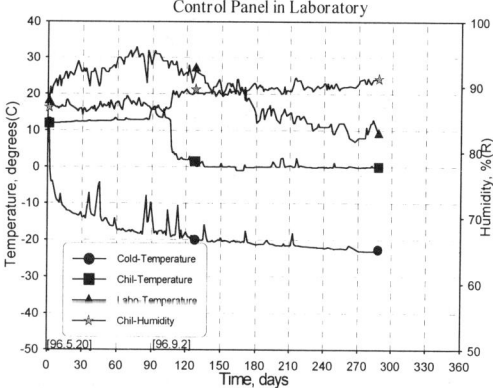

Fig. 3. Variation of Room temperature and humidity after cooling started

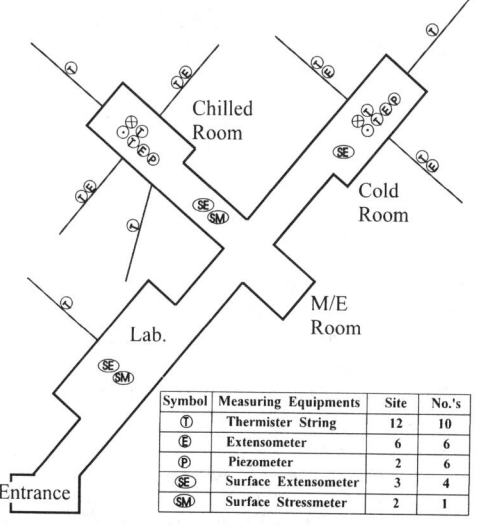

Fig. 2. Installation of monitoring system

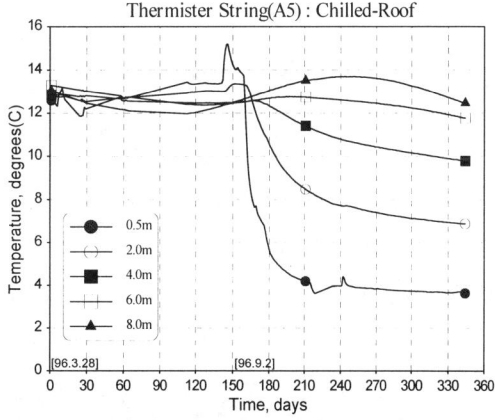

Fig. 4. Temperature distribution within rock mass at the roof of the chilled room

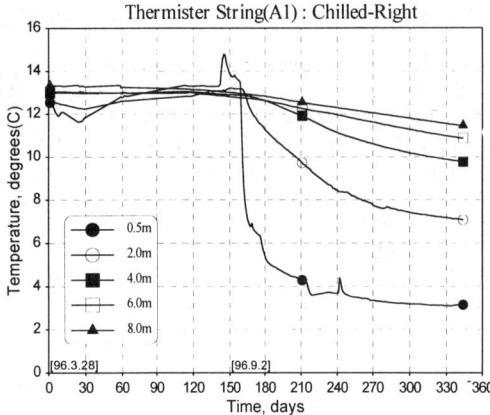

Fig. 5. Temperature distribution within rock mass at the right side wall of the chilled room

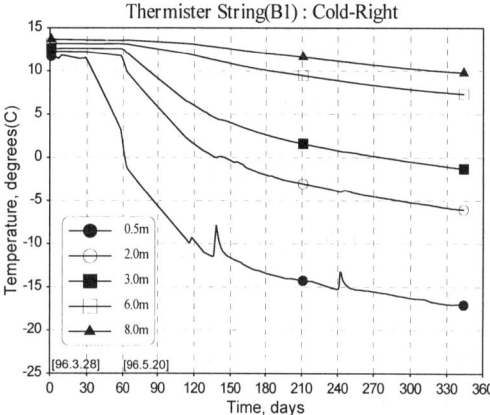

Fig. 6. Temperature distribution within rock mass at the right side wall of the cold room

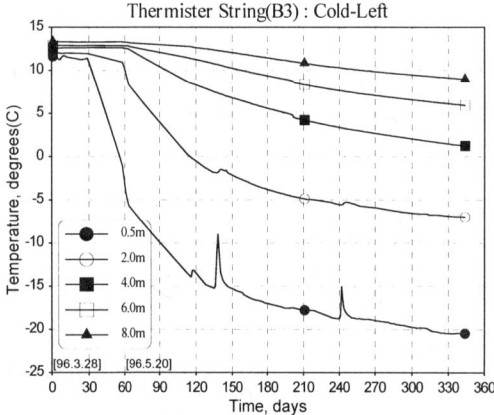

Fig. 7. Temperature distribution within rock mass at the left side wall of the cold room

The temperature at 0.5 m depth in Fig. 4 is dropped to 4 °C and almost stabilized in 7 weeks after cooling started. The temperature at 8 m depth in Fig. 4 shows some influence by outside atmosphere because the distance from this point to the surface is less than 10 m. As shown in Fig. 5, the temperature at the right side of the chilled room, where the cold room is located nearby, is lower by 1~2 °C than that at the left side of the chilled room.

The temperature distribution within rock mass at the left and the right side of the cold room is shown in Fig. 6 and Fig. 7. The temperature at 0.5 m depth went down to -17 °C at the right side and -20 °C at the left side after 8 months. The temperature at the left side of the cold room, where the chilled room is located in this side, is lower by 2~3 °C than that at the right side of the cold room. At the region 8 m apart from wall, temperature dropped by 3~5 °C showing 8~10 °C in about 9 months.

4.3 *Electric power consumption*

The capacity of the evaporator fan and defrost heater in unit cooler for the chilled room is 0.4 KW and 5.5 KW, respectively. And that for the cold room is 0.8 KW and 10.1 KW, respectively. The refrigeration system turns on and off automatically when inside temperature reaches the target temperature. The compressor power of condensing unit for the chilled and cold room is 10 HP and 3 HP, respectively. The defrosting duration is controlled by timer and is now set as 4~6 times per day. The power consumption of each unit has been measured with wattmeters as depicted in Fig. 8. The average power consumption in cooling is measured as 21.9 KW/day for the cold room and 20.4 KW/day for the chilled room, and that in defrosting is measured as 17.5 KW/day for the cold room and 3.4 KW/day for the chilled room. The power consumption rate in cooling the chilled room shows a decreasing tendency compared to that in cooling the cold room. This tendency can be related with the fact that the chilled room temperature reaches the target one more rapidly than the cold room temperature. The difference of power consumption rate for defrosting between the chilled room and the cold room is cause by the difference of timer setting between the two rooms.

Now we are newly monitoring the power consumption of compressor in condensing unit, and so more detailed analysis for the power consumption of the cold storage system will be made together with this monitoring data.

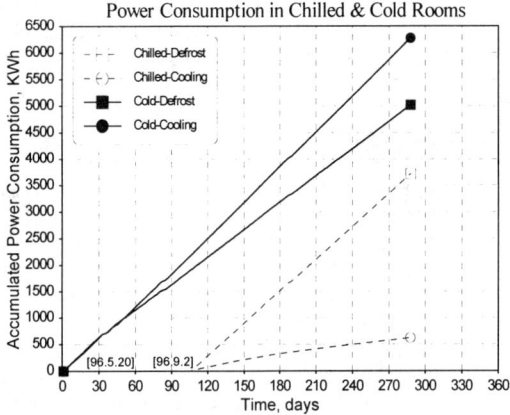

Fig. 8. The accumulated power consumption in cooling and defrosting

5 THERMO-MECHANICAL ANALYSIS

One of the major factors for the design of underground cold storage is estimation of the capacity of the cooling system. Heat flow and thermo-mechanical analysis for this pilot plant of underground cold storage is conducted using FLAC.

5.1 Thermal boundary condition

In analysis the initial temperature of rock mass is set as 12 °C which is the annual average temperature of this region. Annual temperature variation in this region, which shows the range of -1 °C ~ 29 °C, is given as the outer thermal boundary condition. Other thermal properties of rock mass is as follows; Thermal conductivity is 3.1 W/m.°C, specific heat is 800 J/kg.°C, thermal conductivity of ground surface and cavern wall is 9.75 W/m^2.°C and 6.80 W/m^2.°C, respectively. The initial heat flux for cooling the rooms is set as 2.5 KW for the chilled room and 7.5 KW for the cold room considering power capacity of the cooling system. The operating temperature of the chilled and cold room is 0 °C and -25 °C, respectively. A servo-control routine is used in analysis for the control of each room temperature. That is, temperature in each room is checked at every calculation step, and heat flux condition is defined as 'on' or 'off' according as the room temperature reaches the operating temperature or not. The FLAC code used here is 2-dimensional and the analysis is done for each case of the horizontal and vertical section model of the storage cavern.

5.2 Temperature distribution

The analysis is conducted for the long duration of 10 years. Temperature distribution within the rock mass around the cold storage after 1 year and 10 years is shown in Fig. 9 and Fig. 10. The room temperature reached the target value in 2 months later. In 7 months later there was some temperature interference in the region between the chilled and cold room. In 10 years the cooling effect of cold room extended to the distance of about 40 m around the cavern.

5.3 Effect of thermal stress

In thermo-mechanical analysis for the vertical section model including the cold room, initial stress condition is set as K=0.5, where K is ratio of horizontal to vertical initial stress. By cooling the cold room, temperature difference in rock mass around the cold room is about 37 °C and this difference induces thermal stress and deformation. Tensile stress by contraction of rock under cooling is induced around the cold room. This tensile stress occurred especially at the cooling boundary. As the cooling range is expanded with time, the plastic zone by tensile failure is also expanded and tensile cracks are induced in the roof and floor of the cold room. In this analysis, the joint is not considered. The joint which actually exists in rock mass and supports such as rockbolt can have roles on restraining the thermal cracking. There may be, however, a possibility of joint failure by opening and slippage. This kind of rock behavior can be checked by the monitoring system installed in rock mass around the cavern.

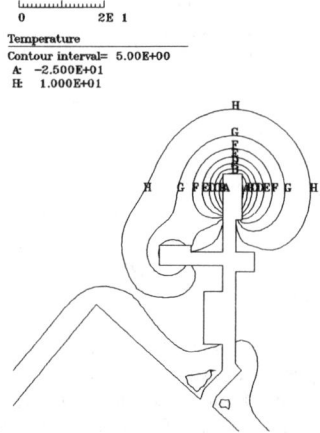

Fig. 9. Temperature distribution after 1 year

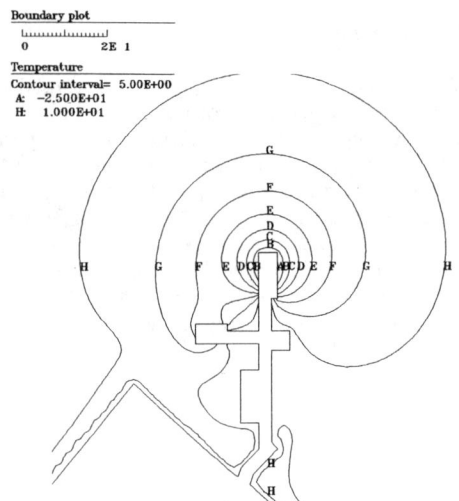

Fig. 10. Temperature distribution after 10 years

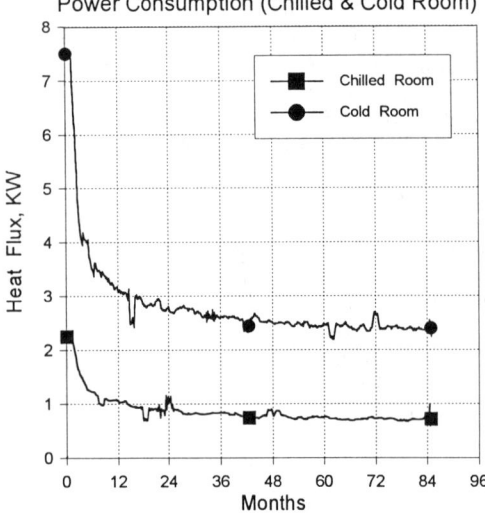

Fig. 11. Heat flux calculated according to the operation of cold storage

5.4 Heat capacity calculation

The heat capacity consumed in operating the cold storage is calculated by analysis as shown in Fig. 11. Analysis shows that the temperature in the chilled and cold room reached the target value in one month and thereafter power consumption rapidly reduced showing half of the initial value in 6 months. In about 2 years after the refrigeration started, the power consumption becomes almost constant. In this analysis efficiency of cooling system is assumed as 100 % and the heat supplied by defrosting is not considered. More detailed analysis will be conducted considering these factors precisely and actual operation data together.

6 CONCLUSION

A pilot plant of the underground cold storage is constructed as an R & D project. The thermo-physical properties of rock are evaluated, and thermo-mechanical behavior of rock mass around cold storage are analyzed by field monitoring and numerical analysis.

Strengths of granite below freezing point were increased. This shows that freezing of rock does not cause instability of cold storage cavern. According to field monitoring, the cold room temperature reached the operating temperature, -25 °C, in 4 months after cooling started. The temperature at 8 m depth from the wall of cold room is 8~10 °C with temperature drop of 3~5 °C in about 9 months.

Thermo-mechanical analysis shows that there is some temperature interference in the region between the chilled and cold room in 7 months, and rock cooling extended to about 40 m around the cold room in 10 years.

More precise analysis will be made with continuous operation and field monitoring, and results can contribute to the refined design and operation of commercial sized underground cold storage.

REFERENCES

Bollingmo, P. 1993. Cold storage plant in rock cavern. *Norwegian Underground Storage*:91- 92.

Broch, E., P. Frivik & M. Dorum 1994. Storing of food and drinking water in rock caverns in Norway. *'94 Int'l Symp. for Grain Elevator and Underground Food Storage*:283-317.

Dorum, M. 1977. Energy economy in Rock Stores. *Rockstore77* 1:73-77.

Kim, H.Y., Y.J. Park & K.C. Nam 1994. Feasibility study for underground frozen meat storage. *'94 Int'l Symp. for Grain Elevator and Underground Food Storage*:319-351.

Park, Y.J., et al. 1996. Construction of pilot plant as a R & D program for underground food storage. *Korea-Japan Joint Symp. on Rock Eng.*:363-368.

Tressler, D.K., & W.B. Van Arsdel 1986. *Commercial food freezing operations fresh foods in the freezing preservation of foods*. AVI. Publ. Co. 3:347p.

Modelling techniques for safety evaluation: Monitoring and interpretation

The automation of MPBX monitoring and the numerical analysis of MPBX displacement

Yong-Bok Jung & Hi-Keun Lee
Department of Mineral and Petroleum Engineering, Seoul National University, Korea

Hyun-Key Jung & So-Keul Chung
Korea Institute of Geology, Mining and Materials (KIGAM), Taejon, Korea

Dong-Hyun Kim
Samsung Heavy Industries Co., Ltd, Korea

ABSTRACT : It is difficult to obtain accurate data, which play an important role in numerical analyses. An automatic MPBX (Multi-Position Borehole Extensometer) monitoring system was developed in this study as a way of obtaining accurate data. This system can be applied to analyses of measured data which include evaluation of the stability, efficiency of construction and the validity of numerical computations. First, the MPBX among field measurement tools was considered. Also, three-dimensional finite element analyses were carried out to verify the influence of geological conditions and excavation sequence on the MPBX displacement.

1 INTRODUCTION

The stability of underground openings is generally affected by the geological conditions of rock mass. The behavior of rock mass is dominated by the mechanical properties and the spatial distribution of discontinuities in rock mass. But those geological conditions cannot be completely determined before construction. Therefore field measurements must be performed to check stability of the underground structure and validity of design and construction (Kim et al.,1991; Dunnicliff,1993; Hanna,1985; Kaiser,1993).

Nowadays, computer-aided numerical analyses are very popular. Given that the accuracy of input data plays an important role in numerical analyses, it is really important to decide how we should get input data and what data we should use from field measurements. An automation of field instrumentation is now being developed to help those tasks. This system can be applied to the evaluation of the stability and the efficiency of construction.

In this study, only the MPBX was considered among field measurement tools. Then three-dimensional finite element analyses were performed to check the influence of geological conditions and excavation sequence on the MPBX displacement.

2 AUTOMATION OF MPBX MONITORING

Displacement is the easiest and the most popular measuring item in underground constructions (Dunnicliff,1993; Brown,1981). Among various types of displacement measuring instruments, MPBX was selected. Developed system can accommodate other instruments which can translate physical signal to electrical signal.

2.1 Specification of the system

The layout of the system is illustrated in Figure 1. This system mainly consists of power supply, main CPU board, I/O, control unit, and storage unit. This version can acquire data from 4 units, which can have maximum 8 channels. Units can be easily added to the system. Sampling rate can be adjusted by S/W and the acquired data are stored in the internal CMOS SRAM and floppy disks for backup.

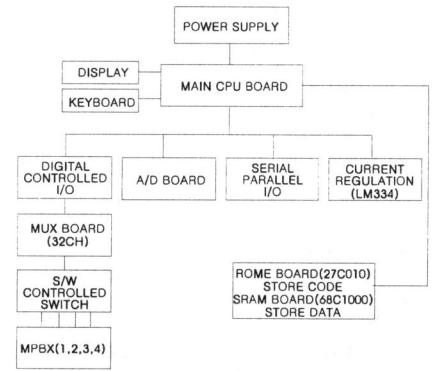

Figure 1. Block diagram of automatic MPBX monitoring system

2.2 Laboratory test

The MPBX used in this system has 4 channels. Internally, the MPBX readout unit measures resistance and converts it into displacement. The relation between resistance and displacement was obtained from a laboratory test. From the test, it has been found that the resolution of the MPBX is 100mv=1mm and the correlation coefficient is 0.99238(Figure 2).

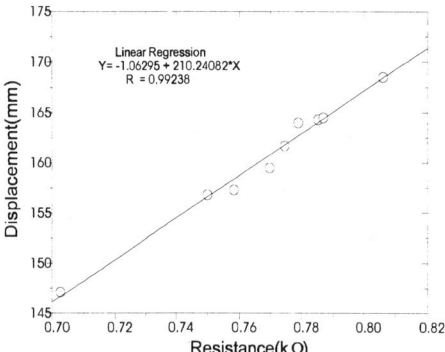

Figure 2. Correlation chart between resistance and displacement

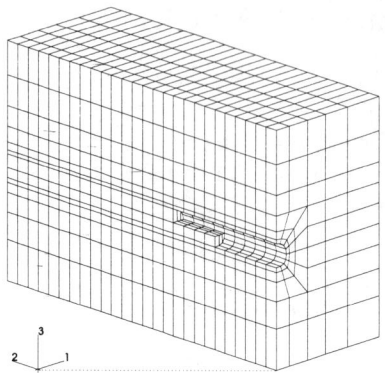

Figure 3. Finite element mesh for numerical analyses

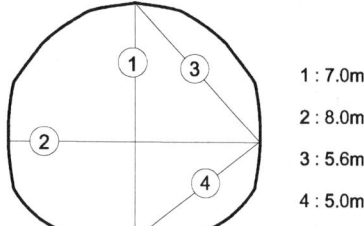

Figure 4. Configuration of convergence measurements

2.3 Field test

A field test was performed at a tunnel on Jinju Highway. Two MPBX were installed near the face. One was damaged by blasting, and the other operated normally. The test was performed for two days. No significant errors were found in measurement when the data acquired by the system and the results from the readout unit were carefully compared. The test proved that the system performed well. However, more field tests and application test are required to prove applicability of the system.

3 THREE-DIMENSIONAL ANALYSES OF MPBX DISPLACEMENT

Three-dimensional finite element analyses were carried out to check the influence of the geological condition and the excavation sequence on the MPBX displacement. Three cases were considered and full face excavation and bench cut were simulated in each case. A commercial finite element analysis S/W, ABAQUS5.4-1 was used in this study(Hibitt et al.,1992). Finite element mesh is illustrated in Figure 3. The tunnel width and the height are 8m and 7m, respectively. Total number of nodes is 2268 and that of elements is 1742. The dimension of model is 40x104x70m.

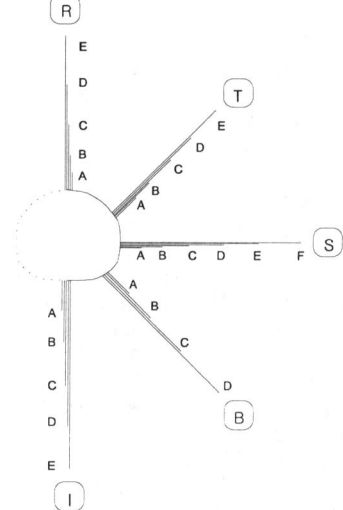

Figure 5. Configuration of MPBX measurements

Measuring positions of convergence and the MPBX are specified in Figure 4 and 5. The location and the distance are summarized in Table 1. Linear elastic behavior is assumed for the sake of simplicity and generalization. In bench cut, upper bench and lower bench advance at the rate of 0.5D. The distance between upper and lower bench is kept during analysis.

Table 1. Depth of each measuring position (unit:m)

	A	B	C	D	E	F
Roof(R)	1.1	3.4	8.0	14.0	23.0	-
Top-right(T)	1.8	5.6	13.0	22.2	34.6	-
Side(S)	1.1	3.4	8.0	15.0	23.6	36.0
Bottom-right(B)	2.0	5.9	13.8	24.5	-	-
Invert(I)	1.4	4.3	10.0	18.0	30.0	-

3.1 Case I - same properties

All elements have same properties in Case I. Young's modulus is 1.08E9Pa and Poisson's ratio is 0.25 and unit weight is 2.6g/cm^3. Excavation is simulated by removing tunnel element group.

3.1.1 Full face excavation

Convergence curves are shown in Figure 6. X axis is the ratio of the distance between tunnel face and measuring position to the tunnel width. Y axis is the normalized displacement(u/u_{max}).

Preceding displacement ratio(PDR) or characteristic displacement ratio(CDR) is summarized in Table 2 (Moon et al.,1995; Barlow et al.,1987).

MPBX displacement is shown in Figure 7. In Figure 7, displacement decreases at X/D=1 because of the displacement components in the direction of face advance. Displacement in the direction of face advance is shown in Figure 8. The reason for the decrease is illustrated in Figure 9. In two-dimensional analysis, displacement in face advance direction cannot be considered because of plane strain assumption. In this three-dimensional analysis, however, it is found that the MPBX displacement was affected by the displacement components in face advance direction.

MBPX displacement vs. anchoring depth is shown in Figure 7(b). The curve in Figure 7(b) is smooth since all areas have same properties.

From the above results, it is reasonable that measuring frequency must be increased when face passes through the measuring location. One measurement a day is not sufficient within a range -1.0<X/D<1.0. If automatic measuring system is introduced, sampling frequency can be easily increased and measuring task can be easily accomplished.

Table 2. Preceding displacement ratio(full face excavation)

Measuring points	1	2	3	4
Preceding displacement ratio(%)	27.9	27.7	29.0	26.9

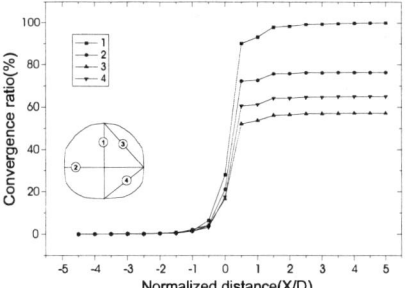

Figure 6. Convergence curve-Case I(full face excavation)

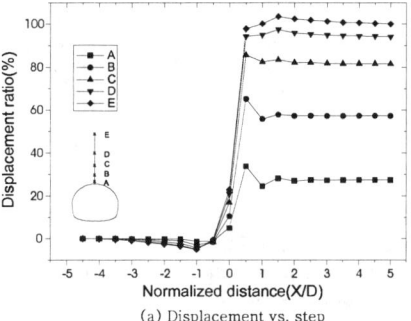

(a) Displacement vs. step

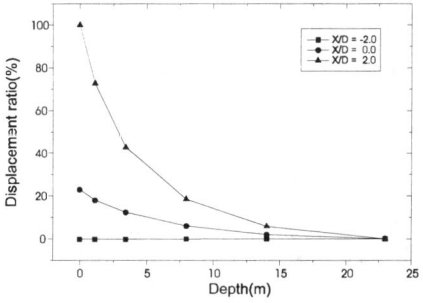

(b) Displacement vs. depth

Figure 7. MPBX displacement curve-Case I(full face excavation)

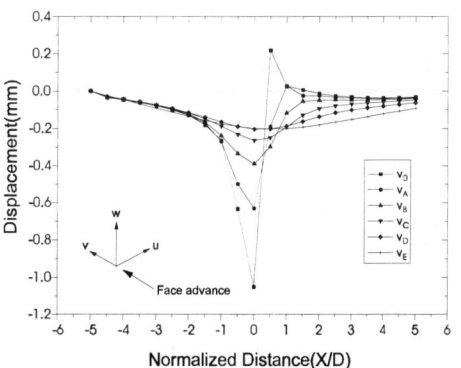

Figure 8. Displacement components in the direction of face advance

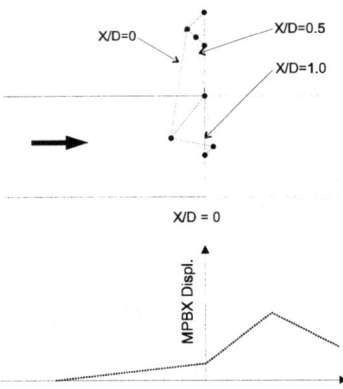

Figure 9. Behavior of two points with respect to the face advance

3.1.2 Bench cut

In bench cut, lower bench follows upper bench by 2D. In this case, the behavior of underground structure differs from that of full face excavation because lower bench plays a role of support.

Convergence curves in bench cut are shown in Figure 10. Further displacement occurred at X/D=2 for removal of lower bench. The deformation of measuring position 2 and 4 is more restricted by lower bench than 1 and 3. Thus there is another increment in displacement after removal of lower bench. While the displacement of measuring position 1 and 3 increases after removal of upper bench. The PDR in case of bench cut is summarized in Table 3.

For MPBX displacement, behavior of roof(R) is similar to that of full face excavation. But in case of top-right(T), side(S), bottom-right(B), and invert, lower bench plays a significant role. This is shown in Figure 11. For full face excavation, displacement converges after X/D=1.0. But displacement increases at X/D=2.0 because of excavation of lower bench.

3.2 Case II - different properties with depth

In general, rock properties vary with depth. To simulate this situation, different properties are assigned to each layer(Figure 12). The properties are shown in Table 4.

Table 3. Preceding displacement ratio(bench cut)

Measuring points	1	2	3	4
PDR(%) Upper bench	26.9	11.4	28.3	13.2
Lower bench	86.1	42.1	81.2	53.7

Table 4. Properties of the model II

	Soil I	Soil II	Weak rock	Hard rock
Young's modulus(Pa)	9.8E6	6.86E7	7.84E8	1.08E9
Poisson's ratio	0.35	0.3	0.27	0.25
Unit weight (g/cm^3)	1.7	2.1	2.3	2.6

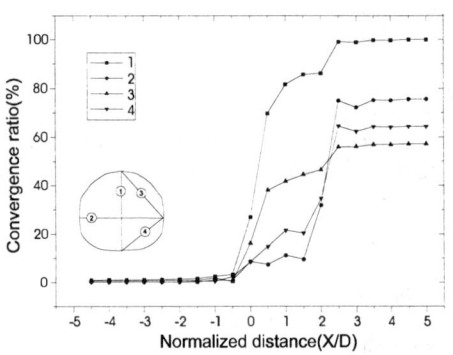

Figure 10. Convergence curve-Case I(bench cut)

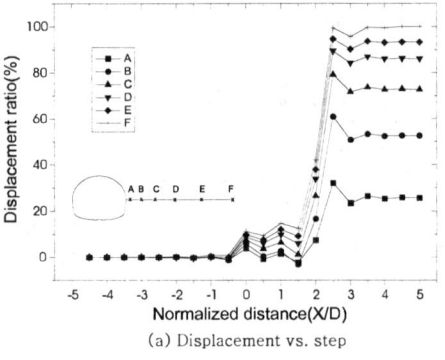

(a) Displacement vs. step

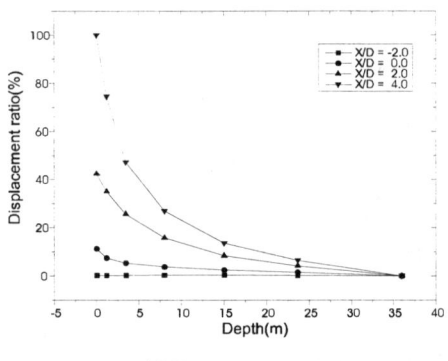

(b) Displacement vs. depth

Figure 11. MPBX displacement curve-Case I(bench cut)

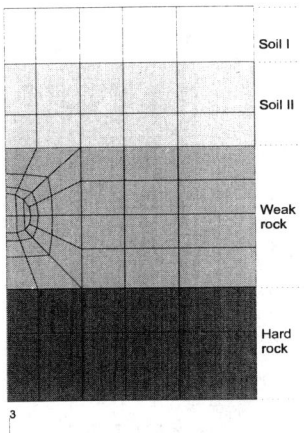

Figure 12. Profile of model II

3.2.1 Full face excavation

Convergence in Case II is similar to that in Case I except for the magnitude of displacement.

The behavior of side is not affected by the different properties. Roof, top-right, bottom-right, and invert pass through different layer. Thus, the behavior is different from that of Case I.

MPBX displacement at roof is shown in Figure 13. In Figure 13(a), C and D are separated from each other, which implies that there is a boundary between C and D. In Figure 13(b), the boundary between different zones can be estimated more clearly. However, this statement is not objective. Further, estimation of boundary is not easy when there is no significant difference in properties. For this reason, boundary is estimated from the slope of displacement vs. depth curve. The statistical values of slope are; minimum 31.0%, maximum 66.8%, mean 43.4% and standard deviation 9.8%. From these data, boundaries between different zones can be estimated.

3.2.2 Bench cut

Convergence is similar to that of full face excavation. The shape of MPBX displacement is similar to that of Case I(3.1).

3.3 Case III - vertical weak zone

To understand the behavior of rock mass which has weak zone, model in Figure 14 is organized. Table 5 shows the properties of model III.

3.3.1 Full face excavation

Roof, bottom-right and invert are not affected by weak zone for they don't pass weak zone. Meanwhile side

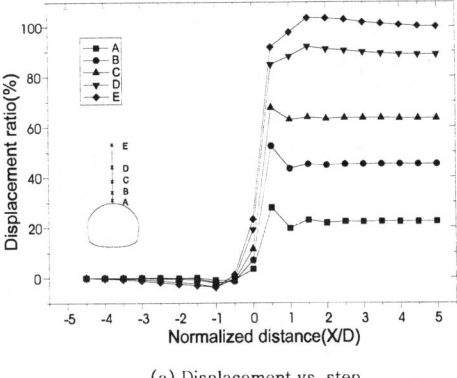

(a) Displacement vs. step

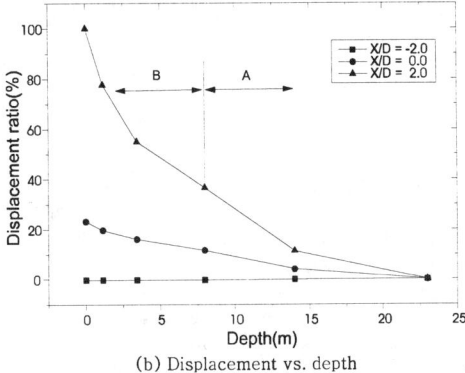

(b) Displacement vs. depth

Figure 13. MPBX displacement curve-Case II(full face excavation)

and top-right pass through weak zone and behave differently. The result at side is shown in Figure 15. The separation of C and D,E,F is shown. The weak zone is estimated by the slope using the same procedure described in 3.2

3.3.2 Bench cut

Convergence is similar to that of full face excavation. The shape of MPBX displacement curve is similar to that of Case I and II.

Table 5. Properties of the model III

	Young's modulus(Pa)	Poisson's ratio	Unit weight(g/cm^3)
Weak zone	6.86E7	0.3	2.1
The other zone	7.84E8	0.25	2.4

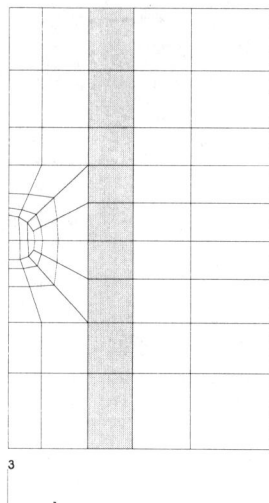

Figure 14. Profile of model III

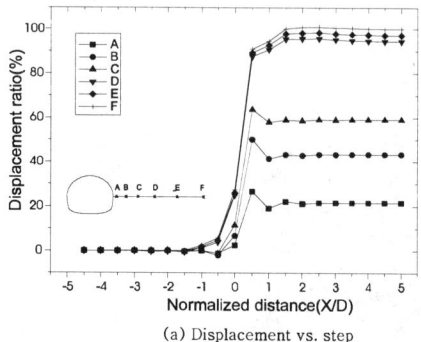

(a) Displacement vs. step

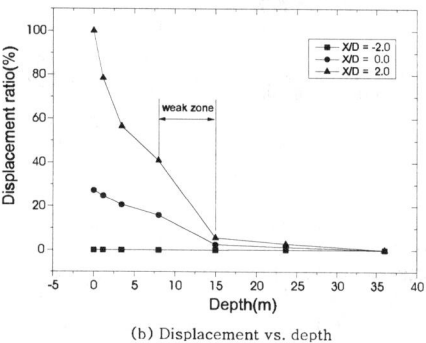

(b) Displacement vs. depth

Figure 15. MPBX displacement curve-case III(full face excavation)

4 CONCLUSIONS

A development and application of automation of monitoring were carried out in order to apply measuring data to design and construction more quickly and accurately. And three-dimensional finite element analyses were performed to check the influence of geological conditions on MBPX displacement.

In this study, numerical analyses were carried out for linear elastic condition. Therefore, other studies are required which consider elasto-plastic condition, time dependent behavior, discontinuities, support, and groundwater.

REFERENCES

Barlow J. P. & Kaiser P. K., 1987, Interpretation of tunnel convergence measurements, *6th ISRM International Congress*, pp.787-792

Brown E. T., 1981, *Rock Characterization Testing & Monitoring - ISRM Suggested Methods*, Pergamon Press

Dunnicliff J., 1993, *Geotechnical Instrumentation for Monitoring Field Performance*, John Wiley & Sons, Inc.

HKS, 1992, *ABAQUS User's Manual*, Version 5.2, Hibitt, Karlson & Sorensen, Inc.

Hanna T. H., 1985, *Field Instrumentation in Geotechnical Engineering*, Trans Tech Publications

Kaiser P., 1993, Deformation Monitoring for Stability Assessment of Underground Openings, *Comprehensive Rock Engineering*, vol.4-II, pp. 607 - 629

Kim, C. H. & Lee, C. I., 1991, Stress analysis and deformation behavior of rock mass around underground cavern by back analysis of field measurement:(I)Analysis on the characteristics of measured convergences, *J. Korean Rock Mech. Soc.*: vol.1-1, pp.75-90

Moon, S. K. & Lee, H. K., 1995, An analysis for the stress redistribution around tunnel face using three dimensional finite element method, *J. Korean Rock Mech. Soc.*: vol.5-2, pp.95-103

Determination of in situ stress using DRA and AE techniques

M.Utagawa, M.Seto & K.Katsuyama
National Institute for Resources and Environment (NIRE), Tsukuba, Japan

ABSTRACT: It is evaluated the possibility of AE and DRA techniques to measure in situ stress, in particular focusing on delay time effect on AE and DRA techniques in the range from one day to four hundreds days. Based on the technique established in the laboratory tests, in situ stress determinations were also conducted using rock cores obtained from a vertically drilled exploratory borehole. Not only axial pre-stress but also lateral one could be estimated from only one uniaxial cyclic loading test. And not only in-laboratory pre-stress specimen but also cored rock could be estimated by AE and DRA method, too. It can be concluded that the vertical geo-stress in this area has a strong effect of the weight of the overlying rock.

1 INTRODUCTION

Information on in-situ stresses is valuable in numerous discipline like mining engineering, civil engineering, petroleum engineering geology. Particularly in rock mechanics, adequate and accurate stress measurement is extremely important to evaluate the stability of rock mass in underground excavations. A variety of methods have been developed and used in attempts to determine in situ stresses in rock mass. Over coring method and hydraulic fracturing method are currently most popular, but both method have certain limitations very costly, time consuming. An alternative method using acoustic emission (AE) technique or deformation rate analysis (DRA) on drilled core rock to determine in situ stress has been proposed by various researchers in the recent past (Kanagawa et al. 1977, Yoshikawa et al. 1981, Seto et al. 1992, Yamamoto et al. 1991).

In this paper we try to evaluate the possibility of AE and DRA techniques to measure in situ stress, in particular focusing on delay time effect on AE and DRA techniques in the range from one day to four hundreds days. Based on the technique established in the laboratory tests, in situ stress determinations were also conducted using rock cores obtained from a vertically drilled exploratory borehole.

But, in case of long delay time, initial-stress can't be accurately estimated because of less clear bend point of this curve. Some parameters have been investigated to influence on the evaluated value of initial stress by DRA. In this paper, the effects of lateral stress, elapsed time after pre-stress loading, and loading rate on the DRA method are described.

2 METHOD

2.1 Acoustic Emission Method

The AE method is one of the methods to estimate the initial-stress of rock mass using cored rock. This method use Kaiser effect of AE. This phenomenon termed the "Kaiser Effect", suggest that previously applied maximum stress might be detected by stressing a rock specimen to the point where there is a substantial increase in AE activity.

2.2 Deformation Rate Analysis (DRA)

Deformation Rate Analysis (DRA) is one of the methods to estimate the initial-stress of rock mass using cored rock. In DRA, more than two cyclic compressive load are applied to a cored-rock specimen. The graph, stress vs difference of axial strain curve in cyclic loading test, have a bend point, as shown in Figure 1. This stress at the bend point expresses in-situ stress of rock mass. DRA is originally proposed by Yamamoto (Yamamoto et al, 1991).

2.3 Test Samples

The rock specimens used in this study were Inada-Granite, Shirahama and Kimachi-Sandstones, Tage-Tuff, and in situ mudstone. The sandstones'

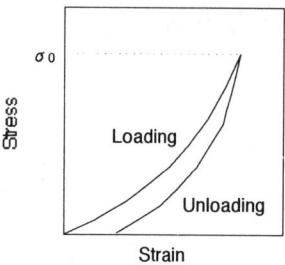

Figure 1 Stress Strain Curve

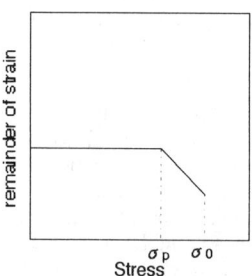

Figure 2 Stress-remainder of strain curve

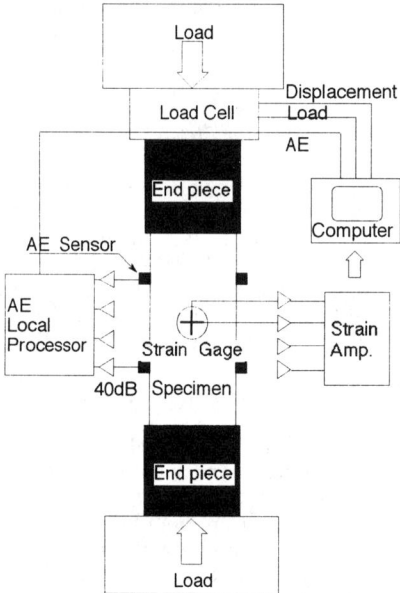

Figure 3 The AE-strain-load measuring system

2.4 Test Equipment

The specimen was loaded by means of a servocontrolled testing machine, MTS 810, under load control. AE sensing and processing was done with four piezoelectric transducers (NF 904DM), an AE analyzer (NF 9600) and AE wave memory (NF 9620). AE sensor's resonance frequency is 550 kHz, and high gain area is from 250 to 550 kHz. Signals from the AE sensors were amplified by a pre-amplifier (Gain is 40 dB) and a post amplifier inside the system (Gain 40 dB). The AE-strain-load measuring system is shown in Figure 3. 4 strain gages were attached to each surface of a specimen to average the inclination of the value of each gage. Load, strain, and displacement data were converted to digital data, and recorded on a floppy disk.

2.5 Test Program

In the tests, at first, uniaxial cyclic loading experiments of rock specimen were conducted to know how accurately artificially applied previous stress can be estimated by AE and DRA. A rock specimen was previously stressed in two or three different directions. Then, after some times being elapsed, the pre-stressed rock specimen was again uniaxially loaded to estimate the previous stress. These rock are taken until 100 m from ground, so the effect of initial geostress is very little.

The loading pattern of cyclic loading test is shown in Figure 4 and 5. Figure 4 is the pattern in the previous loading. Figure 5 is the pattern of loading to estimate the pre-stress value. Elapsed time from previous loading to the test were selected from 1 hour to 400 days.

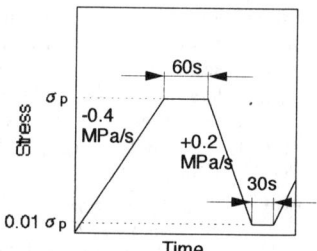

Figure 4 Loading pattern of pre-stress

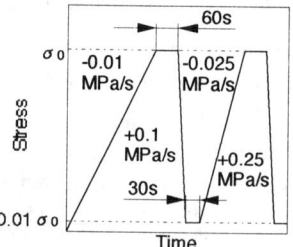

Figure 5 An example loading pattern in cyclic loading

uniaxial compressible strength is about 60 MPa. Granite's strength is about 185 MPa. And Tuff's strength is 40 MPa. The specimens are rectangular shape of which length of square was 30 mm, and height was 60 mm.

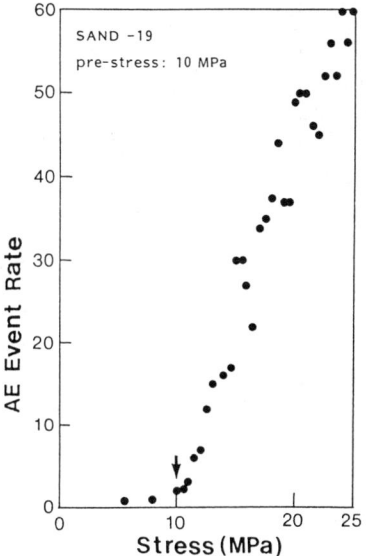

Figure 6 AE events (elapsed time 1 hour)

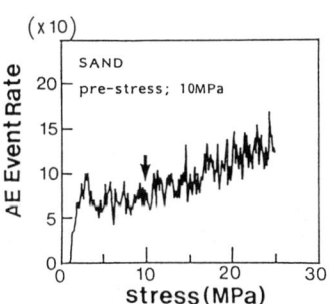

Figure 7 AE events (elapsed time 7 days)

3 RESULT AND DISCUSSION

3.1 Kaiser Effect

Figure 6 shows the typical examples that indicate the existence of Kaiser Effect in a sandstone. Elapsed Time is 1 hour. Kaiser Effect is clearly observed and the estimated maximum stress from the increase point of AE events lies within 5% accuracy.

3.2 The effect of elapsed time on Kaiser Effect

Figure 7 shows the result of the first loading which were done 7 days after the previous loading. Kaiser Effect isn't clearly observed. All data shows in Figure 8. With the increase in elapsed time, AE and DRA evaluation became less clear.

3.3 Estimation by AE on cyclic loading

Figure 9 shows the results of the second, third and fourth loading which is same as figure 7. Kaiser Effect is clearly observed. All data shows in Figure 10. It was found out that AE increase point and inflection point in DRA curve could be clearly recognized at the previous stress level after the second reloading. It was possible to determine the previous stress level within the accuracy of 10% even when the delay time is long and Kaiser Effect in the first reloading is not clear.

3.4 Estimation by DRA

Figure 11 shows the relation between stress and differential strain of the Shirahama sandstone specimen previously uniaxial stressed 1 day ago.

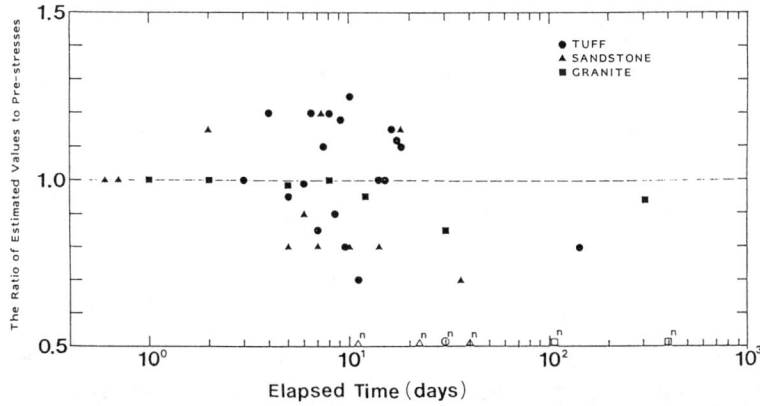

Figure 8 The ratio of estimated value to previous stress as a function of elapsed time (first loading)

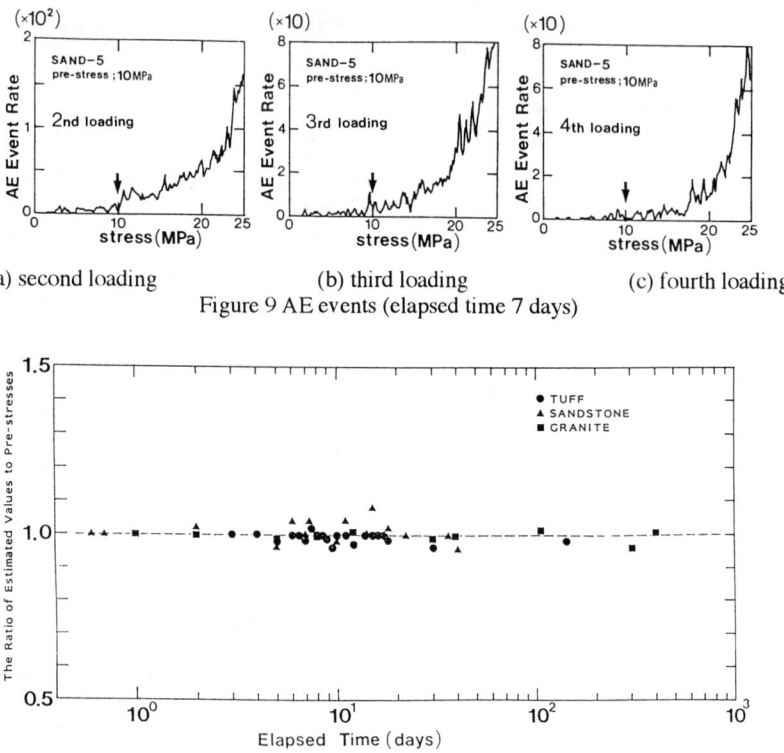

(a) second loading　　(b) third loading　　(c) fourth loading

Figure 9 AE events (elapsed time 7 days)

Figure 10 The ratio of estimated value to previous stress as a function of elapsed time (cyclic loading)

The differential strain was between the second loading and first loading. The previous stress value was 10 MPa. The DRA curve has a clear bend at near 10 MPa.

3.5 The effect of elapsed time on DRA

Figure 12 is a curve of stress and differential strain of the specimen previously stressed 7 days ago. This rock specimen is also Shirahama-sandstone, and the previous stress was 10 MPa. The DRA curve doesn't have a clear bend, while the DRA curve has a clearer bend when delay time is short. With the increase in delay time, the bend point of DRA curve became less clear. The variation of accuracy with elapsed time is shown in Figure 13 for all rock specimens tested in this study. With the increase in elapsed time, error in the evaluated value became bigger.

3.6 Estimation by DRA on cyclic loading

Figure 14 is a curve of stress and differential strain

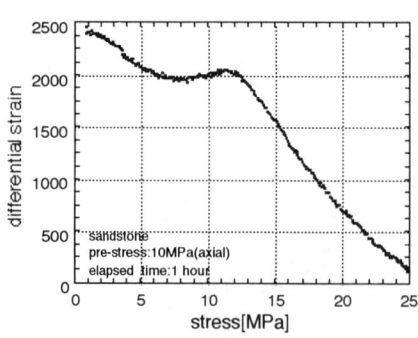

Figure 11 DRA Curve (elapsed time 1 hour)

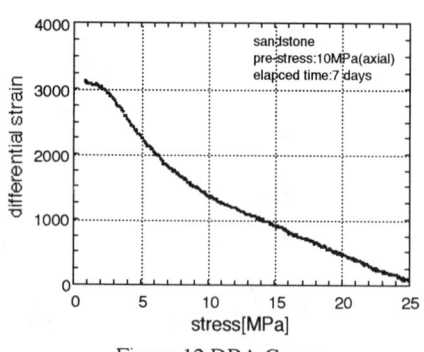

Figure 12 DRA Curve
(elapsed time 7days, First loading)

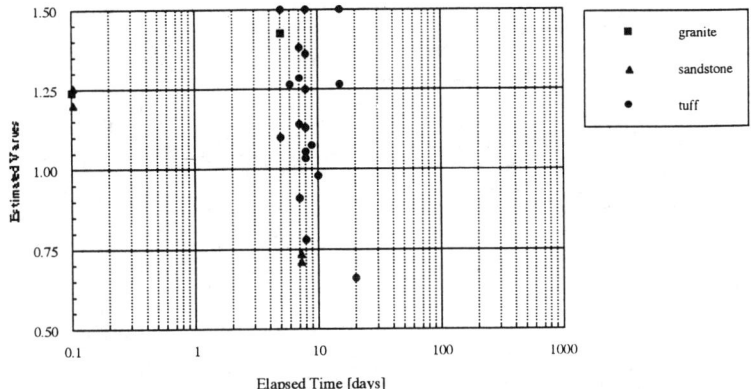

Figure 13 The ratio of estimated value to previous stress as a function of elapsed time (first loading)

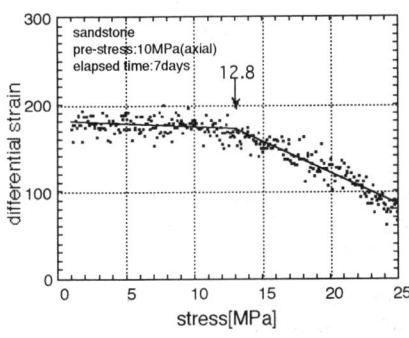

Figure 14 DRA Curve
(elapsed time 7days, Cyclic loading)

3.7 Conclusion in laboratory previous stress test

The laboratory test results indicate that the shorter the delay time is the more accurately the previous stress can be accurately estimated using Kaiser Effect and DRA principle, and that the estimation can be accuracy becomes worse with increase of delay time. But it was found out that AE increase point and inflection point in DRA curve could be clearly recognized at the previous stress level after the second reloading. It was possible to determine the previous stress level within the accuracy of 10%(AE) or 15%(DRA) even when the delay time is long and Kaiser Effect in the first reloading is not clear.

of the rock specimen shown in Figure 15. This strain difference is between the third loading and second loading. The DRA curve has a clearer bend at near 10 MPa of pre-stress value than the one shown in Figure 7. The clear bend of the DRA curve appeared at the pre-stress point on and after the second loading.

3.8 Cored Rock

There was significant correlation between the overburden pressure (estimated from depth and rock density) and estimated vertical stress from AE and DRA technique which are based on the laboratory

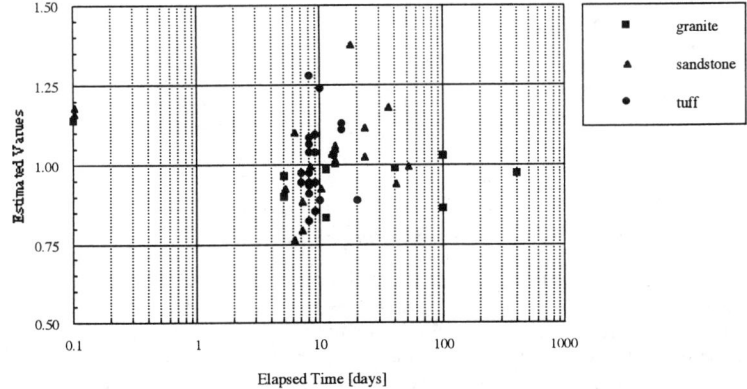

Figure 15 The ratio of estimated value to previous stress as a function of elapsed time (cyclic loading)

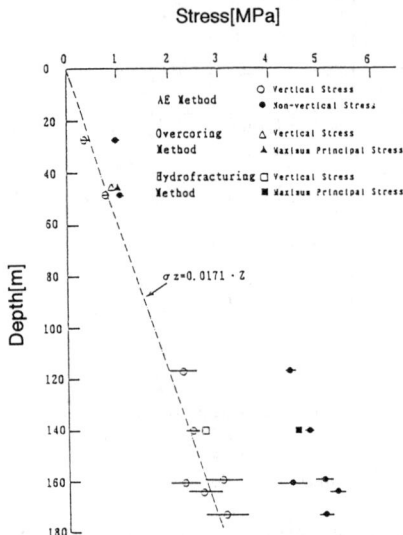

Figure 16 relation between the depth and the vertical stress estimated from AE

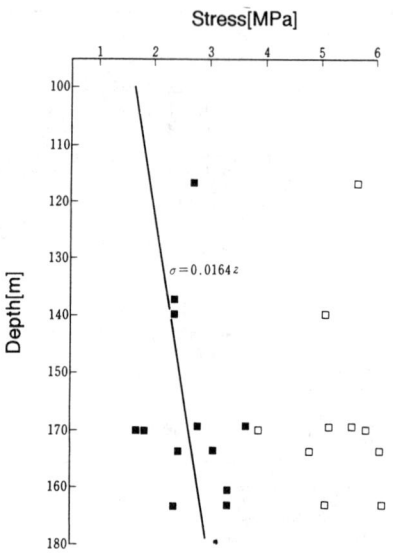

Figure 17 relation between the depth and the vertical stress estimated from DRA

tests. Figure 16 shows the relation between the depth(Z (m)) and the vertical stress estimated from AE. Figure 17 shows the relation between the depth(Z (m)) and the vertical stress estimated from DRA. The dotted line indicated the overburden pressure estimated from the depth and rock density. The estimated vertical stresses from acoustic emission behavior are consistent with the overburden pressure. The borehole was drilled up to the depth of 180 m from the surface five months before the tests, and rock type was sandy shale. And, as it can be seen from figure 18, the results obtained from overcoring method and hydraulic fracturing method also agree with the stresses estimated from AE. The time lag of up to two years didn't deter to evaluate the critical in situ stress condition. Cored rock recollected the in situ stress condition reasonably well, when compared to the results from overcoring method and hydraulic fracturing method.

4.CONCLUSION

The following main results were obtained:
(1) AE and DRA has a clearer evaluation when elapsed time after pre-stress loading, is short. With the increase in elapsed time, AE and DRA evaluation became less clear. Even in that case, however, the clear evaluation by AE and DRA appeared at the pre-stress point on and after the second loading. Therefore, it was possible to estimate the pre-stress even when the bend point of the AE and DRA curve was not observed clearly in the first loading.

(2) Not only axial pre-stress but also lateral one could be estimated from only one uniaxial cyclic loading test.
(3) Not only in-laboratory pre-stress specimen but also cored rock could be estimated from only uniaxial cyclic loading test.
(4) It can be concluded that the vertical geo-stress in this area has a strong effect of the weight of the overlying rock.

REFERENCES

Kanagawa, T., Hayashi, M. and Nakasa, N.(1977), Estimation of spatial geo-stress components in rock samples using the Kaiser effect of Acoustic Emission, Proc. Jpn. Soc. Civil Eng., 285, 63-75.
Yoshikawa, S. and Mogi, K.(1981), A new method for Estimation of the Crustal Stress from Cored Rock Samples: Laboratory Study in the case of Uniaxial Compression, Tectonophysics,74,323-339.
Seto, M., Utagawa, M.and Katsuyama, K.(1992), The Estimation of Pre-Stress from AE in Cyclic Loading of Pre-Stressed Rock, Proc. of Progress in Acoustic Emission VI, 159-166.
Yamamoto, K., Yamamoto, H., Kato, N.and Hirasawa, T. (1991), Deformation Rate Analysis for in situ Stress Estimation, Proc. of 5th AE/MS Activity in Creol. Struct. Mat., 1-13.

Pillar deformation response delay effect in underground mining

A.S.Voznesensky
Moscow State Mining University, Russia

ABSTRACT. The deformation response delay effect in pillars after explosion extraction of other ones in underground mining is studied. A deformation fading creep curve in time period of some days is obtained. In some cases this curve has its beginning and velocity maximum immediately after explosion. In other cases there was a some hours delay registered. This delay preceded the roof fall or the goal stowing technological stage. It was shown this effect can arise in pillar-roof system under rock pressure growing when rock creep parameters vary. These experiments were carried out with the use of the displacement measuring system "Massive" with wireless data transmission developed for underground mine pillar and roof stability control in explosion area. In designed paper the technical parameters of the measuring system, experimental results, possible pillar and roof deformation mechanism and application areas are discussed.

INTRODUCTION

Mine safety is one of the significant problems in underground mining. The roof fall and pillar stability prediction take the important part there. Methods, apparatuses and systems for the measure and control are developed. The difficulties arise in blasting mining when the rock massive condition must be known in some hours after explosion. The operator coming and measuring is not always safe for him. All the methods and apparatuses using the wire data transmission are also not safe in blasting condition. So way of the wireless transmission is one of the safe methods.

1. MEASURING SYSTEM

Measurements of pillar's deformations in underground working at mining methods, providing partial or complete extraction of pillars with the help of explosion, have allowed to register the deformation responses, caused by step loading change on pillars. The measurements were made with the help of system with wireless transmission of the indications "Massive", developed in common in Moscow State Mining University and in "Sibzvetmetavtomatika" (Krasnoyarsk, Russia). The first variant of the system has the data registration on the paper band. The latest variant has the connection with a computer placed on the day surface. The block scheme of the "Massive" system with wireless data transmission from the displacement gauges (DG) to the radio receiver and wire data transmission on the other sites is shown on the figure 1.

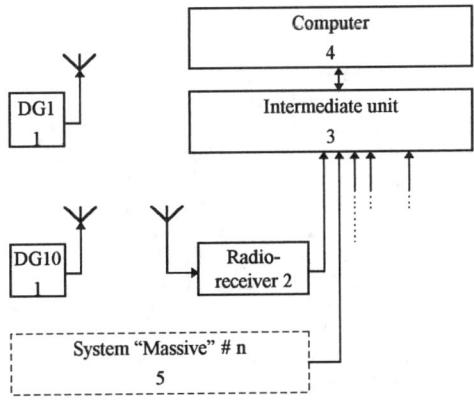

Figure 1. The block scheme of the "Massive" system

This system with wireless data transmission through mining working contains displacement gauges 1 with transmitter, placed in bore-holes, and radio receiver 2, placed in a protected from explosion place. The indications on a wire line from the receiver 2 are transmitted to the intermediate unit 3 or directly on the computer 4, placed on the day surface. The other systems 5 placed in the other mine workings can be

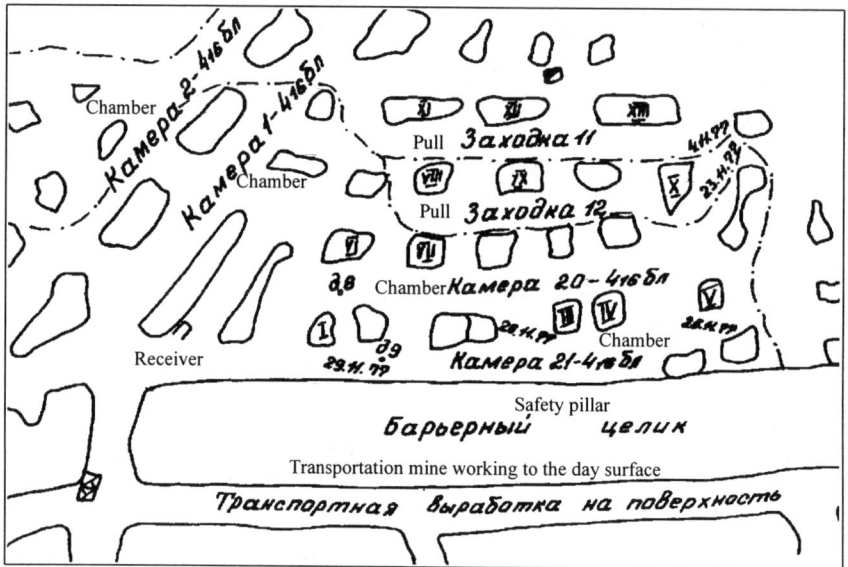

Figure 2. The plan of the experimental site

connected via intermediate unit 4. The system contains 10 radio gauges. Frequency band of the wireless channel is 28-30 MHz. Receiver sensitivity is 0.3 microvolt. Radio transmitter power is 3 mW. Range of a wireless site of transmission is from 50 up to 150 meters depending on the working configuration. Maximum length of the wire site of data transmission is 5 km. Displacement measurement ranges are equal 0-2 mm, 0-6 mm and 0-12 mm with the possibility to put the beginning point at any place of the scale. That allows to carry out a choice for a specific problem. Operating time of the radio gauge in a pulsing mode without change of the power supply is about 6 months. The measurement error at complete term of operation is 6 %. At short-term measurements it decreases proportionally to the measurement time in relation to complete operating time without change of the power supply (6 month). The minimum interval of samples from 5 up to 60 minutes depends on quantity of interrogated radio gauges.

2. MEASURING PROCEDURE

2.1. Experimental site.

The system "Massive" was used for the control of the roof and pillars stability at chamber-and-pillar method on lead-zinc ore mines in Southern Kazachstan. The measurements were made in 416 blocks 7 horizons of mine "Mirgalimsay". The plan of an experimental site is submitted on the figure 2.

The extraction was made by chamber-and-pillar method with roof caving in the receding order.

The primary extraction was made about a year back. Ore in place and the enclosing rock are combined of limestone. There is the layered structure (from thin up to thick) of the ore body. The thickness of the ore body is about 4 meters. A corner of the fall is about 10-25 degrees. A number of large tectonic cracks with amplitudes of displacement 1,5-1,7 m. is revealed. Ore and enclosing rock have an average stability. Depth from a surface is about 300 m. About 10 meter above the main ore layer there was the second ore layer.

Technological roof fall zones accompanied pillar extraction are shown with pitch-dotted lines on figure 1.

Non-technological roof fall occurred in area of pillars I-IV.

2.2. Measuring system allocation

The bore-hole devices were placed in bore-holes by a diameter of 40-42 mm in such a manner that the nearest to the bore-hole mouth support was mounted on it on distance about 60 centimeter, and other support - on distance 2 m from the first one. The radio gauges were placed in pillars horizontally, bore-holes crossed layers.

The receiver-recording equipment together with fittings were placed in the protected from influence of flying rock by explosion place. This site is marked

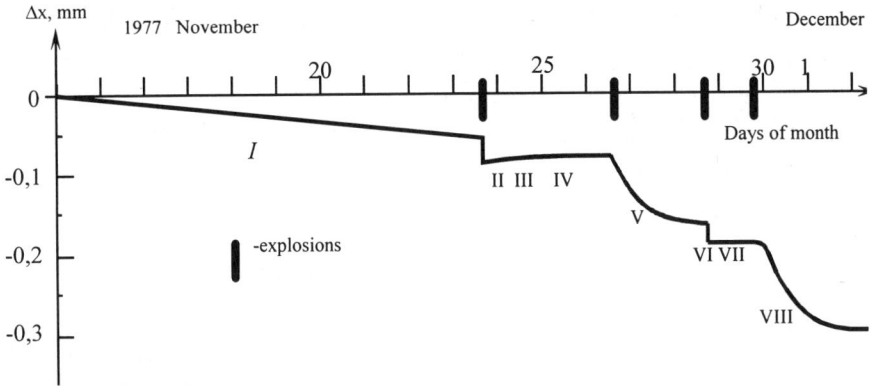

Figure 3. Deformation curve in pillar VII

on the plan by the letter "Ï". The receiver was placed in a metal box and protected from humidity by a plastic bag.

3. RESULTS

The results of supervision for development of cross pillar deformations (gauge 5, pillar VII) are submitted on the figure 3. The time moments of explosion are marked on this drawing by vertical hyphens.

The deformation increased by the jump (site II) on value about 50 microns after explosion of the pillars VIII, IX, X. Then at once after step-loading there was the stabilization of the deforming process. On a site IV the tendency of transition to process, similar the site I is observed. After truncation of the pillar V cross deformations again grow, but this increase is not instant, as on the site II, but gradual, resembling to an exponential function.

The truncation of the pillar II has caused again step change of deformations (site VI) and the deforming stabilization (site VII), and truncation of the pillar I has caused again the deformations increase, resembling fading creep, but this process was detained on a few hours after explosion. The further registration was impossible because of roof fall and pillar' breaking. In this example the pillars' and roof destruction was preceded with the effect of deformation response delay, registered after the last explosion at the pillars' explosion.

4. DISCUSSION

4.1. Pillar and roof deformation model

This phenomenon can be explained by rock creep models. Step loading on pillars arising in result of mining brings a fading creep, which can be allocated from a general deforming curve and can be used for the analysis of pillars' and roof safety.

In the elementary case the model qualitatively describing these processes is the generalized elastic-viscous body (figure 4), described by the equation:

$$\frac{E_1 + E_0}{E_0}\sigma + \frac{\eta}{E_0}\dot{\sigma} = E_1\varepsilon + \eta\dot{\varepsilon}$$

Here A_0 is instant elasticity;
A_1 - detained elasticity;
η - factor of viscosity;
σ, ε - pillars' and roof stress and strain.

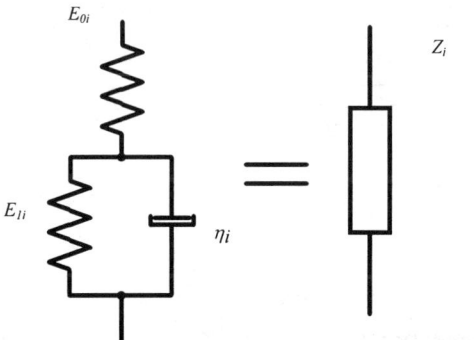

Figure 4. Elastic - viscous body as model of a pillar and roof and its equivalent designation

It is experimentally established, that for rock a rather large area of parameters linearity is observed. However, as for all structured bodies, rock change of viscosity is inherent at rather large stress. That is caused by the structure destruction.

Change of viscosity in the loading process can make of a hundred of time, in some cases tens of a hundred or thousands of time. At the same time the elastic characteristics vary not more than in some times in a wide range of stress. It results in increase of fading creep process time interval. This occurs at a stress value close to destroying one and testifies to destruction of structure and transition in an unstable condition. So for example, for salts this increase reaches up to 100 and more time. For layered pillars with different packs or layers strength it will have an effect first of all on packs with the minimum durability. These phenomena can result in effect, described above.

4.2. Possible mechanism of the "pillars-roof" system deformation

It can be shown, that from all possible cases described effect is possible only in a complex pillars or pillars-roof system in condition of different load on different elements of such system. In this case step loading causes different stress and strain dependencies.

4.2.1. "Two pillars" system

The expressions for stress can be derived for the system of two pillars with a different creep time constants:

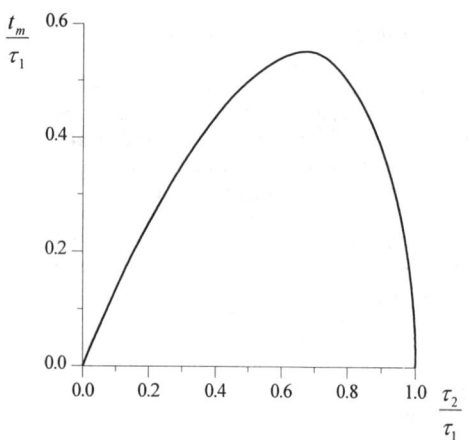

Figure 5. Maximum deformation speed curve delay value t_m depending on time constant τ_2 in relation on τ_1

$$\sigma_i = \sigma_0\left[1-(-1)^i a\left(e^{P_2 t} - e^{P_1 t}\right)\right] \quad (1)$$

Here: i - pillar number (1 or 2);
σ_i - stress changes;
σ_0 - stress step;

$$P_{1,2} = \frac{-3-3\tau_2 \pm \sqrt{9-14\tau_2 + 9\tau_2^2}}{4\tau_2},$$

(if $\tau_1 = 1$);
τ_1, τ_2 - time constants of the pillars 1 and 2;
a - coefficient;
t - current time.

It can be shown that

$$-\frac{1}{\tau_2} \le P_1 \le -\frac{1}{\tau_1}; P_2 \le -\frac{1}{\tau_2}.$$

Function (1) has an extremum. For one pillar σ-curve has a maximum, for the other one it has a minimum. This fact leads to one pillar deformation velocity extremum.

On the figure 5 maximum deformation speed curve delay value t_m depending on time constant τ_2 in relation on τ_1 is shown.

Deformation delay effect is expressed when τ_2/τ_1 ratio has values in middle part of interval from 0 to 1.

4.2.2. "Two pillars - roof" system

The mechanical circuit, being analogue of system from two pillars and roof (figure 6), is shown on figure 7.

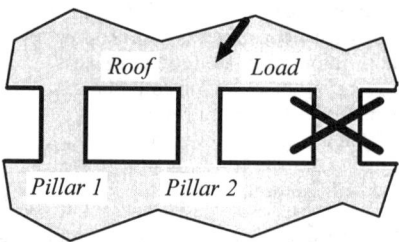

Figure 6. Pillar' and roof loading caused on other pillar extraction

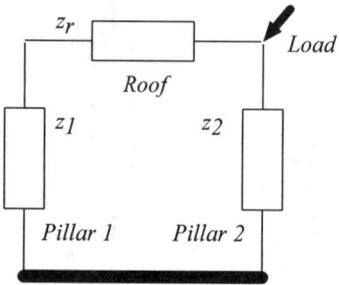

Figure 7. Elastic-viscose body as the model of "two pillars-roof" system.

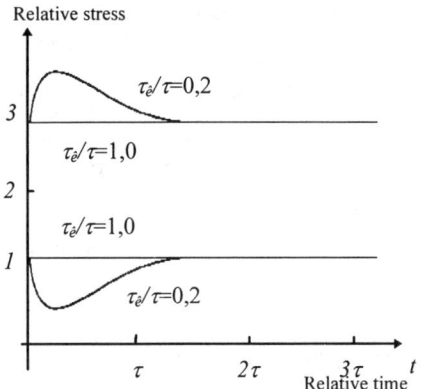

Figure 8. The pillar' stress curves for a different ratio between roof $\tau_{\dot{e}}$ and pillar' τ time constants

On figure 8 qualitative stress dependencies on one and other pillar at different time constant equal to product of viscosity η on detained elasticity Å are shown. For a pillar having greater time constant (and according greater load values coming nearer to destroying ones) the load curve is curved upwards, and for a pillar with smaller constant time - downwards. The change of longitudinal stress will cause respective alterations and cross stress. The deformation speed curve of one of the pillars has an extremum.

4.3. Pillar' and roof stability control method

Such deformation speed extremum time moment in relation to the explosion moment can serve for an estimation of pillar' and roof stability. If the pillars are far from destruction, their time constant are close each other and the deformation response delay effect will be not observed. At approach any of pillars to destruction, the parameters of its elastic - viscous model will vary and in its fading creep curve there will be the delay of the deformation speed maximum in relation to the explosion moment in pillar extraction process.

These results have allowed to develop a method of the pillars' and roof stability control, including measurement of pillar' deformations in the time, when an time interval from the moment of explosion up to the deformations' speed maximum is measured. It will judge instability of rock massive around of mine working at this interval increase higher than critical value.

4.4. Threshold of stability

It is necessary for practical applications to know a threshold value. There is an unstable condition of rock massive if this threshold crossed is. In the table 1 the experimental deformation speed maximum time delay values t_m are shown.

Table 1. Experimental deformation speed maximum time delay values t_m

#	Experiment site	Date	t_m	Condition
1	Panel 416, level 7, Sonkulsay geological ore body	1977, November	9,5	Roof fall
2	Panel 5 , level 19, Central geological ore body	1979, January	18,5	Pillar breaking
3	Panel 5 , level 19, Central geological ore body	1979, January	13,6	Pillar breaking
4	Panel 7, level 19, Central geological ore body	1978, August	0	Stable
5	Panel 7, level 19, Central geological ore body	1978, August	2,25	Stable
6	Panel 416, level 7, Sonkulsay geological ore body	1977, November	0	Stable

On the figure 9 probability densities $p(t_m|\omega_s)$ and $p(t_m|\omega_u)$ of informative parameter t_m and threshold t_{mth} characterized transition from the stable condition to the unstable one are shown. This ones are estimated on sum of Gauss kernel (Parzen estimation) [Fukunaga 1972]. This estimation method belongs to nonparametric methods of statistics and allows to estimate probability densities of various parameter and characteristics of block rock. Such probability densities can have a miltimodal form as since rock blocks dimensions group about defined values.

From the diagrams the good probability density separation appropriate to steady and unstable condition is visible

The threshold is estimated on Byes-criteria for $P(\omega_S)=0.1$ and loss matrix

$$\begin{Vmatrix} 0 & 1 \\ 1 & 0 \end{Vmatrix}.$$

The loss for right decision are equal to 0. The loss for wrong decision are equal to 1.

The loss coefficients can be derived in absolute or relative values. The a-priory probabilities $P(\omega_i)$ can be estimated from reliability theory or from pillar' and roof destruction observation. The threshold characterizing border line between steady and unstable condition depends on all this values.

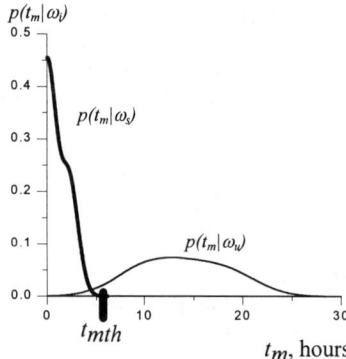

Figure 9. Probability densities $p(t_m|\omega_s)$ and $p(t_m|\omega_u)$ of informative parameter t_m and its threshold t_{mth} characterizing crossing from steady pillar' and roof condition to unstable one.

4.5. Information efficiency criterion of the rock stress state control

Before the control operation the indefiniteness of a rock stress state is characterized by a-priory unconditional entropy $H(\omega)$ where ω means a condition of object of the control. After the control operation by measurement of one or several parameters x the indefiniteness of a condition will be characterized by an average value. This is the complete conditional entropy $\overline{H}(\omega|x)$. Difference of these values:

$$I = H(\omega) - \overline{H}(\omega|x)$$

represents the information quantity received as a result of operation of control. It can characterize quality of monitoring system. However this value depends on a-priory probabilities of a condition of object and on the logarithm basis at calculation of this value. Therefore preferable is the relative value (Voznesensky, 1994).

$$E = \frac{I}{H(\omega)} = \frac{H(\omega) - \overline{H}(\omega|x)}{H(\omega)}$$

Here E is the information efficiency. It characterizes efficiency of the control and system. The value E varies in limits from 0 up to 1. The values, about 0 testify to bad quality of measured parameters and system. Values, about 1 testify to their good quality.

In the table 2 the values of information efficiency E for different values of the a-priori probabilities are adduced. In the table 2 also the threshold values t_{mth} depending on a-priori probability are shown.

Table 2. Influence of a-priori probabilities of unstable and steady condition to information efficiency E and threshold t_{mth}

$P(\omega_s)$	$P(\omega_u)$	E	t_{mth}, hours
0.001	0.999	0.9591	6.09
0.005	0.995	0.9696	5.75
0.01	0.99	0.9610	5.60
0.05	0.95	0.9606	5.23
0.1	0.9	0.9593	5.04
0.15	0.85	0.9579	4.92

6. CONCLUSIONS

1. In designed paper a new method of pillars and roof stability control based on maximum deformation speed curve time delay registration is described. This one has a high value of information efficiency about 0,95. The threshold values of delay are about 5-6 hours.

2. Difficulty in practical realization of this method is requirement of qualitative deformations measuring in blasting area of mining workings.

3. The method can be used for safety control in underground mining.

7. ACKNOWLEDGMENTS

Described results were obtained together with colleagues from the Moscow State Mining University (formerly Moscow Mining Institute), Institutes "Zvetmetavtomatika", (Moscow), "Sibzvetmetavto-matika", (Krasnoyarsk), mining works "Achpoly-metall", who all are gratefully acknowledged.

REFERENCES

Fukunaga. K., 1972. *Introduction to statistical pattern recognition*. School of electrical engineering. Purdue university. Lafayete, Indiana. Academic press, New York and London, 1972

Voznesensky A.S., 1994. *Systems for geomechanical processes control* (In Russian). Moscow State Mining University. Moscow, 1994, 147 pp.

Application of time domain reflectometry to the deformation characterization of rock mass

S.L.Jung & H.K.Lee
Department of Mineral and Petroleum Engineering, Seoul National University, Korea

S.K.Chung
Korea Institute of Geology, Mining and Materials, Korea

ABSTRACT : This paper presents an application of the TDR(Time Domain Reflectometry) to the monitoring of the deformation of rock mass with grouted coaxial cables through laboratory tests. The grouted cable can easily deform together with the rock mass movements, and the deformed cable loses its original capacitance and the reflected waveform produced along the deformed cable consequently represents a change of voltage pulse. Therefore, it is possible to monitor the deformation of rock mass by measuring the changes in these reflection signatures. The behavior of the cables installed into separate boreholes in sedimentary formations was monitored to measure rock mass deformation based on the interpretative techniques developed through laboratory tests.

1. INTRODUCTION

This paper summarizes tests conducted to relate deformation, distance and reflection coefficient amplitude after propagation along coaxial cables as long as 50 m.

Shear test of the cemented mortar containing a specimen of coaxial cable showed that the shear deformation correlated linearly with the reflection coefficient, so the TDR was effective to monitor the displacement of the rock mass. Bending test were carried out in order to determine the influence of the crooked cables on the monitoring of rock mass movements. Controlled crimping and shearing test upon a cable of 50 m long, 12.7 mm diameter showed not only the fact that the reflection amplitudes decreased as the cable length but also the proper crimping depth, width and interval between two adjacent crimps. These relationshipswill allow a quantitative assessment of the usefulness of measuring shearing deformations with long cables.

Two coaxial cables - one 100 m long and the other 175 m long - were installed and grouted into separate boreholes drilled in a sedimentary formation. THe behavior of the cable was monitored with metallic TDR cable tester to measure rock mass deformation based on the interpretativetechniques developed through laboratory tests.

After a brief background section, the equipment and procedure are presented, and the influence of the length on reflected voltage and related issues are discussed.

2. BACKGROUND

2.1 Early usage and advantages

The Time Domain Reflectometry(TDR) approach employed in this study was developed initially as a method to locate discontinuities in coaxial transmission cables(Moffitt, 1964). The concept has been extended to measurement of the properties of the materials in which adjacent conductors are embedded, such as soil water content, and the evaluation of coaxial cable dielectric behaviour. In rock mechanics specifically, the technique has been employed to identify zones of rock mass deformation corresponding to locations of cable failure, and blast effects.

Dowding studied the electromagnetic wave theory necessary to quantitatively relate changes in cable geometry to change in reflected voltage signatures(1988) and tested double shear of very short cables(1 m) in the laboratory(1989). Pierce(1989) made a study of controlled crimping and shearing of a 530 m long, 22.2 mm diameter coaxial cable for comparison with theory and the existing data.

The system is quite economical because of the ease of installation, use of commercially available instruments and cable, and the ease of data acquisition and interpretation. The coaxial cable costs about $12.00(1996 U.S.) per meter and can be grouted in place with a conventional drilling rig water pump. One portable readout unit can be employed for monitoring as many cables as desired and produces either a permanent strip chart record or an analog voltage signal output. These units are mass produced for other industries and therefore are reasonably priced. The reflected signatures can be interpreted directly from the strip chart record or the analog output can be digitized for automated interpretation. Furthermore, the system is capable of being remotely monitored for critical installations such as sinkhole or coalfield subsidence or underground nuclear tests.

2.2 Theory of Time Domain Reflectometry

Time Domain Reflectometry is a remote sensing electrical measurement technique that has been used for many years to determine the spatial location and nature of various objects. An early form of TDR, dating from the 1930s, that most people are familiar with is RADAR. Radar consists of a radio transmitter which emits a short pulse of microwave energy, a directional antenna and a sensitive radio receiver. After the transmitter has radiated the pulse, the receiver then listens for an echo to return from a distant object, such as an airplane or ship. By measuring the time from the transmitted pulse until the echo returns and knowing the speed of light, the distance to the reflecting object may be easily calculated. Detailed analysis of the echo can reveal additional details of the reflecting object which helps in identifying it. The same principles hold for radar, lidar, coax TDR, optical fiber OTDR and broadband impulse radars.

Coaxial TDR is essentially a "closed circuit radar". It involves sending an electrical pulse along a coaxial and using an oscilloscope to observe the echos returning back to the input. A coaxial cable TDR is usually configured as shown in Fig 1. The pulse generator is represented by an open cicuit voltage source, $V_g(t)$, and its source impedance, R_g. It generates a fast rising step function, pulse waveform. This pulse is launched into the coaxial cable. A high impedance oscilloscope is bridged across the input to the coaxial cable. A coaxial cable designated as the "Reference Cable" is connected to the TDR output port. If the ref. cable impedance, R_0, is known, then quantitative results may be obtained from TDR measurements.

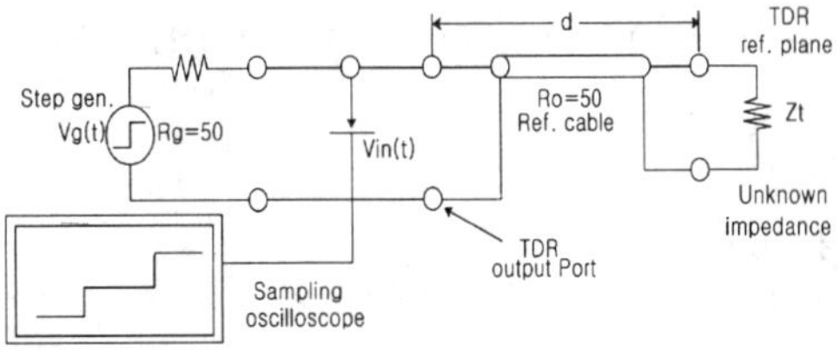

Fig. 1 TDR system configuration.

Usually the reference cable impedance is matched to the generator impedance, i.e. $R_0=R_g$. This is done to prevent reflections which come back into the TDR from being rereflected and corrupting the test results. At the far end of the reference cable is the "Reference Plane". This is the port at which the unknown impedance, Z_t, is attached.

In coaxial TDR the pulse generator sends a pulse through the sampler into the reference coax. The pulse propagates through the coax at a velocity, V_p, and arrives at the far end after a time TD.

$$TD = d / V_p \qquad (1)$$
$$V_p = c / (k)^{1/2} \qquad (2)$$

Where c is the speed of light(3×10^{10}cm/sec) and k is the relative dielectric constant of the coaxial transmission line. If the load impedance matches the coax impedance, $Z_t=R_0$, then the TDR pulse is perfectly absorbed. However if Z_t is not equal to R_0, some of the incident pulse energy will be reflected back to the left towards the generator. This reflected pulse will arrive back at the TDR output port at $t=2TD$. The oscilloscope allows us to visually see the total waveform, $V_{in}(t)$, at the cable input. $V_{in}(t)$ is the algebraic sum of the pulse generator outgoing step pulse and any returning echos present on $V_{in}(t)$ allows us to determine the location and nature of discontinuities within the coax and/or mismatched terminations to the coaxial transmission line.

2.3 Principle of application of TDR to the deformation characterization

Coaxial cable deformities produce electrical circuit discontinuities that can be divided into two catgories: a change in characteristic impedance Z_0(type I), or a change in reactive lumped circuit elements(type II). Cable extension is best modelled as a type I discontinuity while cable shear is best modelled as a type II discontinuity as shown in Fig. 2. When the cable is sheared, the deformation and resultant

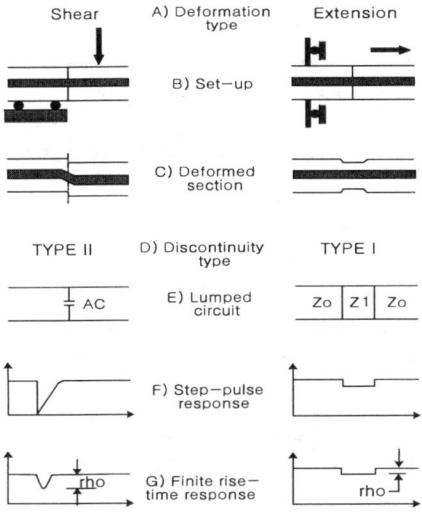

Fig. 2 Relationship between cable deformation and type of reflected signal.

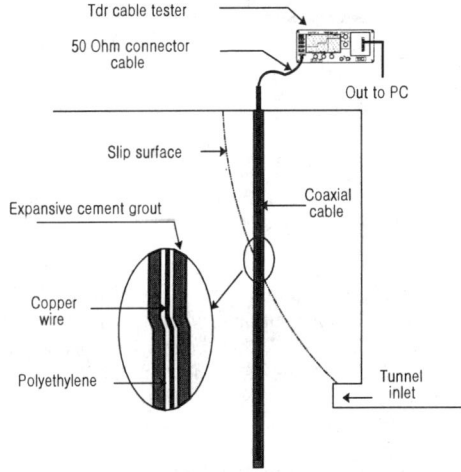

Fig. 3 TDR installation in rock mass showing relationship between cable and instrumentation.

reflection is localized and the deformed crosssection can be modelled by adding an equivalent capacitance to the lumped circuit system as shown in Fig. 2A-G. When the cable is extended, the deformation can be less localized and the necked section could be idealized by a reduction in impedance if the necked section is long compared the wavelength of the voltage pulse.

Either type I or II discontinuities produce a reflected voltage pulse. Travel time between

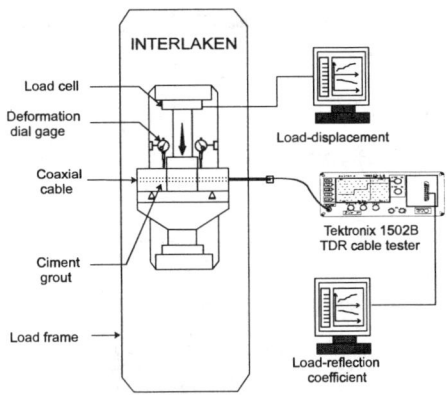

Fig. 4 Apparatus for cable shearing.

initiation and reflection of the pulse allows discontinuity location, while the slope and amplitude of the reflection can be related to specific changes in cable properties. Since individual discontinuities are seperated in space, they are also seperated in time and thus can be analyzed separately. This relation between location and travel time forms ths basis of time domain reflectometry.

As shown in Fig. 3, when the coaxial cable is installed into a borehole drilled in rock mass and grouted with cement mortar, coaxial cable and rock mass are united, so shear deformation or extension of rock mass discontinuity are transferred to the cable and the coaxial cable deforms. Deformation of the coaxial cable makes a change of local capacitance, and reflection signatures shown in cable tester are changed. Therefore, we can analyze a location and the type and amount of deformation of the coax cable by monitoring the reflected voltage signal.

3. EXPERIMENTAL WORK

3.1 Cable testing equipment and data acquisition system

Tektronix 1502B TDR cable tester was used to measure reflected voltage signals through coaxial cables. 1502B itself has a Tektronix SP232 serial extended function module, so it is possible to communicate with external host computer using RS232-C serial protocol. In this study, we made a program using SP232 module that can communicate with the host computer to achieve monitoring and laboratory test.

Coaxial cable used in this study is 12.7 mm diameter and composed of aluminum(outer conductor), copper(inner conductor) and

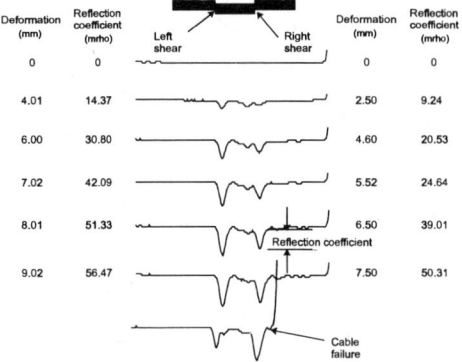

Fig. 5 Reflected voltage signals from shearing 12.7 mm cable to failure.

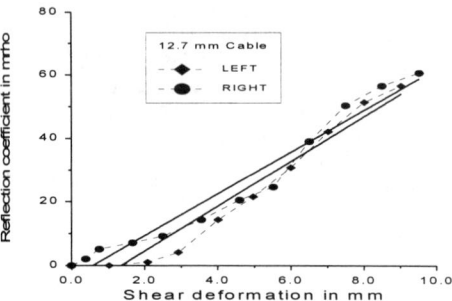

Fig. 6 Relation between reflection coefficient and shear deformation.

polyethylene (dielectric material) and wave propagation velocity in the cable is 0.83.

3.2 Shear test of short cable

In order to measure shear deformation of underground rock mass with coaxial cable, we should know how reflection signatures are changed, when the cable is sheared and the relationship between reflection coefficient and shear deformation.

12.7 mm dia. and 150 cm long coaxial cable was passed through a square pillar($10 \times 10 \times 60$ cm) made of cement mortar. As shown in Fig. 4, the specimen was divided into three parts and supported on each side and the middle part of the specimen was pushed downward. Each part of the specimen represents rock mass block and the location at which shear occurred represents rock mass discontinuities.

Shear deformation was recorded every 1mm with a dial gage located at each side of specimen and TDR reflection amplitude was recorded in digital form by the host PC.

Fig. 5 shows the change of reflection signature during shear deformation. As the shearing is increased, spike of the reflection amplitude increased downward. If the outer aluminum conductor is cut as a result of shearing, the lumped circuit is opened and consequently an upward signal is shown. In Fig.5, a downward signal is shown between the left and right end of the specimen. This is a result of the breaking down of the middle part of the specimen.

Fig. 6 shows the relationship between reflection coefficient and shear deformation. The downward spike of reflection signatures are correlated linearly with shear deformation but the change of reflection coefficient was not

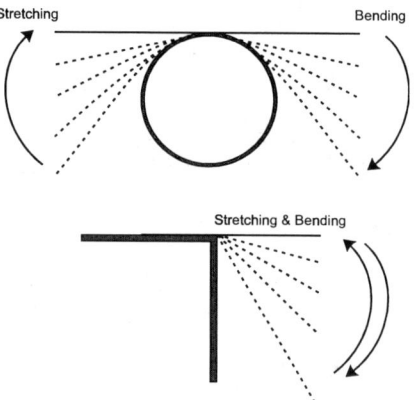

Fig. 7 Appratus for cable bending.

shown until the initial shear displacement is made. Shear deformation, which can make a reflection coefficient that can be perceived becomes different as a type of TDR tester, diameter of cable, setting status of tester, distance between cable tester and location of deformation.

The relationship between reflection coefficient and shear deformation is 6.66 mrho/mm on the left side and 7.04 mrho/mm on the right side. To make the results of a shear test generalize, the test needs to be conducted in different tester settings and distance to location that deformation occurred.

Generally, the coaxial cable used in the place is about $50 \sim 200$ m long and reserved and installed as a rolled form, so it shoud be investigated just how effective on the reflection

3.3 Bending test of short cable

coefficient the crooked cable is. As shown in Fig. 7, the coaxial cable is attatched to a

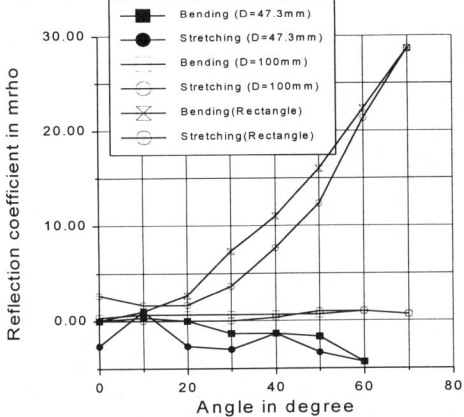

Fig. 8 Relation between reflection coefficient and bending angle.

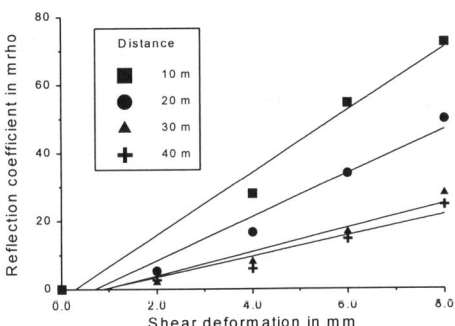

Fig. 9 Reflection coefficient versus shear deformation relationships for long coaxial cables up to 50 m.

cylinder of various diameters, 47.3 mm, 100 mm, 200 mm, 300 mm, and bent and stretched from 0° to 70° and the reflection signatures are recorded at every 10°. The same process was carried out with the rectangular corner.

Fig. 8 shows the relationship between reflection coefficient and the bending angle, where the diameter of cylinder is 47.3 mm, 100 mm and rectangular edges. We can know that crooked cables have no significant effect on the reflection coefficient.

3.4 Shear and crimp test of cable up to 50 m long

This test is carried out in order to compare the data acqusited from controlled crimping and the shearing of a 50 m long, 12.7 mm diameter coaxial cable with theory and the exsiting data from short cables.

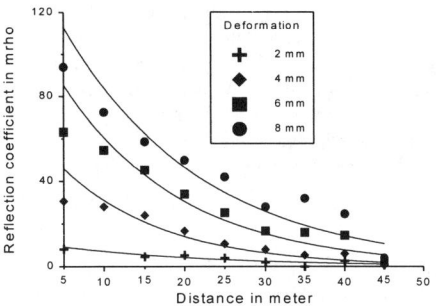

Fig. 10 Reflection coefficient versus transmission distance relationships for four different magnitudes of shear deformation.

The results of a shear test on the short cable showed that the reflection coefficient and shear deformation have linear relationship. This is the same to a 50 m long coaxial cable, but as the distance is increased from the TDR cable tester to the location of the deformation, the impedance of the coaxial cable is increased too, so the amplitude of the reflection voltage pulse is decreased.

A 50 m spool of cable was unrolled and a Tektronix 1502B cable tester and a laptop computer were used to acquire the TDR waveforms. The computer controls the cable tester via the SP232 host application program. The metallic coaxial cable is linked to the cable tester through 1 m precision coaxial cable. The coaxial cable was sheared every 5 m by three jointed vise, and shear deformation was measured by a dial gage. Crimps were produced by squeezing the cable with a set of adjustable vise grips. The depth of the crimp is set by adjusting the separation of the grips. The length of the crimp is controlled by the grip head width of 11 mm and 20 mm.

As shown in Fig. 9, reflection coefficient and shear deformation have a linear relation, but the reflection coefficient is decreased as the distance between the cable tester and the location of shear deformation is increased. When a 8 mm

Table 1. Summary of results for crimped cables.

Dist. Width	Depth(mm)	10 m	20 m	30 m	40 m
11 mm	0.875	6.12	5.33	2.67	2.67
	2.625	36.72	28.00	18.00	16.67
20 mm	0.875	8.16	6.67	3.34	2.99
	2.625	44.88	44.66	23.34	18.67

shear deformation was made to the coaxial cable, the reflection coefficient was 93.88 mrho at 5 m long distance, but decreased to 4 mrho, 4.26 %, at 45 m long. As the location of the deformation becomes more distant, the reflection coefficient decreased, and the width of the downward spike is widened. Fig. 10 is a graph for caculating the shear deformation from the data of reflection signatures acquired at the place. When the reflection coefficient and the distance to the location of deformation are known, We can know the shear deformation using Fig. 10.

Table 1 shows the relationship between reflection coefficient and crimp depth, distance and width. As the crimp width is widened, the reflection coefficient is increased. If the crimp depth is too deep, pulse energy may be absorbed. It is important to consider both crimp width and depth.

3.5 Test case of Hwasoon mine field

The boreholes in which the coaxial cables were installed were 100 m deep at the C4 site and 176 m deep at the F2 site. Before installing the coaxial cables into the boreholes, coaxial cables were sprayed with paint to prevent temporal reaction between the outer aluminum conductor and water in boreholes. In order to weaken the grouting strength, high PH cement was used and cable was installed into boreholes with PVC grouting tube. To prevent from losing the data, waveforms were saved via 6 vertical scale and 6 vertical position.

Fig. 11~Fig. 12 shows TDR data for C4 and F2 site. Upward spike shown at 2.64 m is due to effect of connector with which connection cable linked to cable tester and coaxial cable are jointed. In the data at C4 site, two downward spike were shown at 25 m and 26 m distance. It is due to the defect made at the time when the coaxial cable was installed. As time went on, the reflection coefficient was increased at various depths at both C4 and F2 site. This data can be interpreted by Fig. 10, and can be transferred to the shear deformation at depth.

4. CONCLUSIONS

1. Shear test of the cemented mortar containing a specimen of coaxial cable showed that the shear deformation correlated linearly with the reflection coefficient.
2. The bending test showed that the crooked cables had no significant effect on TDR monitoring.
3. The controlled shearing test of the 50 m long cable showed that the reflection coefficient decreased exponentially as the distance increased.
4. The controlled crimping test suggests a proper crimp depth, width at a certain distance.

REFERENCES

Dowding C. H., Su M. B., O'Connor K., 1986, Choosing Coaxial Cable for TDR Monitoring, Proceedings of the 2nd Workshop on Surface Subsidence Due to Underground Mining, Morgantown, West Verginia, pp. 153~162.

Dowding C. H., Su M. B., O'Connor K., 1988, Principles of Time Domain Reflectometry Applied to Measurement of Rock Mass Deformation, Int. J. Rock Mech. Min. Sci. & Geomch. Abstr. Vol. 25, No. 5, pp. 287~297.

Dowding C. H., Su M. B., O'Connor K., 1989, Measurement of Rock Mass Deformation with Grouted Antenna Cables, Rock Mech. and Rock Engng., 22, pp. 1~23.

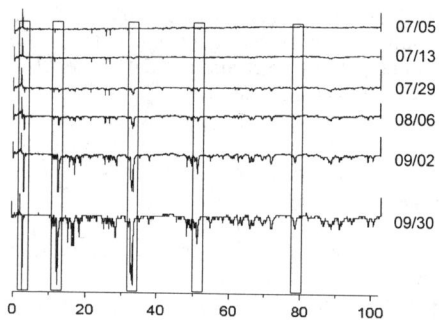

Fig. 11 TDR data for C4 site.

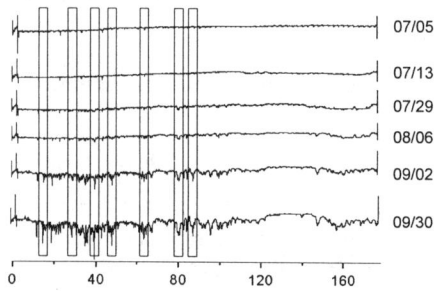

Fig. 12 TDR data for F2 site.

Loosening rock region estimated by field measurements during large underground cavern excavation

Y. Uchita & Y. Hirakawa
NEWJEC Inc., Engineering Technology Development Department, Osaka, Japan

A. Mochizuki
Osaka City University, Japan

ABSTRACT: The behavior of the rock mass was monitored during the excavation of a large underground cavern for a power station in a mountain area in Hyogo prefecture, Japan. The results of three monitoring techniques, which are the observation of discontinuities in the borehole, the measurements of the seismic wave velocity and the air permeability, were reported. It was found that the rock looseness was developed by the expanding aperture of discontinuities and the openings of hidden cracks in the rock mass. The loosening region was monitored in the side depth of 5m from the cavern wall, which is equivalent to one fifth of the excavated cavern width, when the height excavated is equivalent roughly to that width. The region was extended to be more than three fifths of the width when the last stage was achieved. These region was prevented to fail down by the support system.

1 INTRODUCTION

Since the rock in Japan is highly jointed, it is very important in the design and construction of the underground caverns that loosened regions caused by excavation are assessed and worked out countermeasures for their stabilization beforehand. It has generally been understood that the loosening region in a rock mass is caused by direct damage from blastings and developments with openings of hidden cracks and/or increasing of apertures of existing discontinuities with the progress of the excavation. As concerns this, a lots of data have been obtained by use of various monitoring techniques. However, those data described only a part of the rock behavior or behavior at a moment in the excavation, and the actual rock mass behavior is not described yet to give a generalized mechanism of them in the excavating process. Therefor, those data dose not give enough information to design the countermeasures to protect the failure of the cavern wall.

For the purpose of clarifying such rock behavior, six monitoring techniques were adopted at the site of a large underground cavern excavation for power station, such as the observation of discontinuities in the borehole, measurements of the horizontal displacement, the increments of vertical stress, acoustic emissions (AE), the seismic wave velocities between boreholes and the air and water permeability around the boreholes (Uchita 1993,1995),(Ishida 1995).

In this paper, outline of the measuring system and the measurement results obtained will be shown in the following sections.

2 SITE DESCRIPTION

The Ohkawachi pumped storage power station with a maximum output of 1.28MW (0.32MW × 4 units) was constructed since 1988 to 1995. The measurement site is the construction site of a large underground cavern of the power house and located in the middle of Hyogo prefecture, Japan. The underground powerhouse cavern is located at about 280m below the surface of the slope. The geological cross section and the initial stress state is indicated in Fig.1.

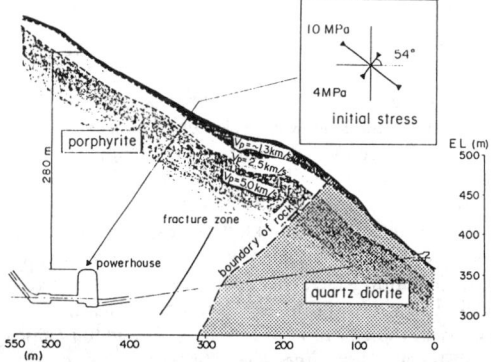

Fig.1 Geological cross section.

Table 1 Mechanical properties of porphyrite.

Mechanical properties	Mean value	numbers of data
Intact:		
Specific gravity	2.75	69
Water absorption	0.34 %	69
Unconfined compressive strength	236.7 MPa	76
Tensile strength	11.8 MPa	49
Young's modulus	76 MPa	71
Poisson's ratio	0.25	31
P wave velocity	5.71 km/s	36
S wave velocity	3.69 km/s	36
Critical strain	0.3 %	69
Rock mass:		
Young's modulus	24 GPa	18
P wave velocity	5 to 6 km/s	—
Shear strength	4.53 MPa	—
Internal friction angle	60.9 deg.	—

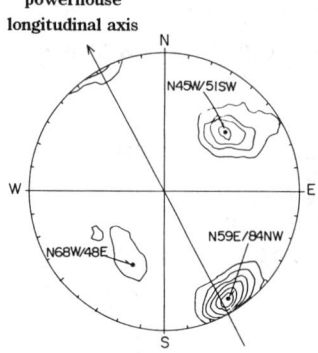

Fig.2 Stereo plots of main joint system. (lower hemisphere).

The rock mass surrounding the cavern is porphyrite of the Mesozoic era, having its property to be fresh and hard. The Q values were about 4 to 20. The mechanical properties of the rock obtained from the laboratory and in-situ tests are shown in Table 1.

There are no weak layers such as a fractured zone or fault. However, three joint sets exist. Two of them have strikes which are approximately parallel to the axis of the cavern. The other set has a strike which is perpendicular to the axis of the cavern (See Fig.2).

The cavern has a width of 24m, a height of 46.6m and a length of 134.5m, as shown in Fig.3. Its cross section has a bullet-like shape. The excavation was started at the top oval portion and the rock mass was removed in the order of ① through ④ indicated in Fig.3. Thereafter, the floor was excavated downward by 10 lifts of about 3m height using the bench cutting method.

The support system of the cavern, based upon the fundamental concepts of NATM, were installed on the excavated wall with the shotcrete of 24cm thick, the fully grouted rock bolts of 5m long and the PS anchors of 10m long with a fixation length of 4m (Yada 1994).

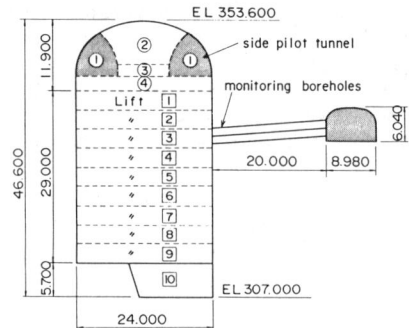

Fig. 3 Cavern cross section, its excavation steps and location of monitoring boreholes.

3 OUTLINE OF MEASUREMENT SYSTEM

The purpose of the measurements is to monitor the behavior of the side wall rock during the excavation process. Therefore six horizontal boreholes with a length of 20m were bored perpendicular to the axis of the cavern from an access tunnel in the tailrace side rock to install many kind pieces of apparatus as shown in Fig.3. The tunnel was previously excavated at the location which was 20m away from the cavern wall. The elevation of the boreholes was approximately at the elevation of lift No.3. The initial values were taken at the time when the excavation of side pilot tunnels ①, shown in Fig.3 was completed.

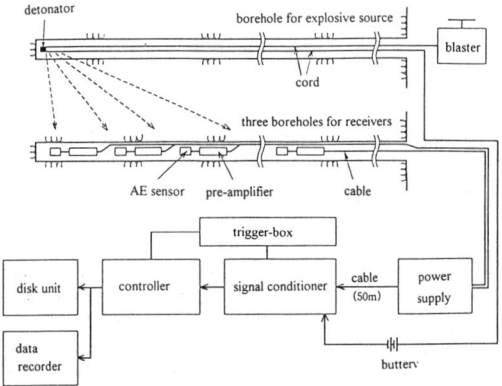

Fig.4 Schematic diagram of seismic wave data acquisition system.

After that, the measurements were carried out at each stage with progressive cavern excavation. The method of these measurements are as follows.
The observation of discontinuities was performed by means of Borehole Television (BTV)(Kamewada

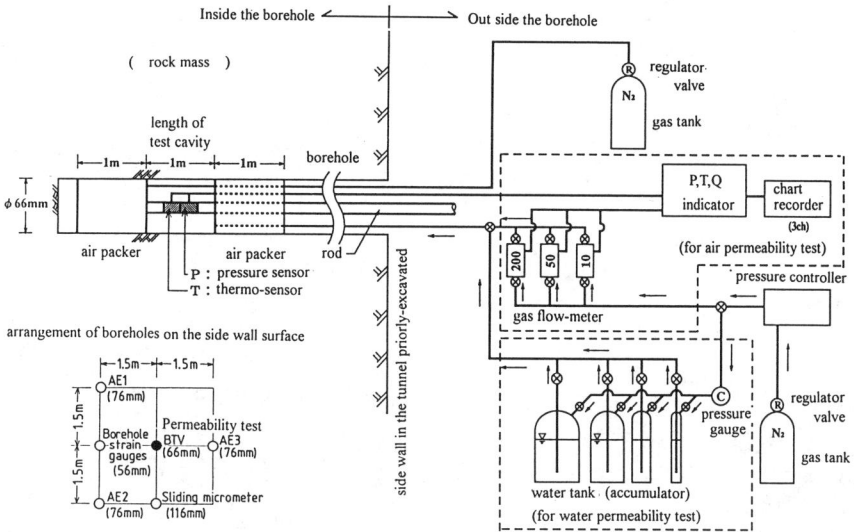

Fig.5 Schematic view of the permeability test system for air and water seepage in the rock mass.

1990), in order to directly detect the location, the direction and opening state of the discontinuities in the borehole which is arranged in the center of six holes. The measurement of the relative horizontal displacement was performed by using of Sliding Micrometer (SMM)(Kovari 1979) which is capable of measuring the relative displacement every 1m section with a high precision.

The measurements of seismic wave velocity and air permeability were carried out to assess quantitatively the loosening region of surrounding rock mass. The seismic wave were measured by using the 24 receivers (AE sensors) buried in three boreholes around the center hole in which an explosive was detonated at 13 points. The longitudinal wave velocities were computed from the measured data. The schematic diagram of seismic wave data acquisition system is shown in Fig.4. The air permeability were also measured by using the permeability test system for air and water seepage in the rock mass as shown in Fig.5, which is capable of measuring the every 1m section formed with double packer unit. Amount of air seepage flow in the rock mass was measured under gas pressure controlled less than 0.1 MPa in the center hole. This test method has more sensitive compared with water permeability test because rock mass surrounding a cavern was usually under unsaturated condition and the kinematic viscosity of nitrogen gas is one sixtieth of that of the water approximately.

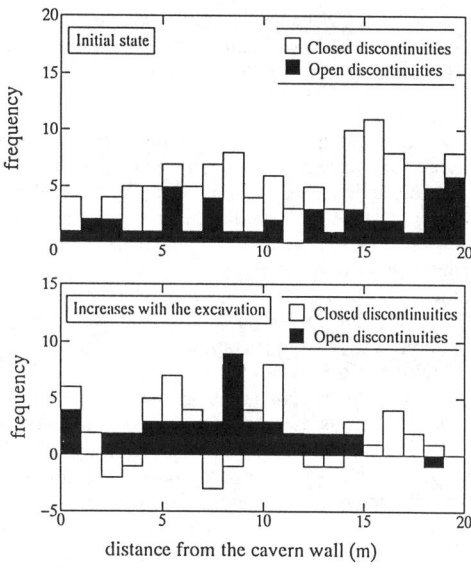

Fig.6 The frequency distribution of discontinuities.

4 ROCK MASS BEHAVIOR DURING UNDERGROUND CAVERN EXCAVATION

4.1 Behavior of discontinuities

The state of discontinuities in relation to the excavation was directly observed by the BTV. The existing discontinuities were defined as open or closed ones by whether these were recognized an aperture width or not. Fig.6 shows the frequency distribution of

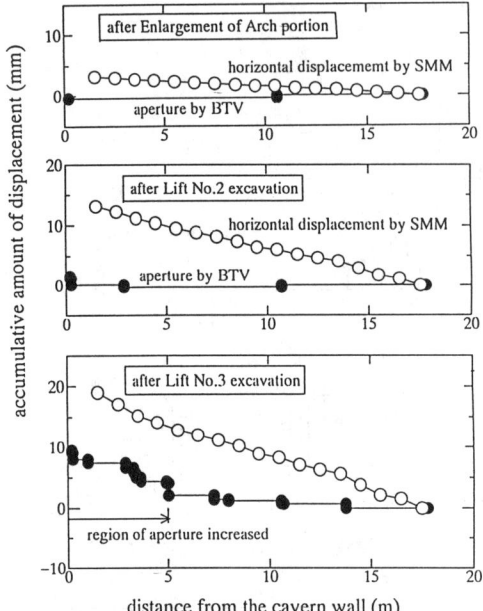

Fig.7 Comparison of accumulative amount between horizontal displacement and aperture increment.

closed and open discontinuities under initial state (before the cavern excavation) and the increased ones caused by the excavation. From the lower figure, it is recognized that open discontinuities in the region from cavern wall surface to 15m increased about two times with the excavation.

Fig.7 shows the accumulative amount of horizontal displacement measured by SMM and of increase apertures with the excavation from arch portion through lift No.3. After the excavation of lift No.2 completed, the accumulative amount of horizontal displacement was 13.8mm, which was calculated as 0.09% in average strain, in the measured section. As compared with this, the accumulative amount of aperture increments was quite small, therefore it is considered that the rock mass behavior in the section was dominated by elastic deformation. After the excavation of lift No.3 completed, the accumulative amount of the displacement and the average strain were 19.1mm and 0.12% respectively. As compared with this, the accumulative amount of aperture increments was 7.9mm. Particularly, aperture increment in the region from cavern wall surface to 5m was 7.4 mm, occupied about 80% of the displacement and 75% of whole aperture increment. Such region where the aperture increment dominated the displacement of the rock mass was definited as loosening region, here. In addition it, the measuring probe for SMM could not be installed after the excavation of lift No.4 because the borehole deformed.

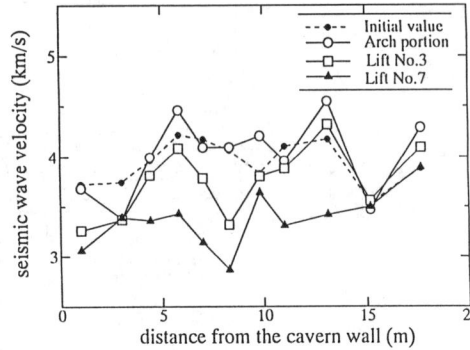

Fig.8 Distribution of seismic wave velocity (Vp).

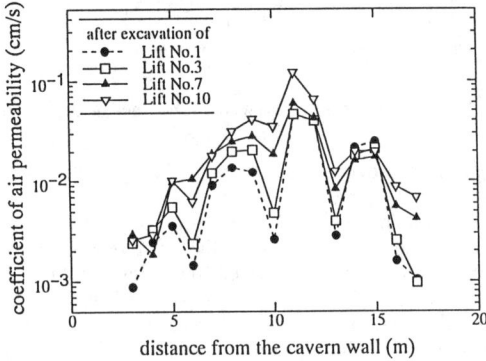

Fig.9 Distribution of air permeability.

4.2 *Distribution of seismic wave velocity and air permeability*

Fig.8 shows the distribution of seismic wave velocity (Vp) which was the mean values in wave lines in the vicinity around the explosive points. The average velocity of the whole initial values indicated 4 km/s and the velocity made an irregular distribution with two peaks. After the excavation of the arch portion, the average velocity increased to make the distribution with four peaks. After the excavation of lift No.3, the velocities decreased at the region of 1 to 3m and of 7 to 8m from the cavern wall. Moreover after the excavation of lift No.7, the average velocity decreased remarkably at the region from 1m through 13m. In addition it, the velocity could not be measured after the excavation of lift No.8 because of cutting the lead wire of many receivers.

Fig.9 shows the distribution of air permeability along the borehole. After the excavation of the lift No.1, the coefficient of air permeability indicated the order of 10^{-3} to 10^{-2} cm/s with the irregular distribution which have maximum value in the middle part of the rock mass and five bottom values. With the progress of the excavation, the permeability increased all over the rock mass except the portion of 14 to 15m from the cavern. Moreover, the permeability at the five portions having bottom value as mentioned above, increased remarkably with the excavation of lift No.3 through lift No.10. In addition it, the packer unit for the permeability test could not be installed after the excavation of lift No.4 and lift No.8 because the borehole deformed, but the borehole was recovered it by means of reboring after the excavation of lift No.7 and lift No.10 respectively.

5 DISCUSSIONS

Fig.10 shows the distribution of increments in the velocity with the enlargement of the arch portion compared with the vertical stress increment (Uchita 1995) measured in the near borehole at the same time. Both of the distributions have a close correspondence. The velocity increments indicated the distribution having four peaks at intervals of 3 to 5m except the portion nearby 3m. Accordingly, it is considered that the velocity express the aperture states in the rock mass, such as the velocity increase suggest existing apertures and micro cracks to close by a concentration of the stress redistributed with the excavation.

Fig.11 shows the comparison of the distributions among the increments in aperture, the increase ratio in the permeability and the reduction in the velocity every representative excavation stages. In this figure, the velocity increase shows downward of longitudinal axis. The main issues from this figure were pointed out as follows.

① Until after the excavation of lift No.2, the rock mass indicated an elastic deformation because the velocity increased only and others did not change almostly.
② The whole measured issues with the excavation of lift Nos.3 through 10 corresponded each other in macro view point.
③ The velocity increment detected the rock mass loosening in the earlier excavation stage, compared with the aperture increment. Because of this, the former expresses three dimensional information in rock mass state around a explosive point, but the latter expresses only a information along a borehole.
④ The increase ratio in permeability, comparing with the aperture increment, is not found the correspondence directly between the both. Because of this, the former expresses the rock mass volume of connected discontinuities around the borehole.
⑤ In comprehensive estimation of measured issues, when the height of the excavated space was nearly equal to the cavern width, the horizontal depth of the loosening region at the monitoring location was about one fifth of the cavern width, and when the height/width ratio of the excavated space was grown

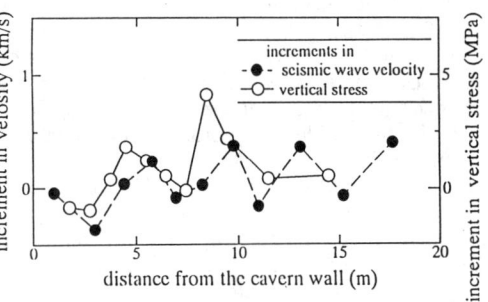

Fig.10 Distribution of increments in the velocity with the enlargement of the arch portion compared with the vertical stress increment.

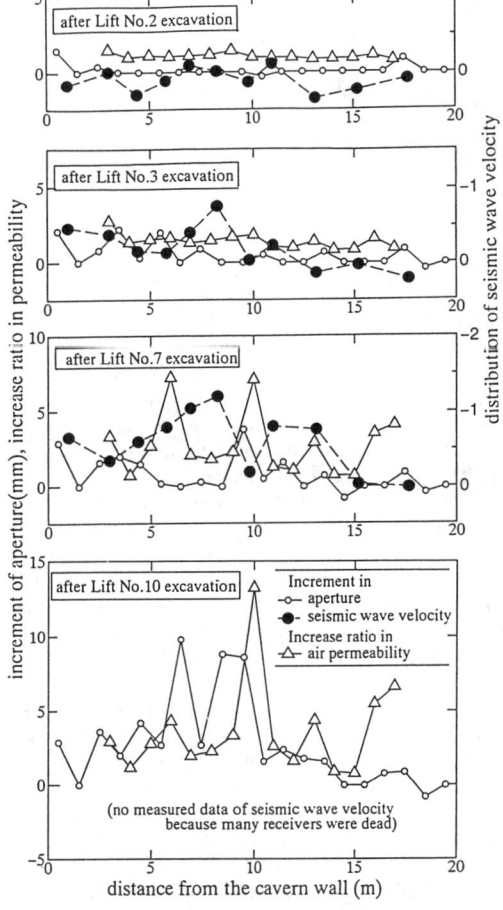

Fig.11 Comparison of the distributions among the increments in aperture, the increase ratio in the permeability and the reduction in the velocity.

to about 2, its depth extended to more than three fifth of the cavern width.

From Fig.10 and Fig.11, the portion in the side wall rock, where the velocities increased with the enlargement of the arch portion, corresponded approximately with the portions at where the aperture and the permeability increased remarkably with the excavation of lower portion. These issues suggest that the vertical stress concentration made an extension of the loosening region with a release in the side wall confinement by the excavation.

Consequently, the extending process of the loosening region with the excavation was considered as follows. The openings of hidden cracks and/or the increasing apertures of existing discontinuities were developed at the portions, where stress concentration occurred, due to the release in the side wall confinement with the progress of the excavation. In the final excavation stage, the loosening region was extended toward a deeper portion with a sliding in the discontinuities which inclined toward the cavern due to the height in the excavated space is longer than its width. These region was prevented to fail down and the stabilization of the cavern wall was kept by the support system as mentioned above.

6 CONCLUSION

Through the measurements and discussions, the following conclusion was obtained.
1) The rock deformation in the horizontal direction indicated elastic behavior in the region where was less than 0.1% in average strain, but it was dominated by increase of aperture width of discontinuities in the region where was more than 0.1% in average strain. Therefore, authors defined the latter as the loosening region in this paper.
2) The horizontal depth of the loosening region at the location of monitoring portion was about one fifth of the cavern width at the time when the excavation height was nearly equal to the cavern width. When the height/width ratio of the excavated space was grown to about 2 after excavation completion, its depth extended to more than three fifth of the cavern width.
3) The seismic wave velocity (Vp) between boreholes indicated clearly decrease at the loosening rock region with progress of the excavation. The increased zones and the decreased zones of Vp at the early excavation steps corresponded approximately to the stress concentration zones and the stress relaxation zones respectively. However, progressing to final excavation stage, the difficulty of the measurements increased because the aperture of the discontinuities increased its width at the loosening region.
4) The air permeability test loading low pressure in the rock mass could be detected the loosening region by increase of permeability. The region increased permeability and its extending process during the excavation indicated good correspondence to the loosening region.

ACKNOWLEDGMENT

During this study, we were grateful to obtain advice and support from Mr.T.Kanagawa of the Central Research Institute of Electric Power Industry (CRIEPI) and Dr.T.Ishida of Yamaguchi University, and comments on the data and kind help in our field work from Mr.A.Yada of the Kansai Electric Power Co.,Inc. and Mr.T.Nakamura of Toda Construction Co.,Ltd. We would like to express our warmest thanks to all of them for their contributions.

REFERENCES

Ishida,T. et al. 1995. Acoustic emission mechanism and rock mass behavior as deduced from in situ measurements during progressive excavation of an underground powerhouse. *8th Congress in International Society for Rock Mechanics*: 593-596. Tokyo.

Kovari,K. & Ch.Amstad. 1979. Decision making and field measurements in tunneling. *25th OYO Anniversary Lecture Meeting*. Tokyo.

Kamewada,S. et al. 1990. Application of borehole image processing system to survey of tunnel. *International Symposium on Rock Joints* :51-58. Loen.

Uchita,Y. et al. 1993. Behavior of discontinuous rock during large underground cavern excavation. *International Symposium on Assessment and Prevention of Failure Phenomena in Rock Engineering*: 807-816: Istanbul.

Uchita,Y. et al. 1995. Rock behavior measured by borehole strain gauges during large underground cavern excavation. *4th International Symposium on Field Measurements in Geomechanics*: 89-96. Bergamo.

Yada,A. et al. 1994. Excavation control of underground powerhouse cavern. *IV CSMR, Integral Approach to Rock Mechanics* :523-534. Santiago.

Visualization of three-dimensional structure of rocks using X-ray CT method

K.Sugawara, Y.Obara, K.Kaneko, K.Koike, M.Ohmi & T.Aoi
Faculty of Engineering, Kumamoto University, Japan

ABSTRACT: The subject of the present paper is to make clear the potential roles of the X-ray CT method in the field of rock mechanics. Principle of the method and the apparatus employed are briefly described with several case examples. The measurements of pore-geometry in rocks, the visualization of the fracture under various circumstances and loading conditions, the measurement of the aperture of open cracks and the analysis of the fluid migration within rock are demonstrated as well as the data-integration methods to display the variations in, and the structure of, the X-ray absorption data in the optimal way for final visual interpretations, aiming at the most efficient way of extracting necessary information.

1 INTRODUCTION

X-ray CT (Computed tomography) is a nondestructive medial imaging technique (Hounsfield (1972)), and its efficiency is widely recognized. Usual photograph of X-ray is a shadow picture to indicate the features of the attenuation of an X-ray beam through a body. On the other hand, the X-ray CT makes it possible to evaluate the attenuation of X-ray at each point within a body, by digitalizing shadow pictures taken from various directions and subsequent image reconstruction by the filtered back-projection method (e.g. Takagi and Shimoda (1991)).

In the field of geomechanics, the X-ray technique was applied to examine the strain localization in sand specimens, by Kirkpatrick et al. (1968), Roscoe (1970), Bransby et al. (1975), Arthur et al. (1977), Scarpelli et al. (1982), Vardoulakis et al. (1982). Although they used X-rays, these studies were not tomographic.

A pioneering paper of the X-ray CT in the field of geomechanics has been presented by Arthur (1971), and followed by many researchers. Tillard-Ngan et al.(1992) and Desrues et al.(1996) were reported on quantitative local measurements of density in sand. On rock specimens subjected to axisymmetric triaxial loading, Vinegard et al. (1991) made clear that both initial heterogeneity and density changes induced by failure mechanisms can be accurately described with the X-ray CT. Verhelst et al.(1995) have concluded, through the study on tensile fracture of the Brazilian test specimens, that the X-ray CT is a promising tool for rock mechanics.

2 PRINCIPLE OF X-RAY CT

The attenuation of X-ray beams by materials is used to reconstruct a cross-sectional image of the object. In the one-dimensional situation as shown in Fig.1, the attenuation of X-ray is expressed by

$$i = i_0 \cdot \exp\{- \int f \, dr \}, \qquad (1)$$

where i : the intensity of X-ray flux after passing the object in air, i_0 : the intensity of X-ray flux at the source, r : the distance from the source, f : the attenuation coefficient depends on the location.

The attenuation coefficient f is a measure for the probability that a X-ray would be attenuated by traversing a unit length of the material, associated with the Compton scattering, the Rayleigh scattering and the photoelectric absorption. This is mainly a function of the electron density of the material and of the energy of the X-ray beam, then the attenuation coefficient f is a constant material property for the X-rays with one single energy level.

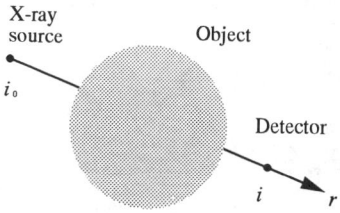

Fig.1 Schematic view of one-dimensional X-ray attenuation.

3 TOSCANER-23200

TOSCANER-23200 is the X-ray CT scanner installed in Kumamoto University, which has been developed for an industrial use by the Toshiba Corporation. Fig.2 shows the system. The X-ray bulb, 300kV/ 2mA, and the detectors of 176 in number are fixed in a horizontal plane, and they can be moved together in the vertical direction, with a specified interval. The specimens are set on the turntable for traversing and rotation, of which the axis is vertical. By the detectors, a few attenuation profiles is recorded for each incidence.

The collimator determines the width of a measured slice, and the cross-sectional image of the slice is reconstructed from a set of profiles by the filtered backprojection method.

The slices can be placed next to each other to reconstruct a three-dimensional image of the rock specimen. In the following measurements, the width of the slice is set in 3 mm. Hence, the radiographic density obtained represents the average over a 3 mm thick slice of the specimen.

The radiographic density map is obtained as a discrete set of numbers supposed to represent the average density over small elements of the picture, termed pixels. In the standard configuration, each pixel represents a 0.29 mm x 0.29 mm square.

The output datum of the attenuation coefficient at each pixel is expressed by the Hounsfield unit : HU, that is defined as -1000 HU in air, and 0 HU in water.

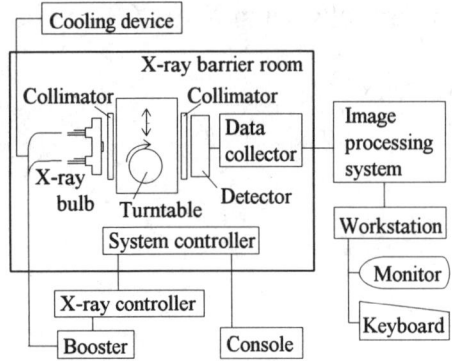

Fig.2 Composition of TOSCANER-23200.

By taking logarithms of eq.(1), the projection p is defined as follows:

$$p = -\log\{i / i_0\} = \int f \, dr \quad (2)$$

and the filtered backprojection method is usually applied to determine the attenuation coefficient at each point within the specimen from the projections of various directions (Takagi and Shimoda (1991)).

The attenuation coefficient which results from the X-ray CT is termed the radiographic density, since it is predominantly proportional to the electron density of the material under inspection. The cross-sectional image of the attenuation coefficient is usually termed the radiographic density map, and the magnitude of the attenuation coefficient is generally expressed by the Hounsfield scale, as described later.

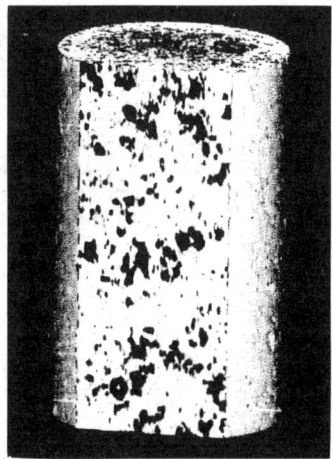

Fig.3 A three-dimensional image of a cylindrical Ryukyu limestone, 19.3 cm in diameter, reconstructed by the X-ray CT.

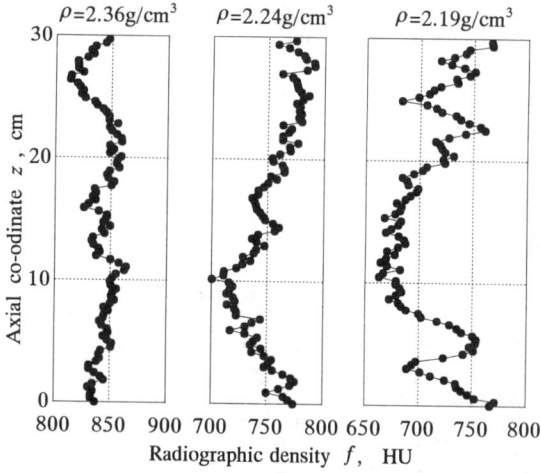

Fig.4 Cross-sectional average of the radiographic density for each slice of the Ryukyu limestone, depending on the axial co-ordinate of the specimen.

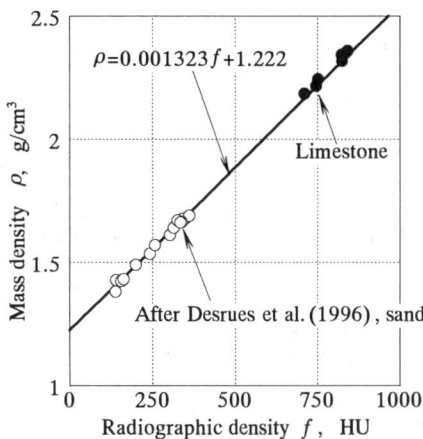

Fig.5 Relation between the average of the radiographic density over of the specimen and the mass density.

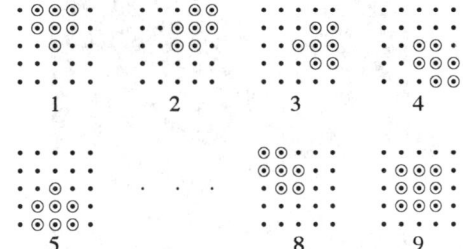

Fig.6 Sub-regions in a 5 x 5 local domain for the edge preserving smoothing.

4 CASE EXAMPLES

4.1 *Measurement of Pore-geometry*

As previously described, the attenuation coefficient which results from the X-ray CT is predominantly proportional to the local mass density of the object. In rocks, it can be linked to the initial heterogeneity, the joint network, the pore-geometry and so on.

Fig.3 shows the density structure of the Ryukyu limestone, investigated using a cylindrical specimen of 19.3 cm in diameter. Clearly, it is indicated that the present system is available for the quantitative measurement of the pore-geometry. Adequate image processing presents the first order pore parameters include pore area, diameter, perimeter orientation, and vertical and horizontal connectivity, as well as the second order parameters such as porosity, pore density, specific surface area, pore shape and tortuosity. These are correlated to the deformability, the strength characteristics and the permeability.

The average of the radiographic density for each slice changes as shown in Fig.4. The fluctuation of the cross-sectional average along the specimen axis is decreasing with increasing the value of the mass density ρ of the specimen. This suggests that the representative elementary volume of the Ryukyu limestone may correlate to the mass density.

Fig.5 shows the relation between the total average of the radiographic density over of the specimen and the mass density of the Ryukyu limestone, defined by the global measurement of the specimen mass divided by its volume, comparing to the data of sand presented by Desrues et al.(1996). It seems that the linear relation given by Desrues et al. is realized in the range from 1.40 g/cm^3 to 2.35 g/cm^3.

4.2 *Edge Preserving Smoothing*

Adequate smoothing is necessary to eliminate the noise, which is un-avoidable in the reconstruction. For rocks, the edge preserving smoothing (Nagao et al. (1978)) is considered to be promising, since it has a function of image sharpening.

In this method, 9 sub-regions are defined in a 5 x 5 local domain as shown in Fig.6, and the variations of the data in the 9 sub-regions are compared. The value of the center pixel in the 5 x 5 local domain is modified by the average in the sub-region having the minimum variation.

The edge preserving smoothing is demonstrated in Figs.7 and 8. The cross-sectional image of concrete in Fig.7 shows that this has the function of image sharpening indispensable for the grain visualization. The cross-sectional images of andesite, in Fig.8, clearly show that an excessive repetition of the edge preserving smoothing results in vanishing of thin cracks. The andesite specimen, in Fig.8, 5.5 cm in diameter and 20 cm in height, has been exposed to the repetitive freezing and melting in water, prior to the measurement, to investigate the mechanism of weathering.

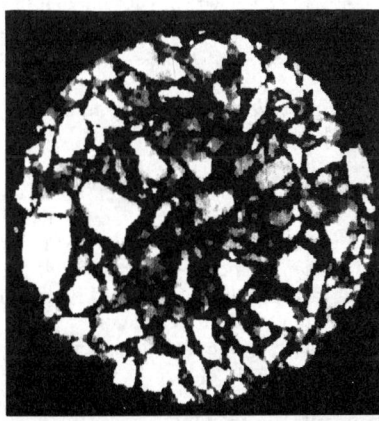

Fig.7 A cross-sectional image of concrete by the edge preserving smoothing.

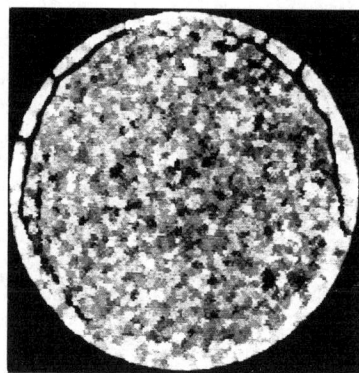

(a) two times smoothing

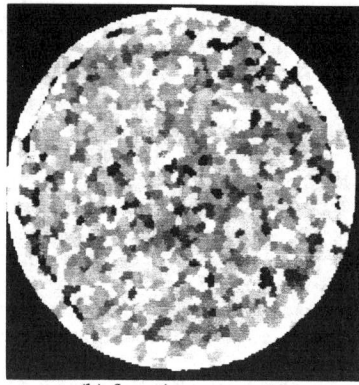

(b) four times smoothing

Fig.8 Cross-sectional images of andesite by the repetition of the edge preserving smoothing, showing the cracks grown by the repetitive freezing and melting in water.

Tangential cracks hidden behind the surface of the specimen is successfully extracted, and it is verified by the X-ray CT that the sufficient amount of water for the tangential crack growth is supplied through radial open cracks generated from the surface.

4.3 Measurement of Crack Aperture

Quantitative measurement of crack-geometry is an interesting and important subject in the application of X-ray CT to rock mechanics. In the following, the method to evaluate the aperture of a single straight dry crack hidden in rock will be presented and the accuracy of measurement is discussed by examining the results of joint model experiments.

The theoretical relationship for the crack aperture evaluation is led from eq.(2). In the case that the radiographic density is given on a line of length L, perpendicular to the crack of aperture w, as shown in Fig.9(a), it is as follows:

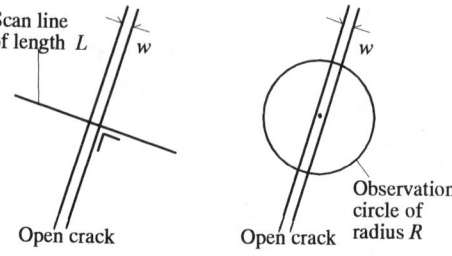

(a) one-dimensional   (b) observation circle method

Fig.9 Schematics of crack aperture measurements.

$$f_{average} \cdot L = f_{solid} \cdot (L - w) + f_{air} \cdot w \quad (3)$$

where $f_{average}$: the average of the radiographical density on the line; f_{solid}: the mean value of the radiographical density of rocks facing each other across the crack; f_{air}: the radiographical density of air, that is -1000 HU. By solving eq.(3) for w, we obtain

$$w = L \cdot \{f_{solid} - f_{average}\} / \{1000 + f_{solid}\}. \quad (4)$$

This is the formula to calculate the aperture from the X-ray attenuation coefficient on a line L, then it can be termed one-dimensional method.

One-dimensional method can be easily extended to two-dimension. The two-dimensional evaluation is performed using the data in a slice perpendicular to the open crack, and two practical schemes will be proposed. One is the stack of the one-dimensional method. In this scheme, the line for measurement

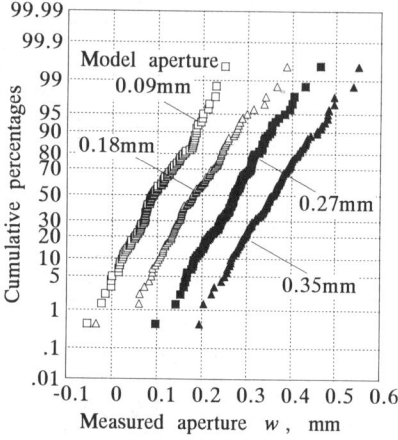

Fig.10 Distributions of the aperture measured by the one-dimensional method, plotted in the normal probability paper.

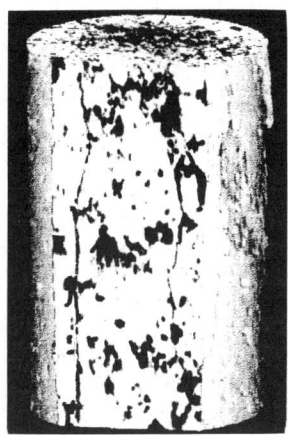

Fig.11 A three-dimensional image of the Ryukyu limestone fractured by the uniaxial compressive loading.

needs to be moved step by step along the crack, for example with an interval of one pixel. The result obtained is the average of aperture in the region covered by the lines for measurement. The other method uses an observation circle. The center of the circle is set in the crack, as shown in Fig.9(b). The crack is assumed rectangular of $2R \times w$, and the mean aperture w is given by

$$w = \pi R/2 \cdot \{f_{solid} - f_{average}\} / \{1000 + f_{solid}\}. \quad (5)$$

In order to examine the applicability of eq.(4), the joint model experiments have been carried out using two flat aluminum-plates. The clearance between the plates is set to 0.09, 0.18, 0.27 and 0.35 mm, and the one-dimensional method has been applied, with a interval of one pixel. The results obtained by the measurements of 110 times for each clearance are summarized in Fig.10, that is the normal probability paper. The linear relations approximately parallel for each other are confirmed, then it can be concluded that the error is obedient to the normal probability law. The standard deviation is about 0.1 mm, and the averages of the measurements coincide with the clearances of the model joint, respectively.

From the model experiments described above, it can be concluded that the one-dimensional method is inaccurate, but the stack can provide the accuracy necessary for the determination of the crack aperture.

4.4 *Visualization of Fracture*

The valuable information on the fracture mechanism of rocks is presented by the pertinent smoothing and the subsequent grey-level thresholding of the image. In the three-dimensional measurements of fracture, the combination of zooming and perspective drawing is effective for the local investigation, as well as the step by step survey using the cross-sectional images demonstrated in Fig.11.

4.5 *Analysis of Transitional Phenomena*

The X-ray CT is available for the investigation of various transitional phenomena, include weathering, degeneration and deterioration, in which the radiographic density maps change with the time elapsed. For the observation of the directing phenomena, the suitable smoothing method needs to be applied as well as the arithmetic operation between the serial

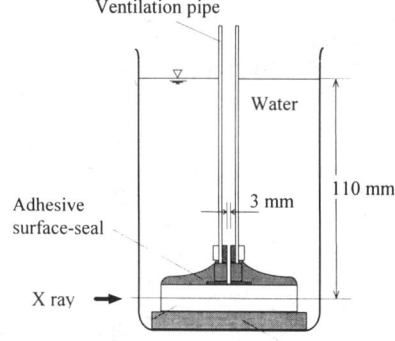

Fig.12 The radial water permeation experiment using a disc of Kimachi sandstone.

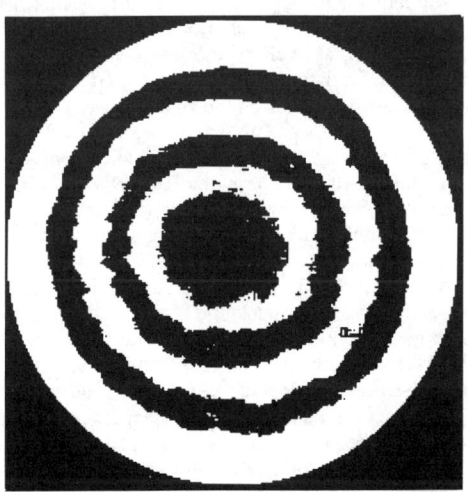

Fig.13 A compound image of the saturation front lines, obtained by means of the subtraction of images.

images, that is usually the subtraction of images.

As an example, the water permeation experiment of the Kimachi sandstone, in Fig.12, is presented with a brief description of processing techniques. A dry sandstone disc, 7 cm in diameter and 1 cm in width, coating both flat end-planes by adhesive seals, has been steeped in pure water to investigate the water penetration of mainly radial direction. To discharge the air preexisted in sandstone specimen. a ventilation pipe is installed at the center of the disc.

Fig.13 is a compound image of saturation front lines, obtained by applying the local region simple averaging, the subtraction of images and the gray-level thresholding. The noise is at first successfully eliminated by the simple averaging, in which the local region is set to 11 x 11, and the value of the center pixel is modified by the mean of the local region, since such an enlargement of the local region is much effective for the extraction of monotonous phenomena like water penetration. The subtraction between the smoothed images and the gray-level thresholding provide the saturation front line of each time elapsed. In Fig.13, the five uneven boundaries of three white rings represent respectively the front lines at 1h, 3h, 6h, 9h and 12h, in order from the outside. The irregularity of the saturation front is closely depending upon the initial heterogeneity of the Kimachi sandstone.

5. CONCLUDING REMARKS

A set of experiments using the TOSCANER-23200 has been presented to indicate the usage of the X-ray CT in rock mechanics. The successful visualizations of the initial heterogeneity of rock, the fracture network and the water migration within rock have been demonstrated, as well as the image processing to display the variations in, and the structure of, the X-ray attenuation data in the optimal way for final visual interpretations, aiming at the most efficient way of extracting necessary information. From these experimental studies, the X-ray CT method has been concluded to be a powerful nondestructive technique for rocks, and available for various subjects in rock mechanics.

The edge preserving smoothing has been confirmed to be effective to visualize the grain boundaries, the pore-geometry, the joint network within rocks and the fracture mechanism of rocks under the various circumstances and loading conditions. The method for the aperture measurement of open cracks hidden in rock has been proposed, and the spatial resolution of the system has been discussed by examining the accuracy of the aperture measurement by the joint model experiments. The enlargement of the area for the local region averaging has been discussed to be useful for the extraction of monotonous phenomena like the water permeation in rocks. Subsequently, the arithmetic operation between the serial images has been presented to indicate how the transitional phenomena in rocks are visualized from the changes of radiographic density maps with the time elapsed.

REFERENCES

Arthur, J.R.E. (1971): New techniques to measure new parameters, Proc. of Roscoe Material Symp. on stress-strain behaviour of soils, G. T. Foulis, Cambridge, pp.340-346.

Arthur, J.R.F., T. Dunstan, Q.A.J.L. Al-Ani & A. Assadi (1977): Plastic deformation and failure in granular media, Geotechnique, 27, 1, pp.53-74.

Bransby, P.J. and P.M. Blair-Fish (1975) : Deformation near rupture surface in flowing sand, Geotechnique, 25, 2, pp.384-389.

Desrues, J., R. Chambon, M. Mokni and F. Mazerolle (1996): Void ratio evolution inside shear bands in triaxial sand specimens studied by computed tomography, Geotechnique, 46, 3, pp.529-546.

Hounsfield, G.N. (1972): A method of and apparatus for examination of a body by radiation such as X- or Gamma-Radiation: London, British Patent No.1283915.

Kirkpatrik, W.M. and D.J. Belshaw (1968): On the interpretation of the triaxial test, Geotechnique, 18, 3, pp.336-350.

Nagao, M. & T. Matsuyama(1978): Edge preserving smoothing, CGIP, 9, pp.394-407.

Roscoe, K.H.(1970): The influence of strains in soil mechanics, Geotechnique, 20, pp.129-170.

Scarpelli, G and D. Muir Wood (1982): Experimental observations of shear band patterns in direct shear tests, Proc. of IUTAM Conf. on defects and failure in granular media, Balkema, pp.473-484.

Takagi, M. and H. Shimoda (1991): Handbook of Image Analysis, Tokyo Univ. Pub., pp.356-371.

Tillard-Ngan, D., J. Desrues, S. Raynaud and F. Mazerolle (1992): Strain localization in the Beaucaire marl, Geotechnical engineering of hard soils-soft rocks, Balkema, pp.1679-1686.

Vardoulakis, I. and B. Graf (1982): Imperfection sensitivity of the biaxial test on sand, Proc. of IUT AM Conference on defects and failure in granular media, Balkema, pp.485-491.

Verhelst, F., A. Vervoort, PH. de Bosscher and G. Marchal (1995): X-ray computerized tomography, Determination of heterogeneities in rock samples, Proc. of 8th ISRM Cong., Tokyo, 1, pp.105-108.

Vinegard, H.J., J.A. de Waal and S.L. Wellington (1991): CT studies of brittle failure in Castlegate sandstone, Int. Rock Mech. Min. Sci. Geomech. Abstr., 28, 5, pp.441-448.

A study on the application of electrical resistivity monitoring technique to detection of seawater intrusion

Tae-Seob Kang, Il-Yeong Han, Jeong-Ho Lee & Soon-Jo Hong
R&D Center, Sunkyong Engineering and Construction Limited, Seoul, Korea

ABSTRACT: It is believed that underground low-level radioactive waste disposal site should be protected from seawater intrusion which might do harm to the underground facilities. Therefore, it is inevitably requested that seawater intrusion monitoring system as well as the other environmental protection schemes be included in the environment monitoring management program for the effective management of underground low-level radioactive waste disposal site. A monitoring technique using electrical resistivity method, which enabled us to get precious information among surface observation holes, was applied to the underground cavern construction site which was quite similar to low-level radioactive waste disposal site in a view point of geological condition for location. Field experiments were designed so that the electrochemical relationship between the resistivity and seawater NaCl concentration might be identified. The test was the way in which NaCl tracer was injected, which was done simultaneously with electrical resistivity measurement at surface into a pressurized horizontal hole at water curtain tunnel and water sampled at a surface observation hole 85m away from the injection hole. During the tracer test, the electrical resistivity showed a similar variation trend of tracer concentration. And the results of electrical resistivity measurements showed a good correlation with the tidal variation. Using the result of experiment, the variation model of apparent resistivity due to seawater intrusion was proposed introducing logistic function of regression analysis. And then the conversion model which infers chloride concentration from surface electrical resistivity monitoring was proposed.

1 INTRODUCTION

Most commonly used methods to characterize seawater intrusion involve collecting a sample of water and analyzing it for targeted species sodium chloride (NaCl) or chloride ion (Cl⁻) in the laboratory. In general, this way of characterizing the seawater intrusion is the only one acceptable to whom it may concern. However, sampling and analysis in laboratory have some difficult problems; time consuming, expensive, discontinuous sampling time, and limitation of information. To overcome these difficulties, many surface and subsurface geophysical methods have been studied. Of all geophysical techniques, electrical resistivity methods have had the most widespread use in groundwater investigations because the groundwater decrease the pore resistivity in contrast with the surround medium (Saunders and Stanford, 1984; Rodriquetz, 1984; Van Overmeeren, 1989; White, 1994; Asch and Morrison, 1989; Bevc and Morrison, 1991).

Especially, the pore resistivity in area saturated with seawater is relatively lower than that in fresh water saturated area. If the relation between resistivity distribution and concentration distribution of subsurface medium is known, it may be possible that the intensity of seawater saturation can be known from the distribution of surface electrical resistivity. Thus, it is the aim of this paper to apply of electrical resistivity monitoring technique to detection of sea water intrusion. To achieve this aim, a salt water tracer test and electrical resistivity monitoring on ground surface were performed. An empirical relationship between seawater concentration and correspondent resistivity change is derived.

2 EXPERIMENTS

2.1 Tracer test

The experiments took place in September, 1996 at the construction site of underground cavern. The top of cavern is located in 30 m below sea level. Water curtains to maintain stable groundwater pressure are installed at 0 m above sea level.

Analysis of geological data inside cavern shows that Cretaceous granite forms a host rock around the cavern. Also according to drilling logs of water curtain holes, tuffaceous rock covers the upper region of

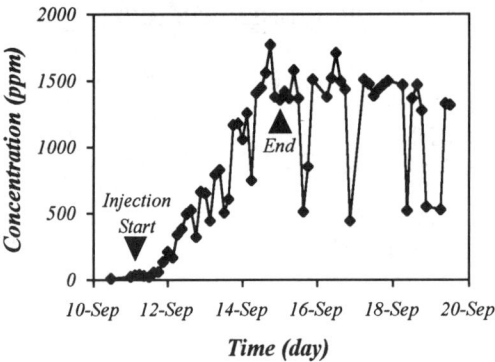

Figure 1. Plots of the variation for the chloride concentration of sampling water

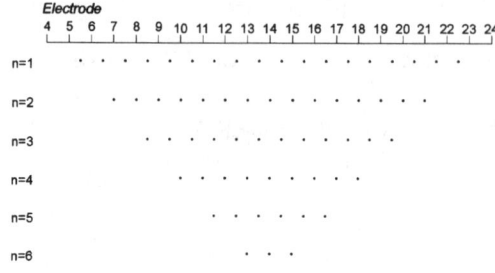

Figure 2. Apparent resistivity pseudosection of a surface Wenner imaging

water curtain holes. In other words, water curtain holes are located in the boundary between tuffaceous rock and granitic rock intruding volcanics.

One of the water curtain holes was designed for fluid injection. Sampling of water was performed at a surface observation hole about 85 m away from the injection hole. Salt water for tracer test was made at mixing tank installed in water curtain tunnel. After mixing and settling to remove residuals, solution was passed through filters and pumped into water curtain hole selected as injection well.

A total of 32 tons of salt water was injected at an injection pressure of 1.8 kg/cm^2 for 96 hours. The injection pressure was fixed taking into account operating condition of water curtain holes. The initial concentration of chloride (Cl$^-$) in solution was about 7000 ppm. Also background concentration in observation hole before injection was 9.7 ppm.

Sampling of water in observation hole was started just after injection at intervals of 3 hours. This resulted in a total of 78 samples and it were analyzed in the site and laboratory. Figure 1 shows the plots of the variation for the chloride (Cl$^-$) concentration in the observation hole for 10 days after the injection of salt water. The maximum value of observation concentration was measured to be 1,770 ppm just prior to the end of injection, and which was about 25 percent of initial concentration. In the distribution, it was shown that the concentration has periodically undergone abrupt drop after an end of injection as well as in the middle of it. This pattern might be caused by the repetition of a rise and drop of water level in observation hole due to tidal effect.

2.2 *Electrical resistivity monitoring*

During the experiment resistivity measurements were made by injecting current at the surface electrodes along the profile line that connects observation hole and ground surface location over injection hole. The equipment used to monitor resistivity variation is STING R1 Earth Resistivity Imaging System and its accessories by Advanced Geosciences Inc.. By using these equipment, it was possible to carry out resistivity imaging with multiple electrodes configuration. A Wenner configuration with 21 electrodes was used to conduct resistivity imaging and the spacing between adjacent electrodes was fixed at 10 m. To maintain the invariant condition of an earthed electrode, the installed electrodes before measurements was unmoved throughout the whole experiment term. Input current of 500 mA was used to enhance signal to noise ratio in all measurements.

An initial data set was taken before salt water injection to obtain the background values. To consider the short term variation by external effect like tide or hydrological change, repeatability measurements of 5 times were performed.

Figure 2 is the apparent resistivity pseudosection of a surface Wenner imaging along the profile line. There was a total of 63 points on pseudosection that could obtain the resistivity value by Wenner configuration using 21 electrodes. The notation from 4 to 24 indicates the location number of electrodes used to conduct monitoring. Originally, a set of electrodes configuration used in this experiment was composed of 28 electrodes. However, because of ground condition, 7 electrodes of the set were not included in measurements.

The measurements were repeated every 3 hours for 10 days at the same time as sampling of water at the observation hole after salt water was injected through water curtain hole. Except for the interval which transmits the data stored in imaging system to PC and reset the system to resume the measurements, a total of 51 measurements was obtained. Throughout this paper the measurement results are presented as apparent resistivity or normalized potential. The normalized potential is the observed voltage divided by the injected current. Also those apparent resistivity values were directly compared with the concentration of sampling water in observat-

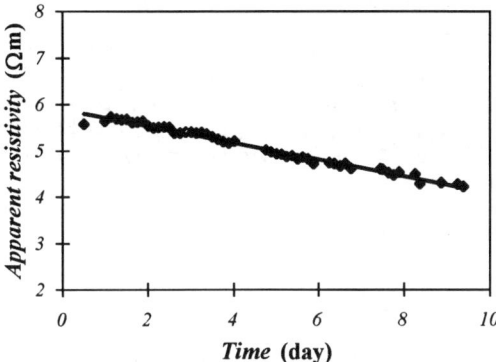
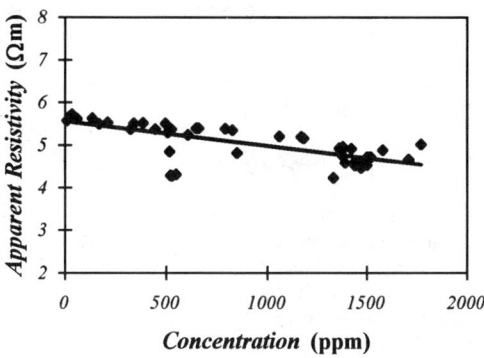

Figure 3. A representative plot of resistivity variation with time

Figure 4. A representative plot of resistivity variation with chloride concentration

ion hole which was obtained on the same time.

On the whole, apparent resistivities range from 2 to 20 Ωm within pseudosection and increase with larger n-spacing or greater depth. There was no large lateral variations in resistivity. In case only that n-spacing is one unit, there shows some indications of near-surface inhomogeneity at certain locations. There were also shown abnormally low resistivities that the measurement values of points on pseudosection including the first electrode (the number 4 of electrode location depicted on Figure 2) were different from them of other points. It might be caused by the unstable earthed condition in that location.

For each of 63 points on pseudosection, the plots of resistivity variation with time during the tracer test of salt water were obtained. Figure 3 shows a representative example of all 63 plots. The decreasing tendency of resistivities with time can be easily ascertained. Because there was no change of condition with the exception of salt water injection, it can be said that the variation was due to the increase of chloride ion in the subsurface groundwater flow.

One of the important features on the resistivity variation is that the apparent resistivities shows a local pattern with periodic change in the same manner of the concentration distribution. It was thought that the fluctuation pattern might be due to the tidal effect. A period of the fluctuation was about 9 hours and the tidal period in the study area ranges from 6 to 6.5 hours. The reason why the fluctuation period of resistivity is greater than that of tide may be understood from the fact that the change of groundwater level due to tide takes place with a certain time lag after inflow or outflow of sea water.

From this phenomenon, to avoid the aliasing of data, it can be known that the measurement interval to perform the resistivity monitoring must be determined within the range which is shorter than half of tidal period. If the measurement is carried out at an interval shorter than half of tidal period, the data will be redundant, but no error is introduced. However, if the measurement interval is longer than half of tidal period, the effect by tide will be distorted. That is, when the measurement points are smoothly connected, they show periodicity that did not exist in the natural condition.

Influence of tide on resistivity variation can be considered as sinuous effect for the decreasing tendency of apparent resistivity due to seawater intrusion. A sinuous effect can be expressed as sinusoidal function. Therefore, when the linear regression function of resistivity variation is obtained, the complete description of resistivity variation may be expressed as linear combination of sinusoidal and regression function.

The linear regression analysis was applied to resistivity measurement data to analyze quantitatively the decreasing level; the objective being to identify the most appropriate parameter for use in model construction of resistivity variation. The mean slope through the whole points on pseudosection was about -0.175 Ωm/day and maximum decrease of 23 percent in apparent resistivity was obtained. The physical meaning of mean slope is the characteristic value for the resistivity change of medium affected on salt water concentration. The latter result is nearly similar to that of experiment using Wenner fixed-spacing by White (1994) even though the setup and objective of experiments were different from each other.

It was shown that the apparent resistivity was inversely proportional to the concentration measured at the observation hole. A representative plot of apparent resistivities and concentrations was given in Figure 4. Using the similar manner in the resistivity change with time, a linear regression was applied to the resistivity data with concentration of sampling water. In the regression results, the apparent resistivities are inverse proportion to concentration with the mean slope of -0.0005 Ωm/ppm. The correlation level during the injection of salt water is much larger than that after an end of injection.

3 DEVELOPMENT OF MODEL

3.1 *Variation model of apparent resistivity*

For the medium be equilibrium in geological and hydrological condition, the measured apparent resistivities can be established as a initial condition of the medium for resistivity monitoring. If the correlation function between resistivity and chloride concentration is known, when the resistivity change due to seawater intrusion takes place for monitoring, chloride concentration can be analogized out of resistivity measurement. The analogized concentration can indicate the extent of seawater intrusion into medium.

In the previous section, it can be known that the apparent resistivities for monitoring are linearly decreased with the increase of chloride concentration. When consider the medium initially saturated with freshwater, the measured apparent resistivity maintains a constant value under no external effects. If the chloride concentration in medium increase from a certain time ($t = t_0$), the resistivity value will decrease. When the chloride concentration reaches an equilibrium state ($t = t_s$), the resistivity will again maintains a new constant value (R_s). If the measured resistivities for monitoring are normalized as dividing by initial value (R_0), then the resistivity variation can be expressed with logistic function.

Logistic function is defined as follows:

$$E(y) = \frac{\exp(\beta_0 + \beta_1 x)}{1 + \exp(\beta_0 + \beta_1 x)}. \tag{1}$$

Equation (1) is nonlinear for β_0 and β_1, but can be transformed into linear function. Let $E(y) = p$ and using the expectation transformation of probability

$$p' = \ln\left(\frac{p}{1-p}\right) = \ln\left(\frac{E(y)}{1-E(y)}\right), \tag{2}$$

inserting (1) into (2) gives

$$p' = -\beta_0 - \beta_1 x. \tag{3}$$

To express properly the monitoring measurements data with time using equation (3), if substitute x for time t, p for $R(t)$, $-\beta_0$ for R_0, and β_1 for β, then equation (3) can be expressed with the first order linear equation as follows:

$$R(t) = R_0 - \beta t. \tag{4}$$

$R(t)$ is the monitoring measurements with time, R_0 the initial apparent resistivity, and β the slope of apparent resistivity change at a specific location.

Equation (4) can be obtained by applying the linear regression with measurements time to the resistivity monitoring data. R_0 is always positive because it indicates the initial apparent resistivity. Although β is positive at a whole monitoring site, it may be negative as a local effect at a certain point on pseudosection.

3.2 *Concentration conversion model*

On the other hand, the relation between the apparent resistivity measurements and the concentration of sampling water can be obtained by applying the linear regression too. The apparent resistivity has a range from initial apparent resistivity (R_0) to value (R_s) in the time when is a equilibrium state of chloride concentration. The regression result in this experiment gives

$$C(t) = \frac{R_0 - R(t)}{0.0005}. \tag{5}$$

$C(t)$ is the variation of chloride concentration with time.

Equation (5) represents the correlation model of apparent resistivity monitoring with the concentration variation of salt water in this experiment. Because of specific characteristic of medium, the coefficient in equation may be different in other medium. Therefore, the on-the-site verification for the various medium is needed to obtain the site-specific coefficient. However, once the above model for a certain site near by sea is obtained, then the extent of seawater intrusion can be inferred by performing monitoring of apparent resistivity.

4 REFERENCES

Asch, T., and Morrison, H. F., 1989, Mapping and monitoring electrical resistivity with surface and subsurface electrode arrays, *Geophysics*, 54, 235-244.

Bevc, D., and Morrison, H. F., 1991, Borehole-to-surface electrical resistivity monitoring of a salt water injection experiment, *Geophysics*, 56, 769-777.

Van Overmeeren, R. A., 1989, Aquifer boundaries explored by geoelectrical measurements in the coastal plain of Yemen: A case of equivalence, *Geophysics*, 54, 38-48.

White, P. A., 1994, Electrode arrays for measuring groundwater flow direction and velocity, *Geophysics*, 59, 192-201.

A.E. source location considering the stress induced velocity anisotropy in rock

Kyu-Sang Lee
Rural Development Corporation, Ansan, Korea

Chung-In Lee
Department of Mineral and Petroleum Engineering, Seoul National University, Korea

ABSTRACT: In this study investigated was the variation of elastic wave velocities during the uniaxial compression tests, and its location analysis program was modified considering anisotropy. Then, the real fracture plane in the rock specimens were compared with those from A.E. source by the modified program. As the stress level increased under uniaxial compressive test, the wave velolcity immediately before the failure increasd 17% in parallel and decreased 26% in normal to the loading direction, respectively. As the stress level increased A.E. source from the modified program coincided nearly with the fracture locations observed in the test specimens for the anisotropic analysis, but not for isotropic analysis owing to velocity variation. It was found out from the analysis of A.E. source that at each different stress level the shear microcracks initiated from the center of specimen and propagated upward and downward.

1. INTRODUCTION

When rock is subjected to uniaxial compression, the microcracks are generated parallel to the loading direction. Owing to the occurrences of microcracks, the elastic wave velocities parallel and normal to the loading direction becomes anisotropic. In some papers, P wave velocity anisotropy is approximately 40 to 50 % immediately before failure (Lockner et al. 1977) ,(Yanagidani et al. 1985). So, the velocity anisotropy should be considered to obtain proper location of source. In this study, was measured the stress induced velocity changes and the analysis program was modified considering velocity variation. And then comprared the A.E. sources with real fracture plane.

2. BASIC THEORY OF SOURCE LOCATION

Direct method and iterative method exist in theory of the source location using acoustic emission. When the acoustic source exists in the internal of detector, the acoustic source can be identified using direct method. When the measured time differences have large errors or the measured velocity fields is anisotropic, the iterative method such as least-square method and simplex method can be used to reduce errors. Herein, the source location is performed using simplex method.

In this study, when the wave propagation velocities at each direction are equal, a simplex-method based solution that was suggested by U.S. Bureau of Mines is used. In the cases of velocity variations at each direction, the following equation is used after the correction of velocity difference at each propagation direction (See figure 1).

$$\sqrt{(a_o - x)^2 + (b_o - y)^2 + (c_o - z)^2} + \delta t_i \cdot v_i + \delta_i = \sqrt{(a_i - x)^2 + (b_i - y)^2 + (c_i - z)^2}$$

where, $\delta_i = v_0 \cdot \delta t_i$
a_0, b_0, c_0 : coord. of 1st hit PZT
a_i, b_i, c_i : coord. of ith hit PZT
x, y, z : coord. of source
v_i : wave velocity to the ith PZT
δt_i : $t_i - t_0$
t_i : wave propergation time to the ith PZT

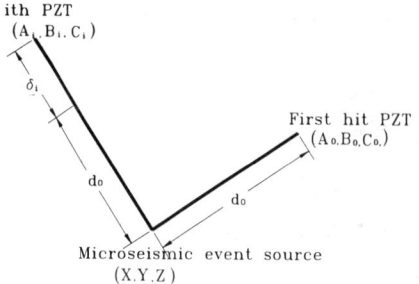

Fig. 1 Graphic representation of distances between microseismic event source and two distinct PZT

3. TESTING METHOD

3.1 Rock Specimens and Physical Properties

In order to make easy attachment of PZT, rock specimens used in the test are made rectangular form by 70mm X 70mm X 130mm.

Physical properties of rocks used in the test are assigned in Table 1.

Table 1 Physical properties of Machon gabbro

Properties	Value
Unit Weight (kN/m^3)	28.1 ± 0.2
Apparent porosity(%)	0.67 ± 0.06
P-wave velocity (m/sec)	4020 ± 60
S-wave velocity (m/sec)	2330 ± 40
Uniaxial compressive strength (MPa)	120 ± 7
Young's modulus (× GPa)	64.6 ± 1.4
Poisson's ratio	0.21 ± 0.03
Brazilian tensile strength (MPa)	8.4 ± 0.1

3.2 Testing Method

In order to examine the velocity change of elastic wave during compression test, the pulse generator which generate elastic wave is adhered to one surface of a rock specimen and PZT is attached to the opposite side. Model 5055PR of Panametrics Inc. is used to generate elastic wave. The generated signals are amplified and stored at the oscilloscope, HP54620A. The stored signals are send to the computer and analyzed later. When the signals are generated in pulse generator, the trigger signals are also generated. These signals are considered as initial time of the wave generation. And then, the velocity is determined by dividing the distance into time at which the first wave arrival is observed.

Four strain gages are attached to the surfaces of the rock specimen to measure volumetric dilation of the specimen. Model 1544 and 1546 strain-meters of B & K Inc. were used to measure the strain. Each strain values were measured and averaged every 2 seconds.

To determine the wave velocity, PZT was attached at 30 mm and 90 mm apart from the top. And then the specimen is stressed to generate acoustic wave by metal-sphere installed at the top of specimen. The measured acoustic emission is temporarily stored at the oscilloscope via the amplifier. And then, is saved at personal computer for the purpose of later analysis.

PZT is used as a detector for the A.E. source location. When stress or force is introduced to PZT, electric voltage proportional to the stress is generated in the surface of PZT. 200 ton capacity universal testing system of Shimadzu Inc., Japan was used generate to fractures. Rock specimen is stressed in uniform velocity, $20kg/cm^2$ per minute.

Electric voltage generated in PZT is amplified by type 2638 wideband conditioning amplifier of B & K Inc. as it is need. The amplified signal was analyzed by type 4429 A.E. analyzer of B & K Inc. where difference of arrival time of the signal can be measured. In A.E. analyzer, if signals higher than 0.5V is detected at any one channel, that particular channel is set to zero. And signals greater than 0.5V arrives at other channel, the arrival time difference between initially detected channel and later detected channel is logged in μ sec.

Schematic diagrams of testing equipments are presented in figure 2.

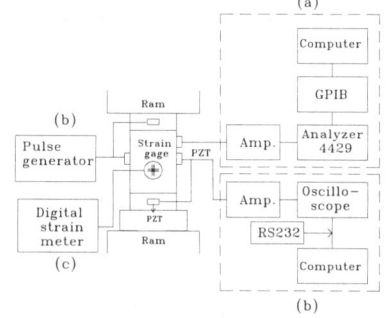

Fig. 2 Block diagram of (a) A.E. source location, (b) P wave velocity, (c) surface strain measuring system

4. RESULTS AND ANALYSIS OF LABORATORY TEST

4.1 Propagation Velocity of Acoustic Emission

The results of laboratory testing to determine velocity of the acoustic emission propagating into rock specimen are presented in figure 3. As shown in figures, the lower wave is A.E. that is measured in PZT installed at 30mm distance from the location at which failure is occurred and the upper wave is A.E. that is measured in PZT installed at 90mm. The first point indicated in the figure is the point of 1.6 μ sec in time axis and the second is the 27.8 μ sec. Therefore, the velocity of A.E. can be readily obtained as 2290 m/sec from the propagation distance of 60mm and arrival time of 26.2 μ sec.

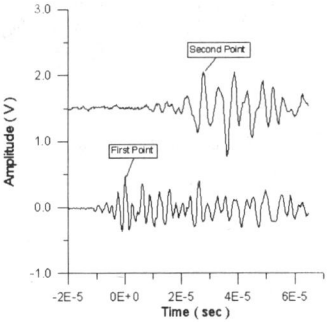

Fig.3 A.E. wave form used in determining the velocity

4.2 Velocity Change

Fig. 4 shows the velocity changes of P wave in the parallel and normal to the loading direction. Velocity parallel to the loading direction increases from 4210m/sec to 4920m/sec, and that of normal decreases to 3120m/sec.

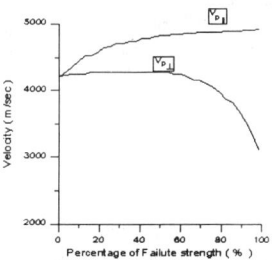

Fig.4 Velocity change under uniaxial compression

The propagation velocity of A.E. used in source location is similar to the velocity of S wave. Tendency of the velocity change of other papers (Nur etal. 1969),(Lockner et al. 1977) obtained by uniaxial compressive test has been reported to have a similar fashion either P or S wave. In this study, therefore, the velocity change ratio of P wave is regarded as that of A.E. wave. Based on the assumption, changes of the propagation velocity with increases of stress level are considered.

4.3 Results of Acoustic Source Location

Because the propagation of crack occurs instantaneously near the failure stress, a number of results of the acoustic source location can't be obtained. So, for the purpose of slow occurrence of the propagation of crack to the specimen, the acoustic source location test is performed with gradual increase of the stress. At about 80 percent of failure stress predicted from uniaxial compressive test, the stress is increased by 5 percent of the failure stress. For 10 minutes, the same stress state is maintained. This process is repeated with gradual increase of stress.

The result of acoustic source location up to 85 % of the failure stress at which the specimen is failured is seen in figure 5. It is shown that crack is developing slowly from the center part of the rock specimen. Occurrence of crack by the partial concentration of stress is observed at the lower side of the specimen.

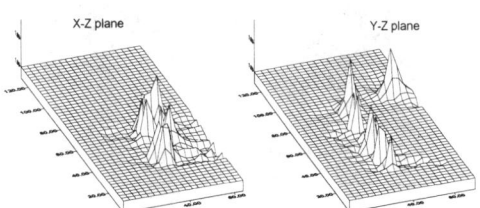

Fig. 5 A.E. sources stressed up to 85 % failure stress

Figure 6 presents the result 85% to 90% of the failure stress. When the results are shown in x-z plane and y-z plane(x-z plane means that x and z coord. is plotted and y coord. is ignored. y-z plane is same), cracks are developed from center to slightly right side of specimen and vertical direction in the center of specimen is observed, respectively. In addition to this, the growing

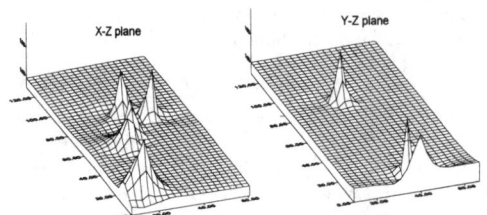

Fig. 6 A.E. sources stressed 85% to 90% failure stress

crack which is normal to y axis at the right upper side of the rock specimen.

Figure 7 shows the result 90% to 95% of the stress level. Showing in x-z plane, the crack observed in figure 6 moves slowly to the upper side of the rock specimen and the crack at the lower side moves to the right direction. Showing in y-z plane, the vertical crack at the center side of rock specimen grows to the lower side, and new crack develops at the upper side of rock specimen.

The result of 95% to the failure stress is presented in figure 8. Observing in x-z plane, the narrow new crack grows at the center side of the specimen. And the fact that the narrow crack is interconnected with the cracks of the upper and lower side. Observing in y-z plane, crack development at slight left side of the center and the lowest side of specimen.

Figure 9 is cumulative plot of the results of

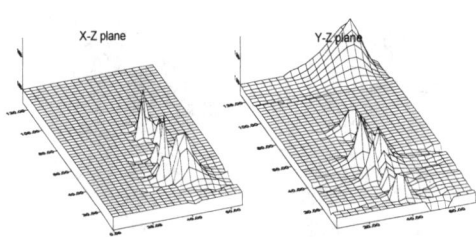

Fig. 7 A.E. sources stressed 90% to 95% failure stress

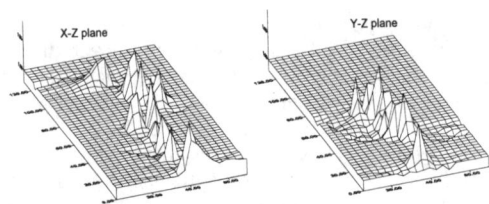

Fig. 8 A.E. sources stressed 95% to failure

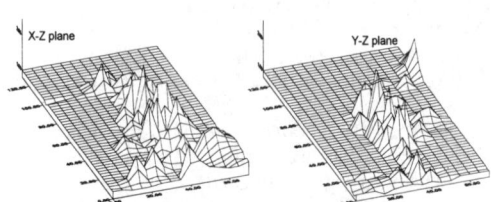

Fig. 9 A.E. sources from initial stress to failure

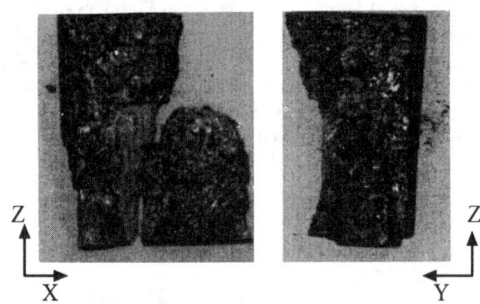

Fig. 10 View of specimen after failure

the acoustic source location. Figure 10 shows an actual failure form of the rock specimen. The location of A.E. sources showed in figure 9 and the plane of real crack shows good agreement.

5. CONCLUSIONS

1. The elastic wave velocity immediately before the failure increases by 17 percent along parallel direction to loading direction and decreases with 26 percent along the normal direction.

2. As the results of anisotropy analysis, the occurring location of the crack obtained by the analysis has agreement with the location of actual crack developed in the rock specimen.

3. When the failure proceeds in the specimen, at the first time the shear crack is occurred by growing of the micro cracks at the center portion of the specimen. When the stress approaches to the failure stress, it is ascertained of cracks to transmit to the upper and lower portion of the specimen.

References

1. Gupta, I. N., 1973. Seismic Velocities in rock subjected to axial loading up to shear fracture.: J. Geophys. Res. 78, pp.6936-6942.

2. Lockner, D.A., Walsh, J.B., Byerlee, J.D., 1977. Changes in seismic velocity and attenuation

during deformation of granite.: J. Geophys. Res. 82, 5374 - 5378.

3. Lockner, D.A., Walsh, J.B., Byerlee, J.D., 1977. Changes in seismic velocity and attenuation during deformation of granite.: J. Geophys. Res. 82, 5374 - 5378.

4. Nishizawa, O. Kusunose, K. 1985. Localization of dilatancy in Ohshima Granite under constant uniaxial stress.: J. Geophys. Res. 90, pp.6840-6858.

5. Nur, A., and G. Simmons, 1969. Stress-induced velocity anisotropy in rock, An experimental study.: J. Geophys. Res., 74, 6667-6676.

6. Rao, M.V.M.S., Sun, X., Hardy, Jr., 1989. An evaluation of the amplitude distribution of AE activity in rock specimens stressed to failure.: Proc., 30th U.S. Rock Mechanics Symposium. Balkema, Rotterdam, 261-268.

7. Riefenberg, J. S. 1989. A simplex-method-based algorithm for determining the source location of microseismic events.: Bureau of Mines. Report of Investigations 9287.

8. Soga, N. Mizutani, H. spetzler, H. Martin III R. J., 1978. The effect of dilatancy on velocity anisotropy in westerly granite.: J. Geophys. Res. 83, pp. 4451-4458.

9. Yanagidani, T., Ehara S., Nishizawa O., Kusunose K., Terada M., 1985. Localization of dilatancy in Ohshima granite under constant uniaxial stress.: J. Geophys. res. 90, 6840-6858.

Crack monitoring in underground construction

S.K.Tewatia
Central Soil and Materials Research Station, Hauz Khas, New Delhi, India

ABSTRACT: The appearance of the cracks and the relative movements of the walls along and across them beyond certain limits, in the underground civil engineering structures, may prove to be potentially damaging. For the safety considerations, the measurement of the magnitude and time rate of deformation of the cracks is essential. At Sardar Sarovar Dam Project on Narmada River (India) an underground power house is being constructed. The Central Soil and Materials Research Station has been doing the work of crack monitoring using demec gauge which involves fixing of 6 special pins (3 on each wall) as shown in figure 1. This method is not very accurate. Moreover demec gauge can not measure the deformation beyond 5 mm. To overcome these limitations a simple 3 pin method has been developed using digital vernier calipers of 0.01 mm least count for measurement of the crack deformations to any desired accuracy.

1 THE 6 PIN DEMEC GAUGE METHOD

This involves fixing of 6 special pins (3 on each wall) on the predetermined periphery of a circle, having a diameter of 250 mm or 500 mm, which (the pins) lie symmetrically about the crack such that the line joining the middle pins on each wall is nearly perpendicular to the crack and any two adjacent pins subtend an angle of 60^0 at the center of circle which lies on the crack (fig. 1).

Due to relative movement of the wall 1 with respect to wall 2, suppose after time, t_1, the points A, B and C on wall 1 move to A_1, B_1 and C_1 respectively. As the wall 1 moves as rigid body and rotation of the wall is ignored therefore AA_1, BB_1 and CC_1 are equal and parallel and components of these, parallel and perpendicular to the crack, give deformation along and across the direction of the crack.

1.1 *Measurement of deformation across the crack*

Original length of BE which is nearly perpendicular to the direction of the crack is known. After certain time intervals t_1, t_2, t_3 etc. EB_1, EB_2, EB_3 etc. are measured and EB is subtracted from them to obtain deformation across the crack at t_1, t_2, t_3 etc. as below

$$y_1 = EB - EB_1 \tag{1}$$

The method suffers from the weakness that EB should have been subtracted from the component of EB_1 along the direction of EB to measure the deformation across the crack after time t_1. But as the deformation along the crack is very less (< 5 mm) as compared to the original length of EB (= 250 mm or 500 mm) therefore EB_1 etc. still remain almost perpendicular to the direction of crack and the method gives good approximation.

1.2 *Measurement of deformation along the crack*

Original length of DA which is inclined at 30^0 to the

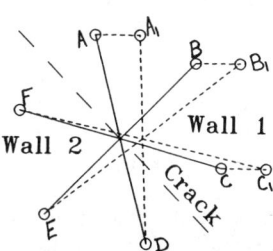

Figure 1. The 6 Pin Demec Gauge Method

direction of the crack is known. After certain time intervals t_1, t_2, t_3 etc. DA_1, DA_2, DA_3 etc. are measured and DA is subtracted from them to obtain deformations along the crack at t_1, t_2, t_3 etc. Similarly after the same time intervals t_1, t_2, t_3 etc. FC_1, FC_2, FC_3 etc. are measured and FC is subtracted from them to obtain deformations along the crack at t_1, t_2, t_3 etc. The average deformation along the crack, x_1, at t_1 is obtained as

$$x_1 = \frac{(DA_1 - DA) - (FC_1 - FC)}{2} \qquad (2)$$

Here, while taking average, minus sign is used because due to the deformation along the crack if DA_1 increases then FC_1 will decrease or vice versa, and therefore $(DA_1 - DA)$ and $(FC_1 - FC)$ have opposite signs. However it is clear from fig. 1 that any displacement across the crack will increase or decrease $(DA_1 - DA)$ and $(FC_1 - FC)$ by equal amount and therefore from equation (2) the effect of displacement across the crack is canceled. As DA and FC are not exactly along the direction of crack but are inclined at 30^0, therefore the deformation shown by this method is less than actual deformation by about 13%, however, good approximation can be obtained by dividing x_1 by $\cos(30^0)$.

2 THE PROPOSED 3 PIN METHOD

Insert 3 pins A, B and C on both sides of the crack as shown in figure 2, so that AB is perpendicular to the direction of crack and BC is inclined at any known angle. Measure original lengths of AB and BC. Suppose after time, t_1, A moves to A_1 and C moves to C_1 so that BA_1 and BC_1 are the new lengths of BA and BC respectively. The basic assumption involved in the method is that the walls move as rigid bodies without any rotation so that AA_1 is equal and parallel to CC_1, and therefore the components of these along and across the crack, x and y (say), are equal. The approximate value of y (= BA_1 - BA, as in 6-pin method) is assumed.

Now in triangle BCC_2: BC and CC_2 (= y) are known in magnitude and direction therefore magnitude and direction of BC_2 can be known. In triangle BC_1C_2: as directions of C_1C_2 (which is parallel to crack) and BC_2 are known therefore angle between them is known. As in this triangle magnitudes of two sides BC_1 and BC_2 and one angle are known therefore third side C_1C_2 (= x) can be determined.

Using this x in right angled triangle BA_1A_2, BA_2 is determined as $(BA_1^2 - x^2)^{0.5}$. New value of y is

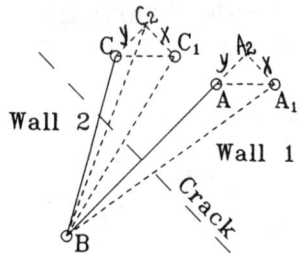

Figure 2. New Method

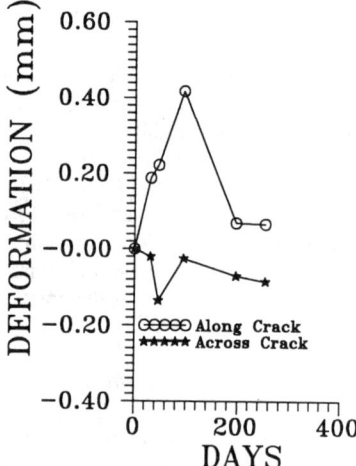

Figure 3. A typical plot of crack monitoring at Sardar Sarovar Project

determined as $(BA_2 - BA)$. As compared to the previous y, this y is closer to the true value. Using this y from triangles BCC_2 and BC_1C_2, new value of x is determined as earlier. As compared to the previous x, this x is closer to the true value. Using this new x, new y is determined from triangle BA_1A_2 as earlier which is further closer to the true value as compared to previous y. Repeating this procedure, after 5-6 iterations, the values of x and y converge to their true values. A very simple software can be developed on these lines for successive iterations (calculations).

To obtain a check on the validity and accuracy of the method and observations, one more pin is inserted on one of the walls. The relative movement (x,y) of the wall is noted as above, separately, by making two sets, each of them consisting of three pins, and comparing the relative movements along and across the crack by each set. If the method, observations and calculations are correct then the two sets should give the same result.

3 CONCLUSIONS

The proposed method involves the fixing of three pins only as compared to 6 pins in the demec gauge method. Using 4 pins a built in check in computer programming can be provided for pointing out the error, if any, in the observations. While demec gauge method is an approximate method the proposed method can be used to any desired accuracy by increasing the number of iterations in the programming. The demec gauge has limitation of measuring crack deformations up to 5 mm but using digital vernier calipers, in the new method, crack deformation of any magnitude can be measured. The methods do not measure the absolute width of crack on the wall surface but the change in width along and across the crack with respect to time is measured. Also the methods do not provide the absolute movements of wall but their relative movements, only, with respect to each other.

4 ACKNOWLEDGMENT

The author is thankful to Mr. S.K.Verma (Chief Research Officer) for a number of fruitful discussions on this topic.

Compact VSP probe for inspection of rock mass quality

Atsuo Hirata
Kumamoto Institute of Technology, Japan

Shintaro Baba
Ohmoto Gumi Co., Ltd, Japan

Tsutomu Inaba
Nishimatsu Construction Co., Ltd, Japan

Katsuhiko Kaneko
Hokkaido University, Japan

ABSTRACT: It is of extreme importance for an analysis with a high degree of accuracy that characteristics of rock around cavern such a tunnel are well known. A new method of rock property investigation by using a compact vertical seismic prospecting system is proposed. This investigation system consists of a probe attached vibration sensor and super elastic alloy plate to tress it against borehole surface, computer, and trigger system.

1 INTRODUCTION

It is a general rule that rock around an underground cavern is damaged up to a point depth in process of excavation, and the extent of damage varies. One of the major causes of damage is above all the blasting at the time of excavation. The others are the yield and degradation of rock due to stress concentration around a cavern caused by earth pressure. The damaged zone around a cavern generated by these causes is usually referred to as a loosened zone. A loosened zone has mechanical characteristics which are different from those of sound rock around a damaged region (Hirata 1991).

A stable analysis of underground structure focuses on a cavern, linings, the loosened zone and its surrounding sound rock areas. It is of extreme importance for an analysis with a high degree of accuracy that characteristics of a loosened zone and those of sound rock are well known. For this reason, devices such as the displacement meter, the strain meter and the earth pressure meter have been developed to be placed underground, and these have a wide range of applications in actual excavations (Jaeger 1972). Once these devices have been planted underground for measurement, however, they are unremovable and thus not reusable for another site. Moreover, 3 to 5 of those devices are normally placed for each measuring section every 10 to 20 meters of an excavation site, and therefore it becomes imperative to gain control over quite a few measuring points, which demands a good deal of time and expenses both for measurements and the collation of measurement results. Yet, the measurement of every 10- to 20-meter area is not enough to maintain and verify the stability of rock around a cavern.

The measurements usually include the displacement of cavern walls by excavation, the relative displacement of rock from the point where rock does not move, the earth pressure that acts on the lining and the force acted on rock bolts. These measurements gradually change influenced by the release of the earth pressure that an excavated rock has been under. It is a foregone conclusion that the changes in measurements of a loosened zone damaged by excavation are the greatest.

There are other indicators to evaluate characteristics of rock. one of them is the velocity of elastic waves that propagate through rock. Since the fastest wave of all is the P wave, the first arrival judgment of it is easy as compared with that of the S wave or surface waves. For this reason, P wave velocity is adopted as a useful indicator to estimate the elastic property of rock (Kaneko 1989). There are a few cases in which P wave velocity was used in investigating. a loosened zone around caverns (Tanaka 1995). But an increase in the complexity of measuring work is inevitable for a fuller investigation of a loosened zone by means of elastic-wave prospecting, and this method is hardly used today in daily measuring.

In this study, therefore, we would like to propose a new method of rock property investigation by using a compact vertical seismic prospecting (VSP) system which we have developed for inspecting rock mass quality. In this method, probes are inserted in boreholes bored into the inside walls of a cavern to place rock bolts as cavern supports, and the vibration of the cavern surface is measured by the impact of a hammer struck on the cavern walls. Since plenty of boreholes are bored for rock bolts over the cavern surface, the proposed investigation method using probes gives us ample information on rock mass quality over broad areas of the inside of a cavern. And the measuring work is greatly simplified with only the insertion of VSP probes and hammer strikes against a rock surface.

2 THE COMPACT VSP PROBE

The compact Vertical Seismic Prospecting (VSP) system consists of probes used to detect the elastic wave, cables, the AD converter, a computer to record wave data, and a trigger system to generate shot mark. The piezoelectric transducer, PZT, with a resonance frequency of 500 kHz, is used as the sensor of vibration. The transducer itself is a 3 mm cube. Vibration induced by hammer strikes against the rock surface is composed of frequency components ranging from 100 Hz to 1 kHz. Vibration designed to be measured is of a frequency range lower than the resonance frequency of PZT. And the sensitivity of PZT to vibrations to be measured is very low so that vibrations are to be recorded after being amplified. Actual signals induced by the impacts of hammer strikes can involve noise of fairly low frequency. This may include noise caused by power supply, cable, etc. Noise may hinder signals being recorded in the memory of the computer as digital data. Thus, high-pass filters are introduced to cut off the low-frequency components, and an electric circuit with a 1 M ohm of electrical resistance is installed between the two terminals of PZT.

Boreholes in which insert VSP probes are bored by driving a drill into the rock while giving a blow to the probes with the tip of the drill bit. The inside surface of a borehole is generally rough and a borehole can wind delicately in a rock mass. The diameter of a borehole can vary rather delicately with rock mass strength. The device for sensing vibration is to be pressed on the inside of a borehole in varying conditions. It is required that a pressing force applied on the rock surface should be kept at a constant level Sensitivity of a sensing device depends crucially on the proper installation of it on the rock surface. We have adopted the method of installing a sensor by using a spring in order to set and remove it at an arbitrary position inside of a borehole. The pressing force applied against borehole surface extents control the sensitivity of a sensing device. The relationship between the pressing force applied and the degree of sensitivity under a constant input is shown in figure 1. Sensitivity increases with the growth of a pressing force when the pressing force is under 10 N. Equivalently, measurements in conditions of a pressing force being kept over 10 N are precise enough without another condition in with only a amount of errors generated by the setting of a sensor.

It is required that the spring mechanism should be easily movable along a borehole whose diameter is 30 to 40 mm, and be able to keep a pressing force over 10 N throughout. The movement of a probe in the borehole gets into difficulties, if the pressing force becomes too large. It would be most ideal to keep the pressing force constantly in the domain of 10 to 20 N. We thus have decided to use a Titanium Ti and Nickel Ni alloy plate for a spring that presses a sensor to the borehole surface. It is well known that Ti-Ni alloy shows super elasticity (Nishida 1989). Super elasticity is the property that metals such as steel have of recovering their shape nonlinearly, in case of a large displacement resulting in permanent deformation. In other words, super-elastic alloy has nonlinearly a high deformation performance that steel does not show, and has also the property of stabilizing its after the deformation has passed a certain stage. A super-elastic alloy plate is thus ideal for a spring with a uniform pressing force.

A super-elastic alloy plate is fixed on a VSP probe as shown in figure 2. Both ends of the super-elastic alloy plate are fixed on the support plate, and the plate is bent outwards. A PZT is attached onto the back side of the spring plate. The deformation caused by the insertion of a probe into a borehole mainly originates at both ends of a spring plate as shown by a broken line in figure 2, and, conversely, the spring plate presses a PZT to the borehole surface at a uniform pressure. A pressing force is pertinent to the length L between the fixed points of a spring plate, the bending height h, the displacement Δh, and the bending stiffness of a spring plate. Bending stiffness is decided by the mixing ratio of Ti to Ni, the thickness t and the width W of a spring plate. In this study, bending stiffness depends on the only variable, the width of a spring plate, the other variables kept constant with a fixed mixing ratio of 49

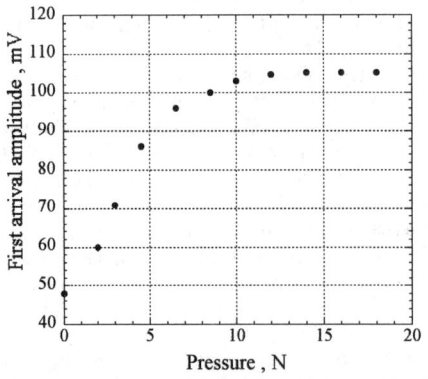

Figure 1 Relationship between pressing force and output of sensing device.

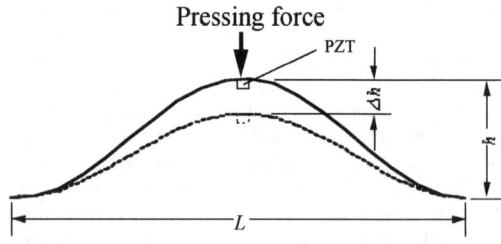

Figure 2 Profile of spring using super elastic alloy and PZT.

% Ti and 51 % Ni and a spring plate with a thickness of 0.25 mm.

A spring plate with a width of 5 mm was first set in the probe. Figures 3 and 4 each show the correlation between the amount of displacement and the pressing force applied in case in which the bending height is variously changed, with L kept constant at 5 cm in figure 3 and 6 cm in figure 4, respectively. Explanatory in these figures shows the length L between the fixed points of a spring plate in cm, the bending height h in cm and the width W of a spring plate in mm, in the order presented. The spring does not show super elasticity and the total amount of displacement becomes low in cases where the bending height is kept low. The displacement pattern of a spring changes to an entirely different mode in which two bulges are formed out of a spring when the deformation is past a certain point. On the other hand, the spring does show enough super elasticity in cases of greater bending heights to provide a uniform pressing force, and the total displacement amount increases to 10 mm. A spring with a width of 5 mm does not show a pressing force of over 10 N. To add to this, a spring is vulnerable to side-to-side movements such as twists because the stiffness of a spring is low.

For these reasons, a spring with a width of 10 mm was used for the VSP probe. The relationship between the pressing force of a spring and the amount of displacement is shown in figure 5. The stiffness of a spring is two times that of a 5 mm-wide spring. Displacement performance increases with the increase of a bending height. As an additional plus, displacement performance increases in case of $L = 7$ cm. But the contact point of a spring with the borehole surface can in this case deviate from the center of a spring plate because L is too large, and the PZT does not touch the borehole surface. It thus turned out that the optimal shape of a spring of Ti-Ni alloy is $W = 10$ mm, $L = 6$ cm and $h = 2$ cm.

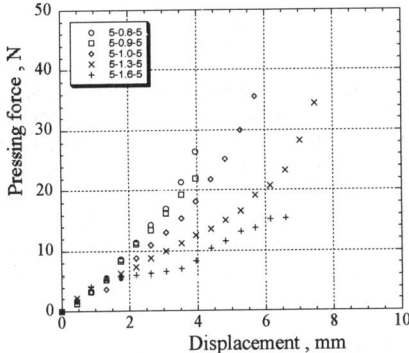

Figure 3 Relationship between the displacement of the spring and the pressing force in case of $L=5$ cm, $W=5$ mm.

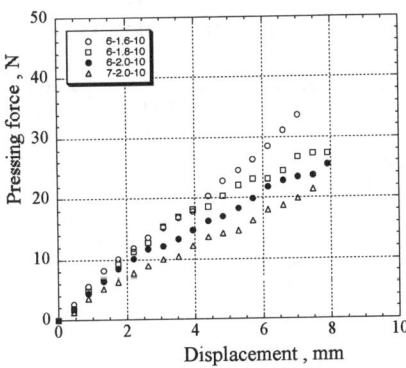

Figure 5 Relationship between the displacement and the pressinf force in case of $L=6$ and 7 cm, $W=10$mm.

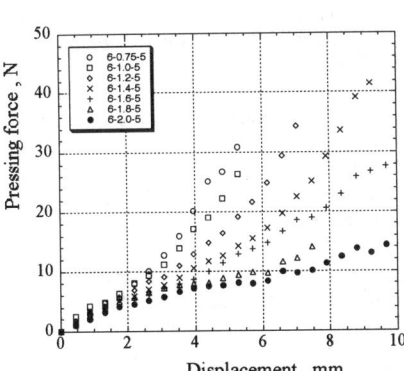

Figure 4 Relationship between the displacement and the pressing force in case of $L=6$ cm, $W=5$ mm.

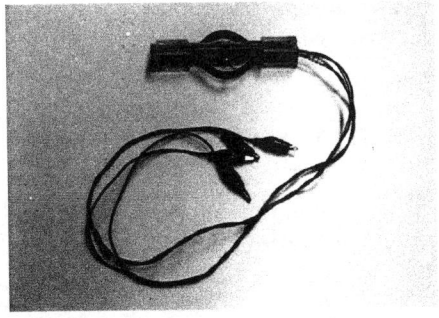

Figure 6 Trial model of the sensing device of the VSP probe with PZT, cables and springs using super elastic alloy plate.

A pressing force of 15 N acts on the borehole surface if a spring is inserted in the borehole under conditions of a displacement of $\Delta h = 4$ mm. A pressing force of 10 N acts on the borehole surface if a change in the diameter of a borehole decreases below 2 mm. On the other hand, if the change increases beyond 2 mm, the pressing force reaches 20 N. If the pressing force becomes still greater, however, the movement of the prove runs into difficulties. It is appropriate that the maximum pressing force should be kept below 20 N.

Figure 6 shows a trial model of the sensing part of the VSP probe that consists of springs, the PZT and cables. The plate on which springs are fixed is 8 mm in thickness. The total thickness of the trial model is 48 mm. We insert a probe in a borehole with a diameter of 40 mm. The total displacement generated by a spring placed on either side of the probe is 8 mm, and a pressing force of 15 N acts on the borehole surface. In this way, a pressing force of 10 to 20 N is maintained on the borehole surface for boreholes with diameters ranging from 36 to 44 mm.

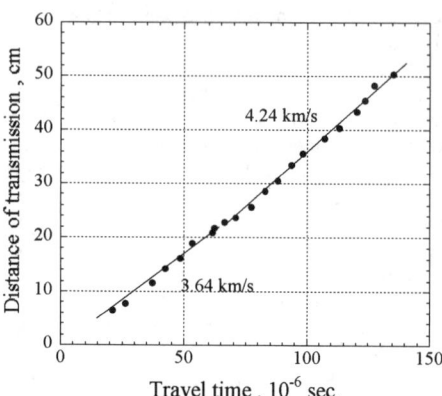

Figure 8 Relationship between travel time and the distance from a shot point to a measuring point in borehole.

3 EXPERIMENTAL DEMONSTRATION

A test specimen made of concrete as a preparation for demonstration is shown in figure 7. This is a right prism (25 cm, 25 cm, 60 cm), and a borehole with a diameter of 42 mm is bored through the entire length. The P wave velocity of the region up to 23 cm from one end of the specimen is 3.64 km/s, and the rest of it shows a P wave velocity of 4.24 km/s. In short, the specimen imitates a real rock structure around a cavern composed of a loosened zone and sound rock.

Vibration induced by the hammering of the specimen was measured by a VSP probe inserted in a borehole. The difference in time between the first arrival of the P wave and a trigger time is counted as propagation time T. The relationship between T and the distance from a shot point to a measuring point in the borehole is shown by open circle in figure 8. Solid straight lines indicate the P wave velocities of the test specimen that are 3.64 km/s and 4.24 km/s. It is evident from figure 8 that the structure of the specimen is precisely measured.

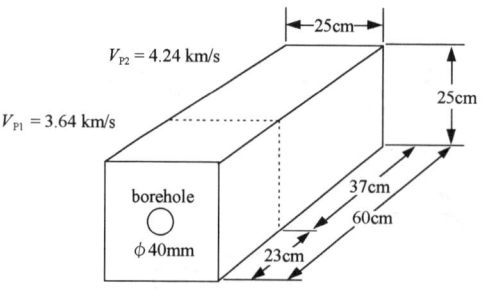

Figure 7 A test specimen made of concrete.

4 CONCLUSION

We have developed a compact VSP system to evaluate a loosened zone caused by excavation work. A super-elastic alloy plate has been adopted as material for a spring needed to press a vibration sensor to a borehole surface in view of the fact that a uniform pressing force is essential for a precise measuring of elastic waves. It has become apparent that our system can provide a constant pressing force, and a high degree of reliability for the measurement of vibration. Although the measuring system that we have developed has not yet been put to actual use at excavation sites, we believe that it has extensive practical applications.

REFERENCES

Hirata, A., T. Inaba and K. Kaneko 1991. Dynamic behavior of tunnel due to blasting. *J. of the Mining and Materials Processing Institute of Japan*. 107, 7:481-487.

Jaeger, C. 1972. Rock mechanics and engineering. *Cambridge university press*.

Kaneko, K., N. Nakamura, A. Hirata and M. Ohmi 1989. Technique for estimation of seismic Q by means of first arrival pulse. *Butsuri-Tansa*. 42, 4:235-244.

Nishida, M., K. Kaneko, T. Inaba, A. Hirata and K. Yamauchi 1989. Static rock breaker using TiNi shape memory alloy. *Proc. of Int. Conf. on Martensitic Transformations*:123-128.

Tanaka, Y., A. Hirata, M. Yamamoto and H. Matsunaga 1995. A experimental study on the damage area of smooth blasting by the electronic delay detonator. *Kayaku Gakkaishi*. 56, 1:26-31.

Geodynamic safety – The main factor in the exploration of mineral resources and the earth surface

A.N.Shabarov, V.V.Zoubkov & N.V.Krotov
State Research Institute of Mining Geomechanics and Mine Surveying (VNIMI), St-Petersburg, Russia

ABSTRACT: This paper presents the results of a series of investigations performed to validate a new approach to ensure safety to the most significant underground structures and surface engineering constructions. In this respect, the insight into geodynamic hazard is of paramount importance.

It is known that many industrial regions over the world are the areas of present-day tectonical activity. Within them the faults in the earth crust become more active. They are the unforeseen dangers for the exploitation of mining enterprises, oil- and gas pipelines, nuclear power stations, the power transmission lines and the other significant structures of national economy. On the other hand, the native behavior of rock mass is still more subjected to the influence of technogenic activity of men, e.g. mineral mining, petroleum- and gas output, ground water pumping-out, etc. As a result, nearly in all the regions where the intensive mineral mining takes place and which previously were regarded as the quiet ones, the seismic events begin to manifestate, for example, the technogenic earthquakes, the number and energy of which are steadily growing.

In these circumstances, better geodynamic safety of major industrial regions all over the world and, first of all, of dangerous structures of technosphere is a high-priority task.

The solution of similar problems is possible with using the method of geodynamic zoning developed in VNIMI, which enables us to reveal among numerous faults existing in the earth crust, those ones which have conserved or even intensify in time their activity.

This method is based on the morphostructure analysis of the earth surface by the topographic maps of various scales with the use of the satellite and air photographs, observations on the geodynamic polygons, the analysis of geological, geophysical and geochemical data and analytical computations. The results of its application have been discussed at the UN ECE Meetings of Experts on Coal and Gas, International Workshops and Symposia in 1994-1996. By decision of UN ECE Committee on Energy this method was recommended for application in consideration of the following matters:

A. Designing, construction and exploitation of solid, liquid and gaseous deposits;
B. Designing and construction of linear structures (railway lines, oil-and gas pipelines, etc.);
C. Selection of the reliable sites for the layout of significant surface and underground structures (electric power stations including nuclear ones, underground gas reservoirs, oil-products storage tanks, burial of harmful industrial wastes, etc.).

Consider the solution of some problems regarding the geodynamic safety in the listed above items.

A. Data on the detection of block structure of the earth crust, active faults, tectonically stressed zones and the assessment of native stress state of block rock mass become at present the basis for the proper cutting of deposits into mine fields and sequential order of their mining.

The most significant element in performing the geodynamic forecasts in mineral deposit mining is the assessment of the native stress state of block rook mass and the calculation of stress field in minable seams with due account of geological disturbances within the deposit region.

It is known that movements along the block contacts initiate the dynamic events in mines. On the other hand, the deposit mining induces these movements and facilitate the formation of new boundaries in block structures. Rather often the contact interaction and movements contribute to total deformation of rook mass, and this contribution may be comparable or even exceeding the deformation of blocks themselves. Therefore, the contact interactions exert a substantial influence both on stress-strain state of rook mass and on the steadiness of this state.

At the Institute VNIMI, the specialists have developed the method for solving a problem on a system of interacting blocks, with its numerical implementation - the software BLOCKS2D. It enables us to make an assessment of stress state of block rook mass with arbitrary conditions on contact planes. Thereby, the blocks themselves consider to be elastic ones, and each of them has its elastic characteristics, and inelastic processes are transferred on the boundary between the blocks. At the contacts between blocks any interaction conditions may be realized: the conditions of smooth contact, constant friction, complete cohesion, etc. The results of calculations are used for construction of forecasting maps of the native stress state.

The availability of forecasting maps of the native stress state of block rock mass makes it possible to make a proper decision at the stages of designing and construction, for example, in mining layout within the range of tectonic disturbances, in selection of sites for drivage of permanent workings and shafts sinking in blocks not subjected to the influence of higher stresses; as well as for well layout at the oil- and gas fields to ensure the constant well yield, for location of hydroelectric stations, nuclear power stations, the burial of chemical and radioactive wastes; for designing of extended products-pipelines and railway lines.

As an example, Fig. 1 presents the results obtained in the geodynamic zoning of the Beipiao coal deposit in China, which is one of most rockburst-outburst-hazardous deposits in the world.

The initial depth of sudden outbursts occurrence is 130 m. The most seismically active appears to be the conjugate zone between the Taidy mine field and the Guanshan mine field. The investigations have shown that the deposit is related to a large fault of rank I. This fault, called afterwards the fault "Beipiao", was revealed for the first time, and it was not formerly indicated on the Chinese geological maps. Its existence was confirmed by the seismologic data and exploration work. Directly within the Beipiao deposit this fault is displaced in an echelon shape in such a way that the retaining wall between the displaced fragments represent the tectonically stressed zone of rank I, which covers two blocks of rank IV - "210" and "212". Just within these limits numerous dynamic events take place.

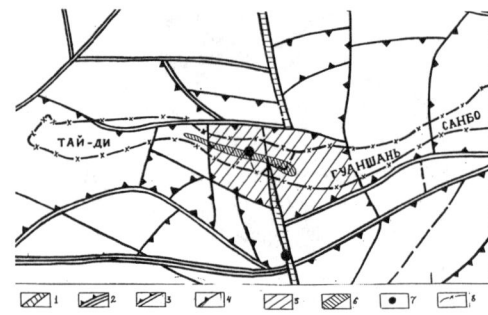

Fig.1. Scheme of geodynamic zoning of the Beipiao deposit (China):
1-4 relative faults of I-IV ranks; 5 - the most stressed blocks; 6 - zone of maximum concentration of foci of dynamic events; 7 - epicentres of earthquakes; 8 - boundaries of mine fields.

It is this fact that is represented by the results of calculation of stress state (Fig .2) of rock mass in the vicinity of the Beipiao fault.

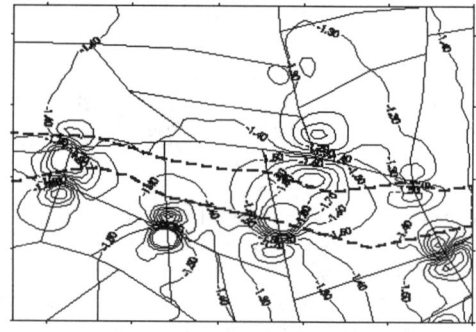

Fig.2.

With using the method for detection of active faults it was assessed the degree of geodynamic hazard of tectonic fractures within the mine fields of the Verkhnekamsk potash deposit (Russia, the Middle Urals).

The tectonical analysis has shown that the area

between the Prikama fault and the diagonal, cross fault is located in a large regional tectonically stressed zone. Besides the Verkhnekamsk potash deposit, in this zone there is also the Kizelov coal deposit, which is known by a great number of dynamic events. In block mass of high order within the Verkhnekamsk deposit the path of stress axes varies, and the axis of maximum stresses takes the sublatitudinal position, the axis of average stresses is subvertical. In such a stress field as the most active should consider the fractures 2-2 and the diagonal, straight-line fracture adjoining it from the south (1b). Relative position of the fracture planes (2-2 and 1b) and stress axes predetermines the development of maximum tangential stresses along their displacements. By the way, in displacement the shearing component prevails that was expressed in relatively small vertical amplitude. The fracture 1b is a left-hand shear, and the fracture 2-2 represents a right-hand shear, i.e. the displacement of internal wings of fractures is directed towards the site of their intersection. This is a typical wedge-shaped pattern, in its top and adjacent area the tectonically stressed zone is formed.

In the intersection zone of the mentioned above faults in January 1995 the technogenic earthquake of magnitude 5 by the Richter scale had occurred at the depth of 5 km. This earthquake caused the rock caving and the earth surface subsidence within the area of 500 hectares (see Fig. 3) with maximum subsidences up to 4 m. Repeated seismic events in this region are the real verification of the present-day activity of tectonical fractures. The tectonically stressed zone is located also within the area of fractures intersection 1b and 4-4 along which the right-hand shearing movement is probable. The stress level of rocks in this zone is lower, but nevertheless it exists.

The next zone of stress concentration is the intersection zone of fractures 5-5 and 1a. The fractures are referred to upthrow fault displacement, and the oncoming movement of hanging wings leads to a substantial rise of tectonic stresses in their intersection zone. In addition, the stress state of rock mass in this zone becomes complicated by a number of small breaks in continuity. The intersection zone of two fractures making the great Prikama fault, also is located in the zone of higher stresses.

In order to assess the native stress state of tectonically disturbed rock mass with using the method of geodynamic zoning, one has constructed a map of block structure within the Chutyr petroleum field and determined the direction of principal stresses. Fig. 4-6 show the calculated model of block structure and the results of calculations of stress state for this structure. To assess the influence of contact interaction along the block boundaries, the mechanical characteristics of rock blocks are preset as equal ones. The boundary conditions at block contacts are preset from the assumption that along the faults of rank II intersecting the deposit, the mode of complete overslipping is possible. This assumption conforms to the situation when due to the fluid pumping-out the opportunity appears for movements along the contacts of blocks positioned within the deposit. The forecasting map of stress state (Figs 4-6) shows the interaction of block structures in compliance with our presentations about the character of potential movements along the contacts of a block structure.

For example, Figure 4 shows the stress distribution with absence of movements along contacts of the mentioned disturbances. If in the deposit mining the movements along contacts took place, e.g. in the

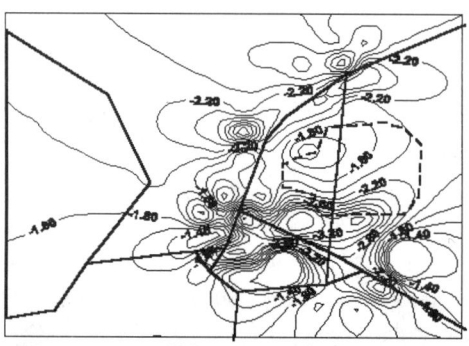

Fig. 3.

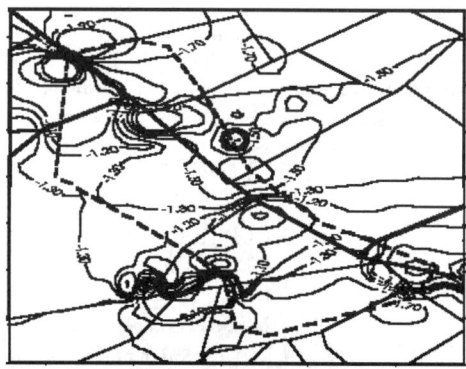

Fig. 4.

block HI, then the stress distribution will be quite different. With partial recovery of the seam pressure the movements along some contacts will be eliminated, and the stress distribution in block structure will be favourable.

B. The method of geodynamic zoning is of great importance in designing and construction of main pipelines. So, based on the analysis of 2000 emergency cases having occurred in the linear part of the main oil- and gas pipelines, it was stated that above 60 % of emergencies occur within the zones of active faults, detected with the use of this method.

C. Detailed study of stress-strain state and gas permeability of enclosing rocks on the geodynamic model, and polygon monitoring observations enable us to provide the long-term stability and reliable maintenance of underground structures. The Institute VNIMI has got an experience and has at its disposal the methodology for prediction and survey observations for the change in geodynamic state of rock mass, initiation of movements in rock strata and on the earth surface with using the satellite and air photographs, geophysical equipment, base stations, etc., that allows to reveal timely the source of generation and mode of rock pressure manifestations, to assess the current state of the environment, individual elements of rock mass at any preset time interval with prediction of this state under varying tectonic situation. In designing and construction of underground structures on the basis of geodynamic model of the region the potentiality appears for selection of the site for their erection, for consideration of any alternative engineering decisions, possible alternate layout, the parameters of rational realization within the range of predicted margins of technogenic events, dynamic disturbances, the mechanical, thermal and structure gradients, energy exchange processes in gas-containing media.

Table 1 below contains the enumeration of basic preventive measures against the geodynamic hazard in designing, construction and exploitation of underground structures.

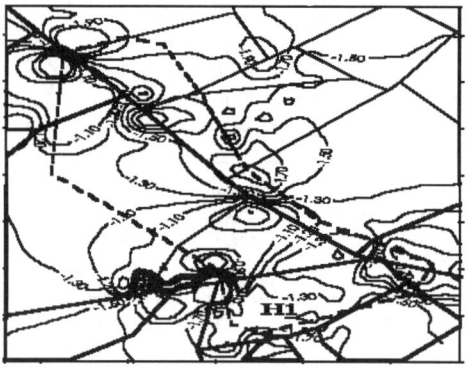

Fig. 5.

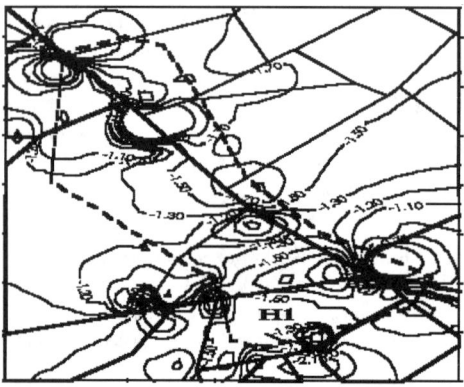

Fig. 6.

Table 1. System for better geodynamic safety of underground structures
(by way of example of the underground gas reservoir (UGR))

Stage	Subject-matter of investigations
Prospecting work	1. Geodynamic zoning of a region. 1.1 Detection of block structure and active faults. 1.2 Assessment of dynamic interaction of blocks. 1.3 Assessment of native stress state. 1.4 Detection of the zones with higher and lower gas permeability in rock mass. 1.5 Development of geodynamic model of the section of UGR.

Table 1. (continue)

Stage	Subject-matter of investigations
Prospecting work	1.6 Prediction of change in geodynamic mode of the UGR section. 1.7 Selection of the site for the UGR layout.
Designing	2. Working out and substantiation of the requirements for geodynamic reliability of the UGRs. 2.1 Investigations of physico-mechanical properties of rocks with due account of rheological processes of creeping and relaxation, dynamic disturbances, mechanical, thermal and structure gradients, energy exchange processes in gas-containing media. 2.2 Assessment of the influence of the external environmental factors. 2.3 Development of the regional and local measures for better protective functions of rock mass for the purpose of long-term stability of the UGRs. 2.4 Analysis of alternate layout of the UGRs. 2.5 Simulation of the UGR operation in preset geodynamic conditions. 2.6 Assessment of the UGR influence on the geodynamic conditions of the section or a region on the whole. 2.7 Development of methods and techniques for control of geodynamic conditions of enclosing rocks.
Construction	3. Creation and assembling of systems for continuous monitoring of geodynamic conditions of the UGR section. 3.1 Development of airphotomonitoring systems for observations of movements and deformations of the earth surface. 3.2 Equipment of geodynamic polygon for observations with using the satellite systems and for geophysical regime observations. 3.3 Working out of suggestions to the Building Code and applicable regulations for providing the geodynamic safety of engineering constructions.
Exploitation	4. Continuous monitoring for geodynamic conditions of mineral resources within the UGR section. 4.1 Observations of the geodynamic polygon. 4.2 Software for the UGR operation. 4.3 Detection of local hazardous zones and unfavourable interaction between the UGR exploitation conditions and stress-strain state of enclosing rocks. 4.4 Application of a package of measures for rock pressure control and elimination of dangerous situations.
Liquidation (abandonment)	5. Assessment of negative consequences (after-effects). 5.1 Sequential order of measures applied for the UGR abandonment with the least negative influence on geodynamic and ecological situation of the region.

Modelling techniques for safety evaluation: Back analysis and others

Displacement back analysis of tunnels in viscoelastic rock masses

Fusheng Zhu
Department of Civil Engineering, Northeastern University, People's Republic of China

Lin Xue
Department of Mechanics, Qingdao Institute of Architecture and Engineering, People's Republic of China

ABSTRACT: This paper presents a boundary element method of displacement back-analysis of rock mass creep parameters based on the creep compliance and the generalized creep compliance concepts. The original stress condition and the creeping parameters of rock mass can be obtained by using the measured displacements during excavating.

1 INTRODUCTION

Available field evidence shows that deformation of underground tunnels is time-dependent. The ground pressure on the lining of a tunnel of a given shape, for example, depends not only on the rock mass properties and the original stress conditions, but also on the time of its installation. Generally, most of those factors affecting the pressure on tunnels have been analyzed, measured and taken into account in any excavation design. Considering the present state-of-art of excavation lining design in creeping rock masses, it is felt that more information is still needed on the basic rock mass creep parameters, most of which can best be done by systematic monitoring of important projects and by back-analyzing the data with help of appropriate numerical methods. The researching works in the back analysis of rock mass parameters show that the material constants can be gotten, not only for isotropic materials, but also for inhomogenous and anisotropic materials based on numerical methods, such as the finite element methods and the boundary element methods. The displacement back analysis has been already proposed by many authors in elastic, elasto-plastic and linear viscoelastic models for rock mass (Shimizu et al. 1983, Sakurai et al. 1983, Hisatake 1986, Maier 1981, Goida 1980, 1987, wang 1988, Zheng et al. 1987).

This paper presents a boundary element method of displacement back-analysis of rock mass creep parameters based on the creep compliance and the generalized creep compliance concepts. The original stress condition and the creeping parameters of rock mass can be obtained by using the measured displacements during excavating. The method proposed is practicable because the appropriate viscoelastic model for the back-analysis, which could not be worked out in other numerical back-analysis methods, can be selected by the identification theorems based on the same concepts mentioned above. The another advantage of the method proposed is that the back-analysis can be worked out without adopting the optimization algorithm as in the conventional methods so that the analysis may be more concise and fast. An application of the proposed method for a tunnel in time-dependent response of the surrounding ground has been discussed in detail to indicate the power of the method.

2 BACK ANALYSIS OF ELASTIC ROCK MASS

In the two-dimensional elastic continuous body, which is assumed to be isotropic and have no body forces, the following relationship, which is based on the boundary element method, between the displacements at a point $p(x, y)$ in the ground and the normalized initial stresses was given as (Shimizu et al. 1983):

$$u_i(p) = \{F_i(p)\}^T \{\sigma'\} \qquad (1)$$

where:

$$\{F_i(p)\}^T = -\{EU_i(p)\}^T [T]^{-1} [n] \qquad (2)$$

The quantity E is the Young's modules, $\{U_i(p)\}$ and [T] are the influence coefficience vector and matrix obtained by integrating the Kelvin's solutions, respectively, [n] is the outward normal of a nodal point on the boundary, and $\{\sigma^0\}$ is called the normalized initial stress, and is defined as $\{\sigma^0\}=\{\sigma_{xx}/E\ \sigma_{yy}/E\ \sigma_{xy}/E\}$.

In tunneling the displacement measurements such as the convergence and the bore-hole extensometer measurements are simple and reliable compared with the stress measurements, so the displacements are considered as input data in the back analysis procedure. We have the following set of equations for a set of measured displacements:

$$\{u\} = [F]\{\sigma^0\} \tag{3}$$

If the relative displacements between two arbitrary points on the tunnel surface are measured, they can be expressed as follows:

$$\{d\} = [A]\{\sigma^0\} \tag{4}$$

where:

$$[A] = [C][F] \tag{5}$$

The [C] is the transformation matrix and matrix [A] can be determined uniquely for a given Poison's ratio.

The equation (5) has the same number of equations as the number of measured displacements, and contains three unknowns. If the number of measured displacements exceeds the number of unknowns, the normalized initial stresses can be determined by a suitable optimization procedure. The normalized initial stresses based on the least squares method is given as follows:

$$\{\sigma^0\}=[[A][w][A]]^{-1}[A]^T[w]\{d\} \tag{6}$$

Where [w] is a weight matrix and $\{d\}$ is the measured relative displacements.

It should be pointed out that the normalized initial stresses are consisted of the initial stresses and the Young's modules, so at least one of these unknowns must be determined in some other way. It is convenient to assume the in-situ state of vertical stress to be equal to the overburden pressure, then the Young's modules and the other components of initial stresses can be calculated easily from the normalized initial stresses.

3 BACK ANALYSIS OF VISCOELASTIC ROCK

In a viscoelastic body all the stresses, strains and displacements are time-dependent. In any time-dependent problems involving governing equations which are linear in the time variable and have time-independent coefficients it is possible to remove the time dependence in the equations by taking a Laplace transform. The constitutive equations for a viscoelastic material in the Laplace plane s become the following algebraic equations:

$$\bar{f}(s)\bar{\sigma}(s)=\bar{g}(s)\bar{\varepsilon}(s) \tag{7}$$

where $\bar{f}(s)$ and $\bar{g}(s)$ are the polynomials in the Laplace plane, which correspond the differential operators f(t) and g(t) in the physical plane.

If we define the fundamental solutions for a viscoelastic region as:

$$u_{ki}=A_1(x,y)J_1(t)+A_2(x,y)J_2(t) \tag{8}$$

where:

$$\begin{aligned}A_1(x,y)&=\frac{1}{2\pi}(\ln\frac{1}{r}\delta_{ki}+r_ir_i)\\ A_2(x,y)&=\frac{3}{2\pi}(\ln\frac{1}{r}\delta_{ki}-r_ir_i)\end{aligned} \tag{9}$$

and

$$\begin{aligned}J_1(t)&=L^{-1}\left[\frac{\bar{f}(s)}{s\bar{g}(s)}\right]\\ J_2(t)&=L^{-1}\left[\frac{\bar{f}(s)}{s(6k\bar{f}(s)+4\bar{g}(s))}\right]\end{aligned} \tag{10}$$

where $J_1(t)$ and $J_2(t)$ are called the creep compliance and the generalized creep compliance, respectively.

Using the fundamental solutions above, together with suitable transformation equations to account for the orientations of the line segments of the boundaries, we can obtain the following indirect boundary element formulation as:

$$\begin{aligned}u_s^i&=\sum_{j=1}^N(A_{1ss}^{ij}P_s^jJ_1+A_{2ss}^{ij}P_s^jJ_2)+\sum_{j=1}^N(A_{1sn}^{ij}P_n^jJ_1+A_{2sn}^{ij}P_n^jJ_2)\\ u_n^i&=\sum_{j=1}^N(A_{1ns}^{ij}P_s^jJ_1+A_{2ns}^{ij}P_s^jJ_2)+\sum_{j=1}^N(A_{1nn}^{ij}P_n^jJ_1+A_{2nn}^{ij}P_n^jJ_2)\end{aligned} \tag{11}$$

where A_{1ss}^{ij} ect., are the boundary displacement influence coefficients, and are the fictitious forces obtained from the corresponding elastic problem and are the measured displacements at the time. If the number of measured displacements exceeds the number of unknowns, the linear algebraic set of the equations (11) can be solved by using the least squares method.

There are five simple linear viscoelastic models, the creep compliance and the generalized creep compliance for these models can be expressed as (Xue et al. 1993):

$$J_j(\tau_i) = x_j[w_j(t_o) - w_j(t_i)] \quad (i = 1, 2, \ldots; j = 1, 2) \quad (12)$$

where x_j (j =1,2) are the algebraic expresses of the elastic constants and $w_j(t_o)$ (j =1,2) are the algebraic express of the viscoelastic constants.

Assume that the displacement measurement starts at any time $t_o \geq 0$, the interval time of measurement is $\tau_1 = t_1 - t_o$ and $\tau_2 = t_2 - t_o$. Let $\tau_2 = 2\tau_1$, then we have (Xue et al. 1992):

$$w_j(t_o) = J_j(\tau_2)/J_j(\tau_1) - 1 \quad (13)$$

and

$$x_j = J_j(\tau_1)/[(1 - w_j(t_o))] w_j(t_o) \quad (14)$$

Once the creep compliance and generalized creep compliance have been obtained by the displacement back analysis, the time-dependent parameters of the viscoelastic rock mass can be expressed easily by using the equations (13) and (14).

4 IDENTIFICATION OF VISCOELASTIC MODELS

If the simple viscoelastic models are chosen in any displacement back analysis, the following theorems of identification for rock mass models can be proved:

Theorem 1 *If $w_1(t_0) = 1$ and $x_2 =$ const. at any time corresponding to t_0, then the rock mass can be described by the Maxwell model.*

Theorem 2 *If $x_1 =$ const. and $x_2 =$ const. at any time corresponding to t_0, and there exists the relation $x_1 \ln w_1(t_0) = x_2 \ln w_2(t_0)$, then the rock mass can be described by the Kelvin model.*

Theorem 3 *If x_1 and x_2 are variables at any time corresponding to t_0, then the rock mass can be described by the Maxwell model or Burgers model.*

Theorem 4 *If $x_1 =$ const. and $x_2 =$ const. at any time corresponding to t_0 and there exists the relation $m \leq 1/3$, then the rock mass can be discerned by the three-parameter model.*

The proof of these theorems have been omitted for simplification, and they can be found in the literature (Xue 1994).

5 VERIFICATION OF THE ALGORITHM AND PROGRAM

Example 1 Consider a circular hole with the diameter R=1 in an infinite plate. Suppose that the rock mass can be modeled by a three-parameter model with the mechanical properties E_1 =700MPa, E_1 =500MPa, η=1.16 $\times 10^5$MPa·d and v=0.3. Assume a set of measured displacements, the creep compliance $J_i(t)$ then can be computed using the back analysis mentioned before (see Table 1), and the viscoelastic ancestral quantities $w_1(t)$, $w_2(t)$ and the elastic ancestral quantities x_1, x_2 are given in Table 2.

Table 1. Values $J_i(\tau)$ of displacement back analysis.

$t_0 \sim t_i$	5~10	5~15	10~20	10~30	20~40
$J_1(\tau) \times 10^{-4}$MPa^{-1}	3.131	4.901	4.555	7.286	4.730
$J_2(\tau) \times 10^{-6}$MPa^{-1}	6.002	9.467	6.112	8.254	3.237

Table 2. Values of x_1, x_2 and $w_1(t)$, $w_2(t)$

t_0	5	10	20
x_1 ($\times 10^{-3}$MPa^{-1})	1.9999	2.0000	1.9999
x_2 ($\times 10^{-5}$MPa^{-1})	2.5641	2.5641	2.5641
$w_1(t)$	0.8057	0.6492	0.4215
$w_2(t)$	0.6263	0.3922	0.1538

It can be found that the values of x_1 and x_1, x_2 are constant and m = 0.17<1/3, so that the model is a three-parameter one by using the theorems 4, as what assumed in the problem.

The computed and given values of the viscoelastic parameters are given in Table 3. The results show that the numerical solution presents a good approximation to the given values.

Table 3. Results of back analysis of mechanical parameters

parameters	E_1(MPa)	E_2(MPa)	η(MPa $\times 10^5$)
given values	700	500	1.16
computed values	679	485	1.13

Example 2 Consider a circular hole with the diameter R=1 in an infinite plate. Assume a set of measured displacements, the creep compliance $J_i(t)$, the viscoelastic ancestral quantities $w_1(t)$ $w_2(t)$ and the elastic ancestral quantities x_1, x_2 are given in Table 4 and Table 5.

The results show that the values of x_1, x_2 and m are variable so the rock mass can not be the Maxwell or viscoelastic solid, but the Burger's model.

Table 4. Results of displacement back analysis.

$t_0 \sim t_i$	10~20	10~30	20~40	20~60
$J_1(\tau) \times 10^{-4}$MPa^{-1}	1.311	2.349	1.960	2.153
$J_2(\tau) \times 10^{-6}$MPa^{-1}	1.146	2.270	2.153	3.954

Table 5. Values of x_1, x_2, $w_1(t)$, $w_2(t)$ and m

t_0	10	20
$x_1 \times 10^{-3} \text{MPa}^{-1}$	7.9385	18.7001
$x_2 \times 10^{-5} \text{MPa}^{-1}$	61.7967	15.7604
$w_1(t)$	0.767	0.855
$w_2(t)$	0.9516	0.812
m	0.776	1.067

6 CONCLUSIONS

1. This paper presents a boundary element method of displacement back-analysis of rock mass creep parameters based on the creep compliance and the generalized creep compliance concepts. The method proposed is practicable because the appropriate viscoelastic model for the back-analysis, which could not be worked out in other numerical back-analysis methods, can be selected by the identification theorems based on the same concepts mentioned above.

2. The displacement back-analysis of rock mass creep parameters based on the mentioned concepts above does not adopt any optimization algorithm as in the conventional methods, and the parameters can been obtained by the simple algebraic calculation, so that the analysis is more concise and fast.

ACKNOWLEDGMENTS

This research has been supported by the National Natural Science Foundation of the People's Republic of China.

REFERENCES

Gioda, G. 1980. Indirect identification of the average elastic characteristics of rock masses. *Int. Conf. Struct. Founds. on Rock.* Sydney, 331-336. Rotterdam: Balkema.

Gioda, G. & S. Sakurai 1987. Back analysis procedures for the interpretation of field measurements in geomechanics. *J. Num. anal. Meth. Geomech..* (11): 555-583.

Hisatake, M. & T Ito 1986. Three dimensional back analysis for tunnels. *Proc. Int. Symp. on ECRF.* 112-116. Beijing: Science Press.

Maier, G & G. Gioda 1981. Optimization methods for parametric identification of geomechanical system. *Num. Meth. Geontech.* (6): 87-90.

Sakurai, S. & K. Takeuchi 1983. Back naalysis of measured displacements of tunnels. *Rock Mechanics and Rock Engineering* (16):173-180.

Shimizu, N. & S. Sakurai 1983. Application of boundary element method for back analysis associated with tunneling problems. In C. A. Brebbia et al.(ed.), *Proc. 5th Int. Conf. Boundary Elements.* Hiroshima, 645-654.

Wang, Z. Y. & H. H. Liu 1988. Back analysis of measured rheologic displacements of underground openings. *Proc. 6th Conf. on Num. Meth. in Geomech.* (4): 1101-1107.

Zheng, Y. R., C. Wang & D. H. Zhang 1987. Back analysis from measured displacement based on elastoplastic theory in strain space. *Proc. Int. Symp. Geomech. Bridges Struct.*, Lanzhou, 505-508.

Xue, L. & Z. F. Yang 1993. The analytical solution of the mechanical parameters of viscoelastic rock mass (in Chinese), *Annual Report of the Engineering eomechanics Lab.* Beijing: Earthquake Publishing House.

Xue, L. 1994. Theorem of identification for viscoelastic model of the rock mass and its applications. *J. Geomechanics (in Chinese)* 16(5):1-10.

Stochastic finite element analysis of underground rock structure using Latin Hypercube Sampling technique

K.S.Choi
Department of Radioactive Waste Technology, NETEC, KEPCO, Taejon, Korea

B.Y.Park
Nuclear Fuel Cycle Research Group, KAERI, Korea

H.K.Lee
Department of Mineral and Petroleum Engineering, Seoul National University, Korea

ABSTRACT : In this study, a stochastic finite element model is proposed with a view to consider the uncertainty of physical properties of rock mass in the analysis of structural behavior on underground caverns. Here, the Latin Hypercube Sampling technique, in which can make up weak points of the Monte Carlo Simulation, is applied for the analysis of underground cavern. To reflect the uncertainty of material properties, multi-random variables are assumed as the elastic modulus and the Poisson's ratio, all of which could be simulated in terms of normal distribution, log-normal distribution, and rectangular uniform distribution. New computer program has been coded and verified through the analysis of verification examples, and its practical applicability has been confirmed by the stochastic finite element technique for the analysis of the underground oil storage cavern in Korea.

1 INTRODUCTION

Underground structures are widely used to overcome space shortages of aboveground structures, to secure a smooth traffic flow or to accommodate important facilities requiring special isolation. With a growing usage of underground structures, there have been increased much concerns about the safety assessment for underground structures. Recent numerical methods for the safety assessment of underground structures are almost based on the deterministic approach technique such as FEM(Finite Element Method), FDM (Finite Difference Method), BEM (Boundary Element Method), and so on. The accuracy of these methods depends on the rationality of input data. However, since the design work for underground structure is generally performed before the stage of excavation, much uncertainties are involved in physical properties for the design. Moreover, since the various in-situ and laboratory tests to be performed at the stage of site characterization are costly, it seems difficult to get enough data for the exact stress analysis. However, most of numerical analyses for underground structures have not been reflected such uncertainties in physical property of rock mass so far, and have been performed on the assumption that all of the characteristic values of rock mass at the site are constant. As the results by such analyses do not reflect the uncertainties in the physical properties of rock mass, they are not simulated in real situations and thus lead to over-design or require drastic changes of design during the construction work which result in economic losses. In order to realize more rational analysis, it is desirable to use the stochastic model which can consider uncertainties of the physical properties in rock mass. But, unfortunately, the study related to the stochastic approach for uncertain structures in progress up to now has been limited to only above- ground structures. In addition, the perturbation theory or the MCS(Monte Carlo Simulation) technique applied to the stochastic finite element analysis field so far requires enormous number of calculation or complex solution procedure in producing more accurate results. The LHS(Latin Hypercube Sampling)(Iman et al. 1984) technique, which is a sort of variance reduction technique capable of maintaining the degree of accuracy while the calculation amount is greatly reduced by preserving strong points and compensating weak points of MCS technique, has been widely used to the analysis of neutron transport in the nuclear engineering field since it was presented by the Sandia National Research Institute of the United States in 1980.

In this study, a stochastic model of numerical analysis that reflects uncertainties of physical properties in rock mass has been developed. In so doing, LHS technique has been applied. To reflect uncertain material properties in rock mass, random variables such as the elastic modulus and the Poisson's ratio have been used, all of which can be applicable through normal distribution, log-normal distribution, and rectangular uniform distribution. The new computer program has been coded and its practical applicability has been confirmed by using the stochastic finite element technique for the analysis of the oil storage cavern in Korea.

2 STOCHASTIC FINITE ELEMENT ANALYSIS

2.1 Generation of random deviates from distribution

In this study, the power residue method has been adopted to generate pseudo-random numbers denoted by $\{u_n\}$, $n=0,1,2,\cdots$. If cumulative distribution function (CDF) of a random variable X is $F_X(x)$ and the cumulative probability of X is given in the form of $F_X(x) = z$, the set of random deviates x defining the distribution of X is represented as the following equation using the inverse method (Abramowitz et al. 1972).

$$x_i = F_X^{-1}(z_i), \quad i=1,2,3,\cdots,n \quad (1)$$

As pseudo-random numbers is in the form of uniform distribution on the interval (0,1), it becomes $z_i = u_i$ when the CDF of pseudo-random numbers and that of desired distribution homologize 1:1. When pseudo-random numbers are not used in sufficiently large size, they may be biased partially between 0 and 1, and are potentially unable to represent the desired distribution. But, if the desired distribution are discretized into m non-overlapping intervals of equal probability and one value from each interval is selected at random, more optimal sampling can be ensured. This means that the sampling is done uniformly on the vertical axis of the CDF. In order to generate distributions of multi-variables, this procedure is repeated for each variable, each time working with the corresponding CDF. The next step involves pairing the selected values. If two variables are sampled independently and paired randomly, the sample correlation coefficient between two random variables shall be considered to review the effect by the sampling fluctuations. In this study, the following correlation factor used widely in the LHS technique is applied (Iman et al. 1982).

$$\rho_{12} = \frac{\sum_{i=1}^{n}\left(S_{x_1^i} - \frac{n+1}{2}\right)\left(S_{x_2^i} - \frac{n+1}{2}\right)}{\left\{\sum_{i=1}^{n}\left(S_{x_1^i} - \frac{n+1}{2}\right)^2 \sum_{i=1}^{n}\left(S_{x_1^i} - \frac{n+1}{2}\right)^2\right\}^{\frac{1}{2}}} \quad (2)$$

where S_{X1} and S_{X2} indicate the random deviates obtained in i intervals of two variables, $X1$ and $X2$. If ρ_{12} is larger, statistical correlation between two random deviates is higher. Therefore, in LHS method, when this value is less than 1, a pair of random numbers is considered to be relatively reasonable.

2.2 Spatial variations of material properties.

The spatial variation of material properties is assumed to be a 2D homogeneous stochastic process in this study. The fluctuating component $g(x)$ of a material property is then assumed to have mean zero $E[g(x)]=0$ and the auto-correlation function $R_j(\xi) = E(g(x)_j \cdot g(x+\xi)_j)$, where $x=\{xy\}^T$ indicates the position vector, j denotes the random variable number, and $\xi = [\xi_x \ \xi_y]^T$ means the separation vector between two points x and $x+\xi$. If the randomness of the spatial variation is isotropic, the auto-correlation function of the spatial variation is supposed to be a function only of the distance $|\xi|$. The following form of an isotropic auto-correlation function is considered for this study;

$$R_j(\xi) = \sigma_j^2 EXP\left[\frac{|\xi|}{d}\right] \quad (3)$$

where, σ_j is the standard deviation of random variable j and d is a positive parameter such that the larger it is, the more slowly the correlation disappears (Yamazaki et al. 1988). If the structure is discretized by m elements, the fluctuating component $g(x)$ is composed of m material property values associated with these m element correlation each other. Their correlation characteristics can be specified in terms of the covariance matrix

$$Cov_j[g^X, g^Y]_{i,k} = E_j[g_X, g_Y]_{i,k} = R_j(\xi_{X,Y})_{i,k},$$

where the subscript j indicates the random variable number and $\xi_{X,Y}$ is the separation between the centroids of elements i and k. Therefore, the final distribution of random variable can be obtained by multiplying equation (1).

$$G_j(x, Cov)_i = Cov_j[g_X, g_Y]_i \cdot [x_u]_j \quad (4)$$

2.3 Finite element analysis

When the elastic modulus or the Poisson's ratio of rock mass are considered as a random variable, the elasticity matrix [c] and the element stiffness matrix [K] of i-th element are represented by

$$[c]_i = [c]_{io}(1+g_i(E, \nu))$$
$$[K]_i = [K]_{io}(1+g_i(E, \nu)) \quad i=1,2,3\cdots,m \quad (5)$$

in which $[c]_{io}$ and $[K]_{io}$ are the elasticity matrix defined by equation (5a) and the element stiffness matrix defined by equation (5c) and (5d), respectively. And, in equation (5), $g_i[E, \nu]$ indicate the distribution function of elastic modulus and Poisson's ratio, which can be obtained by reflecting the $G_j(x, Cov)_i$ calculated by expression (4).

$$[C] = 1/m \begin{bmatrix} E - E\nu^2 & E\nu^2(1+\nu) & 0 \\ E\nu^2(1+\nu) & E(1-\nu^2) & 0 \\ 0 & 0 & Gm \end{bmatrix} \quad (5a)$$

$$m = (1+\nu)[1-\nu-2\cdot\nu^2] \quad (5b)$$

$$[K]_{Tri} = \int_A [B]^T [T_\varepsilon]^T [C][T_\varepsilon][B] dA \quad (5c)$$

$$[K]_{Quad} = \int_{-1}^{1}\int_{-1}^{1} [B]^T [C][B][J] d\xi d\eta \quad (5d)$$

3 STOCHASTIC ANALYSIS FOR THE UNDERGROUND OIL STORAGE CAVERN IN KOREA

A 2D FEM program has been developed by which we could estimate the response variability due to the physical property variation of rock mass. To realize an efficient inverse matrix calculation of global stiffness matrix, the modified Cholesky method and the skyline algorithm have been adopted. In order to verify the program, various stochastic analyses have been performed. Once considering the elastic modulus, the output by this program shows the identity with that of Yamazaki, et al.(1988) Also, for the case of analysis using the elastic modulus as a random variable, the output of LHS method shows the similarity to that of MCS method. In addition, when the stochastic results are compared with deterministic results in many cases, the ranges and the tendency of stochastic output exist within a reasonable scope (Chung & Choi 1996).

In order to confirm the applicability of the program to the field, a stochastic analysis was performed for the underground oil storage cavern which has been constructed in Korea, and analysis model are indicated in Figure 6. The elastic modulus of rock obtained by the lab. tests of the 70 specimens showed the normal distribution with the mean value of about 6.9 x 10^5kg/cm^2 and the standard deviation of 1.7 x 10^5 against mean value, and the Poisson's ratio also showed the normal distribution with the mean value of 0.26 and the standard deviation of 0.054. However, the distribution of the elastic modulus presented from the in-situ test at the 50 locations was irregular distribution shape with the meal value of 5.77 x 10^5kg/cm^2 and the standard deviation of 2.11 x 10^5 kg/cm^2. In this study, the following two cases are examined; the one is the case of which the elastic modulus and the Poisson's ratio are assumed to be the normal distribution(Case 1), the other is the case of which the elastic modulus is assumed to be the rectangular uniform distribution and the Poisson's ratio is assumed to be the normal distribution(Case 2). Then the mean value and the variance of the Poisson's ratio were used as 0.26 and 5.4% respectively, and the mean value and the variance of elastic modulus with normal distribution were used as 5.77 x 10^5 kg/cm^2 and 30% respectively. In the case of elastic modulus with rectangular uniform distribution, 42,121kg/cm^2 was used as minimum value and 1,111,879kg/cm^2 as maximum value. For the analysis of two cases, unit weight of 2.6 t/m^3 and the lateral pressure coefficient K of 1.5 were used from the in-situ test results. These values show a little bit of difference depending on the location of measurement. But in this study, uncertainties of these values were not considered. On the analysis model shown in Figure 1, rock is modeled as the isoparametric quadrilateral node elements of 240, and the stochastic complexity in each element was supposed to be the same. The initial horizontal stress

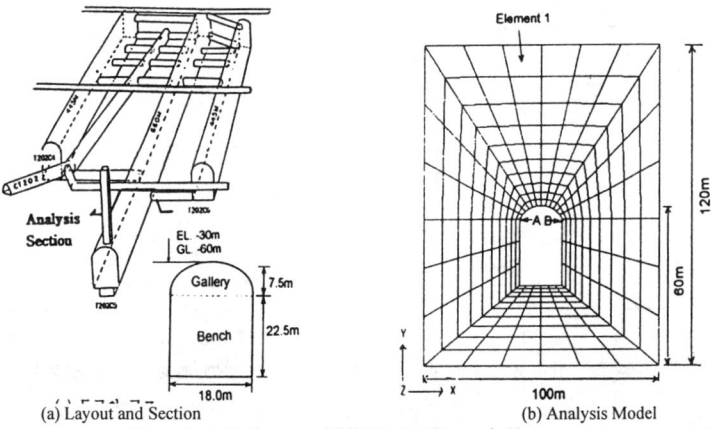

(a) Layout and Section   (b) Analysis Model
Figure 1. A Underground Oil Storage Cavern in Korea

calculated from K and the self-weight of rock was considered for load.

On the analysis model of Figure 1, the distribution form of the elastic modulus sampled at Element 1 is the same as shown in Figure 2(a) means the result of Case 1, and (b) means the result of Case2. As shown in the Figure, the distribution form of the elastic modulus sampled at Element 1 the distribution characteristics defined by the input rationally. In this example, the mean value and the standard deviation of the sampled distribution showed the difference of about 1.0% compared with the input value.

The sampled displacements at several representative points around cavern were examined. The displacements after excavation occur toward inner-part of the cavern except the cavern bottom.

The vertical displacement at the crown point and the horizontal displacement at node for Case 1 are shown in Figure 3, and those for Case 2 are indicated in Figure 4. As shown in Figure 3, the distribution form of displacement around the cavern shows the normal distribution or the log-normal distribution shape. Figure 4 shows the combined shape of the rectangular uniform distribution and the log-normal distribution. In the Case 1, the vertical displacement at the cavern crown is 2.1 ± 0.94 mm and shows the concentrated tendency generally within the scope of $1.0\sim3.5$ mm. The vertical displacement at the mid-point of cavern bottom is 1.54 ± 0.71 mm and shows the concentrated tendency generally within the scope of $0.5\sim2.5$ mm. The horizontal displacement at node A and node B is the same size with 1.16 ± 0.96 mm due to the geometric symmetry about the analysis model. In the Case 2 analysis, the vertical displacement at the cavern crown is 3.3 ± 3.82 mm and shows the concentrated tendency within the scope of $1.0\sim5.0$ mm, while the vertical displacement at the mid-point of the cavern bottom is 2.43 ± 2.82 mm and shows the concentrated tendency within the scope of $0.5\sim5.0$ mm. The variance of displacements at the representative points around cavern is 45% of the mean value in the Case 1, and 115% in the Case 2.

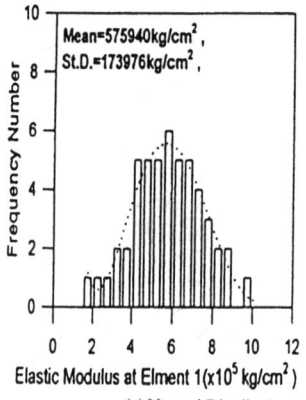 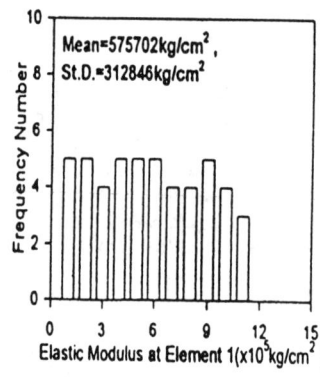

(a) Normal Distribution (b) Rectangular Uniform Distribution

Figure 2. Elastic Modulus Sampled at Element 1

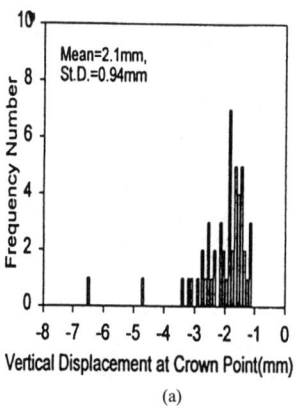 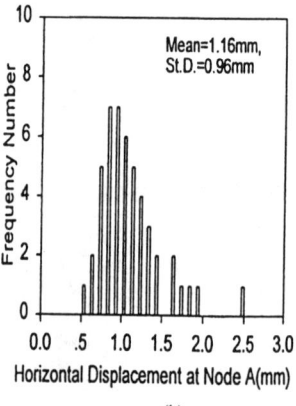

(a) (b)

Figure 3. Case 1 (E=Normal, v=Normal)

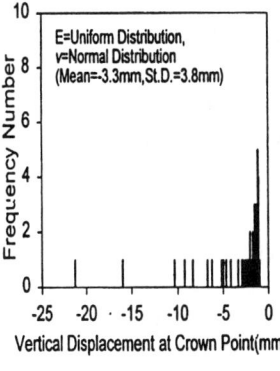
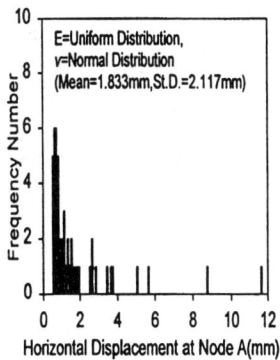

(a)   (b)

Figure 4. Case 2 (E=Uniform, ν=Normal)

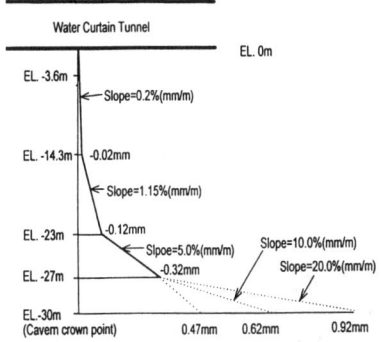

Figure 5. Calibrated Vertical Displacement at Crown Point Based on the Instrumentation Results

In order to verify the feasibility of this stochastic outputs of Figure 3 and Figure 4, we conducted deterministic analyses of many cases in the range of normal distribution and uniform distribution. Accordingly, stochastic analyses of many cases with single random variable are also carried out for the same purpose. As shown in Table 1 and Table 2, the displacement around the cavern shows much difference depending on the magnitude of physical properties and the supposed type of distribution. As for the analyses considering single variable, the largest mean value and standard deviation of displacements around cavern was found in the case of analysis with the uniform distribution of elastic modulus and the deterministic value of Poisson's ratio, while the smallest value was found in the case of analysis with the deterministic value of elastic modulus and the normal distribution of Poisson's ratio. From these results, the elastic modulus has larger sensitivity to the displacement around cavern than the Poisson's ratio. When the stochastic results of Figure 3 and Figure 4 are compared with the results of Table 1 and Table 2, it may be understood that the ranges of displacement distribution and the concentration tendency exist within a reasonable scope.

The basic design of this underground oil storage cavern was finished in 1991. At present, all caverns have been excavated. Because the section of this cavern was too large, the cavern was split into 4 stages (Gallery, Bench 1, Bench 2, Bench 3), and has been excavated by the NATM. The support of this cavern was designed by the empirical method such as Q-system. As the result of the support design, the surrounding cavern was reinforced using the shotcrete and the rockbolt. It is impossible to install the sensor at the exact location to measure the exact deformation during excavation. Therefore, the instrumentation of displacement at crown point was carried out from the observation of the sensor installed between gallery and the water-curtain tunnel. After the excavation of gallery, the convergence displacements were measured during the excavation of Bench parts. Figure 5 indicates the calibrated vertical displacement and cavern crown points based on the instrumentation results. As shown in the Figure, the vertical displacement at the cavern crown has the tendency to increase with depth. In this Figure, 0.47~0.92 mm was calibrated downward after gallery excavation. Accordingly, from the convergence displacements measured at the 2, 3, 4 excavation stages, 0.2 mm after Bench 1 excavation and about 0.1 mm after Bench 2 was measured. The convergence displacements of about 5.5 mm to inner direction between node A and node B, after excavation of Bench 3, was observed. To compare the analysis results and the instrumentation results exactly, the measurement values including the effect of reinforcement and blasting should be utilized. And the three-dimensional analysis model including the water curtain tunnel should be used. But, under the assumption that these influences are involved in the uncertainties in the stochastic analysis, it may be understood that the analysis results exist within a reasonable scope, judging from the direction and the

Table 1. Displacement around the Cavern by Deterministic Approach

Properties	Location		Vertical Displacement (mm)		Horizontal Displacement (mm)	
			at Crown Point	at Center of Bottom	at Node A	at Node B
$E_{max}=1.112\times10^6$ kg/cm^2		$v_{max}=0.427$	-0.675	-0.279	0.543	-0.543
$E_{max}=1.112\times10^6$ kg/cm^2		$v_{min}=0.093$	-1.031	-0.890	0.496	-0.496
$E_{mean}=5.770\times10^5$ kg/cm^2		$v_{mean}=0.26$	-1.879	-1.395	1.029	-1.029
$E_{min}=4.212\times10^4$ kg/cm^2		$v_{max}=0.427$	-17.818	-7.374	14.341	-14.341
$E_{min}=4.212\times10^4$ kg/cm^2		$v_{min}=0.093$	-43.321	-23.496	13.096	-13.096

Table 2. Displacement Distribution Characteristics Obtained by Stochastic Analysis in Case of Using Single Variable in Random Fields

Properties	Location	Vert. Disp. at Crown Pt. (mm)		Hori. Disp. At Node A (mm)		Vert. Disp. At Center of Bottom (mm)	
E	v	mean	σ	mean	σ	mean	σ
Normal D.	0.26	-2.123	0.943	1.163	0.517	-1.576	0.700
Rectangular D.	0.26	-4.149	6.219	1.835	2.114	-3.113	4.620
5.77×10^5 kg/cm^2	Normal D.	-1.853	0.109	1.027	0.018	-1.365	0.187
Normal D.	Normal D.	-2.094	0.937	1.161	0.519	-1.540	0.709
Rectangular D.	Normal D.	-3.305	3.820	1.833	2.117	-2.430	2.820
Deterministic Analysis Results		-0.675~-43.321 (-1.879)		0.543~13.096 (1.029)		-0.279~23.464 (-1.395)	

size of measured displacement. From the analysis result of this example, we can see that the direction and the size of the vertical displacement at the cavern crown point show to be similar to the measurement results, but the measurement results of the horizontal displacement at the cavern side node A and B are larger than the analysis results. To compare the analysis results with instrumentation results actually, it is desirable to add the functions of excavation/removal to the program presented in this study.

4 CONCLUSION

This study suggests a stochastic finite element analysis method capable of considering multi-random variables in the uncertain underground structure by applying the LHS technique. The results of the study have been validated through the verification of the program, and the field applicability has been confirmed by an application to the stochastic analysis of the underground oil storage facility constructed in Korea. The major results of this study are summarized in the followings:

1. The results of the stochastic analysis with the consideration of the elastic modulus of the rock and the Poisson' ratio show that the distribution form of displacement varies significantly with the distribution forms of the elastic modulus and the Poisson's ratio. In the present analysis, when normal distributions are assumed for the elastic modulus and the Poisson's ratio, the resulting distribution form of displacement around the cavern is a normal or log-normal distribution with the variance of approximately 45% about the mean value. When a uniform rectangular distribution and a normal distribution are assumed for the elastic modulus and the Poisson's ratio respectively, the resulting distribution form of displacement shows a combination of uniform rectangular distribution and log-normal distribution with a large variance of around 115%. In addition, the sensitivity analyses with single parameter variation shows that the displacement around the cavern is more sensitive to the change in the elastic modulus of rock than to the change in the Poisson's ratio.

2. Considering the direction and the quantitative magnitude of the displacement measured at the underground oil storage facility, the direction and the magnitude of the vertical displacement obtained from this analysis show good agreement with the measured data. However, for more realistic comparison of results with the data, an addition to the program to consider the effects of the excavation is judged to be necessary. Using the present results, the assessment of distribution characteristics of the movements of underground structures can be achieved simply. Thus, by using the reliability index obtained through the limit state equations formulated from the relationship of the allowable displacement and the allowable strength, the reliability analysis and the optimum support design are deemed possible.

Further studies need to be carried out to consider the numerous uncertainty factors in the geological structure and characteristics of the rock structures, and also the large safety requirement variations due to the function of structure.

REFERENCES

Abramowitz, M. and Stegun, I.A., *Handbook of Mathematical Functions with Formulas, Graphs, and Mathematical Tables,* Dover Pub., 1972.

Chung, Y. S., Choi, K. S., "Development of Stochastic Finite Element Analysis Technique, ", *Korean Society of Civil Engineering*, 1996, pp. 669-681.

Hoek, E. and Brown, E.T., *Underground Excavation in Rock*, Stephan Austin and Sons Ltd., London, 1980., pp. 527.

Hyundai Institute of Construction Technology, "Instrumentation for Underground Oil Storage Cavern and Revised Instrumentation for Excavation Behavior(3)," 92GE003, 1994. pp. 34-43.

Iman, R.L. and Conover, W.J. "A Distribution-Free Approach to Inducing Rank Correlation Among Input Variables," *Commun. Statist.*, Vol.11(3) , 1982., pp. 311.

Iman, R.L. and Shortencarier M.J., "A FORTRAN 77 Program and User's Guide for the Generation of Computer Models," *NUREG/CR-3624 SAND 83-2365*, 1984. pp. 50.Latin Hypercube and Random Samples for Use with

Samrim Consultant Co., "Basic Design Report for ○○ Underground Oil Storage Cavern," Vol. 1, 2, 1991.

Teukolsky, S.A. and Vetterling, W.T., *Numerical Recipes-The Art of Scientific Computing*, Cambridge Univ. Press, 1986, pp. 191-225.

Wold, H., *Random Normal Deviates,* Tracts for Computers 25, Cambridge Univ. Press, England, 1948.

Yamazaki, F., Shinozuka, M. and Dasgupta, G. "Neumann Expansion for Stochastic Finite Element Analysis," *Journal of Engineering Mechanics, ASCE*, Vol.114, No.8, 1988., pp. 1335-1353.

Swellex® in weak/soft rock

U. Håkansson
Atlas Copco, Hong Kong, People's Republic of China

C. Li
Luleå University of Technology, Sweden

ABSTRACT: The increasing use of Swellex in weak/soft rock necessitates the knowledge and understanding of the bolt function under such conditions. An elasto-plastic analysis has been performed based on the "thick" cylinder problem, assuming plane-strain conditions and applying the Mohr-Coulomb criterion. The contact stress between the bolt and the borehole wall is a fundamental factor to achieve a good reinforcement. The effect of the expansion degree of the bolt on the contact stress is paid a great attention in this study. A laboratory test was carried out to install a Super Swellex in a man-made material having σ_c = 4.4 MPa and E-modulus = 2 GPa. The stress distribution, displacement and pull-out force were monitored and measured. The overall objective with the presented analysis is to arrive at Design Guidelines for Swellex use in weak/soft rock. Design parameters such as bolt spacing, bolt length, borehole size and pump pressure are recommended.

1 INTRODUCTION

Swellex is a friction rock bolt which achieves its reinforcement action, in the borehole, by the expansion of a folded steel tube. Swellex has a unique feature in weak/soft rock. The expansion will consolidate the material in the vicinity of the bolt, which will enhance the strength of the rock and create a favourable arching action between the bolts. The use of Swellex in weak/soft rock formations has grown rapidly world-wide during recent years, and consequently the demand to explain the function under such conditions has also increased. The function of Swellex in hard, elastic, rock is well known and has been thoroughly reported earlier (Wijk & Skogberg, 1982). The objective of the present paper is therefore to demonstrate the function of Swellex when expanded in weak/soft, elasto-plastic, rock.

2 THEORY

The theoretical treatment of the expansion of Swellex in weak/soft, elasto-plastic, rock has been reported elsewhere, Håkansson & Li (1997), and therefore only the final equations and results are reported in this paper.

The analysis was made using cylindrical coordinates, assuming plane-strain conditions and applying Mohr-Coulomb failure criterion. Tensile stresses and strains resulting in an elongation of an element are regarded as negative.

The notations and geometry of the problem are shown in Figure 1. A borehole with a radius of r_i, is subjected to an inner pressure of P_i and a far field pressure of P_o. This is the thick wall cylinder problem with the outer radius extends to infinity.

When a Swellex bolt is expanded in weak/soft rock, a zone of plastic (crushed) material is generated in the vicinity of the borehole. The extent of the plastic zone is defined by the radius r_e, see Figure 1, but outside the plastic zone the material remains in an elastic state.

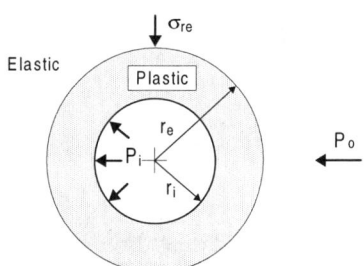

Figure 1. The plastic zone around a borehole.

Under the above assumption the stress distribution in the elastic zone becomes

$$\sigma_r = P_o - (P_o - \sigma_{re})\left(\frac{r_e}{r}\right)^2$$

$$\sigma_t = P_o + (P_o - \sigma_{re})\left(\frac{r_e}{r}\right)^2 \qquad (1)$$

where σ_{re} is the radial stress at the elasto-plastic boundary, see Figure 1.

In order to arrive at a solution for the plastic conditions, a failure criterion must be introduced. The Mohr-Coulomb criterion is often used to describe the strength of weak rock, i.e.

$$\sigma_r = k\sigma_t + a(k-1) \qquad (2)$$

where $k = (1 + \sin\phi)/(1 - \sin\phi)$, $a = C/\tan\phi$
ϕ = friction angle, C = cohesion.

In the plastic zone, the stresses are expressed as

$$\sigma_r = (P_i + a_r)\left(\frac{r}{r_i}\right)^{\frac{1-k_r}{k_r}} - a_r$$

$$\sigma_t = \frac{1}{k_r}(P_i + a_r)\left(\frac{r}{r_i}\right)^{\frac{1-k_r}{k_r}} - a_r \qquad (3)$$

where k_r and a_r are the constants of k and a corresponding to the residual friction angle, ϕ_r and cohesion, C_r. The radial stress at the outer boundary of the plastic zone, σ_{re}, and the radius of the plastic zone, r_e, are given by

$$\sigma_{re} = \frac{2kP_o + a(k-1)}{k+1} \qquad (4)$$

$$r_e = r_i\left(\frac{\sigma_{re} + a_r}{P_i + a_r}\right)^{\frac{k_r}{1-k_r}} \qquad (5)$$

It should be noted that σ_{re}, as given by Eq.(4), represents the limiting radial stress for the hole boundary to enter the plastic condition, i.e. if $P_i \leq \sigma_{re}$ no plastic zone develops. Note also that if the outer pressure, $P_o = 0$, then $\sigma_{re} = C \cos\phi$.

The radial elastic deformation outside the plastic zone, is readily attained by considering static equilibrium, geometrical compatibility, and the generalised Hooke's law under plain strain condition, which leads to

$$u_r = \frac{(1+v)}{E}\left[\sigma_{re}\frac{r_i^2}{r} - P_o(1-2v)r\right] \qquad (6)$$

In order to determine the deformation of the borehole in the elasto-plastic material, it is assumed that the elastic strains are constant throughout the plastic zone, and equal to the strains at the elasto-plastic boundary, and that the total strain is the sum of the elastic and plastic strain.

The following expression for the radial displacement is achieved, Håkansson & Li (1997)

$$u_r = \frac{(1+v)r}{\omega E}\left\{\sigma_{re}\left[2\left(\frac{r_e}{r}\right)^{\omega} + \omega - 2\right]\right.$$
$$\left. - 2P_o\left[\left(\frac{r_e}{r}\right)^{\omega} + (-v)\omega - 1\right]\right\} \qquad (7)$$

where $\omega = 1 + 1/f$ and f is the plastic strain increment.

3 CONTACT STRESS AND REINFORCEMENT EFFECT

When a Swellex bolt is installed and expanded in rock, a contact stress will develop between the bolt and the borehole wall. The reinforcement effect of the bolt depends upon both the contact stress and the frictional properties of the borehole wall. The effect can be expressed by the pull-out resistance of the bolt, given by

$$F_{pull} = 2\pi r_i Lq \tan(\phi + i) \qquad (8)$$

where r_i = the radius of borehole,
L = the "effective" length of the bolt,
q = the contact stress,
ϕ = the friction angle between the bolt and the rock material,
i = the roughness angle of the borehole wall.

The contact stress is closely related to the installation pump pressure. However, not all the pump pressure is transferred to the rock, since a portion of it is sustained as hoop stress, by the Swellex bolt. It is experimentally demonstrated that the pump pressure, needed to inflate a Super Swellex in free air is about 15 MPa. This is called the inflation pump pressure, designated as P_{po}. The pressure transferred to the rock is called the borehole pressure, designated as P_i, which is the difference between the maximum pump pressure and the inflation pump pressure (i.e. $P_i = P_{pmax} - P_{po}$).

The establishment of the contact stress is due to the difference between two deformations, U_r and u_{so}, which is the elastic contractions of the borehole and the Swellex bolt, respectively, see Figure 2. when

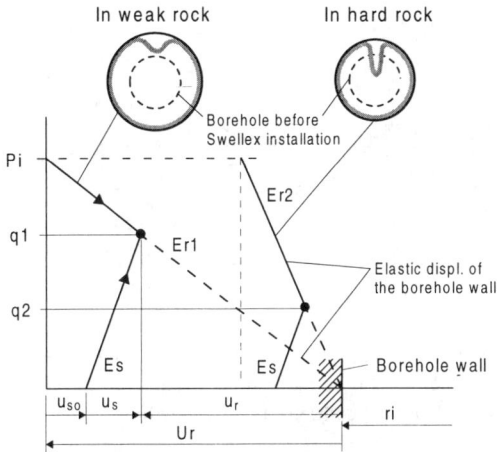

Figure 2. A sketch illustrating the elastic deformations of the borehole and the Swellex bolt.

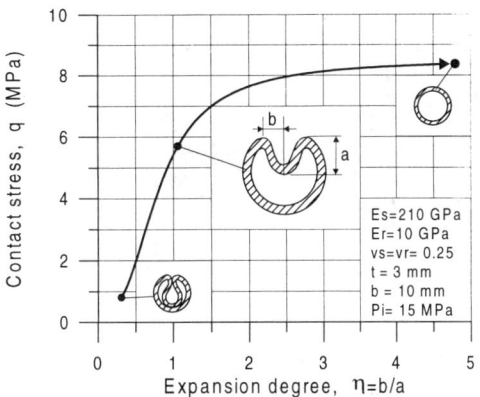

Figure 3. The contact stress vs. the expansion of bolt.

the pump pressure has ceased after installation. U_r is directly proportional to the borehole pressure P_i, while u_{so} to both the borehole pressure P_i and the maximum pump pressure P_{pmax}. The condition to establish a contact stress is that $U_r > u_{so}$.

Three elastic deformation components contribute to the establishment of the contact stress: (1) the contraction of the Swellex bolt after the pump pressure has ceased, u_{so}; (2) the displacement of the Swellex bolt under the contact stress q, u_s; and (3) the displacement of the borehole under q, u_r. It is seen from Figure 2 that the sum-mation of the three displacement components, u_{so}, u_s and u_r, is equal to the displacement, U_r, of the bore-hole wall, i.e.

$$U_r = u_{so} + u_s + u_r \qquad (9)$$

The four elastic displacements are expressed as

$$U_r = \frac{P_i r_i}{k_r}, \quad u_r = \frac{q r_i}{k_r}, \quad u_s = \frac{q r_i}{k_s}, \quad \text{and}$$

$$u_{so} = \frac{r_i(1+v_s)}{E_s(\alpha^2 - 1)}\left[2(1-v_s)P_{pmax} - (1-2v_s)\alpha^2 P_i - P_i\right]$$

where $\alpha = \dfrac{r_i}{r_i - t}$.

Substituting the above into Eq.(9) and taking into account the solution for a fully expanded Swellex bolt by Wijk and Skogberg (1982), we obtain the expression of the contact stress q as follows

$$q = \frac{k_s k_f}{(k_r + k_s)(k_r + k_f)}\left(P_i - k_r \frac{u_{so}}{r_i}\right) \qquad (10)$$

where

$$k_r = \frac{E_r}{1+v_r}, \qquad k_s = \frac{E_s t^3 (2\xi b^3 + 3\pi r_i^3)}{3\xi b^3 r_i^3},$$

$$k_f = \frac{E_s t}{(1-v_s^2)r_i}, \qquad \xi = \frac{\sqrt{1+\eta^2}}{3\eta^3}, \quad \text{and } \eta = \frac{b}{a}$$

The contact stress is not only dependent upon the mechanical properties of the Swellex bolt and rock, but also the expansion degree of the bolt. The variation of the contact stress with respect to the expansion degree, expressed as the ratio of $\eta = b/a$, is graphically illustrated in Figure 3. The contact stress increases with the expansion of the bolt but the "tongue length" of the Swellex bolt has a significant effect on the contact stress.

In weak/soft rock, the Swellex bolt may be over expanded which means that most of the pump pressure is trapped as hoop stress, in the bolt. The borehole pressure P_i will be a variable, depending upon the strength and deformability of the surrounding rock. Considering the above and by combining Eq.(4), (5) and (7), and assuming that $f = 1$, $P_o = 0$ and $u_r = D/2 - r_i$, the borehole pressure is expressed as

$$P_i = \frac{2k_r \sigma_c}{k_r^2 - 1}\left[\left(\frac{D}{d} - 1\right)\frac{E_r}{1+v_r}\frac{1+k_r}{\sigma_c}\right]^{\frac{k_r-1}{2k_r}} - \frac{\sigma_c}{k_r - 1} \qquad (11)$$

where D is the diameter of a fully expanded Swellex.

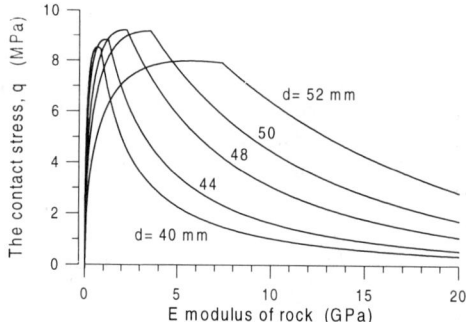

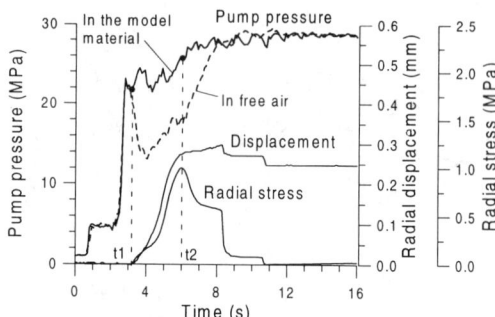

Figure 4. The contact stress vs. the E modulus of rock for a Super Swellex bolt in different boreholes. (Parameter values: D= 54 mm, E_s= 210 GPa, $v_s=v_r$= 0.25, t= 3 mm, b= 10 mm, P_{pmax}= 30 MPa, P_{po}= 15 MPa, σ_c= 3.5 MPa, ϕ= 30°, f= 1, P_o= 0).

Figure 5. Pump pressure and the radial stress and displacement of the model material at a distance of 10 cm from the centre of the Swellex bolt.

d is the borehole diameter and σ_c is the uniaxial compressive strength of the intact rock.

The variation of the contact stress with respect to the E modulus of rock is calculated using Eq.(10) and plotted in Figure 4. In the calculations, the borehole pressure P_i is determined by Eq.(11) if the bolt is fully expanded, otherwise is assumed as a constant of 15 MPa (for the case P_{pmax}=30 MPa). Figure 4 highlights two points: (1) the contact stress decreases with the increase of the borehole size in extremely soft rock (for instance when $E < 2$ GPa), while it increases with the borehole size in hard rock; (2) the contact stress is higher in relatively soft rock compared to hard rock. This second point is even more clearly visually explained by the two displacement lines marked as E_{r1} and E_{r2} in Figure 2.

4 LABORATORY TEST

The purpose of laboratory tests was to verify the performance of Swellex in weak/soft rock. Artificial materials were employed to simulate weak/soft rock in the laboratory tests. The model material was made of sand, cement, bentonite and water. The mixed material was cast in a concrete cylinder. Three displacement sensors and three pressure cells were buried in the material, located at a distance of 10, 25 and 50 cm to the borehole centre, respectively. The properties of the model material after 14 days setting were σ_c= 4.4 MPa, and E = 2.0 GPa. A steel tube of 44.5 mm in diameter was vertically erected in the centre of the cylinder before the material casting in order to make a hole. The tube was pulled out after setting, and a 0.9 m long Super Swellex bolt was inserted into the hole. The bolt was expanded by an electric pump with a maximum pump pressure of 30 MPa. A pull-out test was carried out immediately after the expansion of the bolt.

The two top curves in Figure 5 represent the pump pressures for one Super Swellex expanded in free air (dashed line) and another in the model material (solid line). The difference between the two curves, about 10 MPa, is the pressure transferred to the model material, i.e. the borehole pressure P_i. The two bottom curves are the radial displacement and stress measured at a distance of 10 cm from the borehole centre. The Swellex touched the borehole wall at time t_1, and the stress and displacement started to build up from that time and onwards. A crack was created in the model material at time t_2 when the pump pressure reached about 26 MPa resulting in a decrease of stress. The radial stresses and displacements measured at the distances of 10, 25 and 50 cm from the borehole centre are plotted in Figure 6.

The pull-out test gave a maximum pull-out load of 185 kN for the 0.9 m long Super Swellex. It was observed that the bolt was fully expanded in the material. The expanded section, i.e. the "effective" length, of this 0.9 m long bolt is about 0.7 m. Thus, the pull-out resistance of the Super Swellex in this weak material is calculated to be 264 kN/m.

5 DESIGN GUIDELINES

Johnston (1993) defined weak/soft rock as the group of geotechnical materials for which the uniaxial compressive strength falls approximately in the range 0.5-25 MPa. Yoshinaka et al. (1996) reported

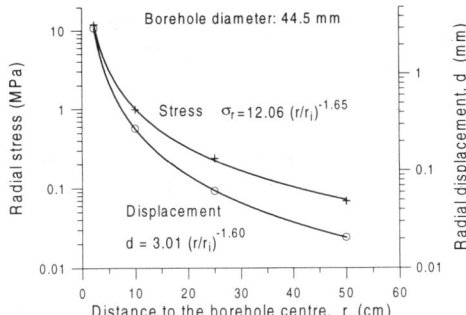

Figure 6. Distributions of the radial stress and displacement within the material.

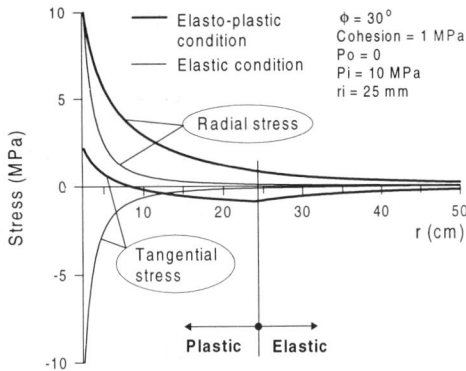

Figure 7. Stress distribution for an elastic and elasto-plastic rock, respectively.

their test results of a group Japanese soft rocks: tuff, siltstone, mudstone and sandstone. The uniaxial compressive strength of these soft rocks ranges from 3 to 12 MPa and the E modulus between 1 and 2 GPa. Weak/soft rock is characteristic of low strength, low deformation modulus, and sometimes high porosity. In most cases, weak rock collapses in the form of ravelling failure.

The objective of bolting in this type of rock is to avoid the disintegration of the rock, through the mechanical interaction between the bolt and the rock. In porous weak rock, compaction of the surrounding rock can improve the mechanical properties of the material through reduction of the porosity and increase in friction. In order to achieve this compaction, the diameter of the borehole and the spacing between bolts should be small. It can be seen in Figure 7 that the radial stress remains relatively high also at a distance of 25 cm from the borehole centre. It seems that a bolt spacing between 0.5 and 1 m is reasonable to achieve an effective interaction with the rock and a good compaction in porous rock.

The length of the bolt depends upon the size of the disturbed zone around the excavation. It is well known that bolts should be long enough to reach the stable ground. In general, the disturbed zone in weak rock is larger than in hard rock. Thus, the length of the bolt should be longer in weak rock. Tanimoto et al. (1981) suggested that the length of the bolt should be 1.5-2 times the width of the plastic zone.

The borehole size directly affects the borehole pressure P_i, and therefore the contact stress q. By using Eq.(11), two diagrams are plotted in Figure 8 and 9 showing the relationship between the borehole size and the borehole pressure. It is seen from the diagrams that the borehole pressure for the full expansion of Swellex increases with the decrease of the borehole size for a given rock. For a given

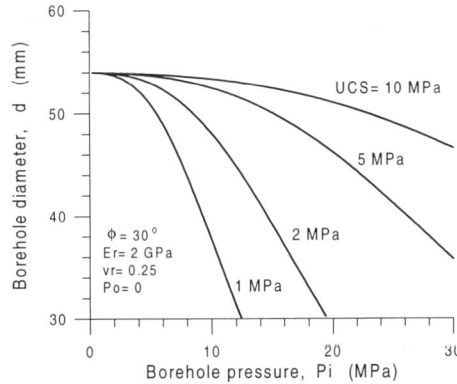

Figure 8. The borehole pressure vs. the borehole diameter for full expansion of the Super Swellex when the strength of rock varies.

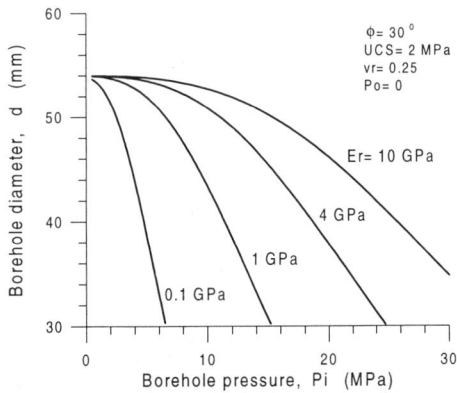

Figure 9. The borehole pressure vs. the borehole diameter for full expansion of the Super Swellex when the deformation modulus of rock varies.

borehole diameter, the borehole pressure increases with the increase both in strength and E modulus. In weak/soft rock, the borehole has to be small in order to establish a relatively large borehole pressure. In hard rock, however, the displacement of the borehole is limited and the borehole pressure is easily built up. Thus in hard rock the borehole should be so large that the expanded Swellex has a short tongue. The basic principle is that the borehole should be small in weak/soft rock, while large in hard rock.

In weak/soft rock, there exist a risk that the Swellex is over expanded. In this case, most of the pump pressure will be trapped in the Swellex. The contraction of the Swellex, u_{so}, will be increas-ed, and therefore the contact stress will be reduced, see Eq.(10). This will result in a low pull-out force. It is concluded that the pump pressure should be smaller in weak/soft rock than in hard rock.

The design guidelines for Swellex bolts applied to weak/soft rock are summarised as follows:

- Swellex type: Super Swellex, Midi Swellex
- Pump pressure: 24 MPa
- Borehole size: $\sigma_c < 1$ MPa, $E < 0.5$ GPa
  $d \leq 44$ mm

  $\sigma_c < 1$~8 MPa, $E < 0.5$~5 GPa
  $d = 44\text{-}48$ mm

  $\sigma_c > 8$ MPa, $E > 5$ GPa
  $d = 48\text{-}50$ mm
- Bolt length: 1.5-2 times the width of the plastic zone.
- Spacing: 0.5~1.0 m
- Bolt installation: A collar sleeve is recommended to prevent the rock around the borehole collar from fracturing.

6 CONCLUSIONS

- The radial stress and deformation, which are caused by the expansion of the bolt, reaches a higher value further away from the borehole in the elasto-plastic material compared to an elastic media. The increased radial stress will enhance the arch-building function between the rockbolts in weak/soft rock. In a porous material the expansion of the Swellex bolt will compact and consolidate the material and thereby decrease the void ratio, leading to an increase of the peak friction angle and therefore an increase of the shear strength of the material.
- The contact stress between the bolt and the borehole wall, and thereby the pull-out resistance, is higher for a lower E-modulus (i.e. as the rock gets weaker). For an extremely low E modulus, however, the contact stress will be very low due to the over expansion of the bolt.
- The borehole size has a significant influence on the contact stress in weak/soft rock, since when the bolt attains it original circular shape, hoop stress in the bolt will terminate further increase of the borehole pressure (and thereby the contact stress). This means that great care must be taken in order to avoid a borehole that is too large.
- The pump pressure should be tailored to the prevailing conditions in weak/soft rock in order to prevent over expansion of the bolt.

Aknowledgement: The authors wish to thank Prof. Bengt Stillborg for reviewing the manuscript and Messrs. Josef Forslund and Ulf Mattila for their assistance during the performance of the test.

REFERENCES

Håkansson, U. & C. Li, 1997. Swellex in Weak Rock Formations. *Swedish Rock Mechanics Symp.'97,* Swedish Rock Mechanics Research Foundation – SveBeFo. 85-106.

Johnston, I.W. 1993. Soft rock engineering. *Comprehensive Rock Engineering: Principles, Practice and Projects, Vol.1,* (J. Hudson et al. eds.) 367-393.

Tanimoto, C., S. Hata. & K. Kariya 1981. Interaction between fully bonded bolts and strain softening rock in tunnelling. *22nd US Symposium on Rock Mechanics, June 29-July 2, 1981, MIT,* 347-352.

Wijk, G. & B. Skogberg 1982. The SWELLEX® rock bolting system. *14th Canadian Rock Mechanics Symposium,* Canadian Institute of Mining and Metallurgy, 106-115.

Yoshinaka, R., M. Osada, & T.V. Tran 1996. Deformation behaviour of soft rocks during consolidated-undrained cyclic triaxial testing. *Int. J. Rock Mech. Min. Sci. & Geomech. Abstr.,* 33(6), 557-572.

# Development of an expert system for safety analysis of structures adjacent to tunnel excavation sites

G.J.Bae, C.Y.Kim, H.S.Shin & S.W.Hong
Korea Institute of Construction Technology (KICT), Seoul, Korea

ABSTRACT: In this study, An expert system called NESASS was developed. NESASS predicts the trend of ground settlements to be resulted from tunnel excavation and carries out a safety analysis for superstructures on the basis of the predicted ground settlements. Using neural network techniques, NESASS learns a database consisting of the measured ground settlements data and infers a settlement trend at the field of interest. NESASS calculates the magnitudes of a evaluation parameter(angular distortion etc.) and, in turn, determines the safety of the structure. In addition, NESASS predicts the patterns of cracks to be formed on the structure using Dulacska's model for crack evaluation. Therefore, in this study, the major factors influencing ground settlement were determined. Subsequently, a database of ground settlement due to tunnel excavation was built. A parametric study was performed to verify the reliability of the proposed neural network structure. A comparison of the ground settlement trends predicted by NESASS with the measured ones indicates that NESASS leads to reasonable predictions. An example is presented in this paper where NESASS is used to evaluate the safety of a structures subject to deformation due to tunnel excavation near to the structure.

1 INTRODUCTION

Recently, there has been a world-wide trend to place various national infrastructure in the underground for the propose of sorting out innumberable problems, e.g. increase in transportation requirement or the urban movement of the population. One of the most essential factors in construction of infrastructure would be safety on construction. Particularly, in the case of excavation in the urban area, not only stability of tunnel itself but also the safety of the adjacent superstructure should be considered because an instances of the adjacent structure damaged due to excavation such as Daego gas blowout and Seoul gas leakage accidents, are increasing constantly. A lot of research on the safety evaluation in excavation has been carried out in the advanced countries. Particularly, Boscardin carried out monitoring a deformation of a superstructure at the underground site in Washington D.C and reviewed factors which are able to evaluate a safety of superstructure, such as angular distortion and defection ratio. In 1984, Attewell, UK, presented the safety evaluation charts of buried pipes using Winkler model according to various construction materials and foundation types. An unexperienced engineers face to a difficult problems because the method of safety evaluation was not presented systematic.
It, therefore, is necessary to develop a system of safety evaluation based on the field conditions in order to perform the reasonable safety evaluation of superstructure.

2 THEORETICAL BACKGROUND

2.1 *Stochastic analysis of surface settlement*

The stochastic analysis of surface settlement suppose the ground to be the aggregate of spheres or disks. When a disk is removed in aggregate of spheres or disks as figure 1, the surface settlement types and size are described by the probability of settlement for each disk.
On the basis of the concept, Gauss normal probability function(GNPF) was derived by Sweet & Bogdanoff(1965) and Schmidt(1969). Equation (1) show a GNPF.

$$\delta_s(0,0) = \delta_{smax}$$
$$\delta_s(X,0) = \delta_{smax}\exp(-X^2/2i^2) \qquad (1)$$

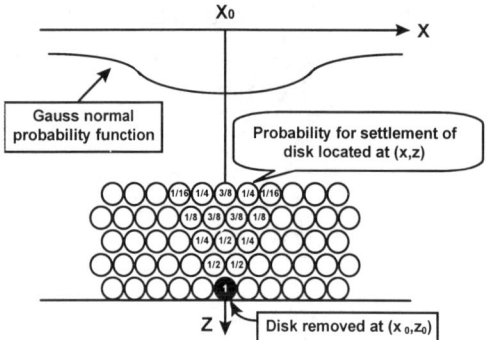

Figure 1. Stochastic settlement theory.

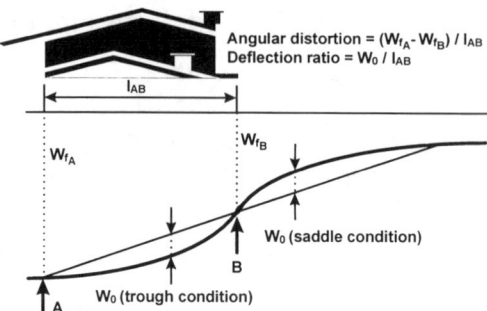

Figure 2. Definition of angular distortion and deflection ratio.

where, δ_{smax} is maximum surface settlement, i is the distance between the centerline of GNPF and inflection point.

Up to now, this equation have suggested to appropriately describe the surface settlement trough(Bae 1989). Thus in order to applied equation (1) to analyze a ground settlement, i-value has determined by consideration of only ground condition and tunnel geometry. But because i-value is affected by a various factors, it is expected that i-value must be determined by consideration of not only the ground condition but also the other factors, such as excavation method, ground water condition, etc.

2.2 *Expert system and neural network.*

Expert system endows computer system with a knowledge of expert for tunneling and have beginners for tunneling could utilize a expert's ability with the computer system. Thus, expert system is a kind of a consulting system using computer. Generally, expert system is consisted of knowledge base(rules and facts) and inference engine, output system, etc. Expert system use various techniques. Among these, in this study, neural network is used. Neural network is simulated the neurons and synapses such as the basic units of human brain. The important factors of neural network are the processor such as neuron and activation function and synapse. The synapse is consisted of connected lines, which can control the weight.

2.3 *Safety evaluation techniques for surface structure.*

2.3.1 *Settlement criteria to prevent building damage.*

Using the settlement criteria, structural safety is estimated. Major distortion factors for surface structures are the angular distortion and deflection ratio. The angular distortion is the level of shear distortion of structures, which is described as the rotation angle between arbitrary two points of structures. The deflection ratio is the level of distortion of structures. Figure 2 show the concept. In this study, it is assumed that the settlement trough is same as the angular distortion. Because the surface structures is not beam, and more flexible structure than beam, hence behave like purely ground.

The structural damage criteria based on the angular distortion and deflection ratio were arranged from various references and presented(KICT 1996).

2.3.2 *Dulacska's model to evaluate a crack phase*

Dulacska's model is used to estimate the crack type and size of structures due to surface settlements. Surface settlement induced in structure foundation has two types. One is trough condition, the other is the saddle condition as figure 3. Also, the crack of wall has two types. One is the σ-crack due to horizontal tension, and τ-crack due to shear deformation(Dulacska 1992). According to the type of structure, the effects of each cracks on structural stability are estimated by various equations.

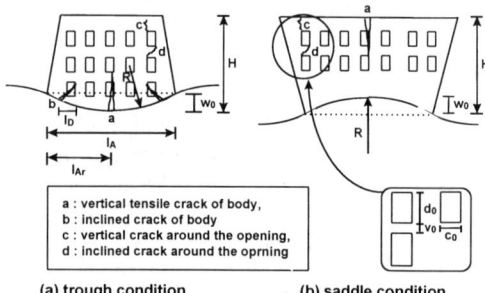

a : vertical tensile crack of body,
b : inclined crack of body
c : vertical crack around the opening,
d : inclined crack around the oprning

(a) trough condition   (b) saddle condition

Figure 3. Crack phase on the wall of structure.

3 DEVELOPMENT OF EXPERT SYSTEM

In this studies, major factors influencing to a ground settlement due to tunnel excavation were set up. And a database consisted of a information of tunnel sites and it's ground settlement trough was built on the basis of the major factors. The neural network and expert system were designed in order to use in a learning of database and predict a ground settlement trough and analyze a safety for superstructures of tunnel sites.

3.1 Determination of the major factors influencing to a ground settlement and representing a settlement trough

In order to building of the robust D/B, reasonable determination of classification factors must be preceded firstly. The classification factors were determined on the basis of major factors influencing to ground settlement. In addition, because a settlement trough was analyzed using a GNPF, settlement trough of a each tunnel site is represent by i and δ_{smax} in a GNPF as table 1.

Therefore, major factors of table 1 were built through the literature study and a counsel of experts for tunneling. In this place, major factors for tunnel geometry are focused on a scale of tunnel, and can be considered for twin tunnel. And major factors for ground condition are involved a kind of host rock mass in order to consider a influence of stratified rock mass, and are considered for a groundwater. Rock and soil layer are classified on the basis of the classification proposed by Attewell and Taylor(1984) and standard specification for construction. And for excavation and support system, condition of support and auxiliary technique are particularly focused.

3.2 Building of a database for the actual monitoring data in tunnel sites

In this study, collecting, sorting and analysis of a actual monitoring data was performed. The data is collected in a part of Seoul subway 3rd, 4th lines. And the database for the monitored 27 fields was built with respect to the major factors. It will be used to learning a neural network. this application of neural network is expect to be a reasonable method to use a monitoring field data to a tunnel designs.

Table 1. Major factors influencing to ground settlement and representing a settlement trough.

Tunnel Geometry	tunnel depth (m)	
	excavation width (m)	
	excavation height (m)	
	tunnel shape	circle, egg, horseshoe, specified
	kind of tunnel	single, twin
	pillar width(m)	
Ground condition	host rock mass	granite, schist
	kind of rock mass (m)	weathered, soft, moderate, hard, very hard
	kind of soil layer(m)	cohesive (stiff, soft), cohesionless (dense, medium dense, loose)
	groundwater level (m)	
	groundwater inflow (l/min-km)	
Condition of excavation and support	support method	rock bolt, shotcrete, steel rib
	excavation method	TBM, shield, drill and blasting, peak
	excavation design	full face cut, divided cut (short bench, long bench, ring cut), temp. invert, single side wall drift, double side wall drift
	auxiliary technique	forepoling, inner grouting, surface grouting, pipe-roof or, horizontal jet grouting
	supporting time	early, proper, delay
	velocity of excavation (m/day)	
	length of working face (m)	
	draining system	drainage, undrainage
Representation of settlement trough	maximum settlement, δ_{smax} (mm)	
	trough width parameter, i (m)	

3.3 Design of expert system, NESASS

NESASS(Neural network Expert System for Adjacent Structure Safety analysis) was designed to build and operate a database for the ground settlement and tunnel information, and learn the database, and predict the ground settlement trough of tunnel sites in interesting using neural network. Moreover, NESASS has the modules of safety analysis, which consist of a safety evaluation of a superstructure using the criteria to prevent a structure damage and a evaluation of crack phase on the structure using the Dulacska's model. In this study, the system structure of NESASS, therefore, was designed as figure 4, and realized using a Delphi and GUI.

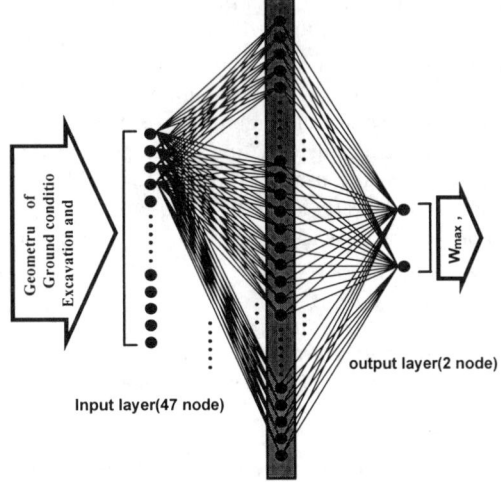

Figure 5. Neural network structure designed in NESASS.

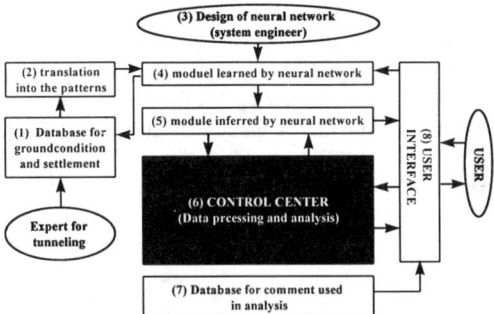

Figure 4. The system structure of NESASS.

In this place, the patterns of neural network involved in NESASS were designed with 47 input nodes, 94 hidden nodes and 2 output nodes as figure 5. It is resulted from the various suitability study of neural network structure. And neural network learns and infers using the value of learning rate(0.2) and coefficient of momentum(0.5) which were set up through a various parametric study(KICT 1996).

4 VERIFICATION OF NESASS FOR TUNNEL CONSTRUCTION PARAMETERS

In order to verify the reliance of the inferred maximum settlement, field instrumental data and laboratory experimental data are compared with results inferred by this system as figure 6. The published equations correlated between tunnel geometric components($Z/2a$) and settlement ratio(S_{max}/S_c) is used(Bae 1989). In figure 6, the correlated equation estimated for the clayey soil is used because sand soil is sensitive to not only tunnel geometric component but ground volumetric changes. In figure 6, Z is tunnel centerline depth, a is calculated tunnel radius. S_{max} is the maximum settlement, S_c is the crown settlement. In this test, the field monitoring data of Seoul subway 4th line are used. The results show the inferred settlement value is lower than the real settlement value. In this result, it is expected that if the tested field is composed of hard rock, and the ground water drop is much smaller than the soil layer. The result of the inferred surface settlement will be closely coincided with the real data.

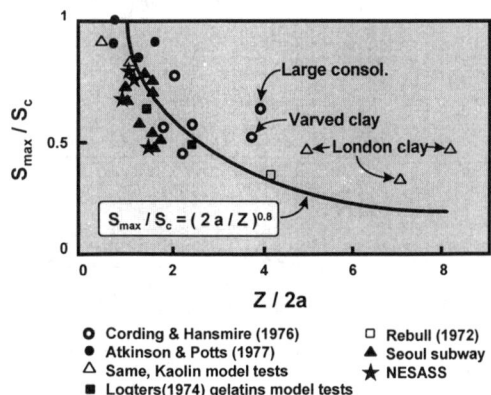

Figure 6. The correlation between settlement ratio and tunnel geometric components.

Figure 7 shows the relationship between settlement width parameter i and tunnel geometric component, and i inferred from NESASS is compared with tunnel geometric components. field monitoring data in a part of Seoul subway 3rd line are used for this inference test. In figure 7, the inferred settlement

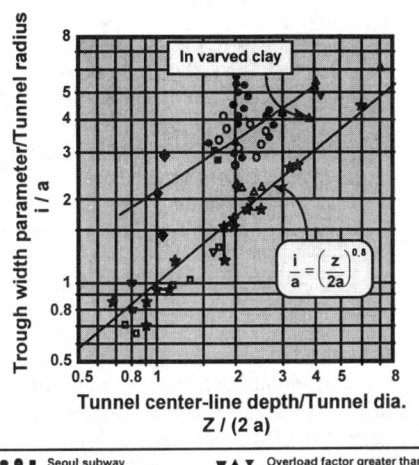

Figure 7. The correlation between settlement width parameter i and tunnel geometric components.

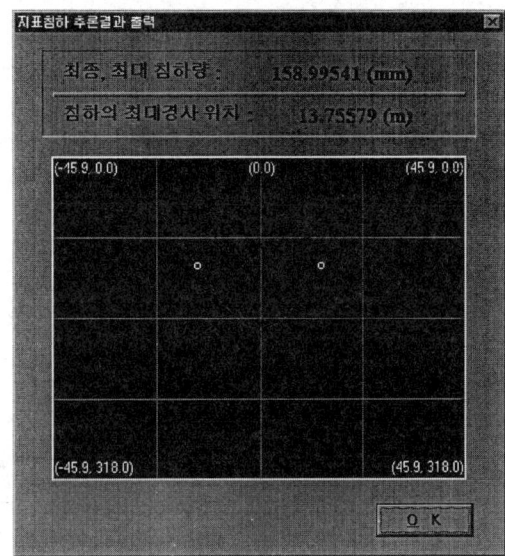

Figure 8. The predicted result of ground settlement trough by NESASS.

trough appears trended relatively bigger than the proposed equation. It is expected that if the lower ground water level and ground condition of clayey soil and weak rock layer is considered, The laboratory experimental output by Cording & Hansmire(1975) for i-value have very similar tendency with i inferred from NESASS.

From the results, It is judged that NESASS can produce reasonable and reliable result for settlement prediction, and have very good adaptability for safety evaluation of surface structures.

5 APPLICATION OF NESASS

5.1 Prediction of ground settlement trough upon a tunnel

Input data for prediction of NESASS used a tunnel information of a horseshoe shape, shallow depth, ground condition of a bad soil only, excavation type of a short bench and ring cut, forepoling. Thus, the predicted result for a input was, as figure 8, a maximum settlement (δ_{smax} =158.995mm). It is 0.003% error rate compared with a actual monitoring result (159mm) in a same tunnel site as a predicted site. The trough width parameter i was predicted in 13.76m by NESASS.

In addition to prediction of the ground settlement trough, NESASS is able to select a similar tunnel site among a database to the applied tunnel site. the selected similar tunnel sites, therefore, resulted in the 320-18k-O in a Seoul subway 3rd line. It's selection is judged properly from a comparison between the maximum value of applied tunnel site and the inferred similar tunnel sites.

5.2 Safety analysis of the structures by NESASS

Using two safety analysis modules in NESASS, the safety of supposed superstructures for the applied tunnel site(δ_{smax} =158.995mm, i =13.76m) are analyzed. thus, the structures are determined on a multistory building of 10m-width, 10m-height, masonry partition and filling wall. The result of a safety evaluation module using the criteria to prevent a structure damage for a superstructure is same as table 2 and figure 9. In this analysis, it is estimated that structure A is stable for shear distortion resulted from a differential settlement, but is very serious for unstable resulted from a bending effect. And, relatively, a large angular distortion is expected in the structure B, C. And instability for the ground settlement is not expected in the structure D.

By a evaluation module of crack phase on the structure using the Dulacska's model, the evaluation of crack phase was performed for the proposed structures(A,B,C,D) resulted in table 3. Particularly, the very large inclined crack(51.902mm) on the lower part of the wall is evaluated in structure A, as figure 10. it is expected that it's crack is resulted from the shear stress generated on the lower edges of the wall due to bending of foundation.

Table 2. Safety analysis result for superstructure.

item \ kind of structure	A	B	C	D
max. dist. From tunnel(m)	-5	10	25	100
min. dist. From tunnel(m)	-5	0	15	90
angular distortion	0	0.0037	0.0057	8×10^{-12}
deflection ratio	0.001	0.0008	0.0004	3.78×10^{-12}
max. settlement(mm)	148	159	88	8.05×10^{-8}
diff. Settlement(mm)	0	36.92	57.247	8×10^{-8}

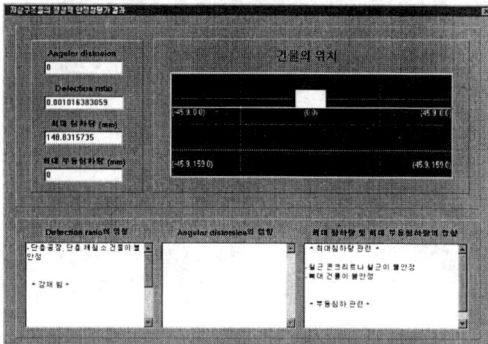

Figure 9. The dialogue of safety evaluation resulted from NESASS.

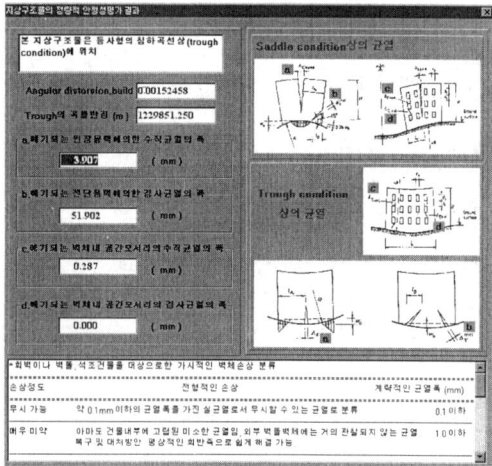

Figure 10. The dialogue of crack phase predicted by NESASS.

6 CONCLUSION AND SUMMERY

The following conclusion have been obtained from this study.

1) NESASS developed using a neural network technique, is expert system for safety analysis of adjacent structures at tunnel excavation sites.

2) NESASS developed by Delphi can be applied to estimate a ground settlement and analyze a safety

Table 3. The prediction of crack phases generated in the wall of superstructure

item \ kind of structure		A	B	C	D
max. distance from tunnel(m)		-5	10	25	100
min. distance from tunnel(m)		-5	0	15	90
deformed condition of foundation		trough	trough	saddle	saddle
crack width (mm)	a-mode	0	0	0	0
	b-mode	51.902	45.549	0	0
	c-mode	0.289	0.017	0	0
	d-mode	0	0	0	0

for the structure on the environment of GUI.

3) The building of database and the application of neural network are expected to a reasonable method to use a monitoring field data in tunnel designs.

4) The NESASS developed in this study will be able to evaluate a crack phase of the structure using Dulacska's model.

5) According to comparison of the ground settlement trends predicted by NESASS with the measured ones, it is expected that NESASS correctly predicted the ground settlement.

6) An example is presented in this study where NESASS is used to evaluate the safety of a structures subject to deformation due to tunnel excavation near to the structure.

7 REFERENCE

Bae, G.J. 1989. A Study on Prediction of Ground Movements caused by Tunneling in Soil. Ph.D. Thesis. Univ. of Yonsei. Seoul. p.159.

Dulacska, E.1992. Soil settlement effects on building. *Development in Geotechnical Engineering 69* Elsevier Science. Amsterdam. p. 447.

Korea Institute of Construction Technology 1996. Development of the Construction Technology for Underground Living Space -Underground Excavation Technology(IV)-. *Technical report*. Seoul / Korea. p. 216.

Schmidt, B. 1969. Settlements and Ground Movements Associated with Tunneling in Soil, Ph.D. Thesis. Univ. of Illinois. Urbana-Champaign. p.224.

Sweet, A. L. and Bogdanoff, J. L. 1965). Stochastic model for predicting subsidence. *Journal of engineering mechanics division*. ASCE. Vol. 91. No EM2. pp. 21 ~ 45.

Safety aspects in tunnelling and salt cavern design

Reinhard Rokahr & Kurt Staudtmeister
Institut für Unterirdisches Bauen, Universität Hannover, Germany

ABSTRACT: The design of underground constructions like tunnels and caverns requires a high amount of idealisations with respect to the rock mass behaviour and the theoretical calculation models. The application of different calculation methods for structural analysis and safety assessment, which are absolutely essential components of the design concept, are discussed.

Within the analysis the non-linear time dependent behaviour of the rock mass and the supporting elements as well as the loading history may play an important role for the layout. Consequently all safety considerations have to include the factor time. The results of calculations serve in defining the necessary geometric and physical boundary conditions, i.e. for example the thickness of a tunnel lining or the necessary minimum pressure in a cavern for storage operations. The objective of the dimensional analysis of construction projects is to guarantee that there is sufficient stability and serviceability with respect to the planned usage.

The paper presents a designing procedure that includes safety considerations for the different influencing factors. Examples show the application in tunnel and cavern design, which reflect the experience of more than twenty years of engineering work in numerous projects.

1 INTRODUCTION

A characteristic of deep lying tunnels constructed according to the New Austrian Tunnelling Method NATM is the fact, that the stress redistribution process during drifting is associated with unavoidable and relatively large deformations at the boundary of the cavity. The outer shotcrete lining has to stabilise the cavity walls and to prevent the collapse of fracture bodies or localised parts of the rock mass. Deformation of the cavity walls should neither be prevented nor to any large extent hindered.

It is never possible to predict in advance the permissible deformation. The term "permissible deformation" is also misleading because it assumes that one knows how large the impermissible deformation is. Impermissible deformations can however only be those for which the cavity wall cannot be adequately stabilised so that localised collapse occurs. On the other hand, if the deformation of the cavity walls does not change the cavity contour, and is thus not associated with any loss of the load bearing ability of the so-called load bearing ring of the rock mass, then the scale of deformation is not so much a safety problem but rather a question of economics.

The classic examples of this are the oil and gas storage caverns in rock salt formations. The theory of supporting force can be applied in its "purest form" in these cases because the existing liquid or gas pressure really is an active pressure and can be kept constant independent of the radial deformations of the cavern wall. The parameter, permissible deformation, does not apply to salt caverns as long as it is proven that no spalling has taken place. Although the characteristic line method, which seems to give a correlation between deformation and supporting pressure, may appear very convincing, it is not suitable for dimensioning.

The supporting force theory is based on the concept that active pressure can be applied to the excavated margin independent of the radial deformation. The value is given by the consulting engineer.

In contrast to the internal pressure acting on the walls of a salt cavern no active pressure is actually applied but rather a shotcrete lining with or without system anchoring. Despite the creep and relaxation properties of shotcrete, the contact pressure between the outer lining and the formation is dependent on the radial displacements over time. It is erroneous to equate system anchoring with active internal pressure because the anchors are usually not prestressed, and system anchoring according to today's level of understanding represents an improvement in the material properties of the rock formation - even though the improvement is practically unquantifiable.

Moreover, there is currently no accepted mathematical relationship between a radially symmetrical, constant supporting force, and the required thickness of a shotcrete lining in connection with system anchoring.

2 SAFETY ASPECTS IN TUNNELING

"The NATM is a lining method to stabilise the tunnel margins using shotcrete, anchors, and other lining materials in connection with measurements to control dimensioning."

This definition has been proposed by Rokahr 1995 during the annual Rock Mechanics Conference in Salzburg.

The important aspect is whether stabilisation of the tunnel walls using shotcrete with or without anchors is necessary and adequate. The NATM can therefore not be defined as a drifting method without safety measures. It is not critical whether the rock is broken up by explosives, part-face heading machines, full-thickness cutting machines, excavators, or in extreme cases, by hand. Dividing up the cross section during drifting is also not a differentiating criterion.

And finally, NATM lining methods do not include steel, cast iron or reinforced steel tubbings. To answer the question about the way controlling the dimensioning it is first necessary to differentiate between dimensioning a deep lying tunnel and a tunnel near the surface.

A characteristic line could be depicted in very simplified form for the area in front of the heading because this is the only place where a reduction in the radially acting stress components could take place.

As shown in figure 1 immediately following the series of shots, the radially acting stress component is nil (zone 2). This value remains nil if no outer lining is installed even though the continuing stress redistribution process resulting from further drifting produces an increasing load on the "perforated disc" load bearing system. The contact pressure between the lining and the formation will only be increased again by installing a shotcrete lining (zone 3). This increase is dependent on several factors, e.g. the deformation behaviour of the formation, the depth of the tunnel, the drifting speed, number and length of the anchors, and the rheological behaviour of the shotcrete.

A stress redistribution process in deep lying tunnels can give rise to deformation components of several decimeters. The installed shotcrete lining - which is hardly in a position to make any significant impact on hindering these deformations - is subjected to this deformation process. This can simultaneously lead to stressing in the shotcrete lining which can reach values exceeding the limits of stability.

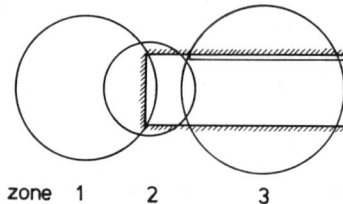

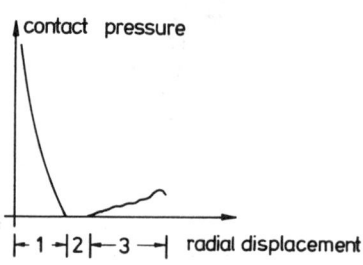

Figure 1. Characteristic line zones classification

If a rock body breaks off during this deformation process, the load on the shotcrete lining only increases if the contact pressure built up between the outer lining and the rock mass at this point in time is smaller than the pressure associated with the weight of the caved mass. Otherwise, it is difficult to imagine why the collapse occurred.

In the case of shallow tunnels, the processes depicted in figure 1 are basically the same. The radially acting stress component at the excavated margin is always nil for a period of time and for short advances per round. The increase in contact pressure between the lining and the formation again arises because of continued drifting. Unlike deep lying tunnels, the deformation that occurs is usually smaller but also leads to stressing of the shotcrete lining. The intention in this case is also to prevent the collapse of localised rock masses, but the extent of these local rock masses is easier to estimate in shallow tunnels. In extreme cases, one must anticipate that the unavoidable settling causes the break out of the rock mass lying above the roof and adds an additional load to the tunnel lining. Initial lower loading at the tunnel face can increase to a much higher value during drifting if the longitudinal load bearing effect is suddenly broken.

On what is the safety now based and how could a safety analysis be carried out? Unlike conventional building, the safety margins between the serviceable condition and the fracture situation are not as easy to determine in tunnel engineering. An unequivocal statement on the safety margin against loss of load bearing ability in a load bearing system - to the extent possible in building - would only be possible if the properties of the construction materials "rock mass" and "shotcrete", as well as their failure mechanisms, were sufficiently well understood. One has to admit, that today a generally accepted calculation model which meets scientific and practical engineering demands does not exist. The recommendations and regulations in tunnel engineering can at best only be considered as aids to decision making - they are "fiction" as far as safety considerations are concerned.

Against this background, the following points should therefore be considered as a proposal of how safety analysis can take place. Firstly, it is obvious that the load bearing behaviour of the formation is of critical significance for an approach of this type. For this reason, the geologists have to provide the engineers with the most detailed possible description of the geological situation, and usually, also an estimate of the material parameters required to describe the load bearing behaviour of the formation.

This is the basis upon which a stability analysis takes place before constructing the tunnel, and can only be considered as an approximation. During the drifting phase, the geologists and the engineers on site are assailed with the almost immoral demand to check the estimated load bearing behaviour of the formation despite knowing that the only method usually available is that of "taking a good look".

The extent to which the actual load bearing behaviour deviates from the estimated load bearing behaviour cannot be assessed because the actual is unknown. Almost all attempts using reverse calculations on the basis of measured dislocations of discrete points on the shotcrete linings - also in connection with deformation measurements in the rock formation - in order to determine the calculation parameters required, have so far failed to yield the results sought. The main reason being that the measured values can only be used to register the integral effects of all parameters and not to determine individual influence factors.

The only means currently available for the quantitative and qualitative estimation of safety is to suggest distorsions of the shotcrete lining on the basis of the deformation measurements of specific measuring points. With the help of a suitable material law, conclusions can be drawn on the state variables and thus the stresses. A possible representation of the level of stressing of a shotcrete lining against the deformation history is given in figure 2. The stress intensity index η is the ratio of the existing state variables to the ultimate state variables in the fracture state.

The question of what combination of roof displacement and convergence or divergence is permissible for the installed shotcrete lining without giving rise to a loss of load bearing ability as a consequence of fracture effects or cracking, can only be answered in the context of the deformation history. This means that it is never adequate to merely assign a limiting value, but rather it is critically important to estimate the permissible deformation increase per time unit or per measuring interval (Zachow 1995). The difference to other building is that in tunnel engineering, construction takes place prior to the calculations. In order to be able to quantitatively estimate the level of stressing of the shotcrete lining - at any time - from the measurements made during construction, it is necessary to take into consideration the rheological behaviour of the shotcrete.

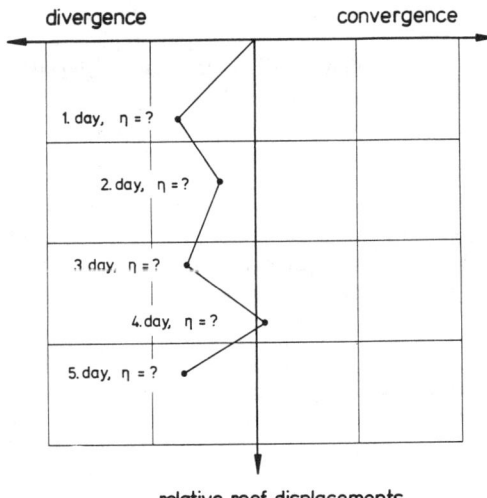

Figure 2. Qualitative representation of the level of stressing against the deformation history

The following two examples are presented to remove any last doubts remaining about the indispensable nature of this requirement. Figure 3 shows the time dependent deformation behaviour of two 30 h old shotcrete test samples subjected to creep tests using different loading histories. Test (1) which used a constant stress of $\sigma_1 = 9$ MPa over the whole test period shows that within 24 hours of starting the test, the elastic-viscous total strain of $\varepsilon_1 = 0.75$ % has already been reached and climbs well above 1% before completely hardening. This clearly shows that it is not possible to calculate the stressing

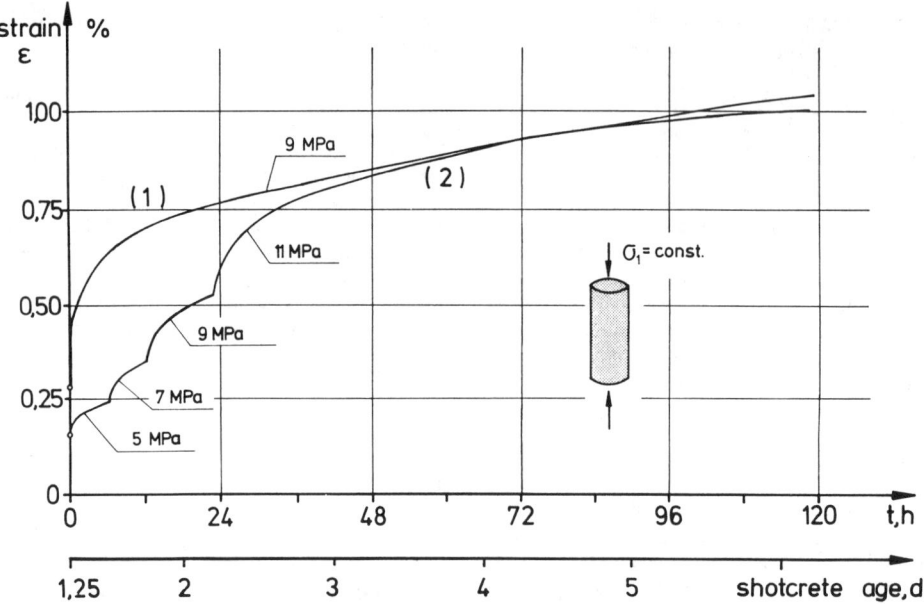

Figure 3. Creep behaviour of shotcrete samples during different loading histories

of a shotcrete lining with parameters which are valid for the material properties of hardened concrete.

Moreover, test (2) shows that when the load is increased in steps where σ_1 = 5/7/9 MPa, much lower total strains result over a comparable period of 24 hours. One can thus conclude that in addition to the level of stressing, the loading history also has a significant influence on shotcrete deformations over time. Another example is shown in figure 4. It depicts the results of two relaxation tests with shotcrete test samples of different age at the start of the tests. The diagram clearly shows the relatively short period of time of only a few hours during which a large part of the originally applied stress relaxes.

If the level of stressing in the shotcrete lining is known, it is possible to estimate the safety against collapse of local parts of the rock mass. The time factor is of critical importance here because the level of stressing is dependent upon time. Failure of the formation load bearing ring is not possible as long as the shotcrete lining has adequate load bearing margins.

The problem still to be solved lies in estimating the load bearing reserves of the formation. When fractures appear in the shotcrete lining - in particular in deep-lying tunnels - this does not mean that there is no safety margin against the collapse of localised rock masses. In this context it is worth emphasising that a crack in a shotcrete lining cannot be equated in its structural effect with a moment hinge because experience has shown that longitudinal cracks occurring in the roof zone are almost without exception sliding fractures and not flexural stress cracks.

The orthogonal force transfer in this case is extremely low so the conditions for a hinge are no longer present. The question of whether such a longitudinal crack represents a critical situation does however depend on the question of whether the load

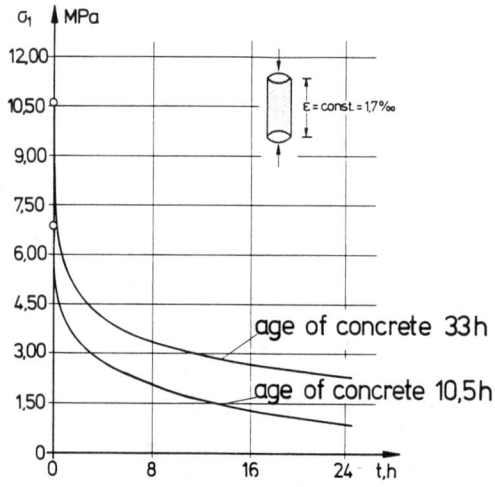

Figure 4. Relaxation behaviour of concrete

pushes from behind or not, and requires a project-specific answer. The only possibility of assessing the safety in this case is to use the deformation speed as an evaluation criterion. When the deformation stops, one can say that a new equilibrium has been established. More detailed conclusions on the safety are currently impossible.

3 SAFETY ASPECTS IN CAVERN DESIGN

Caverns in a rock salt mass designed for the storage of natural gas are generally laid out for optimisation of working gas volume and sufficient withdrawal rates. The factors influencing the stability of an unlined underground cavity are basically divided into the physical effects (formation pressure, internal pressure, temperature), the rock mass stresses, theoretically determined from the above with the help of a calculation model, the resistance (strength) of the rock that acts against the physical effects (material characteristics) and the influence of geological anomalies.

In each influencing factor a number of uncertainties may reduce the load bearing capacity of the rock surrounding the cavity. Significant influences which can result in erroneous evaluations are a false estimation of the formation pressure (lateral pressure coefficients, densities), the deviation of the static system from reality (e.g. excessive idealisation with respect to the geometry or the material behaviour), calculation inaccuracies, the deviation of the assumed rock strength from reality, inappropriate adoption of rock characteristics to the formation characteristics, laboratory tests on non representative core material, time influences and the presence of faults or inhomogeneities that could not be determined from the engineering geological evaluation, or were not taken into consideration in the theoretical calculation model.

Figure 6 shows a typical phase during the operation of a gas storage cavern. For gas withdrawal the internal pressure p is decreased from maximum pressure to minimum pressure in a certain time period. After a following period under minimum pressure conditions the refilling of the cavern starts.

The assessment of the states of stress in the rock mass in the vicinity of the cavern is carried out with the stress intensity index η. This index is calculated on the basis of the results of Finite Element calculations (Staudtmeister and Struck 1991). Among other criteria the values of the stress intensity index are a measure for the determination of the permitted minimum pressure of the gas cavern during operation.

From the level of maximum stress intensity index η immediately after the pressure reduction and the level at the end of time under minimum pressure

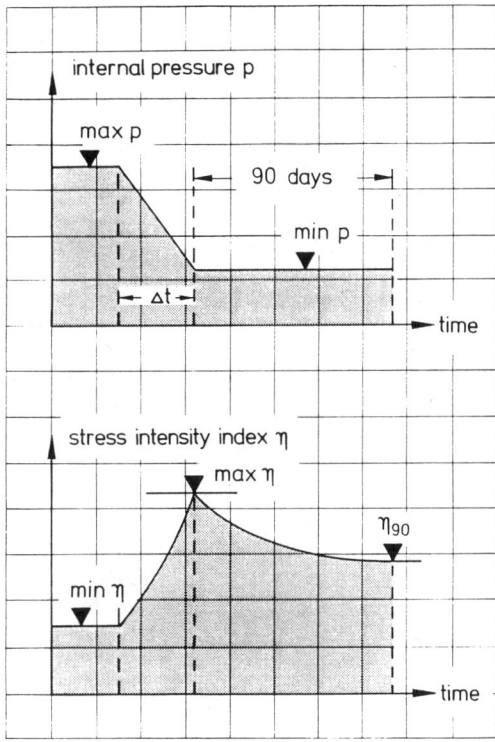

Figure 5. Internal pressure p and stress intensity index η

according to figure 5 the admissible level of min p and the corresponding time is determined. This procedure leads to a dimensioning which excludes any spalling of a part of the cavern periphery. In this case the internal pressure indeed is a fixed supporting pressure which is independent of the deformations of the cavern wall.

The stress intensity index is very similar to the mentioned η-value for shotcrete. It indicates the percentage actually used of the potential stress which the rock mass could absorb during short term stressing. This fracture strength is obtained from uniaxial and triaxial laboratory tests (Staudtmeister and Rokahr 1994).

Much more difficult is the assessment of stability for the „special load case" of a blow out. The question is, how a spalling process at the cavern wall develops over time. The concept is based upon establishing an integral assessment for the stability over time of a certain rock mass zone (Rokahr and Staudtmeister 1993).

The maximum internal cavern pressure is set at a value so that within the whole area of the cavern - in particular in the cavern roof and casing shoe area - fracturing of the rock and an increase in the permeability as a result of micro fracturing can be avoided. Because leaky areas in the rock are difficult to confirm by measuring techniques, and fracturing of the rock is associated with high financial losses for the operator, a comparatively high amount of safety compared to the failure states should be used, Rokahr et al. (1994). The safety against loss of tightness is given by the existence of a closed zone surrounding the cavern, in which an increased secondary permeability is avoided.

As nearly all of the design parameters are dependent on one another, it is initially necessary to select a reference configuration which then can be used to investigate the sensitivity resulting from variations in individual variables which need to be established. In this way, the effort involved in carrying out the predominantly numerical calculations using the finite elements method (FEM) can be reduced.

The discretisation of the calculation models takes into consideration the already mentioned geometrical boundary conditions. The status variables derived from the FE calculations are presented and provide the basis for the determination of the design parameters.

The roof thicknesses and the distances to the adjacent rock formations must also be dimensionally analysed in order to rule out any potential threat to the sealing capacity. Volume losses associated with internal pressure reductions are determined by time dependent convergence calculations taking into consideration realistic internal pressure changes over time.

The calculated volume convergence rate for a gas cavern with a reference depth of approx. 1250 m for various internal pressures is shown in figure 6. The overproportional increase in convergence rate with lower internal pressures is clearly visible.

The rates of cavity closure show a response on the acting deviatoric stresses and have an economic influence as the storage volume shrinks, but they are not a design parameter with respect to safety.

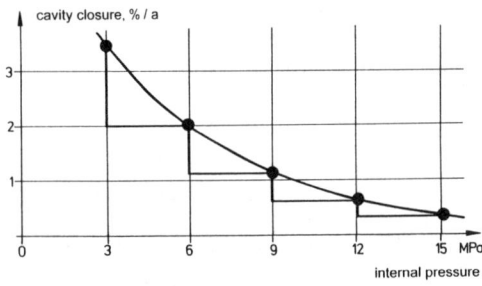

Figure 6. Cavity closure of a gas cavern under different internal pressures

4 CONCLUSIONS

The described procedures show that a high amount of different assumptions is necessary to get a quantitative figure of the safety of the underground load bearing system. In tunnelling as in cavern design the factor time has a decisive importance.
Deformation cannot be a absolute measure for safety. The paper describes a procedure for quantification of the load bearing capacity by a time dependent stress intensity index for the shotcrete lining in tunnelling as well as for the rock salt mass in the vicinity of a cavern.
The level of safety margins can only be fixed site-specific considering all the different influencing factors. They strongly depend on the experience of the consulting engineer.

REFERENCES

Rokahr, R.B. and Staudtmeister, K. 1993. *Rock mechanical study of the load-bearing behaviour of a gas cavern in rock salt after a blow out.* Proc. Seventh Symp. on Salt, April 1992, Kyoto, Japan.

Rokahr, R.B., Staudtmeister, K. and Zander-Schiebenhöfer, D. 1994. *Development of a new criterion for the determination of the maximum permissible internal pressure for gas storage caverns in rock salt.* SMRI Spring Meeting, April 1994, Houston, Texas, USA.

Rokahr, R.B. 1995. *Wie sicher ist die NÖT?* Felsbau 13, Heft 6.

Staudtmeister, K. and Struck, D. 1991. *Design Criteria for the Prevention of Creep Rupture for Gas Caverns in Salt Rock Mass.* SMRI Fall Meeting, Oct. 1991, Paris, France.

Staudtmeister, K. and Rokahr, R.B. 1994. *Laboratory Tests within the Scope of Rock Mechanical Investigations for the Design of Solution Mined Caverns in Rock Salt Mass.* SMRI Fall Meeting, September 1994, Hannover, Germany.

Zachow, R. 1995. *Dimensionierung zweischaliger Tunnel im Fels auf der Grundlage von in-situ-Messungen.* Forschungsergebnisse aus dem Tunnel- und Kavernenbau, Heft 16, Universität Hannover.

A viscoelastic plastic displacement back analysis model for basic parameters of rock mass

Zhenzhong Shen
College of Water Conservancy and Hydropower Engineering, Hohai University, Nanjing, People's Republic of China

Zhiying Xu
College of Civil Engineering, Hohai University, Nanjing, People's Republic of China

ABSTRACT: In this paper, a viscoelastic plastic displacement back analysis model, called flexible tolerance method, is developed for the inversion of the basic parameters of rock mass. The proposed method is wide in application, quick in convergence and high in efficiency, and can be used to solve the complex problem of the displacement back analysis of viscoelastic plastic parameters of rock mass.

1 INTRODUCTION

Because of the complexity of rock mass media and the imperfection of current test methods and techniques, back analysis is an important and efficient procedure for improving the accuracy and reasonability of the basic design parameters of rock mass on the basis of the prototype observation data. The problem of viscoelastic plastic displacement back analysis has been proposed for many years, and some great progress have been made in recent years (Gioda & Maier 1980, Gioda 1985, Sakurai & Takeuchi 1983, Hisatake & Ito 1985 etc.). However, it is an implicit nonlinear problem, and generally speaking, it can only be solved by direct search solution method. So the computation efficiency is not high, and there are many difficulties in actual application, and particularly in solving the large three dimensional viscoelastic plastic problem. This paper considers the characteristics of hard rocks, and uses a seven-parameter viscoelastic plastic model to describe the deformation behavior of hard rocks (Shen 1995), and proposes a new back analysis method for the viscoelastic plastic problems.

2 BASIC PARAMETERS

The basic parameters are determined by the constitutive model of rock mass. Here, a seven-parameter viscoelastic plastic model is recommended, its schematic diagram is shown as in Figure 1, where E_M and η_M are elastic modulus and viscosity coefficient of Maxwell model respectively, E_K and η_K are elastic modulus and viscosity coefficient of Kelvin model respectively, σ_s is yielding stress, E_s and η_s are plastic hardening modulus and viscosity coefficient respectively. Thus, the basic parameters of rock mass include the deformation modulus E_r, Poisson's ratio υ_r and the aforesaid rheologic parameters. If the effect of seepage field is considered, the basic parameters include permeability coefficient also.

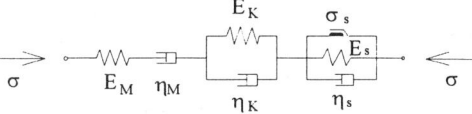

Figure 1. Schematic diagram of the seven-parameter viscoelastic plastic model

For the problem of nonlinear displacement back analysis, when the undetermined parameters are more, the difficulty of back analysis and computation time increase very very fast, and under some condition, the search computation may be not convergent or the result is not unique. Therefore, in order to increase the computation efficiency, the basic parameters, which may be considered separately, should be inversed respectively. For example, there is not close relationship between the deformation modulus and Poisson's ratio and the rheologic parameters of rock mass, so the so-called two-step displacement back analysis method is used here, i.e. the deformation modulus and Poisson's ratio are inversed firstly according to the measuring instantaneous elastic displacement, and then the rest rheologic parameters are inversed by use of the measuring rheological displacement. Usually, the permeability parameter is inversed independently by use of the water level observation data of bore-holes and pumping test data or water pressure test data.

To describe generally, the undetermined parameters are written as the design variable X as follows

$$X = [x_1, x_2, \cdots, x_N]^T \quad (1)$$

where N is the number of the undetermined parameters; x_j represents the each basic parameter, and $j = 1, 2, \cdots, N$.

3 OPTIMIZATION MODEL

Because it is impossible to directly solve the inverse problem of viscoelastic plastic displacement back analysis, it is necessary to establish the inverse optimization model, and to obtain the optimal parameters, which are close to the actual condition. In this paper, the weighted least square method is used to establish the object function of the inverse problem. The object function is shown as following

$$f(X) = \sum_{i=1}^{n} w_i (u_i^c - u_i^m)^2 \quad X \in D^N \quad (2)$$

where n is the number of the observation points; w_i the weight of the observation value of the ith point; D^N the feasible domain; u_i^c and u_i^m are the calculation value and observation value of displacement of ith point. Here, the calculation value of displacement is obtained by the finite element method.

Usually, the variable limits of the undetermined parameters can be given

$$a_j \le x_j \le b_j \quad j = 1, 2, \cdots, N \quad (3)$$

where a_j and b_j are the lower bound and upper bound of the jth parameter.

Therefore, the optimal problem of inversing parameters may be described as following: Under the condition of satisfying the restraint equation (3), find the undetermined design variable X, and make the object function $f(X)$, i.e. equation (2), get to the minimization value.

Because the displacements of observation points are the nonlinear implicit function of the design variable, the aforesaid problem is a nonlinear weighted least square problem. The advantage of the utility of the weighted function includes that the importance of each observation point and its measure precision can be considered. Generally speaking, the measure error can not be avoided, so the weights of the observation points with higher precision should be bigger than that of the one with lower precision. But the sum of all the weights is equal to 1.0.

4 FEASIBLE TOLERANCE METHOD

Generally, the aforesaid implicit nonlinear optimal problem with restraints can be solved only by direct search method. This paper develops a new method to solve that problem, which is called feasible tolerance method here. The method regards the restraint domain as a flexible domain, and on the base of the concept of approximate flexibility, uses the nonlinear simplex method to solve the optimal problem. It can solve not only the optimal problems without restraints, but also the optimal problems with equality restraints or inequality restraints or equality and inequality restraints.

For the problem of displacement back analysis, the equality restraints do not exist usually. So the equation (3) can be written as follows

$$g_j(X) \ge 0 \quad j = 1, 2, \cdots, m \quad (4)$$

where m is the number of the restraint equations.

Constructing a series of common difference tolerance function, represented by $\{\varphi^k\}$, it satisfies the following condition

$$\varphi^0 \ge \varphi^1 \ge \varphi^2 \ge \cdots \ge \varphi^k \ge 0 \quad (5)$$

where φ^k is the given permitted common difference when performing the kth step search, and it is variable. The series is a monotonous descending function series, and when the number of search times increases, it get to zero slowly. The common difference function is a positive function, and given by the vertexes of the simplex. It is defined as following

$$\varphi^0 = 2d \tag{6}$$

and

$$\varphi^k = \min\left\{\varphi^{k-1}, \frac{1}{N+1}\sum_{j=1}^{N+1}\left\|X_j^k - X_{N+2}^k\right\|\right\} \tag{7}$$

where φ^0 is the initial value of common difference function; d the initial length of simplex; X_j^k the jth vertex of simplex in D^N space.

By use of the common difference function, the restraint condition, i.e. equation (4), can be written as

$$\varphi^k - T(X^k) \geq 0 \tag{8}$$

where $T(X^k)$ is the evaluation value of failure restraint, and it represents the failure extents of restraint. It can be calculated by following equation

$$T(X^k) = \left\{\sum_{j=1}^{m} \delta_j g_j^2(X^k)\right\}^{\frac{1}{2}} \tag{9}$$

where δ_j is Heaviside function, and defined as

$$\delta_j = \begin{cases} 0 & \text{if } g_j(X^k) \geq 0 \\ 1 & \text{if } g_j(X^k) < 0 \end{cases} \tag{10}$$

Obviously, when $T(X^k) = 0$, the design variable X^k is in feasible domain, i.e. it satisfies all restraint condition, and otherwise, it does not satisfy all the restraint condition. Because the common difference function has the characteristic given by equation (5), it is able to obtain the optimal solution of the inverse problem by the iteration method in the feasible domain or approximate feasible domain, and the search computation efficiency is higher, because finding the solution of satisfying the restraint condition $T(X^k) \leq \varphi^k$ is easier than that finding the solution of satisfying the restraint condition $T(X^k) = 0$.

The search process of the feasible tolerance method can be described as following.

1. Assume the initial point X^0 and the initial length of simplex d, and calculate the object functions $f(X_i^0)$ for each vertex of simplex, here $i = 1, 2, \cdots, N+1$.

2. $k = 0$.

3. Find out the best point X_l^k and the worst point X_h^k, and calculate the center point X_c^k of all the points except the worst point.

4. Calculate the function $T(X_l^k)$ of the best point, and check whether the discriminant equation (8) is true. If it is true, the point X_l^k is in feasible domain or in approximate feasible domain, and calculate the new point by simplex method and replace the worst point by the new one; if not true, minimize the function $T(X_l^k)$ by simplex acceleration method and obtain a new point to replace the point X_l^k. Thus, the new point satisfies the discriminant equation (8).

5. By use of the same method, perform the minimization search for object functions $f(X_i^0)$.

6. Check whether the following discriminant of convergence is satisfied

$$\varphi^k < \varepsilon \tag{11}$$

where ε is the convergent tolerance. If equation (11) is true, stop search and go to step 8; and if not true, begin the next search and go to step 7.

7. $k = k+1$, go to step 3.

8. Output the result and stop search.

5 EXAMPLE

Figure 2 shows a hydraulic tunnel. Its width is 6.0m and its height is 9.0m. Assume the adjoining rock is the aforesaid rheological body shown as in Figure 1, and its displacement field is given. Thus, the basic parameters of the adjoining rock can be inversed by the proposed method. Here the displacements of the 10 points shown as in Figure 2 are used to inverse the basic parameters of rock mass.

Firstly, the deformation modulus and Poisson's ratio are inversed by used of the elastic displacement

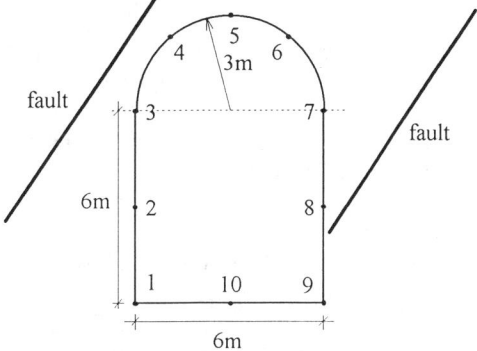

Figure 2. Schematic diagram of a hydraulic tunnel in rock mass.

component. Then, regarding the deformation modulus and Poisson's ratio as known parameters, the rheological parameters can be inversed by use of the rheological displacement. The result is shown as in Table 1, where the search times is $k = 18$, the convergent tolerance is $\varepsilon = 0.005$. And if the simplex method is used, the search times will get to 26. Obviously, the feasible tolerance method is more efficient.

Table 1. The basic parameters of adjoining rock

basic parameter	accurate value	inversion value	relative error
E_r (Gpa)	40.0	40.8	0.2%
υ_r	0.2	0.196	-2.0%
η_M (10^6GPa.s)	500.0	486.0	-2.8%
E_K (Gpa)	60.0	61.3	2.2%
η_K (10^3GPa.s)	80.0	76.7	-4.2%
E_s (Gpa)	30.0	29.2	-2.7%
η_s (10^6GPa.s)	100.0	101.5	1.5%

6 CONCLUSION

The proposed model can be used to solve the complex viscoelastic plastic displacement back analysis problems of geotechnical safety monitoring and feedback design construction. It is wide in application, quick in convergence and high in efficiency. It has not only theoretical but also actual significance.

REFERENCES

Gioda G. & G. Maier 1980. Direct search solution of an inverse problem in elasto-plasticity: identification of cohesion, friction angle and in-situ stress by pressure tunnel tests. *Inter. Numer. Methods Eng.* 15: 1823-1848.

Gioda G. 1985. Some remarks on back analysis and characterization problems in geomechanics. *Proc. of 5th Int. Conf. on Numer. Methods in Geomech.*: 160-174. Nagoya.

Hisatake M. & T. Ito 1985. Back analysis for tunnel by optimization method. *Proc. of 5th Int. Conf. on Numer.Methods in Geomech.*: 1301-1307. Nagoya.

Sakurai S. & K. Takeuchi 1983. Back analysis of measured displacements of tunnels. *Rock Mech. and Rock Eng.* 16(3): 173-180.

Shen Zhenzhong 1995. Deformation analysis and back analysis model of Three Gorges dam and bedrock in construction stages. *Ph.D. dissertation.* Hohai Univ., China.

Full scale tests of steel arch supports

Jong-Woo Kim
Department of Mineral and Mining Engineering, Chongju University, Korea

Hi-Keun Lee
Department of Mineral and Petroleum Engineering, Seoul National University, Korea

ABSTRACT: The results of full scale tests of steel arch supports are reported. Supports tested were both the yieldable and the rigid ones which were assembled by U-sectional and I-sectional beam respectively. The testing program consisted of 1 to 7 point loading conditions to simulate in situ conditions. During tests, deformation behavior of the supports was investigated. It is founded that increasing the number of loading points induces high stiffness and stability of supports. Both vibrating wire gages and curvo-distometer, simultaneous measurer of strain and curvature, were used to calculate load components acting on the support on the basis of the integrated measuring technique. The load components back-calculated by vibrating gage system were in good agreement with the applied ones.

1. INTRODUCTION

The steel arch support system has long been a common tool for tunneling. Of particular importance is the determination of rock pressure acting on steel arches to design an efficient support system. Two methods of support load determination can be considered. The one is direct method to employ various types of loadcells and the other is indirect method to utilize the deformation of the support.

In the indirect method, so-called integrated measuring technique, normal forces and bending moments of the support profile are found out, and then the support load is calculated by numerical differentiation of them. To find them out, Kovàri (1977) inserted cone-type measuring bolts into the profile to measure the change in radius of curvature and chord length by the help of the curvometer and the mechanical extensometer, respectively. Moreover, Choquet(1982) attached small steel balls to steel arch to measure strain of the profile.

On the other hand, scale model test and full scale test can be taken into consideration for physical laboratory tests of support systems. Scale model tests concerning the interaction of ground and roof support were conducted by Everling(1964), who used plaster mixture and steel with low stiffness on the basis of dimensional analysis.

Full scale test has disadvantage in money and man power because of huge test rig, but it can produce more reliable results than scale modeling technique. Horvarth(1971) conducted full scale test of ring support, and Sadler(1984) connected nine steel arches with struts to make three dimensional tests under the concentrated crown loading condition. Recently, Khan and Mitri(1996) reported the results of a testing program comprising five full scale steel arches under 1 to 5 point loading conditions. Their arches, however, were tested by jacking the arch legs against blocking points located along the crown of the arch, which might be in discord with actual state of field.

In this paper, full scale tests of steel arch supports assembled by both U-sectional and I-sectional beams were conducted to secure the integrated measuring technique for support load determination. The testing program consisted of 1 to 7 point loading conditions which were in a symmetric or asymmetric state, in an effort to simulate in situ conditions. The behaviors of steel arches under various loading conditions are described and the back-calculated load components are compared with the applied ones. The test results are, finally, discussed through finite element analysis.

2. THEORETICAL CONSIDERATION

2.1 *Static Analysis*

A steel arch support can be simplified in a semi-circular shape. The connecting part, so-called fishplate, is assumed to be rigid and continuous for convenience. To design rigid arches, axial force N and bending moment M of the support profile should be known. Jukes(1983) derived some equations of N and M for semi-circular arch without arch legs, besides Birön(1983) reported equations for it with arch legs. Their equations are, however, for the structure to be statically indeterminate to the first degree of freedom.

Considering in situ condition, an arch support may be statically indeterminate to the third degree because the boundary condition of the bottoms of arch legs is thought to be fixed rather than pinned, as shown in Figure 1. To solve the statically indeterminate structure, the theorem of Castigliano was employed. Finally the results of structural analysis under crown loading condition are as follows.

$$N = -0.5P\cos\theta - H_A \sin\theta \qquad (1)$$

$$M = 0.5Pr(1-\cos\theta) - H_A(h+r\sin\theta) + M_A \; ; 0 \leq \theta \leq \pi \qquad (2)$$

$$M = -H_A x + M_A \qquad ; 0 \leq x \leq h \qquad (3)$$

where:

$$M_A = \frac{Pr^2(0.048h^3 + 0.125h^2r + 0.108hr^2 + 0.026r^3)}{0.083h^4 + 0.523h^3r + h^2r^2 + 0.785hr^3 + 0.232r^4}$$

$$H_A = \frac{Pr^2(k_1 h^4 + k_2 h^3 r + k_3 h^2 r^2 + k_4 hr^3 + k_5 r^4)}{g_1 h^6 + g_2 h^5 r + g_3 h^4 r^2 + g_4 h^3 r^3 + g_5 h^2 r^4 + g_6 hr^5 + g_7 r^6}$$

k_1=0.072, k_2=0.322, k_3=0.589, k_4=0.420, k_5=0.107, g_1=0.042, g_2=0.392, g_3=1.404, g_4=2.486, g_5=2.348, g_6=1.149, g_7=0.232

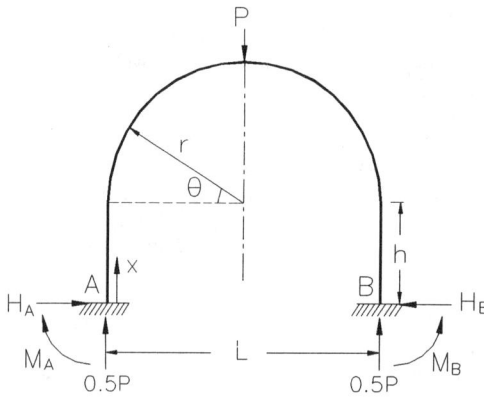

Figure 1. Static model of steel arch support.

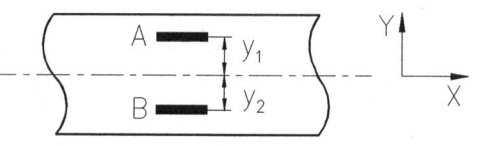

Figure 2. A segment of the beam with two measuring points(after Choquet).

2.2 Integrated measuring technique

Provided that an infinitesimal element of the arch is considered to be in equilibrium under the action of axial force N, shear force Q and bending moment M, the load components of radial direction to the arch can be evaluated as follows(Kovàri 1977).

$$P_i = -\frac{N_i}{r} - \frac{1}{L^2}(M_{i-1} - 2M_i + M_{i+1}) \qquad (4)$$

where i is the number of measuring point, r is the radius of the arch and L is the interval between two measuring points.

On the other hands, N and M of adjacent three measuring points are required, so as to evaluate P_i in equation (4). In regard to this subject, two techniques were reported. The one is proposed by Kovàri(1977), who measured the change in radius of curvature f and chord length l between two measuring bolts in the arch, in turn found out N and M of steel beam from equation (5) and (6).

$$N = (\frac{l}{L} + \frac{8e}{L^2}f)EA \qquad (5)$$

$$M = \frac{8EI}{L^2}f \qquad (6)$$

where e is the eccentricity from the neutral axis of the beam, A is the cross-sectional area, I is the moment of inertia and E is the elastic modulus.

The other is proposed by Choquet(1982) as shown in Figure 2. He marked two measuring points A, B which are located at intervals of y_1, y_2 away from neutral axis, respectively. He also measured two strain components ε_A and ε_B using a pair of 1 mm diameter steel balls, in turn found out N and M of the arch beam from equation (7) and (8).

$$N = \frac{AE(\varepsilon_A y_2 - \varepsilon_B y_1)}{y_1 - y_2} \qquad (7)$$

$$M = \frac{EI(\varepsilon_B - \varepsilon_A)}{y_1 - y_2} \qquad (8)$$

3. TESTING PROGRAM

3.1 Details of steel arches

The yieldable supports and the rigid supports were tested in this study. The yieldable supports were assembled by three Glocken type U-sectional beams, designation U26 beam, and the hook bolts in which overlapping length is 40cm. The rigid supports were assembled by two I-sectional beam, designation GI130 beam, and the wrap type fishplate. The properties and section views of both beams are shown in Table 1 and Figure 3.

Table 1. Properties of U-beam and I-beam

Property		U-beam	I-beam
Height(cm)		12.32	13.0
Width(cm)		14.1	10.0
Moment of inertia(cm^4)	I_x	510	1130
	I_y	530	211
Section modulus(cm^3)	W_x	82	175
	W_y	75	42
Area of section(cm^2)		32.9	44.6
Unit weight(kg/m)		25.8	35.0
Tensile strength(MPa)		510	392

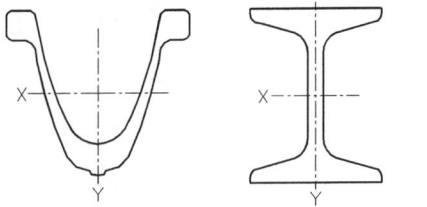

Figure 3. Section views of U-beam and I-beam.

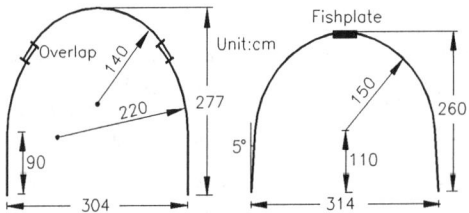

Figure 4. Total views of the 3-piece yieldable support of U-beam and the 2-piece rigid support of I-beam

The yieldable support is of 304cm×277cm arch shape, as shown in the left side of Figure 4. Eleven pairs of vibrating wire strain gages were attached at intervals of 19.13cm along neutral axis of the beam. Of a pair of gages, the one was located 2.93cm apart from neutral axis toward extrados of the arch while the other was located 2.73cm apart from neutral axis toward intrados, as shown in Figure 5. Also, dumbbell-type measuring bolts are inserted into the steel beam at 22.74cm intervals.

In addition, the rigid support is of 314cm×260cm arch shape with their legs splayed at an angle of 5° as shown in the right side of Figure 4. Seventeen pairs of gages were attached at 19.13 cm intervals.

3.2 *Testing facility*

The laboratory steel arch testing facility was designed on the basis that the support would be placed horizontally on the strong concrete floor. It was composed of hydraulic rams, load controller, frame and steel blocks. Loading was applied to the support in radial direction of the arch by seven hydraulic rams. The support and the rams were sustained by frame and steel blocks anchored into concrete floor, respectively. During tests, there were no movement in frame and steel blocks. The schematic view of testing facility with the yieldable support under 7-point loading condition is shown in Figure 6.

The measuring devices used were vibrating wire strain gages, curvo-distometer, loadcells, strain indicator and dial gages. Loadcells were inserted between the arch and the ram to measure the applied load. Arch deformation was measured by dial gages that were placed at the intrados of the arch.

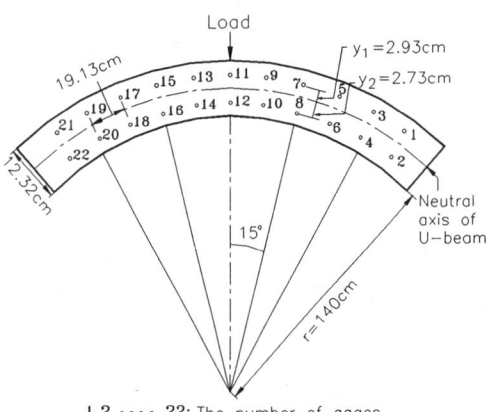

Figure 5. Measuring position of vibrating gages in the yieldable supports.

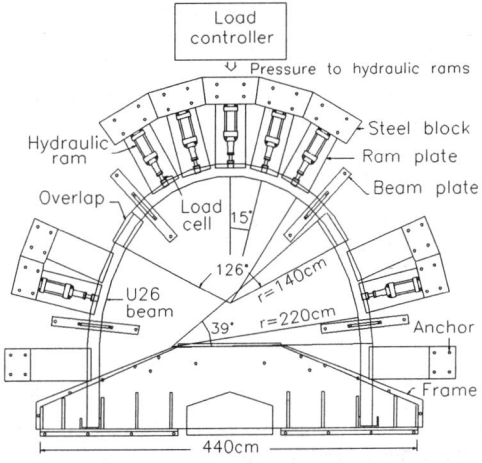

Figure 6. Schematic view of testing facility with 3-piece yieldable support of U-beam.

The curvo-distometer is a designed indicator to measure f and l at one time for small-size arch supports, which was experimentally manufactured in imitation of the curvometer and the mechanical extensometer made by Kovàri(1977).

3.3 Testing procedures

Both legs of the arch were completely fixed with cement mortar of compressive strength 50 MPa in the frame. The hydraulic rams were placed between the arch and steel block, in turn load controller was operated to apply load toward the center of the arch, while applied loads were measured by loadcells. As loading was stepwise increased within the limits of elastic range of the steel beam, deformation of the support was investigated by the measuring devices.

4. RESULTS

4.1 Deformation behaviors

Crown displacements of the yieldable supports and the rigid ones are listed in Table 2 and 3, respectively. These tests were executed in the limits of elastic range of steel beam, so that the displacements were founded to be linearly increased with the applied load. Furthermore, comparing 1-point with 5-point loading condition, it is founded that increasing the number of loading points induces high stiffness and stability of supports. Thus concentrated loading condition would be the worst case in the underground support system.

Crown displacement of the rigid support under 1-point crown loading condition is 2.34 times as large as that of the yieldable support, by simple calculation, while section modulus of I-beam is 2.1 times as large as that of U-beam as listed in Table 1. This originated in the low stiffness of wrap type fishplate connecting two I-beams, in that reason the failure of such supports can be predicted to begin at the fishplate. However, the effect of fishplate was decreased as loading condition was changed into distributed case.

In addition, conventional three-point bending tests were conducted. As a result, the stiffness of I-beam without fishplate was 10.8 ton/mm while that of with fishplate was 0.82 ton/mm.

4.2 yieldable characteristics

The connection part of the two U-beam was composed of a pair of hook-bolts, which were tightened with adequate torque. During 1-point crown loading tests of the yieldable supports, the changes in overlapping length were measured with calipers. The brief results are shown in Figure 7, where the frictional yield load is 5.9 ton.

Table 2. Crown displacements of the yieldable steel arch supports under four different loading conditions.

Test	Geometry	Load (kg)	Displacement (mm)
1-point crown loading		2004	2.51
		4454	5.36
		6094	8.10
		9464	10.86
		12138	13.80
5-point loading		578×5	2.78
		958×5	4.61
		1411×5	6.75
		1648×5	7.83
		2129×5	10.20
7-point loading		739×7	1.27
		1059×7	1.82
		1445×7	2.63
		1768×7	3.27
		2110×7	4.27
		2390×7	5.04
Asymmetric loading		706	0.77
		1589	1.50
		2177	2.02
		2883	2.72
		3471	3.24

Table 3. Crown displacements of the rigid steel arch supports under two different loading conditions.

Test	Geometry	Load (kg)	Displacement (mm)
1-point crown loading		507	1.04
		1088	2.77
		2163	5.76
		2744	7.24
		3882	10.30
5-point loading		684×5	4.93
		1547×5	12.05
		2004×5	14.86
		2467×5	17.56
		3017×5	21.19

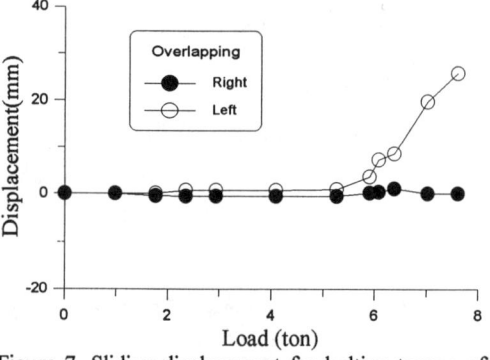

Figure 7. Sliding displacement for bolting torque of 10 kg·m in the yieldable support

4.3 Back-calculation of load components

The strains measured at extrados and intrados of the yieldable support under 1-point crown loading condition are shown in Figure 8, where "Gage No." refers to Figure 5. Strains at crown (Gage No., 11 and 12) represented the highest value while strains at shoulder represented the lowest.

On the other hand, strain data measured were used for back-calculating the applied load by means of equation (4)-(8). The brief results of back-calculation of the yieldable supports under two different loading conditions are shown in Table 4 and 5. At this the back-calculated load components are in good agreement with the applied ones, mean error of which is within ±20%. Thus, this measuring technique is supposed to be applicable in the tunnel support system.

The results of the rigid supports under 5-point loading condition are shown in Table 6. The back-calculated load components except at crown are roughly correspond to the applied ones. This error is owing to the existence of wrap type fishplate. As the fishplate is located at the crown, this support may not be a continuous structure, so that the back-calculated load at crown is far from the applied load.

Table 4. Back-calculated load components of the yieldable support under 1-point crown loading condition.

Applied load (kg)	Position*	Normal force (kg)	Bending moment (kg·cm)	Calculated load (kg)	Calc./appiled
2004	L3	559	76163	2034	1.01
4454	L3	577	163101	4274	0.96
6904	L3	1357	247252	6318	0.92
9465	L3	1302	333446	8743	0.92
12138	L3	1834	431715	11810	0.97

*Legend of Position

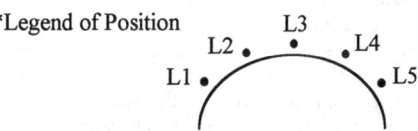

Table 5. Back-calculated load components of the yieldable support under 5-point loading condition.

Applied load (kg)	Position*	Normal force (kg)	Bending moment (kg·cm)	Calculated load(kg)	Calc./applied
1411×5	L1	-2120	60188	1365	0.97
	L2	-1230	98269	1433	1.02
	L3	-1694	105142	1280	0.91
	L4	-189	91582	1423	1.01
	L5	-806	48484	1232	0.87
2129×5	L1	-3365	93068	2227	1.05
	L2	-1720	148425	2002	0.94
	L3	-2165	158271	1788	0.84
	L4	-131	137465	2143	1.01
	L5	-1150	73934	2112	0.99

Table 6. Back-calculated load components of the rigid support under 5-point loading condition.

Applied load (kg)	Position*	Normal force (kg)	Bending moment (kg·cm)	Calculated load(kg)	Calc./applied
2004×5	L1	-2848	-67383	1565	0.78
	L2	918	24503	2353	1.17
	L3	229	5513	-5087	-2.54
	L4	-183	15927	2200	1.10
	L5	-4042	-77184	2319	1.16
3017×5	L1	-4364	-87598	2935	0.97
	L2	-45	44718	3047	1.01
	L3	459	14701	-9117	-3.02
	L4	-780	36142	3682	1.22
	L5	-5558	-108426	3593	1.19

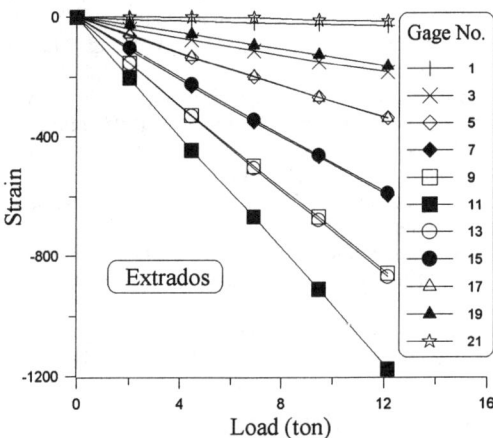

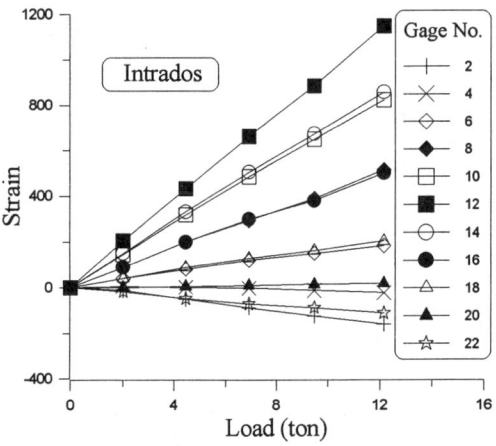

Figure 8. Strains of the yieldable support in extrados and intrados under 1-point crown loading condition.

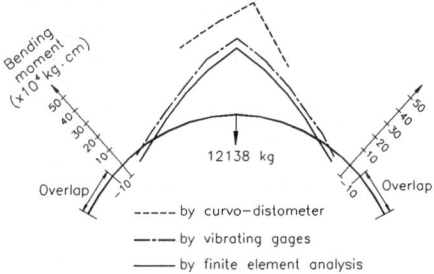

Figure 9. Comparison between test results and finite element analysis of the yieldable support under 1-point crown loading condition.

4.4 *Finite element modeling*

Finite element analysis of structural program SAP90 was conducted to verify the results of full scale tests. The yieldable support of U-beam was discretized into 82 three-dimensional frame elements. The 1-point crown loading condition was investigated, where both ends of legs were fixed in the same way of the full scale test. The results of analysis such as displacements, normal forces and bending moments were in good agreement with the test results.

Figure 9 shows the distribution of bending moments along the arch, where results of analysis are compared with test results. Between test results, the output from strain data measured by vibrating gage system was more correspond with analysis results than the output from the change in radius of curvature and chord length measured by curvo-distometer. It is thought to originate in inaccuracy of measuring device.

5. CONCLUSION

The full scale tests of two different steel arch supports produced a number of available results to design an efficient tunnel support system. The yieldable support assembled by U-sectional beam and hook bolts is found to withstand more load than the rigid support with the wrap type fishplate. It is also founded that the number of loading points induces high stability of support system. The back-calculated load components on the basis of integrated measuring technique are in good agreement with applied loads.

ACKNOWLEDGEMENTS

The authors wish to acknowledge the financial support of the Korea Institute of Geology, Mining and Materials(KIGAM). The authors are grateful to Dr. So-Keul Chung and Dr. Hee-Soon Shin for their encouragement and assistance. Thanks are also due to Mr. Boo-Hwan Kim for the highest assistance.

REFERENCES

Birön, C. & Arioglu, E. 1983. Design of supports in mines, *John Wiley & Sons*:3-87

Choquet, P. 1982. A failure criterion of steel arch supports for the interpretation of in-situ monitoring results. *Proc. 14th Canadian Rock Mechanics Symposium*:116-123.

Everling, G. 1964. Model tests concerning the interaction of ground and roof support in gate-roads, *Int. J. Rock Mech. Min. Sci.* 1:319-326

Horvath, J. 1971. Calculation of rock pressure in shaft and roadways of circular section, *Int. J. Rock Mech. Min. Sci.* 8:239-276

Jukes, S.G., Hassani, F.P. & Whittaker, B.N. 1983. Characteristics of steel arch support systems for mine roadways., Part 1. Modeling theory, instrumentation and preliminary results, *Mining Science and Technology* 1:43-58.

Khan, U.H., Mitri, H.S. & Jones, D. 1996. Full scale testing of steel arch tunnel supports. *Int. J. Rock Mech. Min. Sci.* 33:219-232

Kovàri, K., Amstad, Ch. & Fritz, P. 1977, Integrated measuring technique for rock pressure determination. *Proc. Int. Sympo. on Field Measurements in Rock Mechanics, Zurich*: 289-316.

Sadler G.W. et. al. 1984. Testing of roadway support equipment, *The Mining Engineer*: 237-245

Back calculation of initial stress state from incremental displacement measurements

Youn-Kyou Lee & Chung-In Lee
Division of Civil, Urban and Geosystem Engineering, Seoul National University, Korea

ABSTRACT: By applying the finite element formulation which is capable of handling the geometrically altered structure in a successive manner, it was shown that the magnitude of initial stress field has the linear relationship with incremental displacements. Based on this relationship, two dimensional back analysis code having the capability of dealing with multi-stage excavation problem was built and verified. With this code, the initial stress state prior to excavation can be back-calculated from the incremental displacement measurements taken at any excavation stage.

1 INTRODUCTION

It has been recognized as an essential part in tunneling to monitor the deformation behavior of a tunnel for the purpose of assessing the adequacy of the design and checking the possible instability during construction. Of the various monitoring techniques available, displacement measurements which provide information on the overall movements of the rock mass around the tunnel have accepted to be the most useful. In the past, the main purpose of the displacement measurements was to assess the safety of a tunnel based on the observation of magnitudes and rates of deformation. However, the introduction of modern tunneling method such as the New Austrian Tunneling Method has led to the need for the development of a rapid and precise method which can evaluate the field measurements and feed them back to the construction process. The back analysis has shown up as one of the promising solutions to this requirement.

Recently, efforts have been directed towards developing methods to back-analyze rock mass properties and insitu stress state from monitoring data. Many back analysis techniques have been proposed and some of them were successfully applied to solving the practical problems. The direct methods are based on the iterative procedure searching the optimum rock mass parameters by minimizing the difference between the measured and predicted values(Gioda & Maier, 1980). In the inverse methods, deformability and state of initial stresses of rock formation are estimated by a formulation which is the reverse of that of the ordinary stability analysis(Sakurai & Takeuchi, 1983). The method proposed by Feng & Lewis(1987) is similar to Sakurai & Takeuchi's method in that the state of initial stresses can be directly calculated from the measured displacements, but different in that a flexibility relation, which represents the sensitivity of each predicted displacements to each components of initial stresses, should be established in advance by using the ordinary stress analysis algorithm. Feng & Lewis' method can be applied to non-linear problems, too. In most back analysis method including those methods reviewed so far, optimization processes are performed using the least square approach.

Although there may be some advantages and limits in each back analysis method, the fact that the quality of the back-calculated parameters strongly depends on the accuracy of monitoring data is true of all the methods. For example, it is the total displacements including those occurred ahead of installing instruments that are necessary for the back analysis based on the inverse approach. However, it is common that only a part of total displacement is measured due to many constraints in driving a tunnel. Even though many attempts have been made to estimate the total displacements from this kind of field measurements by the use of regression analysis, no attempt has been successful in incorporating all ground conditions and excavation features, yet. In performing back analysis, therefore, there has been a possibility that the estimated parameters can be greatly distorted through misinterpreting the observed displacements. In order to mitigate this difficulty, Akutagawa et al.(1991) used incremental displacements measured in a particular excavation step for back-calculating the initial stress field. Their algorithm is identical to Feng & Lewis' when it is applied to a single stage excavation problem. In excavating a tunnel which has a large cross section, the section is divided into several sections and excavated successively in consideration of safety and for the convenience of equipment accommodations. Under this circumstance, the measurements of incremental displacements is much easier than those of total displacements. Although Akutagawa et al. did not describe the explicit formulation procedure, the finite element model capable of handling the multi-step excavation efficiently is essential for computer implementation of their method.

The objective of this paper is to discuss a new and efficient back analysis formulation which gives the same result as Akutagawa et al.'s. Based on the incremental finite element formulation for determining stress and deformation resulting from successive geometrical alteration in structure (Ghaboussi & Pecknold, 1984), it is shown that the initial stress state of elastic medium has linear relationship with the incremental displacements occurring in any excavation steps. The 2-dimensional back analysis code was constructed by the use of this formulation. In order to verify the code, a simple illustrative example is given.

2 FORMULATION OF BACK ANALYSIS

In a standard incremental formulation for nonlinear analysis, the incremental equilibrium equations for i-th analysis step are expressed as,

$$K_i \Delta U_i = \Delta F_i + R_i \qquad (1)$$

where K_i = global stiffness matrix of model, ΔU_i = incremental displacements vector, $\Delta F_i = $ incremental external force vector ($F_i - F_{i-1}$), and R_i is residual force vector calculated by,

$$R_i = F_{i-1} - I_{i-1} \qquad (2)$$

I_{i-1} in Eq. (2) is the consistent nodal force vector computed from stress state σ_{i-1} of $(i-1)$th excavation step. In case of multi-step excavation problem, I_{i-1} is the assembly of contribution from each element which is active in analysis step i. Thus,

$$I_{i-1} = \sum_j \int (B_i^j)^T \sigma_{i-1}^j dv_i^j \qquad (3)$$

in which B_i^j is strain-displacement matrix of j th element in i th analysis step. Σ symbol stands for the sum of contribution from each element, but will be omitted for the clarity of expression in the remainder of the paper.

Eq. (1), therefore, can be restated as,

$$K_i \Delta U_i = F_i - I_{i-1} \qquad (4)$$

The right hand side of Eq. (4) is the load vector whose components are all zero except for the components pertaining to nodes on the newly created excavation boundary. Assuming that no artificial support forces are applied on the excavation boundary in each excavation step, F_i should be zero vector and, therefore,

$$K_i \Delta U_i = -I_{i-1} \qquad (5)$$

By successive applications of Eq. (5), incremental displacements occurring in any excavation can be easily calculated. Numerical excavation algorithm employed in this paper is based on Eq. (5).

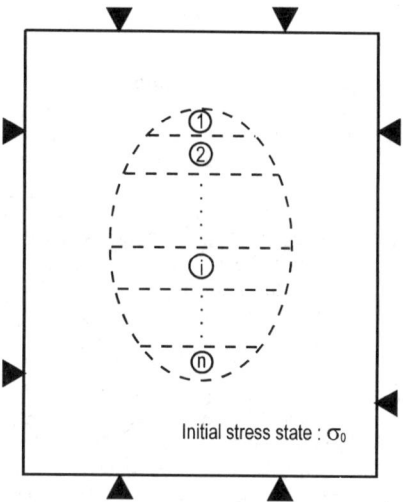

Figure 1. Model for sequential excavation.

Consider a situation that a portion of domain is divided into n regions and that portion is to be excavated in sequence from region ① to region ⓝ (see Figure 1), and further assume that the initial stress state prior to excavation is uniform over the domain and its magnitude is $\sigma_0 = \{\sigma_{x0}, \sigma_{y0}, \tau_{xy0}\}^T$. As the same final displacements are produced irrespective of number of excavation steps, if the final configurations of a elastic model are the same(Ghaboussi & Pecknold, 1984), ΔU_i can be regarded as the difference of solutions resulting from two one-step excavation problems, namely,

$$\Delta U_i = U_i - U_{i-1} \qquad (6)$$
$$= -K_i^{-1} \int B_i^T \sigma_0 dv_i + K_{i-1}^{-1} \int B_i^T \sigma_0 dv_{i-1}$$

where U_{i-1} and U_i are induced by simultaneous excavation of the first $(i-1)$ and i regions, respectively. Recalling the assumption that initial stress state σ_0 is constant over the region under consideration, Eq. (6) gives the linear relationship existing between ΔU_i and σ_0, which is the same result as Akutagawa et al.'s(1991). This means that the initial stress state can be back-calculated from incremental displacements measured during any excavation stage. This statement can be written by the simplified expression

$$\sigma_0 = C_i^{-1} \Delta U_i \qquad (7)$$

where

$$C_i = -K_i^{-1} \int B_i^T dv_i + K_{i-1}^{-1} \int B_i^T dv_{i-1}$$

The matrix C_i is known as the flexibility matrix which represents the sensitivity of each predicted displacements to each components of initial stresses. The coefficients of C_i depend on geometrical shape of excavation and material constants, i.e. elastic mudulus and Poisson's ratio. The importance of Eq.(7) is in that σ_0 can be estimated from incremental displacements measured during particular excavation stage. In practice, the measurements of incremental quantity are much easier than those of absolute quantity.

It should be noted, therefore, that the construction of matrix C_i is the central task in performing the back analysis adopting the method outlined above. If total k measurements are used for back-calculating σ_0, C_i will take the form,

$$C_i = \begin{bmatrix} c_i^{11} & c_i^{12} & c_i^{13} \\ c_i^{21} & c_i^{22} & c_i^{23} \\ \vdots & \vdots & \vdots \\ c_i^{k1} & c_i^{k2} & c_i^{k3} \end{bmatrix} \qquad (8)$$

The first, second and third columns of Eq. (8) simply mean the collection of displacement increments caused by i th excavation in measuring positions, when the imaginary initial stress states prior to excavation are assumed as $\sigma_0 = \{1,0,0\}^T$, $\{0,1,0\}^T$, and $\{0,0,1\}^T$ respectively. Each element of C_i can be easily determined with the aid of numerical methods such as finite element method having the capability of dealing with sequential excavation problems. In this research, this task is carried out systematically by sequential excavation algorithm based on Eq. (5).

Under the assumption that the material properties are known, Eq. (7) shows that minimum three incremental displacement measurements are required for back-calculating three unknowns, $\sigma_{x0}, \sigma_{y0}, \tau_{xy0}$. If more than three incremental displacements are used as the input data for back analysis, the results should be optimized by appropriate optimization processes such as the least square approximation. If the elastic modulus(E) is not known, normalized initial stresses $\{\sigma_{x0}/E, \sigma_{y0}/E, \tau_{xy0}/E\}$ can be obtained by assuming $E=1$.

Provided that the finite element mesh for back analysis is constructed in such a way that the measuring points coincide with the nodal points of the mesh, the incremental displacements for back analysis can be x or y components at any nodes, or relative values between any two nodes. In practice, the relative incremental displacements, the change of distance between two measuring points, are commonly measured by the use of such instruments as convergence meter and multi-point borehole extensometer. In case of relative incremental displacements being included in the input data, the elements of flexibility matrix corresponding to these relative values are evaluated by appropriate transformation of absolute incremental values at any two nodes. For example, the incremental change of distance(Δl) between two nodes, i and j in Figure 2, can be computed from absolute incremental displacement components at the two nodes (Δu_i, Δu_j, Δv_i, and Δv_j) by

$$\Delta l = (\Delta u_j - \Delta u_i)\cos\theta + (\Delta v_j - \Delta v_i)\sin\theta \qquad (9)$$

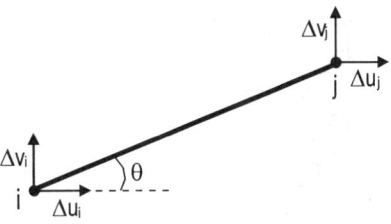

Figure 2. Incremental displacements at both ends of a measuring line.

It should be noted that angle θ in Eq. (9) is implicitly assumed to be constant during the sequential excavations. In real situation, however, θ can be deviate from initial value with on-going excavations, even though the magnitude of the change is not significant. This means that some errors can occur in back-calculating initial stress if the measured values are the length changes such as tunnel convergence measures between any two points; on the other hand, the flexibility coefficients corresponding to those measured values are calculated based on Eq. (9), where θ is considered to be constant and taken from layout of monitoring station. This topic is discussed again in the following section through illustrative numerical example.

3 CONSTRUCTION OF BACK ANALYSIS CODE

Based on the theoretical background described in previous section, two dimensional finite element code, which can back-calculate initial stress state from incremental displacement measurements, was developed. In order to fabricate the flexibility matrix of Eq. (8), this code has the ability of dealing with the initial stress loading and the sequential removal of elements. The algorithm for successive deletion of elements is based on Eq. (5). This algorithm is similar to that of ordinary nonlinear incremental finite element analysis, but the changes of global stiffness matrix and load vector with analysis step going on are due to the removal of some elements.

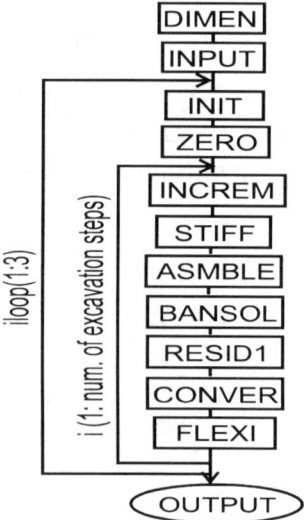

Figure 3. Flow chart of back analysis code

In implementing the code, whole mesh are grouped into as many subregions as needed to facilitate the numerical treatment of sequential excavations. Figure 3 shows the simplified flow chart of the code. The outer loop is repeated three times in order to evaluate three columns of flexibility matrix by assuming the imaginary initial stress states, (1,0,0), (0,1,0), and (0,0,1) in turn. The inner loop continues to the last excavation step in which incremental displacement measurements were conducted. The functions of each module shown in Figure 3 are as follows:

INIT : compute $-I_0$ corresponding to initial stress states, (1,0,0), (0,1,0), and (0,0,1) respectively
ZERO : initialize state variables and arrays
INCREM : evaluate $-\int B_i^T \sigma_{i-1} dv_i$, and assemble global incremental load vector
STIFF : compute the element stiffness matrix
ASMBLE : assemble global stiffness of present excavation step
BANSOL : find ΔU_i , $\Delta \sigma_i$ and update U_i , σ_i
RESID1 : evaluate consistent nodal force $\int B_i^T \sigma_i dv_i$
CONVER : compare total load vector with consistent nodal force
FLEXI : fabricate flexibility matrix and then calculate σ_0

4 ILLUSTRATIVE EXAMPLE

To verify the code constructed, the simple finite element model shown in Figure 4 was selected. This mesh consists of 35 eight node quadrilateral elements prior to excavation. Both horizontal and vertical displacements are prescribed to zero at all nodes along the boundary. Elastic modulus(E) and Poisson's ratio(ν) were assumed to be 3000MPa, 0.25 respectively. Moreover, uniform initial stress field before the start of excavation was assumed as $\sigma_{x0} = -10 MPa$, $\sigma_{y0} = -5 MPa$, $\tau_{xy0} = 2 MPa$.

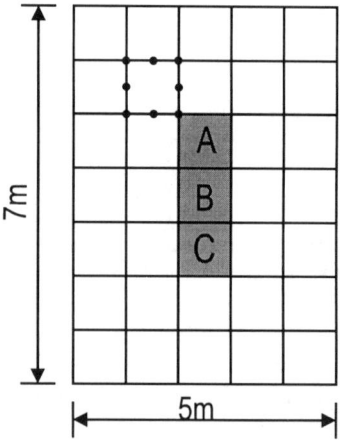

Figure 4. Finite element mesh for back analysis.

Before performing back analysis, we conducted three step ordinary finite element analysis by removing three elements A, B, and C in sequential manner. In each step, incremental displacement measurements were taken at three nodes a, b, and c of element A (Figure 5). Results are tabulated in Table 1.

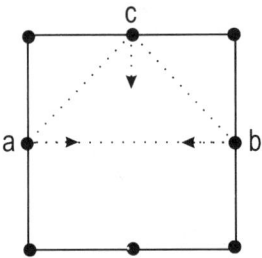

Figure 5. Assumed measuring positions.

Table 1. Incremental displacements induced in each step.

Excavation step	Node	Δu (m)	Δv (m)
1	a	0.278657E-2	-0.716051E-3
	b	-0.280746E-2	0.841315E-3
	c	0.759988E-3	-0.975565E-3
2	a	0.151745E-2	-0.535311E-3
	b	-0.841787E-3	0.512196E-3
	c	0.169306E-3	0.114675E-3
3	a	0.196914E-3	-0.651364E-4
	b	-0.936512E-4	0.197914E-3
	c	0.475691E-4	-0.253108E-4

Δu :horizontal component, Δv :vertical component

Using the incremental displacement measurements shown in Table 1, it is demonstrated that the initial stresses can be back-calculated. Some combinations of displacement measurements are assumed. First, three absolute incremental measurements, that is, horizontal values at nodes a, b and vertical value at node c, are selected. In this case, Eq. (7) for excavation stage 3 can be rewritten as,

$$\begin{pmatrix} -0.129645E-4 & -0.312755E-5 & 0.258158E-4 \\ 0.129645E-4 & 0.312755E-5 & 0.258158E-4 \\ -0.119546E-5 & 0.745309E-5 & 0.000000E+0 \end{pmatrix} \begin{pmatrix} \sigma_{x0} \\ \sigma_{y0} \\ \tau_{xy0} \end{pmatrix}$$

$$= \begin{pmatrix} 0.196914E-3 \\ -0.936512E-4 \\ -0.253108E-4 \end{pmatrix} \quad (10)$$

from which the exact values, namely $\sigma_{x0} = -10\,MPa$, $\sigma_{y0} = -5\,MPa$, and $\tau_{xy0} = 2\,MPa$, are obtained.

Although not shown, the correct results are also produced in the cases of excavation stages 1 and 2.

Under the assumption that the directions of measuring lines do not vary during excavation process, three relative incremental displacements can be calculated by the use of the Eq. (9), along the measuring lines, a-b, a-c, and b-c in Figure 5. The coefficients of three flexibility matrices for each excavation stage can be easily computed with the same procedure. All these flexibility equations give correct solutions. Especially for excavation stage 3, the flexibility relation is

$$\begin{pmatrix} 0.259290E-4 & 0.625510E-5 & 0.000000E+0 \\ 0.184801E-4 & -0.344584E-5 & 0.450648E-4 \\ 0.184801E-4 & -0.344584E-5 & -0.450648E-4 \end{pmatrix} \begin{pmatrix} \sigma_{x0} \\ \sigma_{y0} \\ \tau_{xy0} \end{pmatrix}$$

$$= \begin{pmatrix} -0.290565E-3 \\ -0.774418E-4 \\ -0.257702E-3 \end{pmatrix} \quad (11)$$

As the last case, the real change of distance between two points is considered as the measurement value, but the flexibility matrix is constructed by the use of Eq. (9) under the assumption of constant θ. By assuming constant θ, the formulation of back analysis becomes more simple. Real distance change is the difference between the distances before and after executing the excavation of particular analysis step. The displacement measurements recorded in the actual underground work are usually of this kind. If the distance changes of three pairs of points, (a,b), (a,c), and (b,c), during the excavation stage 3 are regarded as measurements, the following flexibility relation is obtained.

$$\begin{pmatrix} 0.259290E-4 & 0.625510E-5 & 0.000000E+0 \\ 0.184801E-4 & -0.344584E-5 & 0.450648E-4 \\ 0.184801E-4 & -0.344584E-5 & -0.450648E-4 \end{pmatrix} \begin{pmatrix} \sigma_{x0} \\ \sigma_{y0} \\ \tau_{xy0} \end{pmatrix}$$

$$= \begin{pmatrix} -0.289833E-3 \\ -0.769182E-4 \\ -0.257833E-3 \end{pmatrix} \quad (12)$$

It should be noted that Eqs. (11) and (12) have the same flexibility matrix. Eq. (12) gives $\sigma_{x0} = -9.9817\,MPa$, $\sigma_{y0} = -4.9588\,MPa$, and $\tau_{xy0} = 2.0073\,MPa$. The result is close to the actual solution. This result shows that the assumption of the constant θ during the excavation sequence is reasonable for the practical use.

5 SUMMARY AND CONCLUSION

An efficient back analysis algorithm was presented for the estimation of initial stress state from incremental displacement measurements measured during any excavation stage. By applying the sequential excavation principle in the context of finite element method, it was shown that the initial stress state of elastic rock mass prior to excavation has a linear relationship with incremental displacements produced in any excavation step. Based on the equation that represents this relationship, two dimensional back analysis code having the ability of treating multi-stage excavation problems was written and verified. Illustrative example showed that the code can back-calculate initial stress field exactly if the accuracy of displacement measurements is guaranteed.

Considering the fact that the measurements of tunnel behavior usually are started in some later stages due to many constraints of tunneling work, the use of incremental displacements for back analysis is of practical importance. Although its application is limited to rock mass condition showing elastic behavior, the developed code is advantageous for practical use because absolute or relative displacement increments measured in any excavation stage can be used as input data.

REFERENCES

Akutagawa, S., J.L. Meek & E.T. Brown 1991. The back analysis of in-situ stresses in a multiple stage excavation problem. *Proc. 7th Int. Confer. on Computer Methods and Advances in Geomech, Cairns*:937-942.

Feng, Z. L. & R.W. Lewis 1987. Optimal estimation of in-situ ground stresses from displacement measurements. *Int. J. Numer. & Anal. Methods in Geomech.* 11:679-686.

Ghaboussi, J. & D.A. Pecknold 1984. Incremental finite element analysis of geometrically altered structures. *Int. J. for Num. Meth. in Eng.* 20: 2051-2064.

Gioda, G. & G. Maier 1980. Direct search solution of an inverse problem in elastoplasticity: identification of cohesion, friction angle and in situ stress by pressure tunnel tests. *Int. J. Numer. Mech. Eng.*15:1823-1843.

Sakurai, S. & K. Takeuchi 1983. Back analysis of measured displacements of tunnels. *Rock Mech. & Rock Eng.*16(3):173-180.

Environmental and Safety Concerns in Underground Construction, Lee, Yang & Chung (eds)
© 1997 Balkema, Rotterdam, ISBN 90 5410 910 6

An integrated back-analysis system for monitoring underground openings

N. Shimizu & K. Nakagawa
Department of Civil Engineering, Yamaguchi University, Ube, Japan

S. Sakurai
Department of Civil Engineering, Kobe University, Japan

ABSTRACT: Back analysis has been recognized as an effective tool for the observational design method. In this paper, back-analysis methods for continuous and discontinuous rock mass behavior are integrated for monitoring underground openings.

1 INTRODUCTION

The observational method improves the design and construction methods by using field measurements during construction period (Terzaghi and Peck 1967). It has become a common method for constructing underground openings. Back analysis has been recognized as an effective tool for realizing the quantitative observational method. Actually, back-analysis methods for field measurement results could offer such advantages as

1. the modification of assumed design parameters through a comparison with the back-analyzed design parameters, i.e., initial stress, deformation modulus, and strength parameters, etc;
2. the verification of the numerical (or analytical) model assumed at the time the initial design was performed;
3. feedback of measurement results during the construction period to quantitatively improve the design;
4. a better understanding of the mechanism of rock mass behavior by improving the numerical model to match reality.

Research concerning back analysis is still on going, however, and there are few methods of back analysis which can completely provide the above advantages. In rock mechanics problems, modeling the transition from continuous to discontinuous behavior of rock masses is the key to successfully enrolling the back analysis in the observational method.

The purpose of this paper is to integrate the back-analysis methods proposed by the authors for dealing with both continuous and discontinuous rock mass behavior.

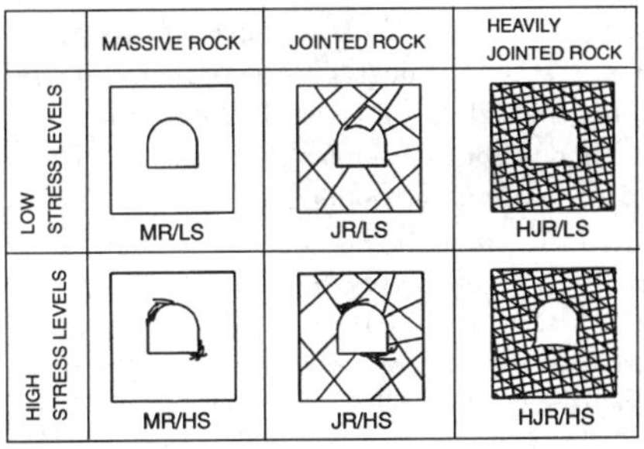

Figure 1: Types of mechanical behavior in different rock mass conditions (Hoek et al. 1995)

2 TYPES OF ROCK MASS BEHAVIOR

Hoek et al.(1995) described the various types of rock mass behavior around underground openings associated with the frequency of discontinuities and the in situ stress levels as shown in Figure 1.

A massive rock subjected to low in situ stress levels (type MR/LS) exhibits a linear elastic response. A heavily jointed rock subjected to both low (HJR/LS) and high (HJR/HS) in situ stress conditions fails by unraveling into small interlocking blocks and by sliding on discontinuities and being crushed into rock pieces, respectively. Both cases, types HJR/LS and HJR/HS, are modeled by pseudo or equivalent continuum models if the behavior of the rock mass is well controlled by the appropriate support systems.

In the case of a massive rock with relatively few discontinuities, subjected to low in situ stress conditions (JR/LS), rock blocks released by intersecting discontinuities fall or slide due to gravity loading.

A massive rock with no or relatively few discontinuities subjected to high in situ stress conditions (MR/HS and JR/HS) behaves by spalling, slabbing, crushing, and splitting. In addition, progressive failures may occur in the rock mass.

This paper deals with types MR/LS, HJR/LS, and HJR/HS, and type JR/LS by using continuum and discontinuum models, respectively, on back analysis. Types MR/HS and JR/HS are excluded here.

3 BACK-ANALYSIS METHOD BASED ON CONTINUUM MECHANICS

The behaviors illustrated as types MR/LS, JR/LS, and HJR/HS in Figure 1 are treated with the continuum mechanics model. A back-analysis method can be formulated based on continuum mechanics.

3.1 Outline of the formulation

As is well known, *FEM* can easily be applied to problems involving non-homogeneous, anisotropic, and non-linear characteristics of rock mass, but it is not suitable for problems in unbounded domains. On the other hand, *BEM* is much more appropriate for homogeneous, linear elastic problems in infinite domains, but it is not as easily applied to non-homogeneous and non-linear problems as *FEM*.

If each method is applied to the area for which it is best suited, it must be most promising for use in the analysis of underground excavation problems. That is, *BEM* should be applied to areas far from the underground openings where the ground may be homogeneous, while *FEM* should be applied to areas around the openings where the ground may be non-homogeneous and non-linear.

In this section, a back-analysis method based on continuum mechanics is formulated by a coupled *FE* and *BE* method (Shimizu and Sakurai 1989).

displacement equations

The entire region around the underground openings to be analyzed is divided into *FE* and *BE* regions, as shown in Figure 2.

The equation for the *FE region* is represented by the following matrix form:

$$\left\{ \begin{array}{c} \mathbf{F}_s \\ \mathbf{F}_f \end{array} \right\} = \left[\begin{array}{cc} \mathbf{K}_{ss} & \mathbf{K}_{sf} \\ \mathbf{K}_{fs} & \mathbf{K}_{ff} \end{array} \right] \left\{ \begin{array}{c} \mathbf{u}_s \\ \mathbf{u}_f \end{array} \right\} \quad (1)$$

where $\{\mathbf{F}.\}$ and $\{\mathbf{u}.\}$ are the nodal forces and displacements of the finite elements, respectively. [K.] is the stiffness matrix. Subscripts $\cdot_f$ and $\cdot_s$ stand for the *FE region* and boundary S between the *FE* and *BE regions*, respectively.

The equation for the *BE region*, on the other hand, is obtained as the following equation.

$$\{\mathbf{F}_b\} = [\mathbf{K}_b]\{\mathbf{u}_b\} \quad (2)$$

where $\{\mathbf{F}_b\}$ and $\{\mathbf{u}_b\}$ are the nodal force and the nodal displacement of the boundary elements, respectively. $[\mathbf{K}_b]$ is the stiffness matrix of the *BE region*.

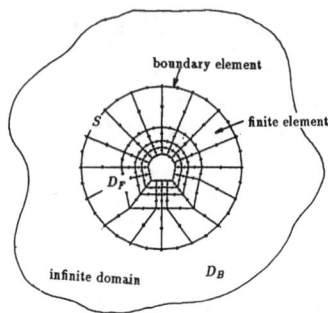

Figure 2: Finite element region and boundary element region

Considering the condition of continuity in displacements and the force equilibrium on S, the equation for the entire region is obtained from Equations (1) and (2) as follows:

$$\left\{ \begin{array}{c} \mathbf{0} \\ \mathbf{F}_f \end{array} \right\} = \left[\begin{array}{cc} \mathbf{K}_{ss} + \mathbf{K}_b & \mathbf{K}_{sf} \\ \mathbf{K}_{fs} & \mathbf{K}_{ff} \end{array} \right] \left\{ \begin{array}{c} \mathbf{u}_s \\ \mathbf{u}_f \end{array} \right\}. \quad (3)$$

For excavation problems of underground openings, force vector $\{\mathbf{F}_f\}$ is given by the following equation under the assumption of a constant initial stress around the excavation area:

$$\{\mathbf{F}_f\} = [\mathbf{P}]\{\sigma_0\} \quad (4)$$

where $\{\sigma^0\} = \{\sigma_{11}^0 \ \sigma_{22}^0 \ \sigma_{12}^0\}^T$ is the initial stress vector and $[\mathbf{P}]$ is the matrix depending on geometry only.

Separating the displacements into measured ones $\{\mathbf{u}_m\}$ and others $\{\mathbf{u}_u\}$, Equation (3) is written as follows:

$$\left[\begin{array}{c} \mathbf{Q}_1 \\ \mathbf{Q}_2 \end{array} \right] \{\overline{\sigma_0}\} = [\mathbf{K}^*] \left\{ \begin{array}{c} \mathbf{u}_m \\ \mathbf{u}_u \end{array} \right\} \quad (5)$$

where $\{\overline{\sigma_0}\} = \{\overline{\sigma_{11}^0} \ \overline{\sigma_{22}^0} \ \overline{\sigma_{12}^0}\}^T = \{\sigma_{11}^0/E \ \sigma_{22}^0/E \ \sigma_{12}^0/E\}^T$ is called the normalized initial stress. E is the deformation (or elastic) modulus of the rock mass.

Eliminating $\{\mathbf{u}_u\}$ from Equation (5) gives

$$\{\mathbf{u}_m\} = [\mathbf{F}]\{\overline{\sigma_0}\} \quad (6)$$

where $[\mathbf{F}]$ is the flexibility matrix derived from $[\mathbf{K}^*]$, $[\mathbf{Q}_1]$, and $[\mathbf{Q}_2]$.

back-analysis procedure

The deformation modulus and the initial stress of rock masses can be determined by minimizing the discrepancy between the measured and the calculated displacements. The discrepancy is defined here by the residual sum of squares, namely,

$$R = (\{\mathbf{u}_m\} - \{\mathbf{u}_m^*\})^T (\{\mathbf{u}_m\} - \{\mathbf{u}_m^*\}) \quad (7)$$

where $\{\mathbf{u}_m^*\}$ is the measured displacement during the excavation of the underground openings and $\{\mathbf{u}_m\}$ is the displacement calculated by Equation (6).

To minimize Equation (7), an appropriate numerical optimization technique is adopted.

If we assume a linear elastic homogeneous isotropic model as the back-analysis model, it is easy to obtain the elastic modulus and the initial stress of the rock mass as described below.

1. Applying the least squares method to Equation (7) under the assumption of an appropriate value for Poisson's ratio, the normalized initial stress $\{\overline{\sigma_0}\}$ is obtained as follows:

$$\{\overline{\sigma_0}\} = ([\mathbf{F}]^T[\mathbf{F}])^{-1}[\mathbf{F}]^T\{\mathbf{u}_m^*\}. \quad (8)$$

2. The vertical component of the initial stress is assumed to have the same values as the overburden pressure, namely,

$$\sigma_{22}^0 = \gamma H \quad (9)$$

where γ and H denote the average unit weight of the rock mass and the overburden height of the opening, respectively. The elastic modulus and the initial stress are obtained from the normalized initial stress as follows:

$$E = \gamma H / \overline{\sigma_{22}^0}, \quad \sigma_{11}^0 = \overline{\sigma_{11}^0} E, \quad \sigma_{12}^0 = \overline{\sigma_{12}^0} E. \quad (10)$$

3. If the rock mass suffers from overstress and a plastic zone occurs around the openings, modulus E obtained by the above procedure is no longer the real elastic modulus. It should be evaluated as the deformation modulus representing the equivalent elastic modulus. The real elastic modulus, E_{real}, can be estimated by solving the following nonlinear equation (Sakurai et al. 1985):

$$E_{real} = \frac{E \sin \phi}{\left\{\left(\frac{2\overline{p}}{\varepsilon_0 E_{real}} - 1\right) \sin \phi + 1\right\}^{\frac{1-\sin\phi}{\sin\phi}} - \frac{\varepsilon_0 E}{2\overline{p}}(1 - \sin \phi)} \quad (11)$$

where E and $\overline{p}$ are the back-analyzed deformation modulus and the maximum principal initial stress obtained by Equations (8) $\sim$ (10), respectively. Friction angle ϕ and critical strain ε_0 (Sakurai 1981) of the rock mass are assumed here. The cohesion of the rock mass is obtained by $c = \varepsilon_0 E_{real}(1 - \sin\phi)/(2\cos\phi)$. Also, the extent of the plastic zone can be estimated based on the results of the back analysis. The detailed procedures are described in the references (Sakurai et al. 1985).

4. By comparing the back-analyzed parameters, i.e., initial stress, elastic modulus, and cohesion, with the parameters used at the initial design stage, the design parameters can be verified under the construction process. If the design parameters are over- or under-estimated, the original design should be changed to match the actual rock behavior.

(a) Second Bench

(b) Final stage

Figure 3: Oil storage caverns

5. In order to assess the stability of the underground openings on the basis of the strain control procedure (Sakurai 1981), the strain distribution is calculated by the ordinary BE-FE coupled method.

A similar procedure to the one described above can be developed for more complex geological conditions (Sakurai 1993).

3.2 Practical application

The back-analysis method described in Section 3.1 was applied for realizing a rational and quantitative observational design method for multi-parallel oil storage caverns in Japan. In this project, ten parallel caverns were excavated 50 m apart from the center of each cavern (Fukuhara and Hasegawa 1995). The length, height, and width of each cavern are 555 m, 22 m and 18 m, respectively (Figures 3 and 4). To conduct a back analysis of the measured results, personal computers were set up at the construction site. Figure 5 shows some results of the back analysis for assessing the stability of the caverns; i.e., the plastic zone around the caverns was evaluated at the arch excavation stage, and then, the maximum shear strain distribution was predicted for assessing the

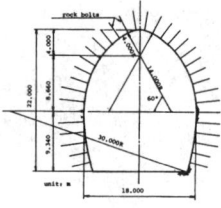

(a) Section of cavern

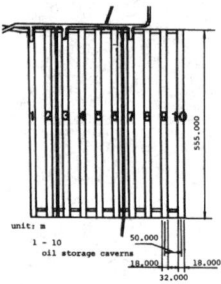

(b) Plan view of site

Figure 4: Profile of caverns

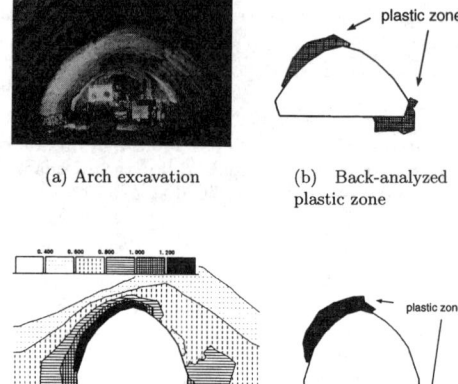

(a) Arch excavation

(b) Back-analyzed plastic zone

(c) Maximum shear distribution predicted by back analysis (final stage)

(d) Plastic zone predicted by back analysis (final stage)

Figure 5: Results of back analysis

stability at the final excavation stage together with the plastic zone.

4 BACK-ANALYSIS METHOD BASED ON DISCONTINUUM MECHANICS

In this section, the back-analysis modeling is described for the behavior illustrated as cases JR/LS in Figure 1. This modeling is formulated based on a discontinuum model (Shimizu et al. 1996).

The method requires only measured displacements and geometric conditions of a discontinuous rock mass, i.e., the location and the direction of the discontinuities, as input data. The mechanical constants and in situ stress conditions of the rock mass are not required in order to apply the method.

4.1 Outline of the formulation

assumption

The assumptions of the method are as follows:

1. the rock mass is composed of a number of blocks surrounded by discontinuities;

2. continuous displacement and small strain occur within each block, and displacements of rock block are formed by rigid-body parallel translation and rotation, and strain;

3. rock mass deformation shows steady-state time dependent behavior.

the state equation for displacements

The displacement at point $p(x,y)$ in the rock block can be expressed by the following equation (Figure 6):

$$u^i(p) = a^i(p)\, d^i \tag{12}$$

where

$$\left.\begin{array}{rl} u^i(p) &= \left\{\begin{array}{c} u_x^i(x,y) \\ u_y^i(x,y) \end{array}\right\} \\ d^i &= \{u_{Rx}^i\ u_{Ry}^i\ \omega_{Rz}^i\ \varepsilon_x^i\ \varepsilon_y^i\ \gamma_{xy}^i\}^T \\ a^i(p) &= \left(\begin{array}{cccc} 1 & 0 & -(y-y_0) \\ 0 & 1 & x-x_0 \\ x-x_0 & 0 & y-y_0 \\ 0 & y-y_0 & x-x_0 \end{array}\right) \end{array}\right\}. \tag{13}$$

Index ^{i} is denoted by the ith block ($i = 1, 2, \ldots N$; N: the total number of blocks). u_x^i and u_y^i are displacement components in the directions of x and y, respectively. u_{Rx}^i and u_{Ry}^i are rigid-body parallel translations, and ω_{Rz}^i is a rigid-body rotation at an arbitrary point (x_0, y_0) about axis z. ε_x^i, ε_y^i and γ_{xy}^i are strain values in the ith block. d^i is called the deformation parameter.

Assuming that the rock mass behaves in a steady state creep deformation, parameter d^i is governed by the following equation:

$$c_{j0}^i d_j^i(t) + c_{j1}^i \dot{d}^i{}_j(t) + \ldots + c_{jn}^i \overset{(n)}{d}{}^i{}_j(t) = g_j^i(t). \tag{14}$$

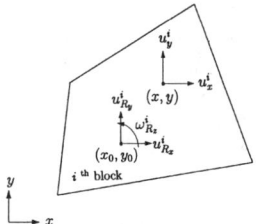

Figure 6: Deformation parameters of a rock block

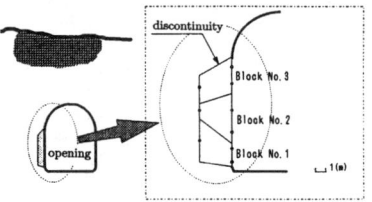

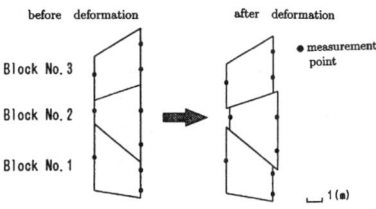

Figure 7: Rock blocks along the wall of underground opening

Coefficient c^i_{jk} is represented by the time dependent feature of d^i_j. $\dot{d}$ is denoted as the derivative of d with respect to time t. $g^i_j(t)$ represents a function related to the excavation procedure.

For simplicity, the formulation is described below for the case in which $n = 1$. In discrete time Δt (constant), $\dot{d}^i_j(t) \simeq (d^i_j(t+\Delta t) - d^i_j(t))/\Delta t$, and Equation (14) becomes as follows:

$$x^i_{j,\tau+1} = \theta^i_j \, x^i_{j,\tau} \qquad (15)$$

where $x^i_{j,\tau} = d^i_{j,\tau} - d^i_{j,\tau-1}$, $d^i_{j,\tau} = d^i_j(t)$, $t = \tau \cdot \Delta t$, and $\theta^i_j = c^i_{j1}/(\Delta t \, c^i_{j0} + c^i_{j1})$.

θ^i_j is called a creep parameter and it is composed of c^i_{j0} and c^i_{j1}. τ is the number of steps at time t. In deriving Equation (15), $g^i_j(t)$ is assumed to be constant. This means that no impact (i.e., excavation or loading) acts on the rock mass during rock deformation. For example, this study deals with rock deformation from the last excavation to the next excavation.

Equation (15) leads to the following *state equation* for all rock blocks.

$$X_{\tau+1} = B(\theta) X_\tau \qquad (16)$$

where $X_\tau = D_\tau - D_{\tau-1}$ and $D_\tau = \{d^1_\tau{}^T \; d^2_\tau{}^T \; \cdots \; d^N_\tau{}^T\}^T$. Matrix $B(\theta)$ is composed of creep parameter θ^i_j.

the observation equation

Equation (12) leads to the following equation:

$$\overline{F}_\tau = \overline{K}_\tau \, D_\tau. \qquad (17)$$

Equation (17) forms in a similar manner to the backward DDA method (Shi 1993).

Subtracting equation $\overline{F}_{\tau-1} = \overline{K}_{\tau-1} \, D_{\tau-1}$ at time step $\tau - 1$ from Equation (17) yields the following equation:

$$\overline{Y}_\tau = C_\tau \, X_\tau + V_\tau \qquad (18)$$

where

$$\left. \begin{array}{rcl} \overline{Y}_\tau & = & \overline{F}_\tau - \overline{F}_{\tau-1} \\ C_\tau & = & \overline{K}_{\tau-1} \; (\simeq \overline{K}_\tau) \end{array} \right\} \qquad (19)$$

V_τ is represented by an observation error.

Equation (18) represents the relationship between incremental deformation parameter X_τ and observation vector $\overline{Y}_\tau$ at time step τ. X_τ is called the state vector. $\overline{Y}_\tau$ comes from measured displacements. Equation (18) is adopted as *the observation equation* for the measured displacements of the rock mass.

the system equation

State equation (16) and observation equation (18) are summarized as follows:

$$\left\{ \begin{array}{rcl} X_{\tau+1} & = & B(\theta) \, X_\tau \\ \overline{Y}_\tau & = & C_\tau \, X_\tau + V_\tau \end{array} \right. \qquad (20)$$

Equation (20) is a system equation used to estimate both X_τ and θ from the observation vector $\overline{Y}_\tau$ at time step τ.

It should be noted that parameter θ has to be estimated as well as state vector X_τ. This system forms *the adaptive filtering*. In order to solve the equation, the extended Kalman filtering can be applied.

4.2 Numerical simulation

To demonstrate the method, a numerical example for predicting the deformational behavior of rock blocks around the wall of an underground opening is shown here (Figure 7).

In this case, Block No. 1 is assumed not to move, whereas Block No. 2 and Block No. 3 exhibit rigid-body parallel translation. The number of unknown parameters is 27 in this example, i.e., six deformation parameters and three creep parameters for every three blocks.

Estimates for the deformation parameter are obtained with good accuracy by this back-analysis method. The predicted final values converge within a few percentage points of accuracy even for scattered measurement data.

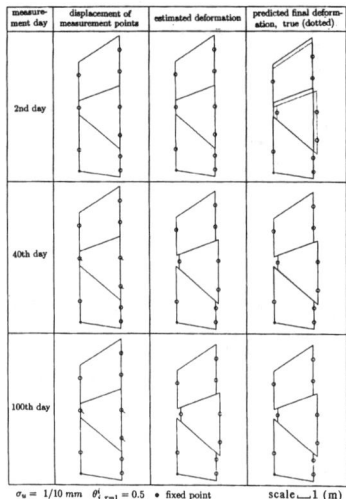

Figure 8: Estimation of movements of blocks and prediction of the final state

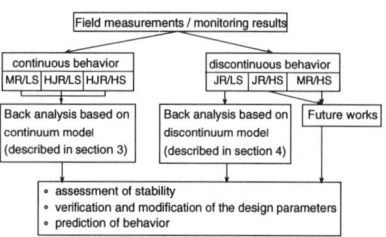

Figure 9: Procedure for integrated back analysis

Figure 8 shows estimated and predicted movements of rock blocks by using back-analyzed deformation parameters. The simulation was conducted by using displacements which were measured every two days. It is seen that this method can predict the final deformation with fairly good accuracy on the fortieth day. The detailed results are discussed in the reference together with other examples (Shimizu et al. 1996).

5 PROCEDURE FOR INTEGRATED BACK ANALYSIS

The procedure for the back analysis which integrated the methods for continuous and discontinuous behavior of a rock mass is summarized in Figure 9. The outlines of both continuous and discontinuous back-analysis approaches are described in Sections 3 and 4. The selection of the appropriate approach (continuous or discontinuous) is based on experience and/or engineering judgment. We should be familiar,therefore, with both the advantages and the limitations of the numerical models. However, once the back-analysis model and approach have been selected, the results of the back analysis are then obtained through an optimum computational method without a trial and error process in the procedure demonstrated in this paper.

6 CONCLUSION

The goal of the back analysis of field measurements and monitoring is to help engineers improve the design and construction methods under the observational method so that they will match with real rock mass behavior. The same contents as this are seen in the observational modeling approach (Kaiser 1995).

The method described in this paper may not be enough to represent and understand the behavior of rock masses as changing from continuum to discontinuum. Further research on the integrated back-analysis method would assist the observational method for underground openings.

ACKNOWLEDGEMENTS

The authors wish to thank Ms. H. Griswold for proofreading this paper.

REFERENCES

Fukuhara, A. and M. Hasegawa, 1995. An observational method for excavation of large underground caverns. In *Observational Method of Construction of Large Underground Caverns in Difficult Ground Conditions, The 8th Int. Cong. on Rock Mechanics*, pp. 87–97.

Hoek, E., P. K. Kaiser and W. F. Bawden, 1995. *Support of underground excavations in hard rock*. Balkema.

Kaiser, P. K., 1995. Observation modelling approach for design of underground excavations. In *Observational Method of Construction of Large Underground Caverns in Difficult Ground Conditions, The 8th Int. Cong. on Rock Mechanics*, pp. 1–7.

Sakurai, S., 1981. Direct strain evaluation technique in construction of underground openings. In *Proc. 22nd U.S. Sympo. Rock Mech.*, pp. 278–282.

Sakurai, S., 1993. Back analysis in rock engineering. *Comprehensive Rock Engineering*, **4**:543–569. Pergamon Press.

Sakurai, S., N. Shimizu and K. Matsumuro, 1985. Evaluation of plastic zone around underground openings by means of displacement measurements. In *Proc. 5th Int. Conf. Numerical Methods in Geomechanics*, vol. 1, pp. 111–118, Nagoya.

Shi, G., 1993. *Block system modeling by discontinuous deformation analysis*. Computational Mechanics Publications.

Shimizu, N., H. Kakihara and K. Nakagawa, 1996. A back analysis method for predicting deformational behavior of discontinuous rock mass. In *The 2nd North American Rock Mechanics Symposium: NARMS'96*, pp. 2001–2008.

Shimizu, N. and S. Sakurai, 1989. Back analysis formulated by the boundary element - finite element coupled method for monitoring multi-parallel underground openings. In *Proc. Int. Conf. on Engineering Software*, New Delhi.

Terzaghi, K. and R. B. Peck, 1967. *Soil Mechanics in Engineering*, pp. 627–632. John Wiley, 2nd edn.

Back analysis of linear and nonlinear deformational behaviors for multiple cross sections in discontinuous rock masses of a large underground power house cavern

S. Akutagawa
Research Center for Urban Safety and Security, Kobe University, Japan

S. Sakurai
Department of Architecture and Civil Engineering, Kobe University, Japan

ABSTRACT : This paper gives a brief summary of a back analysis procedure applied for characterization of deformational mechanism of discontinuous rock mass of a large underground powerhosue cavern. The method employed is capable of isolating the deformational characteristics due to nonelastic movements along rock joints. The back analysis is conducted for three cross sections for which displacements were measured and their behaviors are compared.

1 INTRODUCTION

Often encountered in a monitoring procedure of an excavation process of a large scale underground power house cavern, is unpredictable and undesirable movements of rock masses, which are due largely to nonlinear deformation along rock joints. It is, however, difficult to estimate, prior to construction, deformational characteristics of the jointed rock mass, even if the best is made as to prior on-site investigation of geological conditions, and so on.

To facilitate a quick and reasonable method for identification of the state of rock mass deformation with the presence of multiple rock joint sets, the authors have developed a back analysis procedure taking into account non-elastic strains which occur during excavation (Sakurai et al, 1994, Hojo et al, 1997).

In this paper, the proposed back analysis method is applied for characterization of deformational mechanism of discontinuous rock mass of a large underground powerhosue cavern. The back analysis is conducted for three cross sections for which displacements were measured and their behaviors are compared with each other and also with what have already been identified for this site by other studies (Uchita et al, 1996).

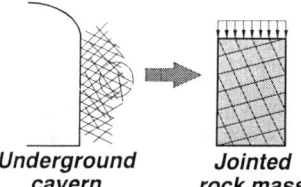

Figure 1. Representative volume of a jointed rock mass

2 METHOD OF BACK ANALYSIS

2.1 *Compliance of a jointed rock mass*

Figure 1 represents a general view of a rock mass with multiple, in this case *two*, joint sets by which deformational characteristics of the rock mass are controlled. Provided that density and distribution of rock joints be sufficiently large and uniform, compliance of the rock mass, C, may be given as

$$C = C_e^r + C_p^r + C_e^{j1} + C_p^{j1} + C_e^{j2} + C_p^{j2} \qquad (1)$$

where superscripts r and j represents association with intact rock and joints, while numbers 1 and 2 indicate joint set numbers. Subscripts e and p represent elastic and plastic (or, nonlinear in general) components.

Of these compliances on the right hand side, those associated with elastic behavior of either rock or rock joints may be obtainable, though some scattering may be unavoidable, prior to excavation by lab and field tests. The nonlinear compliances of rock joints, originating from joint slip and dilatancy, however, are the most difficult to estimate *a priori*. These compliances are, in most cases, hardest to be defined with confidence, yet the most critical indices to control nonlinear nature of rock mass deformation and to assess the stability of the cavern. This is the reason that the back analysis used in this approach defines the nonlinear joint slip as the main unknown parameters to be determined from displacements measured during construction. The detailed description of the back analysis procedure in general may be found in Sakurai et al (1994) and a brief mathematical formulation of the method used in this paper is given in Hojo et al (1997).

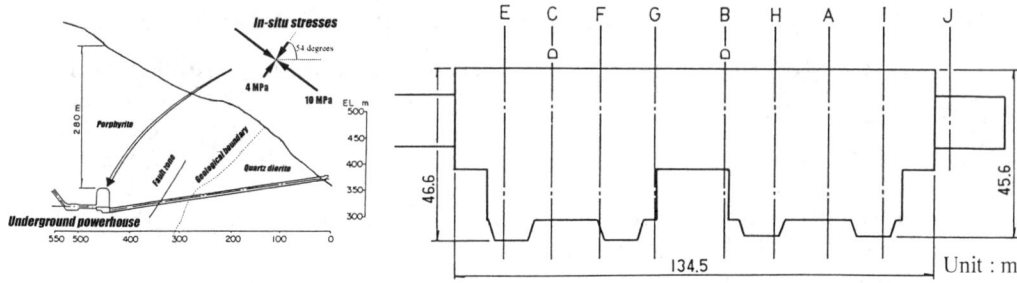

Figure 2. Site location of Okawachi underground power house cavern

Figure 3. Side view of the cavern[Katayama et al. 1992]

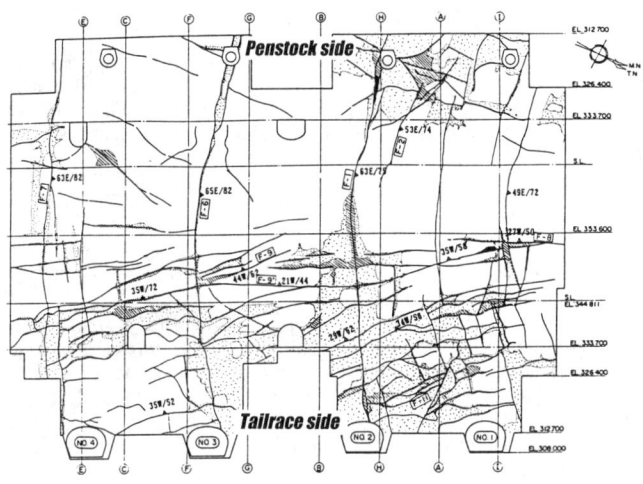

Figure 4. Results of geological survey [Katayama et al. 1992]

3 APPLICATION

3.1 Site and rock mass characterization

An underground powerhouse cavern constructed in Okawachi-cho, Hyogo Prefecture, Japan, has an excavation volume of about 120,000m³ and dimensions of 24m (width), 46.6m (height) and 134.5m (length). The site, as illustrated in Figure 2, has an overburden of about 280m and situated in porphyrite of the Mesozoic era.

The side view of the cavern is given in Figure 3. Of the monitoring sections from A to J, three sections were chosen as main monitoring sections in which extensive measurements were taken for displacements, rock bolt forces, axial forces in PS anchors, etc. In this paper, the displacement measurements taken in Sections A, B and C are considered to identify deformational mechanism of those three cross sections. The geological condition, see Figure 4, on the tail race side (the right hand side of the cavern in Figure 2) was generally worse compared to that on the pen stock side (the left side).

From field observation, three major joint sets were identified for the site, of which the two have their strikes almost parallel to the axis of the powerhouse. In this study, only these two joint sets were considered in the state of plane strain. The mechanical and geometrical properties obtained for intact rock specimen and rock joints are given in Table 1. Dilatancy angle defined for the joint planes of this rock mass was set to the same value as given by Yoshida et al (1996).

Table 1. Mechanical and geometrical properties of intact rock specimen and rock joints

Category	Property description	Value
Intact rock	Elastic modulus : E	75GPa
	Poisson's ratio : ν	0.25
Joint set 1	Dip angle	50°
	Joint spacing	20cm
Joint set 2	Dip angle	-40°
	Joint spacing	33cm
Both joint sets	Normal stiffness : K_n	30 GPa
	Shear stiffness : K_s	7 GPa
	Dilatancy angle : Θ	10°

3.2 Excavation sequence and measurements

As shown in Figure 5, the excavation of the cavern was conducted in several steps. The arch portion was excavated first in 4 steps, which was followed by excavation of main body in 9 benches.

Displacements measured by extensometers are shown in Figure 6 for sections A and C.

3.3 Results of back analysis

Maximum shear strain distribution identified by back analysis are given for the final stages in sections A and C, and for just after bench 5 for section B, in Figure 7. The high concentration of maximum shear strain is observed from the top-right corner down to the cavern bottom on the right hand side. Some of these behaviors is due to natural stress concentration given from the state of in-situ stress and orientation of joint sets. However, nonlinear deformation due to slip displacements along joint planes are believed to be present to some extent as well.

In general, maximum shear strains are greater on the right hand side than in the left, which was expected from the results of geological investigation. For section B, the state after bench 5 was calculated, because the displacement increase after that was very small. The concentration of maximum shear strain is seen particularly at the top right corners for all sections, where stress level becomes considerably high because of geometrical conditions and direction of in-situ stresses. Both for sections A

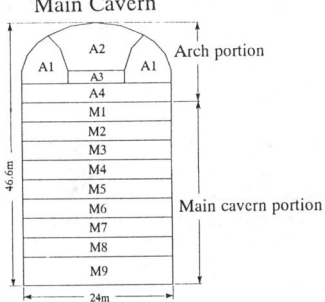

Figure 5. Excavation sequence

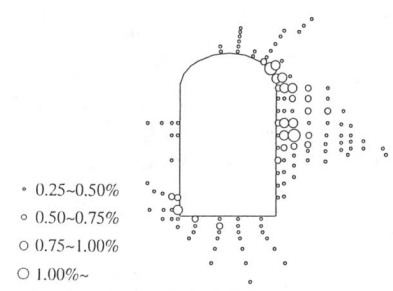

(a) Section A after bench 9

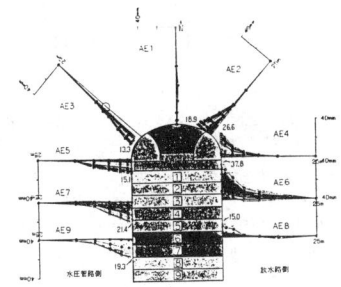

(a) Section A

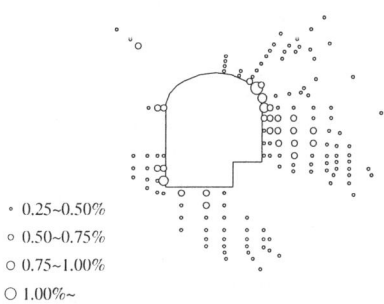

(b) Section B after bench 5

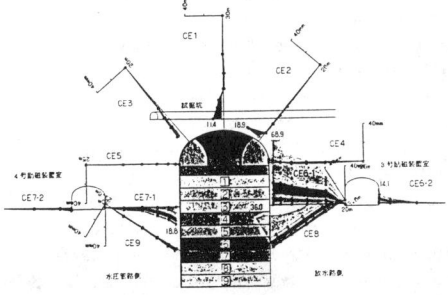

(b) Section C
Figure 6. Measured displacements [Katayama et al. 1992]

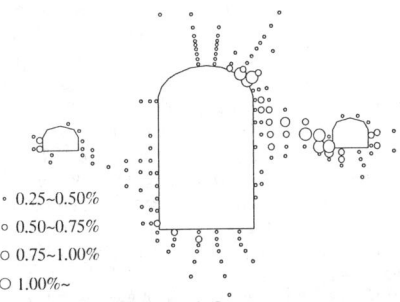

(c) Section C after bench 9
Figure 7. Maximum shear strain distribution

near wall surface and decrease at depth. However, for section C, another peak of strain concentration is seen near a sub-cavern, which also suffered high stress concentration and large joint deformation (Uchita et al, 1996). Though, there are some differences in strain distribution due to geometrical differences in each cross section, the general behaviors of deformation for this cavern is regarded nearly as plan-strain.

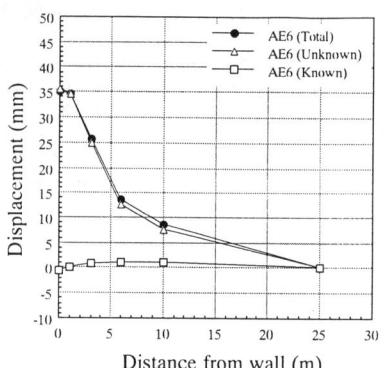

(a) Extensometer AE6 for section A

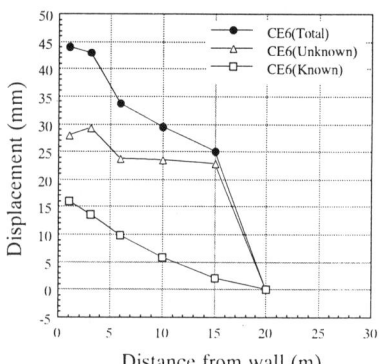

(b) Extensometer CE6-1 for section C
Figure 8. Displacement partitioning

Finally, the displacements measured by extensometer AE6 and CE6-1 are shown in Figure 8, in which the displacement caused by *Known* and *Unknown* factors are also indicated. *Known* displacements indicate those that can be computed before excavation by using known parameters. *Unknown* displacements indicate those which cannot be computed before excavation, but can be defined by back analysis. *Known* and *Unknown* displacements therefore may be simply added to yield the total displacements. For the displacement measured by AE6, almost all of it is attributed to the effect of nonlinear joint movement. For the displacement measured by CE6-1, greater contribution of joint slip is seen near the sub-cavern and the wall surface, which was also confirmed by the distribution of maximum shear distributions. Known displacement in this case are present which are naturally distributed along the extensometer though its contribution to total displacement is not comparable to what was achieved by joint slip movement.

4 CONCLUSION

A back analysis procedure developed for a nonlinear problem, involving excavation within a jointed rock mass, has been applied to monitor displacement and strain fields for multiple sections of an underground powerhouse cavern. It has been demonstrated that nonlinear deformational behaviors, due to existence of rock joints, could be quickly identified from measured displacements and necessary support measures promptly taken to assure safety of the cavern.

ACKNOWLEDGMENT

The authors wish to express their gratitude to the Kansai Electric Power Co., Inc. of Japan for providing the data. Assistance from Mr. M. Kakihara, a graduate student, and M. Takeyama, a senior student of Kobe University, in computer simulation, is also greatly appreciated.

REFERENCES

Katayama, T., A.Yada and Y.Hirakawa (1992) Observational method for excavation of underground cavern for Okawachi Pumped Storage Power Project. Denryoku Doboku, No.237, 97-107.

Sakurai, S., S. Akutagawa and O. Tokudome (1994). Back analysis of non-elastic strains based on minimum norm solution, *Journal Geotechnical Engineering, JSCE*, No.517/III-31, pp.197-202.

Yada, A., A. Hojo & S. Sakurai (1994). Excavation control of underground powerhouse cavern. *Proc. the 1994 ISRM Int. Symp. Integral Approach to Applied Rock Mechanics*, Santiago. 523-534.

Yoshida, H., H. Horii and Y. Uchida. (1996). Analysis of the excavation of underground power house at the Okawachi power station by Micromechanics-Based Continuum model and comparison with measured data, *Journal of Geotechnical Engineering, JSCE*, No.547/III-36, pp.39-56.

Hojo, A., M.Nakamura, S.Sakurai and S.Akutagawa. (1997) Back analysis of non-elastic strains within a discontinuous rock mass around a large underground power house cavern, *Proc. of The 36th U.S. Rock Mechanics Symposium, NYROCKS '97*, (in press)

Uchita. Y., T.Yoshida, Y.Hirakawa and T.Ishida. (1996) Rock behavior measured by means of borehole strain gauges during a large underground cavern excavation, *Journal of Geotechnical Engineering, JSCE*, No.554/III-37, 19-30.

Theoretical study on deformational behavior and reinforcing effect by bolting for tunnels in soft rock

Y.J.Jiang & T.Esaki
Institute of Environmental Systems, Kyushu University, Fukuoka, Japan

Y.Yokota
CTI Engineering Co., Ltd, Fukuoka, Japan

ABSTRACT : A theoretical method is proposed to predict the deformational behavior of tunnels in soft rocks and to give the appropriate design of tunnel support. Large deformational phenomenon of tunnels is considered to be due to expansion of the surrounding medium, and the representative charts for predicting the occurrence of the plastic flow zone are shown. Then, according the post-failure behavior of soft rocks and bolting pattern, the ground-bolts interaction modes are presented to describe the mechanical stabilization of the rocks around tunnels by bolting. The effect of rock bolts on stability of tunnels is investigated carefully by the detailed parametric analyses. It is clarified that there is a proper combination between bolt length and installing density according to the ground condition where rock bolts could exhibit the maximum active effect on controlling the deformation of tunnels and the expansion of plastic zones in the surrounding medium.

1 INTRODUCTION

Deformational behaviour and stability of tunnels during excavation depend greatly on the mechanical characteristics of the surrounding medium. Almost one third of land in Japan is covered with tertiary sedimentary rocks of volcanic origin and many highway and railway tunnels have been already or planned to be constructed in such rocks. As a result, the number of tunnels suffered large deformations has been increasing and it is necessary to clarify the mecha-nism of expansive phenomenon of rocks and to design the suitable supports in such rocks. Although there are some proposals from experiences or theoretical approaches in literature, it is difficult to say they are appropriate and complete.

This paper is to present an alternative theoretical method for the prediction of the deformational behavior of tunnels in soft rocks and the suitable design for supports. The large deformation phenomenon of tunnels(generally termed as expansion) is considered to be due to the occurrence of plastic-flow zone in the surrounding medium and the representative charts are quantatively shown for the prediction. According to the post-failure behavior of soft rocks and bolting pattern, the ground-bolts interaction modes are proposed to describe the mechanical stabilization of the rocks by bolting. The effect of rock bolts on controlling the deformation of tunnels and the expansion of plastic zones in the surrounding medium are investigated by the detailed parametric analyses.

2 EMPIRICAL FEATURE OF SOFT ROCKS AND ITS CONSTITUTION

It has been made clear from laboratory tests that soft rocks represent the strain-softening feature and dilatancy when stress exceeds the peak strength under low confining pressure. The transition condition from brittle failure (i.e. strain - softening) to ductile failure (perfect - plasticity) for sedimentary rocks, as shown in Figure 1, is approximately defined by means of the relation between the confining pressure(σ_3) and principal stress difference ($\sigma_1 - \sigma_3$) as equation(1):

$$\sigma_1 - \sigma_3 = 3.4\sigma_3 \tag{1}$$

Most rocks show strain-softening behaviour clearly if the confining pressure be below the brittle-to-ductile transition pressure. In this paper, the constitution of strain-softening behaviour is approximated as an assembly of three line segments, as shown in Figure 2. Parameters h and f are plastic Poisson's ratios, at a stress decending region and residual region in stress-strain curve respectively, which can be represented by dilatancy angle ϕ^* based on Mohr-Coulomb failure criterion and non-associated flow rule. The residual strength σ_{co}^* and brittle modulus α are the measure of the degree of strength loss occurring in the post-failure process. The brittle modulus α is defined by

$$\alpha = 1 + \int d\varepsilon_1^f / \varepsilon_1^e \quad ; \quad \varepsilon_1^e = (\sigma_c - 2\mu\sigma_3)/E \tag{2}$$

In three stages before peak strength and in the post-

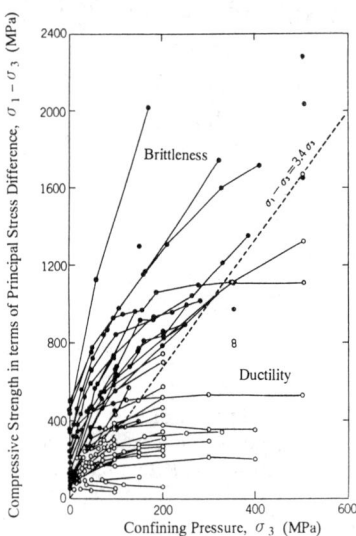

Figure 1 Transition of brittle failure to ductile failure according the confining pressure for sedimentary rocks (after Jiang, 1993).

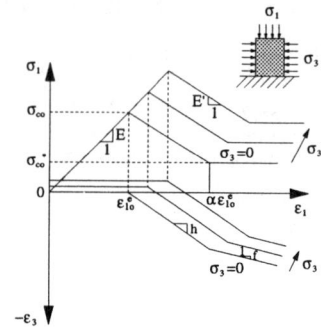

Figure 2 Strain-softening behaviour of soft rocks.

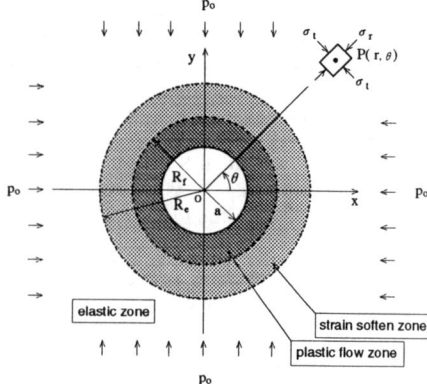

Figure 3 States about a tunnel in soft rocks and notations.

failure process, the constitutive relations can be given as follows:

$$\sigma_1 = \begin{bmatrix} E\varepsilon_1 & (\varepsilon_1 \leq \varepsilon_m) \\ (1+\bar{E})(K_p\sigma_3 + \sigma_{co}) - E'\varepsilon_1 & (\varepsilon_m \leq \varepsilon_1 \leq \varepsilon_r) \\ \sigma_{co}^* + K_p\sigma_3 & (\varepsilon_r \leq \varepsilon_1) \end{bmatrix} \quad (3)$$

where

$$\varepsilon_m = (\sigma_{co} + K_p\sigma_3)/E \; ; \quad \varepsilon_r = \alpha\varepsilon_{10}^\varepsilon + K_p\sigma_3/E$$

$$\bar{\sigma}_{co} = \sigma_{co}^*/\sigma_{co} \; ; \; \bar{E} = E'/E \; ; \; K_p = (1+\sin\phi)/(1-\sin\phi)$$

From many previous experiments, the major mechanical parameters of rocks can be plotted as a function of the uniaxial compressive strength of rocks. The relations between friction angle, ϕ, brittle modulus, α, plastic Poisson's ratios, h and f, residual strength, σ_{co}^* and uniaxial compressive strength, σ_c can be fitted to the following expressions (Jiang, 1993):

$$\phi = 38.28\sigma_c^{-0.004} \; ; \quad \alpha = 1.33\sigma_c^{0.153}$$
$$h = 1.88\sigma_c^{0.136} \; ; \quad f = 1.41\sigma_c^{0.035} \quad (4)$$
$$\sigma_c^* = 0.65\sigma_c^{0.8}$$

3 JUDGEMENT OF OCCURRENCE OF PLASTIC-FLOW ZONE

It has been recently recognized that the expansion was mainly due to the squeezing of the surrounding media associated with the plastic flow. Especially when tunnels be excavated in soft rocks with the strain-softening characters under low competency factor of ground S_{rp} which is defined by a ratio of the uniaxial compressive strength to the initial ground pressure, the deformation and stability of tunnels are largely controlled by the area of the plastic-flow zone because it occurs surround tunnel wall.

For estimating the occurrence of the plastic-flow zone, Jiang(1993) introduced an analytical solution, considering circular tunnels in soft rocks in a hydrostatic state of initial ground pressure(P_o), as shown in Figure 3. Strain soften zone and/or plastic flow zone would be caused in the surrounding rocks if the redistributed stresses exceed the strength of the material owing to excavation. The extent of two plastic zones is defined by the plastic radii, R_e and R_f, respectively. Yield initiation of rocks is assumed to occur following a linear Mohr-Coulomb failure criterion. The competency factor of ground S_{rp} is assumed as a synthetic parameter of evaluating the ground condition. Then the limit value of S_{rp} for estimating the occurrence of the plastic-flow zone, $[S_{rp}]$, is given by following equation:

$$[Srp] = \left[\frac{\sigma_{co}}{P_o}\right] = 2\left\{1 - \frac{(1-\chi)(1+K_p)}{h+K_p}\left(1 - \frac{2(1-\zeta^{1-K_p})}{(1-\alpha)(K_p-1)}\right)\right\}^{-1} \quad (5)$$

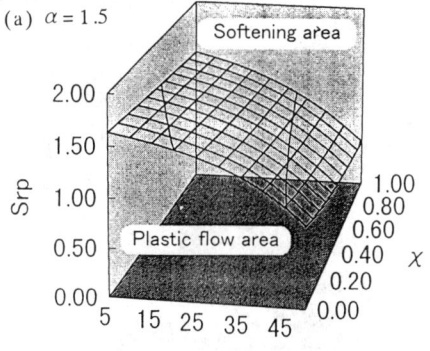
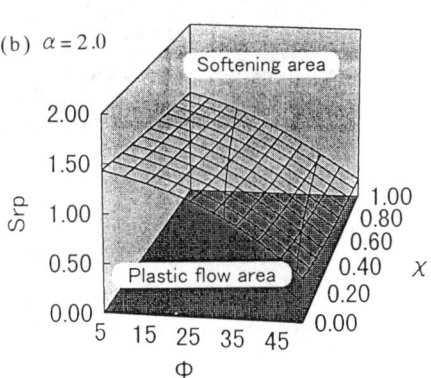
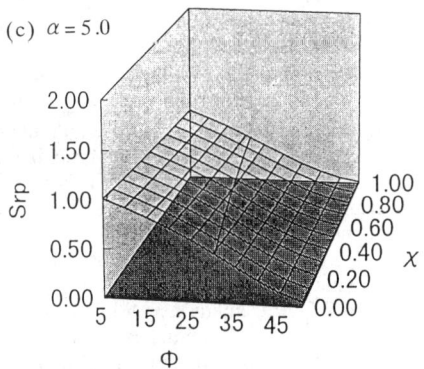

Figure 4 Representative charts for adjudging occurrence of the plastic-flow zone around circular tunnels.

Here parameters χ and ζ mean ratios of the residual strength to the uniaxial compressive strength and R_f to R_e, respectively, and defined by

$$\chi = \sigma_{co}^* / \sigma_{co} \qquad \zeta = \left[\frac{2}{\alpha(1+h)+(1-h)} \right]^{\frac{1}{1+h}} \qquad (6)$$

If the competency factor of ground S_{rp} satisfies

$$S_{rp} < [S_{rp}] < 2 \qquad (7)$$

the plastic-flow zone will occur around tunnels.

If the uniaxial compressive strength of the material is given, another characteristic parameters can be introduced from the empirical equations(4).

From the equation (5), the representative charts for predicting the occurrence of the plastic flow zone are demonstrated in Figure 4.

4 ANALYTICAL MODELING OF ROCKBOLTS INTERACTED WITH THE GROUND

A number of fundamental analysis with respect to the supporting and stabilizing effect of rock reinforcement by bolting have been presented since the early 1940s (Reed et al. 1993), in which the function of rockbolts is often considered to be an equivalent uniform internal pressure acting on the tunnel wall (Rabcewicz & Golser 1973, Hoek & Brown 1980, Labiouse 1991, etc.). Recently, there are some new analytical proposals to illustrate the rockbolt-ground interaction in association with the deformation of the ground(Indraratna & Kaiser 1990a, 1990b). However, whether the rockbolt-ground interaction will vary with the mechanical properties of rocks and rockbolt-ground interface and how to carry out the optimum design of bolt pattern according to the ground condition have not been quantitatively illustrated yet in the above mentioned proposals.

Grouted rockbolts deform with ground if no sliding occurred between the rockbolts and ground. Because of the difference of deformability of bolts and ground, the rockbolts cause the restrained force to support the ambient rock by the bonding shear stress occurred around the boundary interface between the rockbolts and ground. As illustrated by Tao et al.(1983) with considering the equilibrium of the grouted rockbolt relative to the surrounding rock, there is a neutral point along the rockbolts where the shear stress equals zero and the axial stress attains to a maximum. The location of the neutral point is given by

$$\rho = \frac{L}{\ln(1+L/a)} = \frac{L_1 - a}{\ln L_1 - \ln a} \qquad (8)$$

where ρ is the radial distance from the center of tunnel to the neutral point, a is tunnel radius, L is bolt length, $L_1 = L + a$, respectively, shown as in Figure 5.

We assumed that the bolt pattern be axisymmetric, consisting of identical bolts with equal spacing along the tunnel axis and around the circumference. A minute element is used for equilibrium analysis (Figure 5). The tangential stress and the shear stress parallel to the bolt surface are assumed to be distributed uniformly.

Note that throughout this analysis, compressive stresses are assumed to be positive. Therefore, the equilibrium equations of the bolted-element are obtained as

$$\frac{d\sigma_r}{dr} + \frac{\sigma_r - (1-\beta)\sigma_t}{r} = 0 \qquad (9)$$

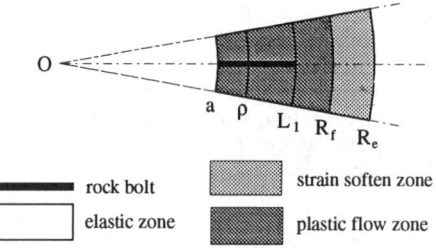

Figure 5 Relationship among plastic radii, bolt length and neutral point ρ.

$$\beta = \pi d_b \lambda a / L_z L_T \quad (10)$$

where L_T and L_Z are the circumferential bolting interval and longitudinal interval. d_b is a bolt diameter, λ is a friction coefficient between rockbolts and ground, β is a dimensionless coefficient called the bolt density parameter, whose value varies between 0.05 and 0.20 for most cases (Indraratna & Kaiser, 1990). On the reinforcing length of the bolt inside the neutral point to the tunnel wall, the shear stress is opposite to that acting on the anchor length outside the neutral point, so the bolt density parameter β in the equation (9) must take a negative sign.

At the condition of hydrostatic initial stress, the tangential stress σ_t and the radial stress σ_r become the maximum principal stress and the minimum principal stress, respectively. So the Mohr-Coulomb failure criterion can be written in the two stresses as

$$\sigma_t = \sigma_c + K_p \sigma_r ; \quad K_p = (1 + \sin\phi)/(1 - \sin\phi) \quad (11)$$

In the plastic flow stage, equation (11) is represented in the residual strength σ_c^*, and in the strain soften stage, the uniaxial compressive strength is reducing linearly with the increase of the axial strain(maximum principal strain ε_1), and the equation (11) can be represented by

$$\sigma_t = \sigma_c' + K_p \sigma_r \quad (12)$$

where σ_c' is decided by the negative incline E' and the maximum principal strain ε_1 (Figure 2).

Combination of the failure criterion equation (11) and the equation (9) leads to a modified equilibrium equation:

$$\frac{d\sigma_r}{dr} + \frac{[1-(1-\beta)K_p]\sigma_r}{r} = \frac{(1-\beta)\sigma_c}{r} \quad (13)$$

As illustrated in Table 1, nine cases of the rockbolt-ground interaction modes are categorized according the extent of two plastic zones relative to the location of the neutral point, ρ, and the boundary of the bolted zone, L_1. These modes can also be divided into following three main groups:

Group 1 (Case 1 to Case 3) : two plastic zones occur because the strength of ground is lower and the length

Table 1 Nine analytical cases of the rockbolt-ground interaction modes.

Cases	Categorization of plastic zones and bolt length
1	$a < \rho < L_1 < R_f < R_e$
2	$a < \rho < R_f < L_1 < R_e$
3	$a < R_f < \rho < L_1 < R_e$
4	$a < \rho < R_f < R_e < L_1$
5	$a < R_f < \rho < R_e < L_1$
6	$a < R_f < R_e < \rho < L_1$
7	$a < \rho < L_1 < R_e$
8	$a < \rho < R_e < L_1$
9	$a < R_e < \rho < L_1$

of bolts remains shortly inside the plastic zones.

Group 2 (Case 4 to Case 6) : two plastic zones are developed as the same with the group 1 and rockbolts are installed to the elastic zone.

Group 3 (Case 7 to Case 9) : there is no occurring of the plastic flow zone in the surrounding rocks.

The solutions of stress and strain in the plastic zones are obtained by combining the equilibrium equation (13) and the boundary conditions at the elasto-plastic boundary and at the tunnel wall. A detailed derivation and the exact analytical solutions for each case have been presented by authors(Jiang et al. 1997).

5 EFFECT OF GROUTED ROCKBOLTS ON TUNNEL STABILITY

The effect of the rockbolts is influenced by the shear strength of grouting and/or bonding strength between bolts and ground, bolt density and bolt length. In order to illustrate the influence of the grouted rockbolts on the extent of the plastic zones and the deformational behaviour of tunnels using the proposed analytical model for the ground-bolts interaction, a detailed parametric study and its consideration are carried out. In the case study, it assumes that the initial ground pressure P_o is 2.5MPa and tunnel radius is 5.0m. The friction coefficient λ is assumed as 0.4, diameter of bolts is 32mm. Mechanical parameters of rocks, such as (ϕ, α, h, f and σ_{co}^*), are introduced from the empirical equations (4) according the uniaxial compressive strength for a given competency factor of ground, S_{TP}.

(1) Influence of grouted bolts on plastic radii and tunnel wall strain

Figure 6 illustrates the variation of the analytically predicted plastic radii and the wall strain of the model tunnel with bolt length and a range of the competency factor of ground up to 1.0. It is evident that no obvious influence on the tunnel behaviour arises if the relative bolting distance normalized by tunnel radius, L_1/a, is

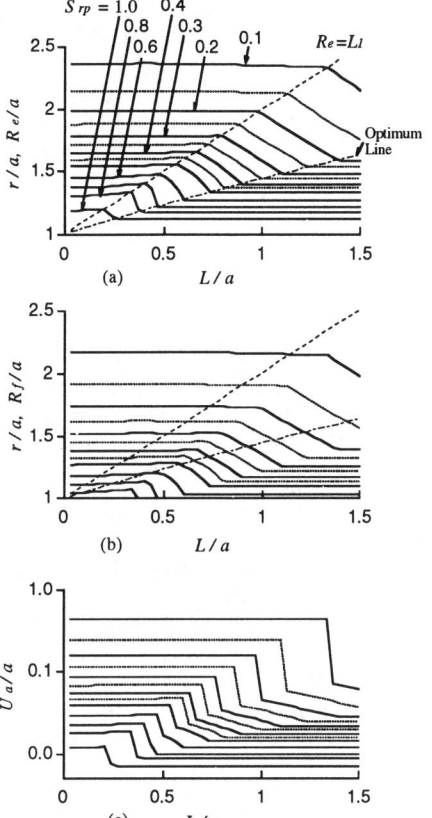

Figure 6 Influence of the relative bolt length, L/a, on plastic radii and tunnel wall strain, the bolt density parameter $\beta = 0.4$.

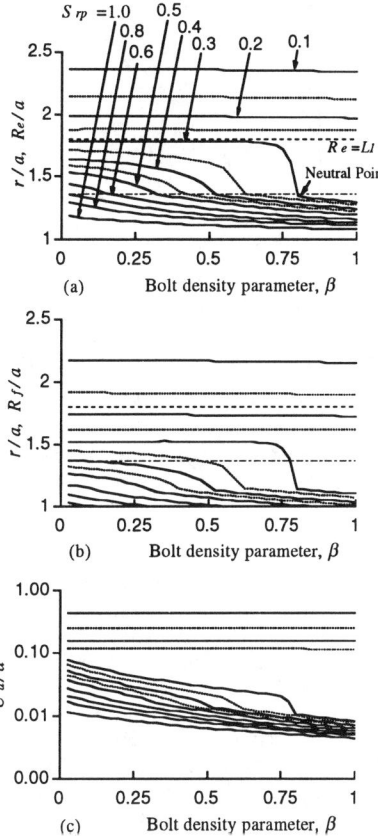

Figure 7 Influence of the bolt density parameter, β, on plastic radii and tunnel wall strain, the relative bolt length $L/a = 0.8$.

less than R_e/a which is corresponding to the case of the unsupported tunnel. The pronounced reduction of the plastic radii and the wall strain arises with the increase of bolt length can be expected when bolts are installed to the elastic zone and the competency factor of ground be in the range of 0.4 to 0.8. However, more effect on controlling the deformational behaviour of tunnels is not expected even if extending bolt length over the optimum line shown in the figure.

Influence of blot length on restricting the tunnel wall strain shows the same trend. In the case of $S_{rp} > 0.8$, the tunnel wall strain can be controlled within 1% when the relative bolt length L/a ranges 0.2 - 0.4, but the more controlled effect can not be expected even if the length of bolt is longer than that ranges. In the case of $S_{rp} = 0.3$, it is hardly to reduce the extent of the plastic zones when L/a is smaller than 0.8. Even if the surrounding rockmass of tunnel is reinforced by the bolts whose length is the same as the tunnel radius, the tunnel wall strain can only be restricted to 5%. It means that if the ground condition is too weak, such as the squeezing ground ($S_{rp} < 0.3$), the plastic radius and the tunnel wall strain are difficult to be controlled even if the bolt is installed longer than the tunnel diameter, due to the extreme relative deformation at the boundary between bolts and ground, a ground-improvement method should be taken into account.

Figure 7 illustrates the influence of the bolt density parameter on the tunnel behaviour according to the competency factor of ground. In the case that the competency factor S_{rp} is lower than 0.4 or larger than 0.8, the controlling effect of the bolt density parameter can not be expected.

From above analytical study, it is clarified that the rockbolts exhibit the reinforcing effect prominently on the tunnel stability if the ground condition is characterized with $S_{rp} = 0.4 \sim 0.8$.

Furthermore, since the bolt length and the bolt density have prominent dependent effect on the tunnel behavior, there is a proper combination between these two parameters for the ground condition for the optimum design of the rockbolts.

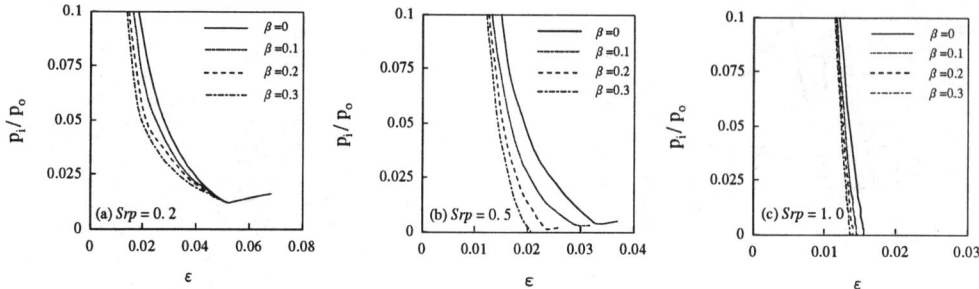

Figure 8 Influence of bolt density parameter β on ground characteristic curve($L/a = 1.0$).

(2) Effect of bolts on the ground characteristic curve

Figure 8 presents the typical ground characteristic curves with consideration of the influence of the bolt density parameter β for different ground conditions ($S_{rp} = 0.2$, 0.5 and 1.0). The installation of rockbolts improves the apparent material properties of the soft rocks, thereby restricting its deformability and the extent of plastic zones. However, if the strength of rockmass is too low, in particular the residual strength decreased suddenly after the peak stress, the excessive relative deformation between the ground and the rockbolts would cause the shear failure and sliding at the ground-bolts boundary interface. This trend is notable especially in the squeezing ground. For example, the ground characteristic curve in the chart shifts to the left with the increase of the bolt density parameter, in the condition of $S_{rp} = 0.2$, while no obvious effect could be fined on the location of the lowest point on the curves which defines the minimum load for support design(Jiang, 1993). On the other hand, in the case of $S_{rp} \geq 1.0$, these is no occurrence of the plastic flow zone around the tunnel, so the ground characteristic curves don't appear the lowest point. The result from the case of $S_{rp} = 0.5$, which corresponds to the proposed region of $S_{rp} = 0.4 \sim 0.8$, illustrated that not only the form of the ground characteristic curve but also the lowest point on the curve can be effectively controlled by the grouted blots.

6 CONCLUSION

In this study, an alternative theoretical method for the prediction of the deformational behavior of tunnels in soft rocks and the influence of fully grouted rockbolts on tunnel stabilization is proposed with the advantage of considering the post-failure behavior of rocks. The large deformation phenomenon of tunnels is considered to be due to the occurrence of plastic-flow zone in the surrounding medium and the representative charts for the prediction are shown.

According to the post-failure behavior of soft rocks and bolting pattern, nine cases of ground-bolts interaction modes are proposed to describe the mechanical stabilization of the rocks by bolting. The effect of rock bolts on controlling the deformation of tunnels and the expansion of plastic zones in the surrounding medium are investigated carefully by the detailed parametric analyses. It is clarified that there is a proper combination between bolt length and installing density according to the ground condition that the competency factor of ground S_{rp} ranges from 0.4 to 0.8.

The presented analytical approach and some results based the detailed parameter study are being verified by the experimental modeling and field applications.

REFERENCES

Hoek,E. & Brown,E.T. : *Underground Excavations in Rock*, The Institution of Mining and Metallurgy, London, 1980.

Indraratna, B. & Kaiser, P.K. : Analytical model for the design of grouted rock bolts, *Int. J. for Numerical Methods in Geomechanics*, Vol.14, pp.227-251, 1990.

Indraratna,B. & Kaiser,P.K. : Design for grouted rock bolts based on the convergence control method, *Int. J. Rock Mech. Min. Sci. & Geomech. Abstr.*, Vol. 27, No.4, pp.269-281, 1990.

Jiang,Y.J. : Theoretical and experimental study on the stability of deep underground opening, *PhD. thesis*, Kyushu University, 1993.

Jiang,Y.J., Esaki,T. and Yokota,Y. : The mechanical effect of grouted rockbolts on tunnel stability, *Proc. of EUROCK'96*, Italy, pp. 893-900, 1996.

Jiang,Y.J., Esaki,T. and Yokota,Y. : The mechanical effect of grouted rock bolts on tunnel stability, *J. of Geotechnical Engineering, JSCE*, No.561/III-38, pp.19-31, 1997.

Labiouse,V. : Rockbolting in the rock-support interaction analysis, *7th Int. Cong. on Rock Mech.*, pp. 1321-1324, 1991.

Rabcewicz,L. and Golser,J. : Principles of dimensioning the supporting system for the New Austrian Tunneling Method, *Water Power*, March, pp.88-93, 1973.

Tao, Z.Y. & Chen, J.X. : Behavior of rock bolting and tunneling support, *Proc. of the Int. Symp. on Rock Bolting*, pp.87-92, 1983.

Detection of underground cavity by inverse calculation

Baek-Soo Suh & Kwon-Ik Sohn
Kangwon National University, Chuncheon, Korea

Byung-Doo Kwon & Hyun-Kyo Jung
Seoul National University, Korea

ABSTRACT: Geotomography is widely used to interpret datas of underground structures. In this study seismic and electrical geotomography is tried to map underground cavity. In seismic geotomography, finite element method is used to calculate the first arrival time and LSQR[3] inversion method is applied to calculate slowness matrix. In electrical inversion method, model experiment is conducted and experimental data are compared with that of theoretical inversion data. And a new method for electric resistivity tomography(ERT) is developed for geophysical inverse problems by adapting the sensitivity analysis.

1 INTRODUCTION

Many geophysical interpretation needs high resolution and tomographic imaging is very important task in recent geophysical data processing. In order to meet this requirement, seismic LSQR inversion method and some sensitivity tomography algorithms are developed in this study. In seismic inversion method, forward method such as ray tracing method, shooting method and finite difference method are used to calculate the first arrival time. In this study, finite element method is used to investigate tunnel which is one of the complex inhomogeneities. The electric resistivity tomography is one of the materials in underground space by measuring the disturbance of the electric potential distribution due to the material there. The potential outputs of the potential electrodes are dependent on the resistivity distribution as well as the configuration of the electrodes. Hence, from the potentials of the potential electrodes, the resistivity can be deduced if the electrode configurations are already known. In this paper, a new algorithm, that adapts a general optimization method and the sensitivity coefficients computed using the finite element method as the gradient vector for variable update, is presented. And theoretical inversion data are compared with that of experimental model data.

2 SEISMIC INVERSION CACULATION

2.1 Inversion calculation

In two dimensional model, the total traveltime of a ray from source to receiver in borehole seismic prospecting is defined as follows

$$t_k = \int_{R_k} \frac{1}{V(x,z)} dl$$

$$= \int_{R_k} s(x,z) dl \quad (1)$$

where v(x,z) is velocity of the medium
s(x,z) is slowness
R_k is raypath.

Eq.(8) can be described by a simple form of matrix equation as follows:

$$\hat{D}\,\hat{s} = \hat{t} \quad (2)$$

where $\hat{D}$ is the raypath distance
$\hat{s}$ is slowness.

SVD(Singular Value Decomposition) method is applied to Eq.(12), and the model parameter equation can be obtained as follows:

$$\hat{s} = (\hat{V}\hat{Q}^2\hat{V}^T)^{-1}\hat{V}\hat{Q}\hat{U}^T\hat{t} \quad (3)$$

where, U : N×N matrix of eigenvectors
V : M×M matrix of eigenvectors
Q : N×M diagonal eigenvalue matrix

2.2 Inversion of theoretical tunnel model

Theoretical tunnel is designed similar to the field tunnel model. The size of model is 18m in x-direction, 45m in z-direction. The x-direction model size is 18m but the receivers are arrayed 15m in x-direction because it can remove the reflections. The mesh size is 0.6m and the number of meshes are 30 and 35 in x- and z-directions, respectively. And the theoretical model can be shown in Fig.1-(b).

33 sources and 33 receivers are arrayed every 1m in one borehole and another borehole. Main frequency of source is designed 833 Hz and tomogram is mapped calculating 33 x 33 datas. The density of mother rock is supposed to 2.7 g/cm^3 and that of tunnel is 0.027 g/cm^3. We assume that tunnel is filled with air. The P and S wave velocity of mother rock are supposed to 4.0 km/sec and 3.0 km/sec, respectively, and those of tunnel are 0.4 km/sec and 0.03 km/sec, respectively.

Finite element method is used to calculate the first arrival time in this model and inversion calculation was done. LSQR inversion method is applied to calculate slowness matrix, and tomogram is mapped by above results as shown in Fig.1-(a). Data interpretation is very important as well as data calculation and data aquisition. The shape and size of tunnel are shown very differently by changing mapping method of geotomography. Sometimes it is possible to us making a misinterpreting tunnel data because of different mapping method of geotomography.

3 ELECTRICAL INVERSION CACULATION

3.1 Finite element analysis of ERT

The relation the constant conduction current density J and the stationary electric field intensity E in the isotropic electric conductivity σ is given, from the Ohm's law, by the following equation:

$$J = \sigma E = -\sigma \nabla \phi, \qquad (4)$$

where ϕ is a electric scalar potential. Applying the principle of conservation of charge over a volume, the following current continuity equation is obtained:

$$\nabla \cdot J = \frac{\partial \rho}{\partial t} \delta(x_s)\delta(y_s)\delta(z_s) = I\delta(x_s)\delta(y_s)\delta(z_s), \qquad (5)$$

where ρ is the charge density specified at a point in the Cartesian x-y-z space by the Dirac delta function and I is the total current injected at the point (x_s, y_s, z_s). From (4) and (5), the following field governing equation is obtained:

$$-\nabla \cdot \{\sigma(x,y,z)\nabla \phi(x,y,z)\} = I\delta(x_s)\delta(y_s)\delta(z_s). \qquad (6)$$

The potential $\phi(x, y, z)$ is calculated from the $\tilde{\phi}(x, K_y, z)$ of the K_y space using the inverse transformation. If K_y is discretized into a series of linear elements, and $\tilde{\phi}(x, K_y, z)$ is assumed to be varied linearly in each elements, $\phi(x, y, z)$ is represented as following equation.

$$\phi(x_0, y_0, z_0) = \frac{2}{\pi} \sum_{k=1}^{L} \int_{K_k}^{K_{k+1}} \tilde{\phi}(x_0, K_y, z_0) \cos(K_y y_0) dK_y$$

$$= \frac{2}{\pi} \sum_{k=1}^{L} D_k(y_0) \tilde{\phi}(x_0, K_y^k, z_0), \qquad (7)$$

where L is the number of linear elements, and

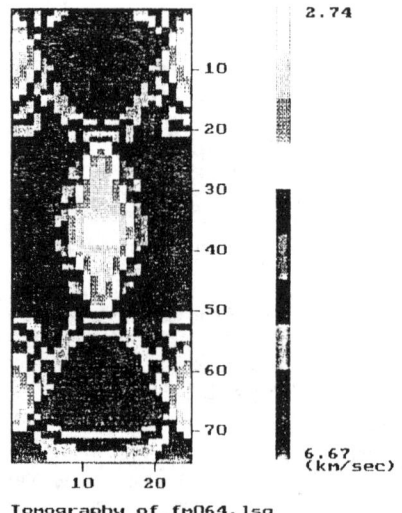
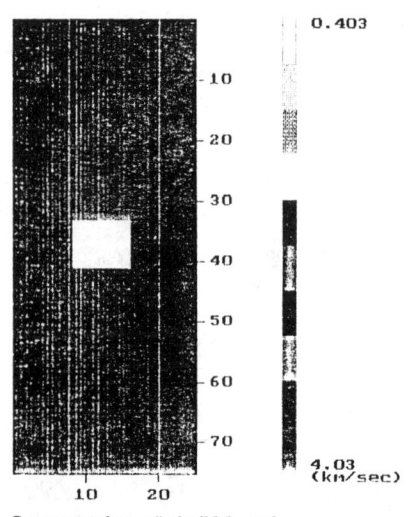

Fig.1 The results of inversion cavity model (a) and theoretical cavity model (b)

$D_k(y_0)$ is the integration value in k-th element.

In ERT systems, the physical quantity that is to be found is the electric resistances or conductivities which give the measured potential distribution for the given current electrodes.

The potential distribution is a function of resistances (or conductivities) only if the electrodes and injected current are fixed. In the system, since making a many boreholes is expensive, usually only two boreholes are made. And in order to get as much data as possible the positions of the current electrodes are changed to many place. Hence the resistance distribution to be found is that makes the following objective function zero:

$$F = \frac{1}{2} \sum_{p=1}^{N_c} \sum_{j=1}^{n} \left(\phi_j(r_j) - \phi_{j,0}(r_j) \right)^2, \quad (8)$$

where n is the number of potential electrodes and N_c is the number of current electrode position, and $\phi_{j,0}(r_j)$ and $\phi_{j,0}(r_j)$ are computed with a given resistance distribution and measured potentials at the point r_j with the p-th current electrode position, respectively. After the sensitivity coefficients are calculated, the conductivity vector is updated using gradient type optimization algorithm as follows.

$$[\sigma]_{new} = [\sigma]_{old} + \sigma \cdot \frac{dF}{d[\sigma]} \quad (9)$$

where σ is the step length and it is approximated by linearizing the objective function.

3.2 Model experiment

Model experiment has been conducted which is made of 1cm thickness Acryl. The size of three dimensional water tank is $20cm \times 120cm \times 60cm$ which is filled with sodium chloride water(Fig.2). The electrodes are arrayed each 4 cm intervals and the number of electrodes are 21 at each side. The material of electrode is 4mm diameter carbon and numbers of 42 electrodes are connected outside terminal to protect water shake. The solution of water tank is filled with city water and the conductivity of solution is 7.22×10^3Mho/m. As an experimental cavity model, acryl model of $8cm \times 8cm \times 120cm$ size is used. The conductivity of acryl is 10^{-14}Mho/m and the cavity model is filled with air. Another cavity model is made of cupper which is the same size compared with that of acryl cavity model. The cupper cavity model is filled with water and the conductivity of solution in cavity model is 3.61×10^3Mho/m, that is one half of conductivity of water tank. At first measuring, 5mA current is input in two current electrodes of No.1 and No.2 current electrode. One of potential electrode is fixed at one position and measured potentials changing position by another electrode. Next time, two current electrode is moved to No.2 and No.3 current electrode and measured the potentials using above method. Fig.3 shows the difference between the theoretical inversion potentials and experimental measuring potentials when current electrodes are located at No.10 and No.11 electrode.

3.3 Numerical example

The suggested algorithm is applied to a simple numerical example which has a rectangular block of 75x75 [mm] with electric resistivity ρ =50[Ωm] in the homogeneous semi-infinite plane of electric resistivity ρ=100[Ωm] as shown in Fig.3, where all the electrodes are located in

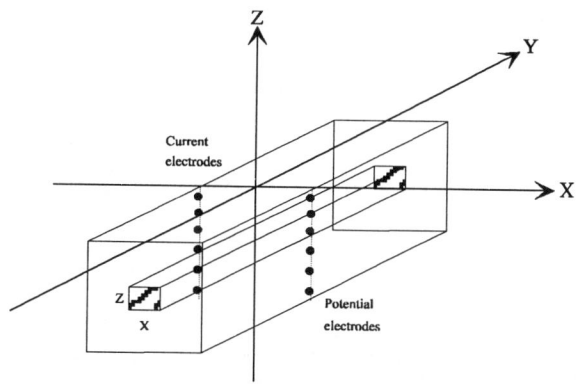

Fig.2 Configuration of model experiment.

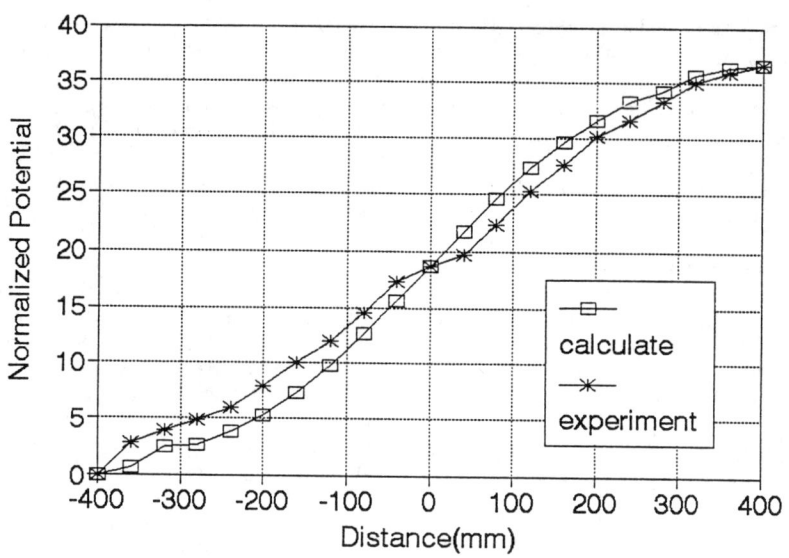

Fig. 3 Comparison of theoretical inversion data and experimental data

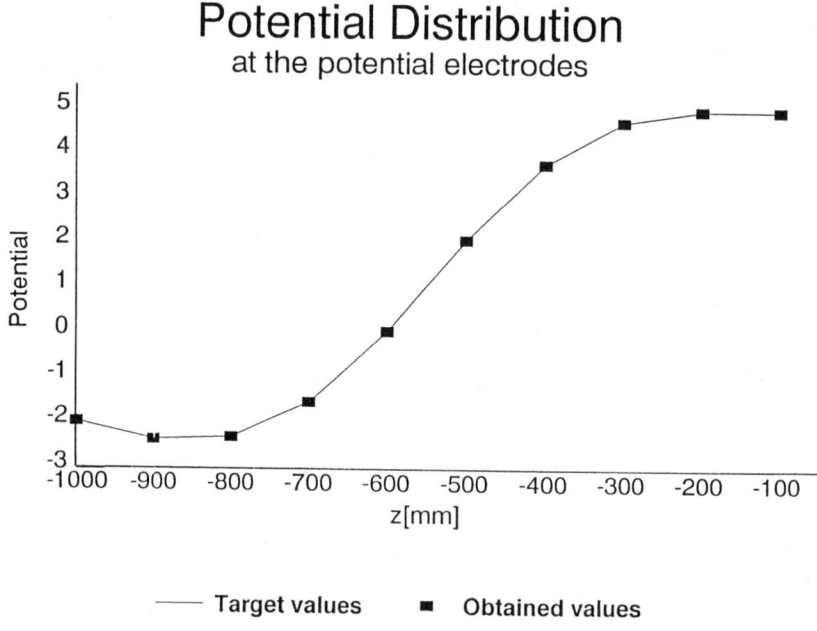

Fig. 4 Comparison of the potential distribution between computed and target values

equal space at y=0. The number of the pairs of the current electrodes are 4 and +1 and −1[A] are applied at the upper and lower electrodes, respectively. The number of the potential electrodes are 10. As the region to be tested, the region of 175 × 175[mm] including the above inhomogeneous block is taken and divide into 25 small rectangular and the resistivities of the small rectangular blocks are taken as variables.

For the inverse deduction of the resistivities from the measured potentials, the tested region id divided into 9 small blocks, and the resistivities of the blocks are taken as variables. And comparison of the potential distribution between computed and target values are shown in Fig.4.

4 CONCLUSION

(1) It has been thought that finite element method needs a lot of memory dimension and calculating time to solve wave equation problems. But developing of computer technology, tunnel detection problem which is one of complex inhomogeneities can be solved using this method and the result is thought to be satisfactory.

(2) All procedure of prospecting such as aquisition, processing and interpretation of datas are very important. Sometimes anomalies can't be recognized in tomogram which is final step of interpretation. To find exact target, several numbers of tomogram have to be mapped using various color bar versus wave velocity.

(3) A new algorithm for the mapping of the electric resistivities in the subspace is developed by combining the finite element method and sensitivity analysis. The computing time for sensitivity coefficients could be remarkably reduced by introducing the adjoint variable. The developed algorithm was applied to a numerical example that has a simple subspace structure, and a good image was obtained without any smoothing (filtering) technique. And calculation accuracy of this method was confirmed compared theoretical inversion data with that of experimental data.

REFERENCES

1) Cooke, D.A. and Schneider, W.A., Generalized linear inversion of reflection seismic data, *Geophysics*, Vol. 48, pp. 665-676(1983).

2) Daily W. and Owen W., "Cross-borehole resistivity tomography", *Geophysics*. Vol. 56, No. 8, pp1228-1235(1991).

3) Dey A. and Morrison H.F., "Resistivity modeling for arbitrarily shaped two-dimensional structures", *Geophys. Prop.*, Vol. 27, pp 106-136(1979).

4) Dresen, L., Locating and Mapping of cavities at shallow depth by the seismic transmission method, *Proceedings of DMSR 77*, Vol.3, pp.214-223(1977).

5) Gersztenkorn, A. and Scales, J.A., Smoothing seismic tomograms with alpha-trimmed means, *Geophysical Journal*, Vol.92,pp.67-72 (1988).

6) Lines,L.R. and Treitel, S.,A review of leastsquare inversion and its application to geophysics problem, *Geophys. Prosp.* Vol 32, pp.159-186(1984).

7) McCann,D.M., Application of cross-hole seismic measurements in site investigation surveys, *Geophysics*, Vol.51,pp.914-929(1986).

8) McMechan, G.A., Seismic Tomography in boreholes, *Geophy. Journal of the Royal Astr. Society*, Vol.74, pp.601-612(1982).

9) Suh,B.S. and Hyun,B.K., Tunnel detection using seismic geotomography, Vol.3, *Journal of Korean Society for Rock Mechanics.* pp.50-53(1993).

Back analysis of subsidence above old open stopes in an Indian hard rock mine by numerical simulation

Achyuta Krishna Ghosh, Amalendu Sinha & Dhiresh Govind Rao
Geomechanics Division, Central Mining Research Institute, Dhanbad, India

ABSTRACT: Occurrence of surface subsidence above old, shallow open stopes in Badia Section of Mosaboni Copper Mine was noticed in October/November, 1994. As mining in this patch was completed more than 25 years ago, no correct information could be gathered regarding the latest topography of the area and it could not be inferred really how much ground movement took place here because of this subsidence.

To find out an answer to this question, a Boundary Element Model study of this subsidence was carried out simulating progressive failure in rock mass. Comparison of the findings of the simulation results with the field observations indicated the model to be realistic and acceptable. Thus the magnitude of maximum subsidence was estimated to be about 0.46 m and a methodology was worked out to evaluate subsidence proneness of the ground above other shallow open stoping areas in this mine. The problem, the approach, the exercise and the findings has been discussed in this paper.

1 INTRODUCTION

Occurrence of surface subsidence above old, shallow open stopes in Badia Section of Mosaboni mine - the oldest running Indian copper mine located in Singhbhum Thrust Zone of Bihar, was first reported on October 31, 1994 at 9 a.m. IST. Collapse took place in underlying open breast stopes, worked between 1940 and 1970 to extract 1.8 to 4.5 m wide orebody, dipping at about 38° due N83°E, leaving irregular *in situ* pillars at random to support the hangwall.

Subsidence continued for two weeks covering an area of about 0.106 km^2 on surface. Simultaneous change in water mark of on the bank slope of a pond (Tank A in Fig. 1) located in the northern half of the subsided zone clearly indicated that the ground level on its east has moved down compared to the western part. Main crack noticed on surface extended over a length of about 200 m in the direction of strike of the orebody (Fig.1). Horizontal and vertical displacements across surface cracks usually varied from 0.5 to 3 cm, and from 0 to 5 cm respectively, but respective maximum values observed over short lengths were 10 and 11 cm. A pothole also formed in close vicinity of an abandoned vertical shaft located at the up-dip extreme of underlying stopes.

The area of subsidence rests above exhausted open breast stopes at a depth of 37 to 237 m, worked in six levels. After mining of these stopes more than 25 years ago, much change has taken place in surface topography. The area which was earlier a jungle of rocky, rugged terrain having no inhabitation, was later levelled and used for cultivation and dwelling. No information was available about the topography of this locale prior to this subsidence. Nor any subsidence was reported earlier in this mine. Consequently, magnitude of ground movement actually taken place here could be neither known nor guessed.

Here soil cover is 1.5 to 3 m, underlain by hard rocks, namely mica schist, granite schist and quartz chlorite biotite schist, the last one being the host rock of chalcopyrite ore as well. Rock mass around these stopes is hard, jointed, and not much weathered.

As the top soil cover is nominal and rock mass near surface is not significantly weathered, it is clear that this subsidence was a result of progressive failure of rock mass initiating from underground stopes. Moreover, rock mass here appeared to be very much similar to that of the present workings in this mine. Hence it was planned to try to simulate this event of subsidence by the 2D boundary element modelling (BEM) technique, which was already standardised, proved and extensively used with success to evaluate rock mass stability and design of support in current workings in this mine, in conjunction with the principle of progressive failure in rocks, for realistic estimation of its extent and magnitude. It was decided to compare the findings of the simulation exercise with prominent, quantifiable observations in field to verify its validity before accepting its results.

Accordingly, a 2D BEM exercise was conducted after relevant laboratory and field investigations. Its results matched well with field observations. The task and findings have been discussed in this paper.

2 OBSERVATIONS

Observations of various field and laboratory investigations made in this connection, are as follows:

2.1 Basic Mining Information

Basic information of the stopes just below the subsided area, are as follows:

Stope	Depth (m)	Strike Length (m)	Length along Dip(m)	Extraction Thickness (m)	Mining Period
above 1L	37-50	320	30	2.0	1964-69
1L-2L	50-88	320	62	1.8-2.4	1942-56
2L-3L	88-127	350	65	2.2-2.4	1948-56
3L-4L	127-164	410	55	2.2-2.4	1949-59
4L-5L	164-202	300	60	2.2-2.4	1952-62
5L-6L	202-237	195	67	1.4-3.0	1956-59

Stope drives at levels were heightened to 4.5 m. There width varies from 2.5 to 12 m or even more.

As gathered from mine plans, extraction in these stopes was beyond 85% with rest ore being locked-up in 276 *in situ* vertical pillars. About 200 pillars were arranged along both sides of stope drive at each level while 76 pillars were scattered in stopes. These pillars, whether along stope drives or inside stopes, were irregular in shape and spacing. Some of them were too small to provide long-term support, while a few were mismatchingly large. Average length, breadth and cross-section of all pillars in this area were 6.58 m, 4.16 m and 27.38 m^2 respectively. They were grouped as per their dimensions as tabulated below:

Dimension	Number
Cross-section (m^2)	
4 to 10	27
10 to 16	46
16 to 20	24
20 to 27	65
above 27	114
Pillar width (m)	
1.0 to 2.5	50
2.5 to 3.0	34
3.0 to 4.0	50
above 4.0	142
Pillar length (m)	
2.0 to 4.0	32
4.0 to 5.0	48
5.0 to 6.0	66
6.0 to 6.5	23
above 6.5	107

2.2 Field Observations

From visits to the stopes below the subsided zone it was revealed that collapse took place in open stopes above 3L. Voids of these stopes seemed to be fully filled with large boulders from collapsed hangwall. But no visible symptom of high stress concentration or impending failure was noticed inside stopes below 3L or in raises and drives in their vicinity. No inflow of water was noticed anywhere. Rock mass above 3L and below appeared to be same and similar to that in other parts of this mine.

Joint characteristics and groundwater condition in this area are as follows:

Joint Characteristics	
Number of sets	2 to 3
Spacing	5 to 30 cm
Surface	unidirectional, wavy, rough to smooth
Alteration	unaltered, 3 to 5 mm thick hard quartz filling at a few locations
Aperture	opening of foliation planes and other joints at some places
Groundwater Condition	Dry, no water inflow

2.3 Laboratory Investigations

Physicomechanical properties of intact rocks, as determined in the laboratory from the samples collected from different levels, are listed here.

Site	Density (t/m^3)	Compressive Strength (MPa)	Tensile Strength (MPa)	Strength Ratio
1L (S)	2.67	80.27	9.62	0.1198
3L (S)	2.82	46.03	10.30	0.2239
3L (N)	2.75	107.36	5.89	0.0548
6L (N)	2.73	65.36	12.37	0.1892
Mean	2.74	74.76	9.57	0.1469
S.D.	0.06	26.86	2.70	0.0751

3 ROCK MASS PROPERTIES

Based on the field and laboratory studies, the rock mass was classified according to various engineering classification systems as tabulated below:

Classification System	Index	Value	Class
Bieniawski (1973)	RMR	59-65	Fair to Good
Barton (1974)	Q	8-11	Fair to Good
Laubscher (1990)	RMR	49-54	Fair

Compressive strength of the rock mass, derived from the average compressive strength of intact rock, (Laubscher, 1990) was 32.15 MPa ≈ 32 MPa. Following the statistical approach used earlier in this mine (Rao et al, 1995), rock mass tensile strength was determined to be 2.31 MPa. Angle of internal friction and cohesion of rock mass were estimated to be 30° (Barton et al, 1974) and 9.2 MPa (Jaeger & Cook, 1979) respectively.

Though broadly the characteristics of rock mass in this region was similar to that of other portions, viz North Badia, Central and North sections in the mine, and values of compressive strength and angle of internal friction were found to be same as those used in earlier 2D BEM simulations of the same mine, its tensile strength and Young's Modulus were comparatively lower (Ghosh et al, 1995; Rao et al, 1995; Ghosh, 1996).

4 VIRGIN *IN SITU* STRESSES

As values of virgin *in situ* stresses were not available for the given location, they were estimated from the following empirical relationship established for this mine among its virgin *in situ* major horizontal stress (S_H), virgin *in situ* vertical stress (S_V) and depth (D, in m) (Ghosh et al, 1987):

$$S_H/S_V = 0.95 + 263.5/D$$
considering $S_V = d.g.D$
where, d = rock mass specific gravity, and
g = acceleration due to gravity.

Thus the stress equations finally accepted here are,
S_H (in MPa) = 0.03474 × D + 7.42503
S_V (in MPa) = 0.02780 × D

5 PRELIMINARY ANALYSIS

A preliminary analysis was done on the basis of the information obtained or generated so far. Had those 276 *in situ* pillars in stopes were of uniform 6.58 × 4.16 m cross-section and spaced at regular intervals in 0.106 km^2 area, average area to be supported by each pillar would have been about 3870 m^2. Presuming mean stoping depth of 137 m, estimating vertical load on a pillar from the relationship given in the earlier section and taking rock mass strength as 32 MPa, these pillars are inadequate to support the total vertical load i.e. dead load of hangwall in stopes.

In reality, hangwall in these stopes were somewhat self-supporting, and so they did not collapse for such a long time. Although rocks here are jointed, as they are very hard as well the possibility of accumulation of strain energy in pillars with time resulting in their abrupt failure instead of gradual collapse cannot be ruled out. Rather, such a long gap between cessation of mining and subsidence and the suddenness of the event in spite of shallow depth, seem to be in corroboration with this idea. It is also obvious that these pillars were of little significance from the standpoint of long term stability of these stopes.

6 NUMERICAL SIMULATION

6.1 *Formulation*

As extraction thickness in the concerned stoping area was nominal compared to its length and breadth, and as the *in situ* pillars are insignificant in resisting any subsidence, 2D BEM technique with fictitious stress elements was used to simulate the given condition.

Rock mass properties and virgin *in situ* stresses used in modelling have already been discussed. The first model was formulated for mining of a 3 m thick excavation, dipping at 38° and extending from above 1L (depth = 37 m) to 7L (depth = 278.5 m), according to the section provided by the mine management.

As no data was available about the extent of bulking of collapsed ground in stopes, bulking factor was assumed to be 1.25, same as that of Indian gritty sandstone in a caved goaf (Dutta et al, 1986).

6.2 *The Analysis*

The result from the first model indicated that extensive failure would occur in hangwall in stoping areas above 2L and would extend upto the surface (Fig.2).

In the second step it was considered that some part of the immediate hangwall of stope in failure zone would physically come down and fill the void of stope immediately below it. Taking bulking factor as 1.25, height of the collapsed zone would be 4 times the mining thickness i.e. 12 m (Fig. 2). Accordingly, the excavation geometry was modified in the second model. Though it was considered that the void was full with broken rocks, the fill would definitely allow small amount of vertical movement (max. 0.225 m) as indicated by the result (Fig. 3). The zone of failure would extend further along the stope hangwall almost to 3L (Fig. 4).

Excavation geometry for the third model was designed in the same way as the second one (Fig. 4). It was found that at this stage, the limit of failure zone along the stope hangwall coincides exactly with the extent of its collapsed zone i.e. the zone with increased excavation height (Fig. 5). This implies that there would not be any fresh collapse in the immediate hangwall of stopes. Trend of vertical displacement above the mined out area, and subsidence and horizontal displacement profiles at surface are presented in Figs. 6 and 7 respectively.

The results indicate that maximum movement of 0.795 m would occur normal to the dip of orebody. Maximum subsidence and horizontal movement of 0.46 m and 0.68 m respectively would occur on surface between 160 to 180 m, and at 300 m respectively in the dip direction of orebody from the site above the uppermost limit of mining. Major vertical and horizontal ground movements of ≥0.35 m and ≥0.6 m respectively would take place simultaneously on the surface at a distance between 140 and 350 m from the same site in the same direction (Fig. 8).

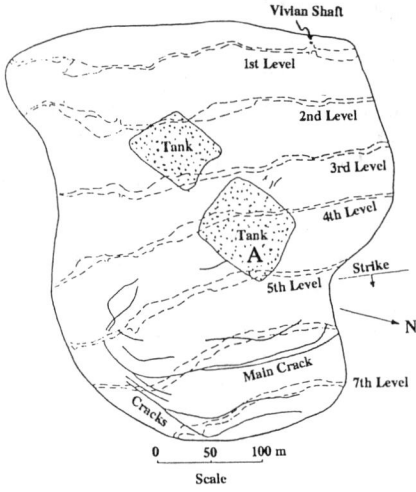

Fig. 1 : Plan showing locations of tanks and cracks on surface

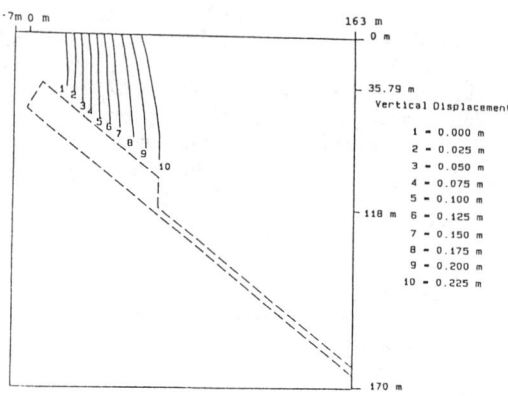

Fig. 3 : Vertical displacement contours above collapsed hangwall

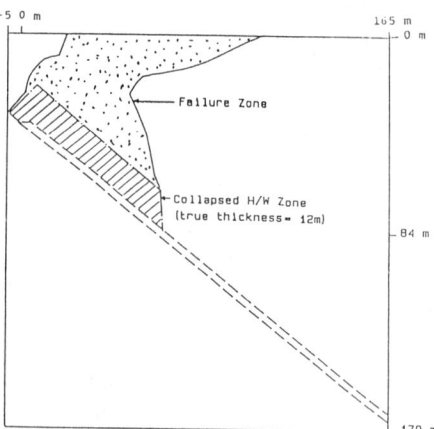

Fig. 2 : Primary failure zone in stope hangwall (Model-I)

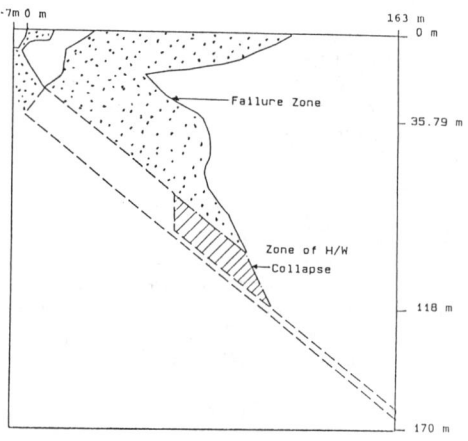

Fig. 4 : Primary failure zone in stope hangwall (Model-II)

6.3 Interpretation of Results

From the findings of the above analysis following conclusions may be drawn:
1] collapse in stopes would not extend beyond 3L;
2] while primary failure zone in surface would be limited to about 100 m from the point above the uppermost limit of mining, vertical displacement would be limited here;
3] referring to Fig. 5, there would be a small patch of stable zone which would divide the primary failure zone on the surface in two unequal parts above the uppermost limit of mining; in the part which is smaller, only a small portion (as shown in Fig. 5) would have mining void below it and would thus be prone to potholing;

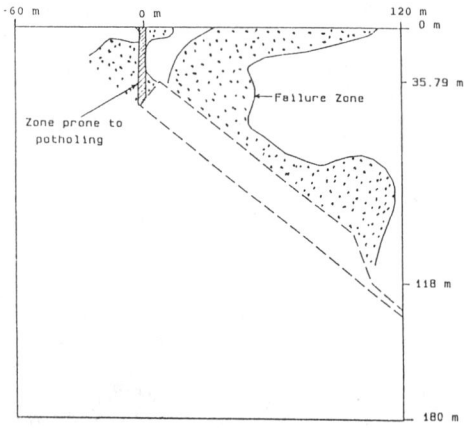

Fig. 5 : Primary failure zone in stope hangwall (Model-III)

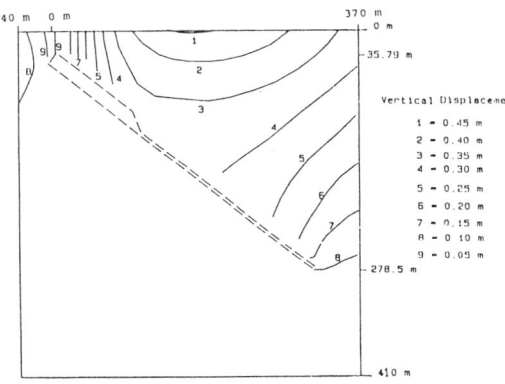

Fig. 6 : Vertical displacement contours in hangwall of stoping area (Model-III)

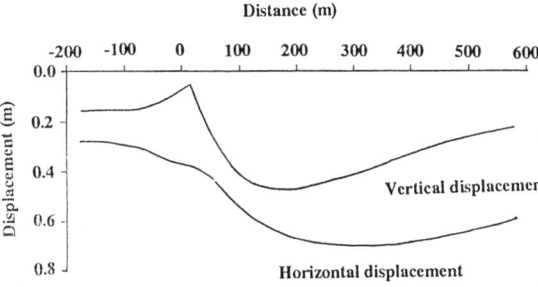

Fig. 7 : Magnitudes of horizontal and vertical displacements at the surface (Model 3)

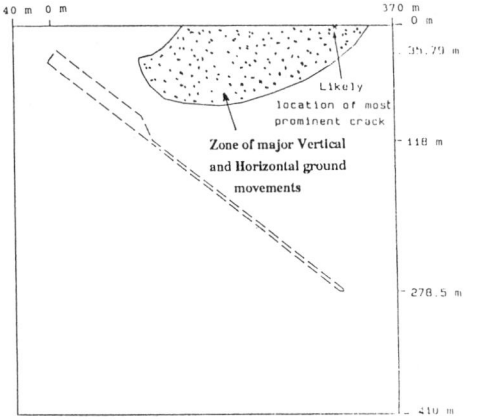

Fig. 8 : Zone of major vertical and horizontal ground movements

4] surface cracks should be observed at sites where both, vertical and horizontal displacements are high i.e. in the zone 140 to 350 m away in the direction of the dip of the opening from the site above the uppermost mining limit; and

5] most prominent cracks should appear on surface at locations of maximum horizontal displacement, that is around 320 m away from the site above the uppermost mining limit in the direction of the dip of the orebody.

7 COMPARISON OF SIMULATION RESULTS WITH FIELD OBSERVATIONS

Finally the results of simulation exercise was compared with actual field observations to assess the validity of the former. Followings are the outcome of the comparison:

* Occurrence of subsidence over a considerable time period and in stages clearly indicates that failure in rock mass took place progressively, the first stage being the major one; the simulation was also carried out considering progressive failure in rock mass, and the failure indicated in the first stage of simulation was the prime one.

* Most prominent cracks were observed on surface about 300 m away from the location above the topmost mining limit, the distance being measured along the dip direction of orebody; the modelling results indicated the distance to be about 320 m, the deviation being 6.66% only.

* A pothole appeared in the field near the uppermost boundary of mining; in the modelling results also the zone has been identified as prone to potholes.

* Because of the lack of any measured data, exact location of maximum subsidence could not be precisely identified; but according to the local people maximum subsidence took place in the vicinity of the pond (Tank A in Fig. 1), about 170 m away from the site above the uppermost mining limit, the distance being measured along the direction of dip of the orebody; on the other hand, from the findings of simulation this distance was estimated to be 160 to 180 m.

8 INFERENCE

From the comparison of numerical simulation results and actual field observations it could be inferred that the findings of the model were satisfactorily matching with the facts. In this modelling exercise, the event of subsidence above Badia section in Mosaboni copper mine was simulated rationally and realistically and other findings of this study can be reliably accepted to estimate relevant unknown factors, like magnitude of subsidence at different locations. This methodology may be applied to simulate, predict and estimate subsidence in hard rock mines provided the cover of top soil is nominal and rock mass properties and *in situ* stress conditions are known and proved.

REFERENCES

Barton, N., Lien, R. & Lunde, J. 1974. Engineering classification of rock masses for the design of tunnel support. *Rock Mechanics* 6(4):189-239.

Bieniawski, Z.T. 1973. Engineering classification of jointed rock masses. *Trans. S. Afr. Inst. Civil Engg.* 15(12):335-244.

Dutta, D., De, P., Bhattacharya, J., Ghosh, A.K. & Ghose, A.K. 1986. Longwall face support design - A microcomputer model. *Indian J. Mines, Metals & Fuels* XXIV(3):97-107.

Ghosh, A.K., Sinha, A. & Raju, N.M. 1987. Rockbolts design guideline for Mosaboni and Surda mines. *CMRI Report (unpublished)*.

Ghosh, A.K., Sinha, A., Rao, D.G., Jain, P.N. & Chatterjee, K. 1995. Design of rock reinforcement system for horizontal cut-&-fill stoping at great depth in Mosaboni mine. *Proc. First National Conference on Ground Control in Mining, Calcutta, India* :169-191.

Ghosh, A.K. 1996. Application of numerical simulation techniques in ground control problems - Facts and fallacies. *Proc. Second National Conf. Ground Control in Mining, Calcutta, India* :357-376.

Jaeger, J.C. & Cook, N.G.W. 1979. *Fundamentals of rock mechanics.* Chapman & Hall:96

Laubscher, D.H. 1990. A geomechanics classification system for the rating of rock mass in mine design. *S. Afr. Min. Metall.,* 90(10):257-273.

Rao, D.G., Sinha, A., Ghosh, A.K. & Jain, P.N. 1995. Estimation of rock mass properties for numerical modelling of an underground mine at great depth. *Proc. 8th International Cong. Rock Mech. (ISRM), Tokyo, Japan* :Vol. 3.

Evaluation of long cable tendon load distribution using Computer Aided Bolt Load Estimation (CABLE™)

W.F.Bawden, M.Moosavi & A.J.Hyett
Department of Mining Engineering, Queen's University, Kingston, Ont., Canada

ABSTRACT: The paper investigates the mechanism of load distribution along fully grouted cable bolts. It is based on a finite difference algorithm that links small sections of cable together, the behaviour being known from short embedment length laboratory tests. The load distribution is determined for both cases of continuous and discontinuous (crack opening) rock mass displacements. Higher axial load will be developed for a rock mass that behaves as a discontinuum with deformation concentrated at few fractures, than for one which behaves as a continuum. These results suggest that the stiffness of fully grouted cables is not simply a characteristic of bolt and grout, but also of the ground deformation. The obtained results from modified geometry and conventional cables are compared which shows that under identical conditions, higher cable capacity can be obtained using the former. Based on these differences, some practical recommendations are presented.

1 INTRODUCTION

Using bolts as a means of reinforcement has been shown to be one of the most effective ways of sustaining the integrity of excavations in rock masses. As mentioned by Hoek and Brown (1980), in design of underground excavations, the main aim is to help the rock mass to support itself. This is more critical near the excavation boundaries where the rock mass, which is usually jointed, is further damaged by blasting. Fully grouted cable bolts are used widely throughout the mining industry for this purpose. Cable bolts are flexible elements consisting of seven high strength steel wires stranded as specified in the ASTM 416-80 standard. The flexibility of this element makes it possible to coil, transport underground and to install in very long lengths from fairly confined working areas. To obtain different mechanical and frictional properties from the same material, the geometry of the cable can be altered. A chronological development of cables with enhanced properties is discussed by Windsor (1992) and the most popular ones are shown in Figure 1.

A comprehensive laboratory test programme was conducted to study the bond capacity of different cable types and to capture the bond failure mechanisms in models that can be used for reinforcement design. The result of this study was the development of models which can properly simulate the frictional-dilational response of the cable-grout interface due to the axial pullout, the details of which are discussed in earlier publications (Moosavi (1997), Hyett *et al.* (1995) and Hyett *et al.* (1996)) and not repeated here for brevity. This paper discusses the extrapolation of those models for calculation of load distribution along long cable lengths (which is the primary interest of the mine engineer). Finally, a parametric study emphasizes some of the more important points for reinforcement design and highlights some of the differences between conventional and modified geometry cables.

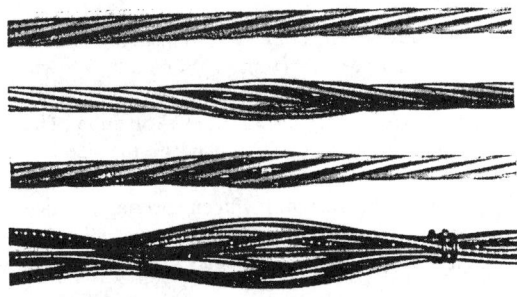

Figure 1: Different cable types; a) conventional, b) Garford bulb, c) nutcase and d) birdcage.

2 LOAD DISTRIBUTION ALONG CABLE BOLTS

Untensioned fully grouted bolts are usually placed around the stope well before the excavation of the ore. Therefore, the mine induced deformations activate the bolt. Following Farmer (1975), the equilibrium of a fully grouted rock bolt may be written as:

$$A_b \delta\sigma_x = -F_s \delta x \quad (1)$$

or

$$\frac{d\sigma_x}{dx} = \frac{-F_s}{A_b} \quad (2)$$

where F_s is the shear force due to bond per unit length and A_b is the cross sectional area of the bolt (Figure 2). For an elastic bolt:

$$\sigma_x = E_b \frac{du_x}{dx} \quad (3)$$

Substituting (3) into (2) results in

$$\frac{d^2 u_x}{dx^2} = \frac{-F_s}{A_b E_b}. \quad (4)$$

Let us assume that the shear force due to bond for a unit length of bolt is a linear function of the relative slip between the bolt and the rock:

$$F_s = k(u_r - u_x) \quad (5)$$

with k as shear stiffness of the interface per unit length (in N/mm^2). Combining equations (4) and (5), the distribution of displacement along the bolt is described by the second order non-homogeneous linear differential equation:

$$\frac{d^2 u_x}{dx^2} - \frac{k}{A_b E_b} u_x = -\frac{k}{A_b E_b} u_r. \quad (6)$$

To solve for this differential equation the rock mass displacement function (u_r) should be known. Rock displacement generally decreases with distance from the surface of the excavation. The form and rate of this will depend on the size and shape of the opening, and the strength and structure of the rock mass. In engineering practice u_r can be routinely determined using either an extensometer or a borehole camera.

Assuming the distribution of displacement within the bolt varies quadratically, the difference form of equation (6) may be written as:

$$\frac{\dfrac{u_x^{i+1} - u_x^i}{x^{i+1} - x^i} - \dfrac{u_x^i - u_x^{i-1}}{x^i - x^{i-1}}}{\dfrac{x^{i+1} - x^{i-1}}{2}} - \frac{k}{A_b E_b} u_x^i = -\frac{k}{A_b E_b} u_r^i \quad (7)$$

where the superscript i refers to the ith nodal point. After rearrangement this can be written as:

$$-\frac{A_b E_b}{x^i - x^{i-1}} u_x^{i-1} + \left(\frac{A_b E_b}{x^i - x^{i-1}} + \frac{A_b E_b}{x^{i+1} - x^i} + k\left(\frac{x^{i+1} - x^{i-1}}{2}\right)\right) u_x^i$$
$$-\frac{A_b E_b}{x^{i+1} - x^i} u_x^{i+1} = k\left(\frac{x^{i+1} - x^{i-1}}{2}\right) u_r^i \quad (8)$$

or more concisely,

$$-c^{i-1} u_x^{i-1} + \left(c^{i-1} + c^i + k\left(\frac{x^{i+1} - x^{i-1}}{2}\right)\right) u_x^i - c^i u_x^{i+1} = k\left(\frac{x^{i+1} - x^{i-1}}{2}\right) u_r^i \quad (9)$$

with

$$c^i = \frac{A_b E_b}{x^{i+1} - x^i} \quad (10)$$

Equation (9) comprises a matrix of n simultaneous equations, which when assembled in

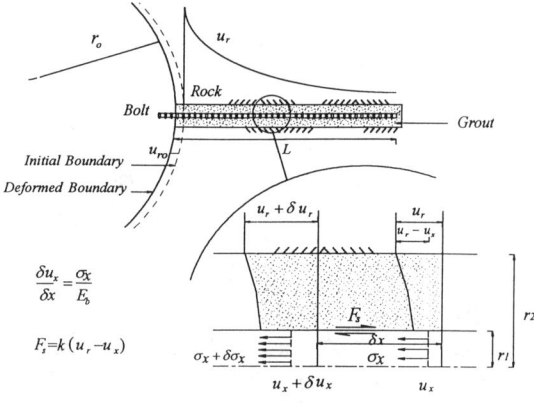

Figure 2: Stress distribution along a fully grouted bolt.

the correct order is tridiagonal in form with n unknown values of u_x.

Accounting for different boundary conditions (plated and unplated cable) and realizing that K_u^i is just dF_p/du_a for the i_{th} element, which is the slope of the load-displacement graphs (and calculated analytically by the developed models), then the problem can be solved in a piecewise manner to simulate the bond behaviour of a fully grouted cable bolt (Hyett et al. (1996)).

3 PARAMETRIC STUDY

A comparative study is performed to emphasize on the different response of conventional versus 25mm Garford bulb cables. In these simulations a 500mm bulb spacing is assumed for the entire cable length.

3.1 *Effect of rock mass displacement profile*

When cables are installed in a blocky rock mass the developed load in the cable would be maximum at the location of the fracture which is opening (upper row in Figure 3). Obviously the load decreases away from the fracture towards the two ends which are free (*i.e.* there is no faceplate attached to the cable). A comparison between the left-hand plots (conventional) and right-hand ones (Garford bulb) indicate that much higher loads can be mobilized in the latter. Obviously cable bolts can not take any loads higher than their ultimate capacity, (260 kN for a 0.6" cable), but these results are only meant to emphasize on the differences in t response.

Next, consider the case of multiple fractures. Middle row plots in Figure 3 show the effect of three fractures, the combined displacement on which is equivalent to that of the single fracture above. Again, the maximum load development occurs at the discontinuities, although the maximum load is considerably less than when the displacement was concentrated on a single discontinuity.

Finally, consider the case when a much greater number of discontinuities exist and therefore the strain distribution in the rock can be assumed to be continuous, (*i.e.* as in using a continuum numerical model to simulate failure of a fractured rock mass). Assuming that the profile of rock displacement is given by ($u_{ro}/(1+x)$) where u_{ro} is the rock displacement at the face, and x is the distance from the face, the load distribution along the cable is shown in lower row of the Figure 3, again for different values of displacement at the face.

These three simulations indicate that the stiffness of a fully grouted bolt is not only a function of the bolt and the grout properties, but also depends on the rock mass displacement profile. In other words, load development in a cable depends critically on whether displacement is concentrated at a single fracture or is more evenly distributed on multiple or many fractures. The results are also higher for modified geometry cables since the presence of the bulge generates higher radial dilation in the sample, causing higher normal pressure to the cable and consequently, higher axial load in cable.

3.2 *Effect of mining induced stress change*

After installation of cable bolts and excavation of the ore, the state of stress around the cables changes. This, combined with the fact that the bond strength (which is the limiting factor in many failures) is frictional, makes cable bolt capacity susceptible to mining induced stress changes. Kaiser et al. (1992) studied this effect and emphasized that the effect of stress change on cable bolt behaviour is most important in low modulus rocks. In fact, the stress change phenomenon explains why the cable bolting practice which was very successful in cut and fill operations for a long time was often unsatisfactory in open stoping operations (Bywater and Fuller, 1983). The installed cables at the back (in cut and fill operations) perform very well under stress increase situations but in the hangingwall (open stoping), they undergo stress decrease and loose bond capacity. Figures 4 shows the load developed in a 10 m cable installed in a 10 GPa and 100 GPa rock respectively. In a 10 GPa rock a stress decrease of 30 MPa can result in a 75% reduction in the capacity in a conventional cable, so that for a 10m long cable only 4t is mobilized. This drop in capacity is only 25% for a 100 GPa rock. In other words the combination of a poor quality rock mass and mining induced stress decrease can result in completely ineffective cable behaviour. The use of modified geometry cables can significantly improve cable performance in stress decrease situations. For comparison, in a 10 GPa rock, a modified geometry cable can utilize about 13t even for a 30 MPa stress decrease. Obviously this drop in capacity is even less for a 100 GPa rock.

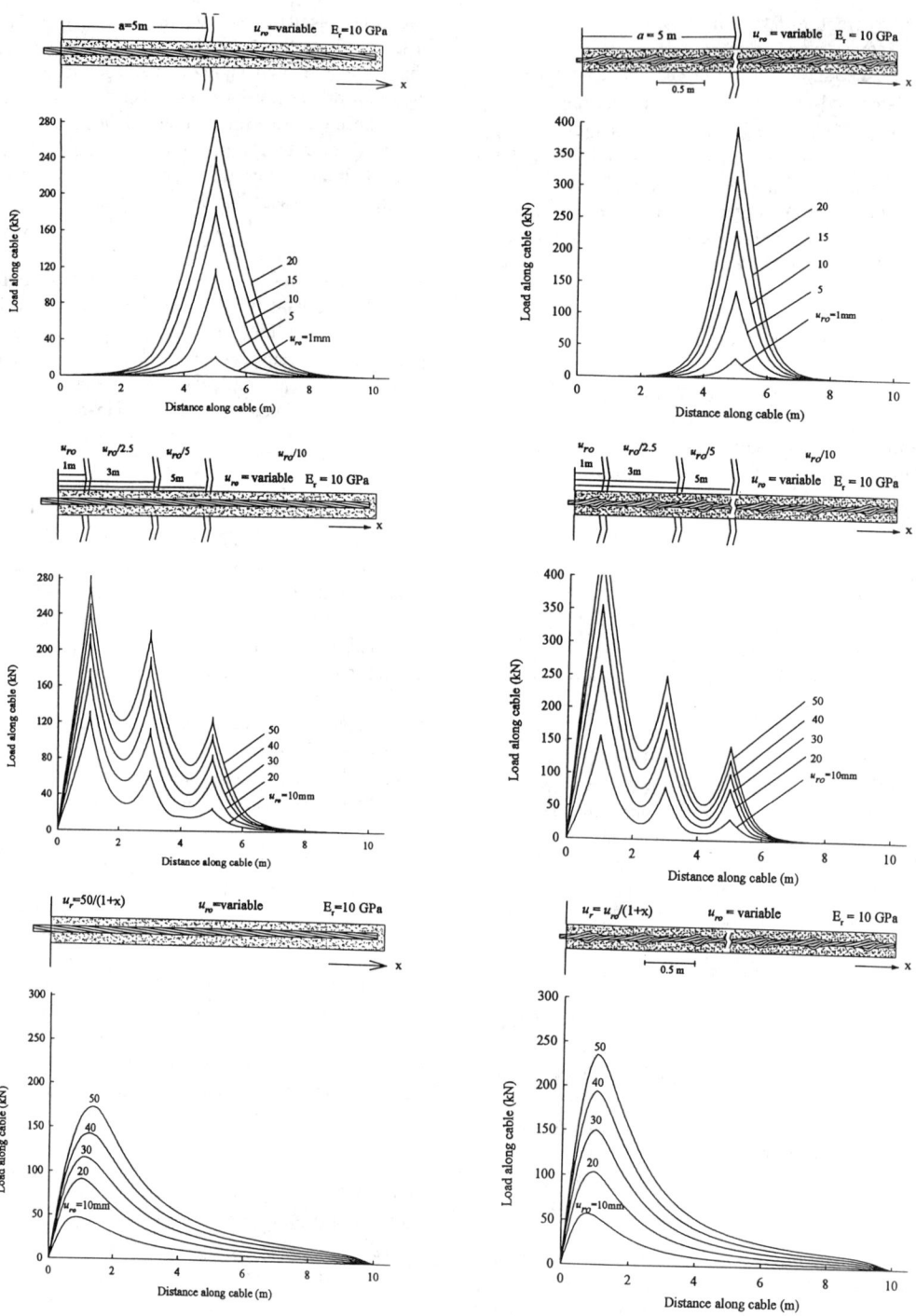

Figure 3: Load distribution for different rock mass displacement profiles in conventional (left) and modified (right) cables.

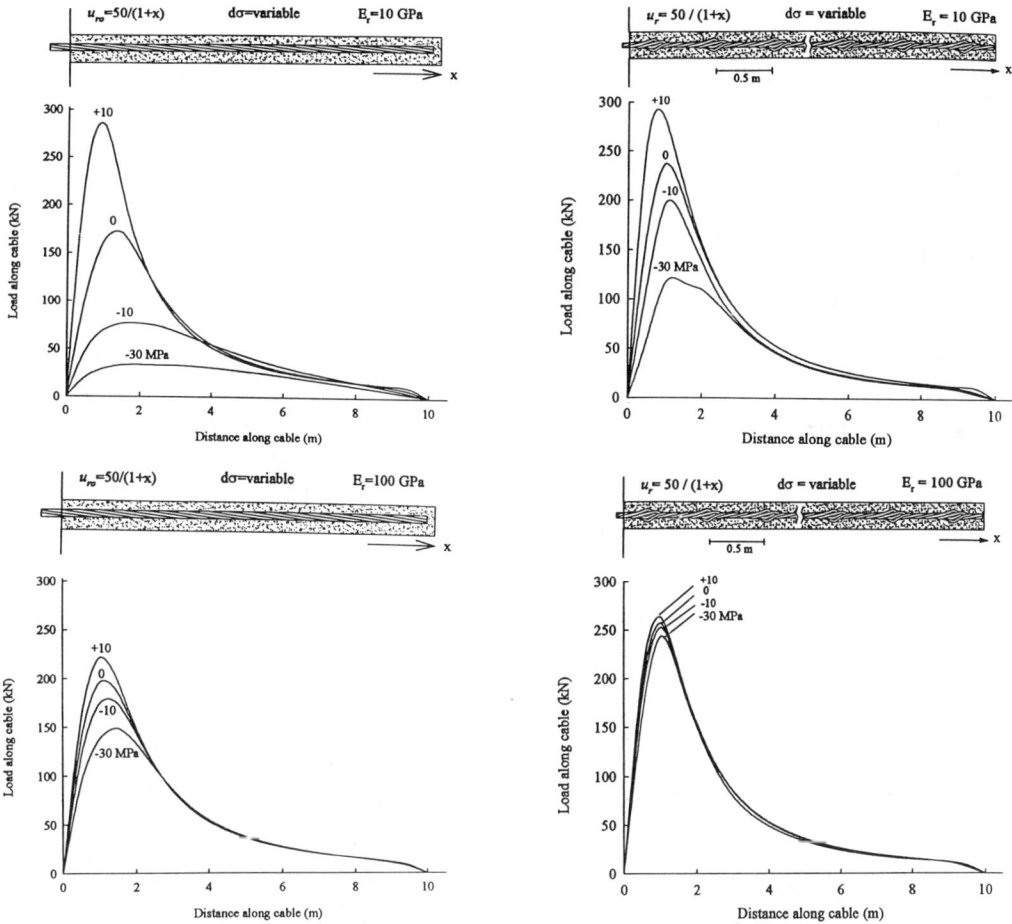

Figure 4: Effect of stress change on the load in conventional (left) and modified (right) cables.

3.3 Effect of faceplate attachment

Where access is available and particularly for situations where mine personnel will be exposed beneath cable bolted ground, attaching a faceplate is strongly recommended. This is also one possible solution to overcome a sloughing failure mechanism. Figure 5 shows the effect of the faceplate as the distance of the crack from the free face (a) is decreased. The plate is especially effective when the fracture is located close to the collar.

In comparison, the load is higher for modified geometry cables but there is a significant difference in the load results when the crack is very close to the face (a=0.25m). In this case the location of the first bulb relative to the crack position becomes very important. Case "A" in Figure 5 represents the situation in which there is a bulb in front of the crack (between the excavation face and the crack) while "B" simulates a case with no bulb in front of the crack. As expected, in the second case there is no difference between a conventional and a modified geometry cable. This effect emphasize the importance of bulb location, particularly when the rock mass is highly laminated. Having a bulb close to the free face in effect acts like a plated cable. In circumstances with no access to attach faceplates, the use of modified geometry cables should be considered as an alternative to plating.

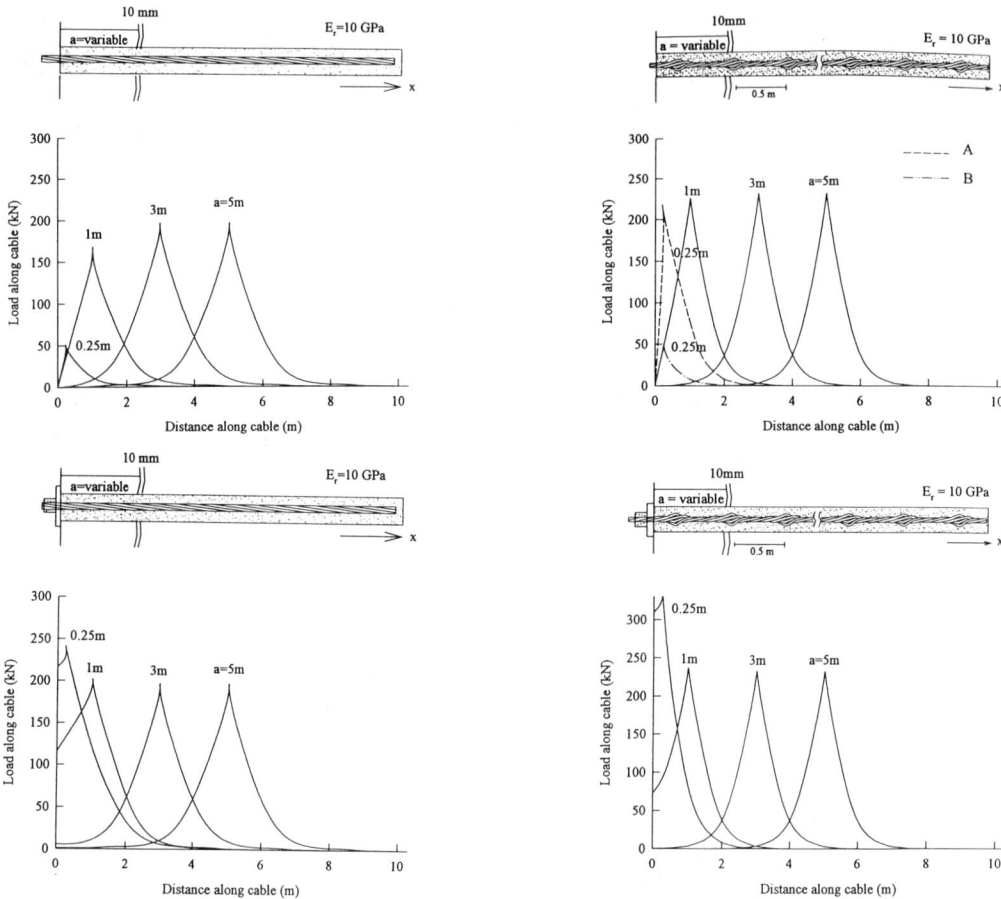

Figure 5: Effect of crack location and plating on the load in cable.

4 CONCLUSIONS

The paper has focused on the load distribution along fully grouted conventional and modified geometry cable bolts. The presence of a deformed structure on the cable, either a Garford bulb or a nutcase, improves the load mobilization response of cables by generating higher radial dilation, which in turn creates higher pressure acting normal to the cable. It seems that by using these types of cables sufficient normal pressure at the cable-grout interface should be able to be mobilized even in poor rock mass conditions (low modulus rock in a jointed environment). For the same reason, it is expected that mine induced destressing should not have as significant an impact on the mobilized load for modified cables as for plain strand. Practically, it has been demonstrated that the use of these cables can result in considerable improvement in reinforcement behaviour including better dilution control and reduced costs due to expanding the borehole pattern (Anderson and Grebenc, 1995).

The importance of the rock mass displacement profile (continuous *vs.* discontinuous) on mobilized load in cable was investigated. If all other parameters are equal, it was shown that a bolt installed in a discontinuum will be much stiffer (more effective) than one in a continuum. This was pointed out earlier by Sakurai and Kawashima (1992) in their physical models and similar differences were recorded earlier in some field observations. Bjornfot and Stephanson (1983) concluded that, in hard rocks where the rock mass is crossed by a limited number of joints, shear stress along the bolt (or axial load) reaches its maximum at the intersection with a major discontinuity. They also pointed out that for specified bond stiffness, higher axial loads are developed if the rock mass displacements are concentrated at a few points rather than continuously distributed along the bolt length. In other words, the effect of reinforcement appears to be underestimated in a continuum approach. This also clarifies why conventional cables have been found more effective in blocky (good quality) rock masses than in heavily jointed (poor quality) conditions.

The predicted mobilized load in cables proposed here will be calibrated based on field studies using a newly invented cable bolt instrument (called SMART cables). The details of the instrument and some initial field results are presented in an accompanying paper (Hyett and Bawden (1997), same conference).

ACKNOWLEDGEMENT

Funding from MRD, INCO, BARRIC Gold Corporation, HBM&S and NSERC is greatly acknowledged.

REFERENCES

Anderson, B. and Grebenc, B. (1995), "Controlling dilution at the Golden Giant Mine". *12th mine operators conference*, Timmins.

Bjornfot, F. and Stephanson, O. (1983), "Interaction of grouted rock bolts and hard rock masses at variable loading in a test drift of Kiirunavaara Mine, Sweden". Stephanson, O. (ed), *Proc. Int. Symp. Rock Bolting*, Abisko, pp 377-395.

Bywater, S. and Fuller, P.G., (1983), "Cable support of lead open stope hangingwalls at Mount Isa Mines Ltd.", *Proc.of the Int. Symp. on Rock Bolting*, Abisko.

Farmer, I.W. (1975), "Stress distribution along a resin grouted anchor". *Int. J. of Rock Mech. and Min. Sci. Geomech. Abstr.* Vol 12, pp 347-351.

Hoek, E. and Brown, E.T., (1980), " Underground excavations in rock", Institution of mining and Metallurgy.

Hyett, A.J, Bawden, W.F., MacSporran, G.R., and Moosavi M. (1995), "A constitutive law for bond failure of fully grouted cable bolts using a modified Hoek cell". *Int . J. of rock mechanics & geomechanical abstracts,* Vol 32, No.1, pp 11-36.

Hyett, A.J., Moosavi, M and Bawden, W.F. (1996), "Load distribution along fully grouted bolts, with emphasis on cable reinforcement". *Int. J. of Num. & Anal. Meth. in Geomech.* Vol. 20, pp 517-544.

Kaiser, P.K , Yazici, S. and Nose, J. (1992), "Effect of stress change on the bond strength of fully grouted cables". *Int . J. of rock mechanics & geomechanical abstracts,* Vol 29, No.3 pp 293-306.

Moosavi, M. (1997), "Load distribution along fully grouted cable bolts based on constitutive models obtained from modified Hoek cells". Ph.D. Thesis, Queen's University, Kingston, Ontario, Canada.

Sakurai, S and Kawashima, I. "Modeling of jointed rock mass reinforced by rock bolts". *Rock Support in Mining and Underground Construction,* pp 547-550, 1992.

Windsor, C.R., (1992), "Invited lecture: Cable bolting for underground and surface excavations", *Rock support in mining and underground construction,* Balkema.

The S.M.A.R.T. cable bolt: An instrument for the determination of tension in 7-wire strand cable bolts

A.J.Hyett, W.F.Bawden & P.Lausch
Department of Mining Engineering, Queen's University, Kingston, Ont., Canada

M.Ruest, J.Henning & M.Baillargeon
Complexe Bousquet, Barrick Gold Corp., Canada

ABSRACT: Design tools have not kept pace with other developments in cable bolting (e.g. improved grouting procedures, modified geometry cables etc.). One of the major problems in the use of fully grouted cable bolts is the lack of a reliable method to directly monitor the tension that develops due to rock mass displacements. Existing gauges are mounted external to the cable itself and therefore potentially interfere with the bond strength in the measurement area. This can make the results extremely difficult to interpret.

This paper describes a novel instrumented cable bolt. The technology is suitable for conventional and modified geometry cables. Twin strand cable bolts can also be instrumented. Since the hardware is internal to the cable, it does not interfere with the bond strength. Results from laboratory calibration and an initial field trial are presented.

1 BACKGROUND

In their comprehensive handbook on cable bolting Hutchinson and Diederichs (1996) introduced the cable bolting cycle (Figure 1), "a cyclical, iterative process which should be worked through a number of times as mining progresses to ensure that the cable bolting process is well tuned". It involves three principal components: *Design, Implementation* and *Verification*. Whereas approximately 70% of the text was devoted to *Design* and 25% to implementation, less than 5% was concerned with verification. The verification process involves an evaluation of the effect of cable bolts on rock mass stability, based on a combination of *Observation* and *Instrumentation*.

For rock bolts, strain gauges adhered to the bolt surface have been used quite effectively (Freeman, 1978; Xueyi, 1983) to measure the strain and hence load developed along the length. However, the complex helical geometry of a cable and the tendency for debonding to involve an untwisting mechanism, complicate the load determination problem for a 7-wire cable. Also, since the vast majority of cable bolts are fully grouted, strain measurement devices adhered to the outside of the cable will affect the bonding process and hence the overall behavior of the cable bolt.

A number of tension measuring devices, all of which attach to the outside of the cable, have been developed. In Canada the *Tensmeg* (Choquet and Miller, 1988) comprises a spiral resistance wire wound into the flutes of the cable.

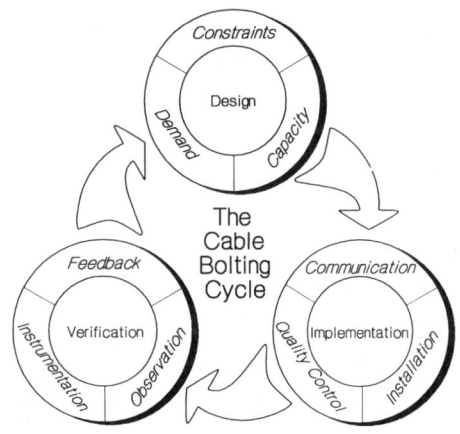

Figure 1: The cable bolting cycle (after Hutchinson and Deiderichs, 1996)

Based on tensmeg results from Winston Lake mine, Maloney et al. (1992) commented that "cable strains and, hence, loads developed in a somewhat random fashion". Both Hutchinson (1992) at Ansil Mine and Goris et al.(1991) at Homestake mine have reported mixed experience with *Tensmeg* cables. In Australia, a number of instruments designed to measure cable bolt load have been developed by CSIRO (Windsor & Thompson, 1992). The Resistance Wire Cable Strain Cell (RWCSC) was used successfully to determine the load distribution that developed in cable bolts at Mount Isa Mine and indicated that the cables were attaining significant loads (up to 250kN). However, these instruments were never made commercially available in Canada. As pointed out by the authors, an inherent problem with these instruments is that load (or more accurately strain) is measured over a finite base length and inclusion of multiple instruments on the same cable to determine the load distribution becomes expensive. Overall, the number of documented cases for which instrumentation has been successfully used to verify cable bolt design are limited.

2 SMART CABLE CONCEPT

If the extension or stretch ($d^i - d^{i+1}$) between two known locations (L^i and L^{i+1}) along a 7-wire strand cable can be measured, the strain (often referred to as elongation) may be written:

$$\varepsilon = \frac{(d^i - d^{i+1})}{L^i - L^{i+1}}. \quad (m/m)$$

For 0.6" low relaxation 7-wire strand (ASTM A416) the corresponding tension is
$F = k\varepsilon$ (kN),
where for the elastic response:
$k = 25000$ kN/m/m ($0 < F < 225kN$)
and for the strain hardening response after yield:
$k = 600$ kN/m/m ($F > 225kN$ $\varepsilon < 3.5\%$).

Thus the average load in the cable can be calculated from the strain between adjacent anchor points, and by using multiple anchor points the load along the whole cable bolt is determined. The instrumented cable described in this paper employs this basic principle. It will be referred to as **S**tretch **M**easurement to **A**ssess **R**einforcement **T**ension or *SMART* technology.

3 SMART CABLE DESIGN

An MPBX incorporated within a cable bolt.

In the manufacture of a SMART cable a miniature multi-point borehole extensometer (MPBX) is constructed within the kingwire of a 7-wire strand cable. To minimize size, spring loaded stainless steel wires rather than rigid rods were used. In the current SMART cable design, 6 wires are secured to the kingwire at specified distances from the instrumentation head where they are spring loaded. Tests have proven that this design is suitable for cables up to 20m in length.

The displacement of the 6 spring-loaded wires caused by stretch of the cable is measured using potentiometers in much the same way as for a commercial MPBX readout head. The advantage of using linear potentiometers rather than strain gauge technology, is that the high-level output (0-5V full range is used) can be easily measured using a low cost readout device. Furthermore, the output is less susceptible to the underground environment.

For a regular MPBX, the cost of a potentiometric readout head is approximately $2000. Since cable bolts are often installed in locations where safe access cannot be guaranteed during mining, it was realized that a low cost disposable electrical readout head was necessary. Hence a major breakthrough in this research has been the development of a reliable potentiometeric readout head. As a further benefit, since the size of the readout head is dramatically reduced it is possible to fully encapsulate it within the cement grout at the collar of the borehole. This leaves only the lead wires emerging from the borehole which are much easier to protect than an exposed head especially in areas such as open stopes and drawpoints.

A series of tests have been used to calibrate the SMART cable. The simplest of these is a tension test. In this test a cable with three anchor points, the locations of which are indicated in Figure 2 was loaded in tension. The applied force (F) and the corresponding response of the SMART cable are shown. The magnitudes of displacement, even for a 1m length of cable, are easily measured to high resolution using potentiometers.

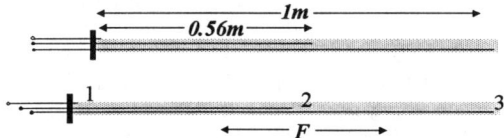

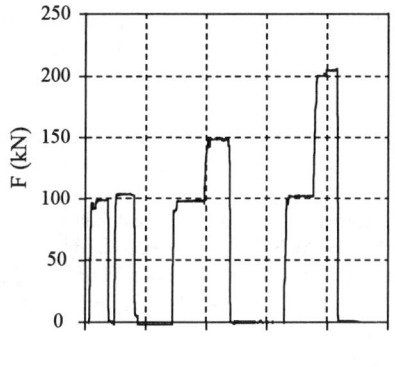

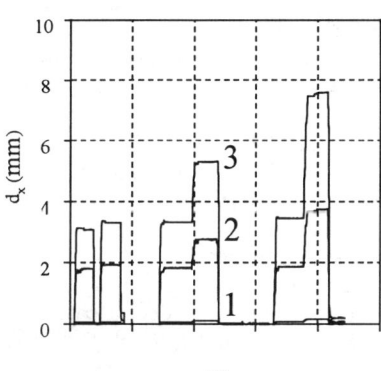

Figure 2: The response of the three anchor points located at $x_1=0.05m$ $x_2=0.56m$ and $x_3=1.0m$, when the cable was subjected to loads up to $200kN$. As expected the anchor point furthest away from the readout ($x_{readout}=0$) exhibits the highest displacement.

4 APPLICATION

9-1-15 Hanging wall at Bousquet # 2 mine.

An experiment was conducted at Barrick Gold Corporation's Bousquet #2 mine to verify the performance of the hanging wall cable bolts and, in particular, to determine whether the design could be improved to reduce costs. Before the experiment, doubt was expressed as to whether the cable bolts used in such situations were in fact being significantly loaded. The experiment was conducted in the H/W of the 9-1-15 stope, which is a primary stope in a relatively new mining block.

Cable bolt layout

The cable bolt design for the 9-1-15 stope specified 8 cable bolt rings (designated 151-158) recessed from the 9-1 H/W access, each comprising a fan configuration of 9 cables ranging from $18m$ to $27m$ in length. The spacing between rings was $2m$ (Figure 3). A single bulge cable with a 12" bulb spacing was placed in each 2.5" ($63mm$) hole, and were grouted from toe to collar using a 0.35 w:c ratio grout.

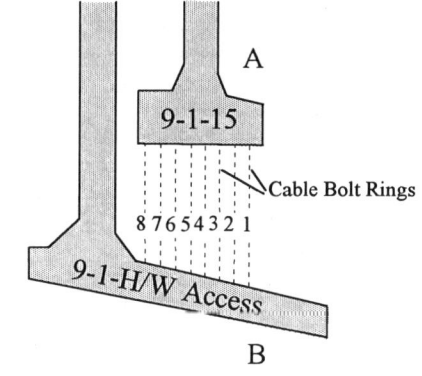

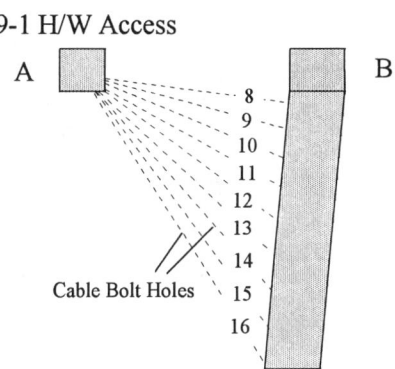

Figure 3: The cable bolt layout for the 9-1-15 stope in plan (LHS) and section (RHS).

A combination of eight SMART cables and two MPBXs were installed. The SMART cables were substituted for regular cables (Figure 4). Four of the readout heads were completely immersed

Figure 4: Installing a SMART cable.

within the grout for protection. Due to operational problems, the two MPBXs were installed close to one another towards one side of the experiment.

The 9-1-15 stope was mined in essentially three stages between November 28th and December 11th (Figure 5). An MPBX located between rings 154 and 155 and coaxial with hole 10 measured relatively insignificant displacements during blasts 1 and 2. However following the main production blast, 231mm of displacement was measured at the furthest anchor point, with over 100mm of this being concentrated between 4m and 7m from the hanging wall. Similar results were obtained from the MPBX located in hole 155-12.

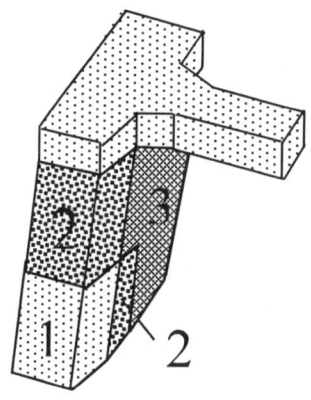

Figure 5: The mining sequence for the 9-1-15 stope.

SMART cable results

Since fully grouted cable bolts behave as a passive reinforcement system, tension develops in response to the rock mass displacements. Therefore, based on the MPBX results, high loads were expected in the cable bolts.

Figure 6 shows the displacement of each of the six anchor points relative to the readout head over the duration of the experiment for three of the SMART cables (153-10, 153-12, 154-10). For the readout head used, the stroke of the potentiometers was limited to 63mm. Due to the length of the cables and the high rock mass displacements, when the production blast was taken, this limit was exceeded for the anchor points closest to the stope. (To solve this problem potentiometers are now being used with a stroke length of 125mm).

During each blast the electronic readout heads allowed the SMART cables to be monitored at 10 second intervals for a 2hr period using a data acquisition system. The detailed response of cable 153-12 to blast 2 is shown in Figure 7. Immediately after the blast the anchor points moved significantly, indicating an increase in load in the cable. During the following 60 mins the anchor points continued to move as the rock mass readjusted to the stress redistribution associated with excavation. These time-dependent movements (which should not be confused with the measurement resolution which is 0.01mm) sometimes affected the whole length of the cable (event A), and sometimes just a particular segment (events B and C). It is evident that the rate at which the anchor points were moving gradually decreased over the monitoring period.

The distribution of anchor point displacement along the length of the cable is shown in Figure 8. Like the MPBX data, the highest displacements occur for those anchor points closest to the stope. However whereas the MPBX describes how the rock is moving, the SMART cable measures how the cable is stretching. If the cable were perfectly bonded to the rock then both would give identical results. However the grout bond is not rigid, and therefore it must undergo some slip as tension is mobilized in the cable. Consequently, the displacements recorded by the SMART cable were less than those for the MPBX. The difference between the two represents the bond slip at the cable grout interface. In this experiment a true comparison between a SMART cable and an MPBX was not possible due to differences in location.

The corresponding average load at the midpoint between anchor points was calculated as described in Section 4 (Figure 9). Since the range

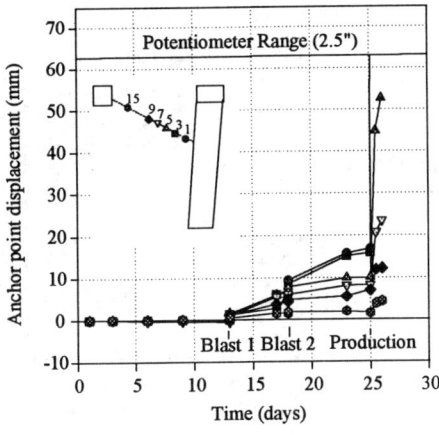

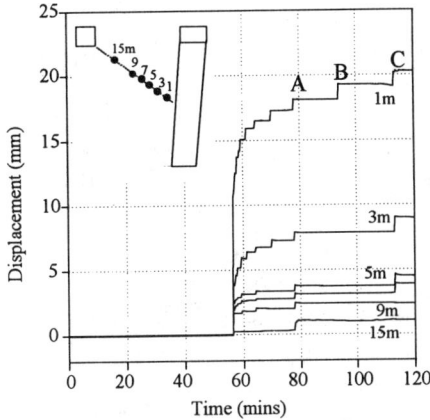

Figure 7. The detailed response of cable 153-12 to Blast 2. Data was recorded at 10s intervals using a data acquisition unit.

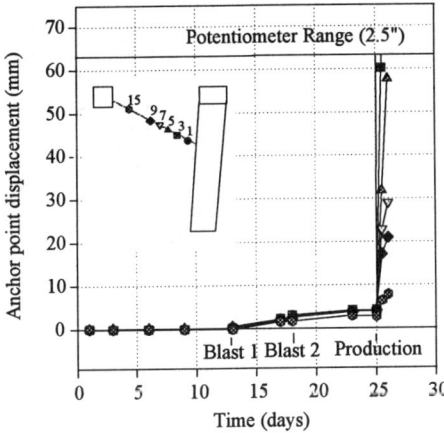

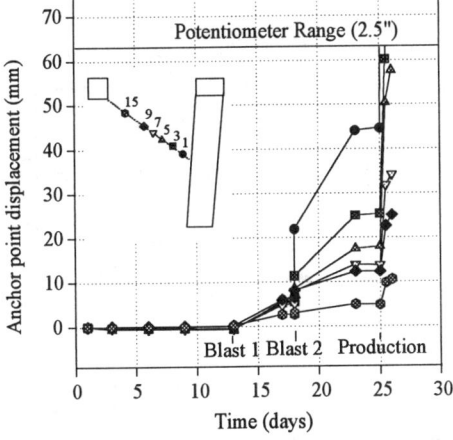

Figure 6: Response with time of the SMART cables 153-10(top), 153-12 (middle) and 154-10.

of the potentiometers was exceeded during the production blast, the detailed load distribution in the immediate hanging wall (based on anchor points 1 and 2) could not be determined (dashed line drawn to known boundary condition F=0 at distance zero). The response of the three SMART cables to blasts 1 and 2 is controlled by their proximity to the excavation created. The hanging wall adjacent to 153-12 was exposed by blast 2. This is reflected by the build up of load in the first few meters of the cable. After the production blast when the stope is fully mined out, the load appears to increase further back along this cable, possibly as the zone of significant rock mass deformation becomes increasingly "deep seated" due to the wider span. In comparison, cable 154-10 which is more remote from the initial blasts, shows virtually no response until the production blast, when it becomes very significantly loaded at a depth of 6*m* from the toe. In fact the response from all three cables (and the MPBX) at this stage appears to indicate a significant amount of opening across a structure that was mapped at 5-6*m* from the hanging wall contact.

CONCLUSIONS

A new instrument that measures the load distribution that develops along a cable bolt has been developed. The instrument is fully enclosed inside the cable so that the bond performance of the cable is not inhibited. A disposable electronic readout unit allows readout using a hand-held unit and data collection using a data acquisition unit.

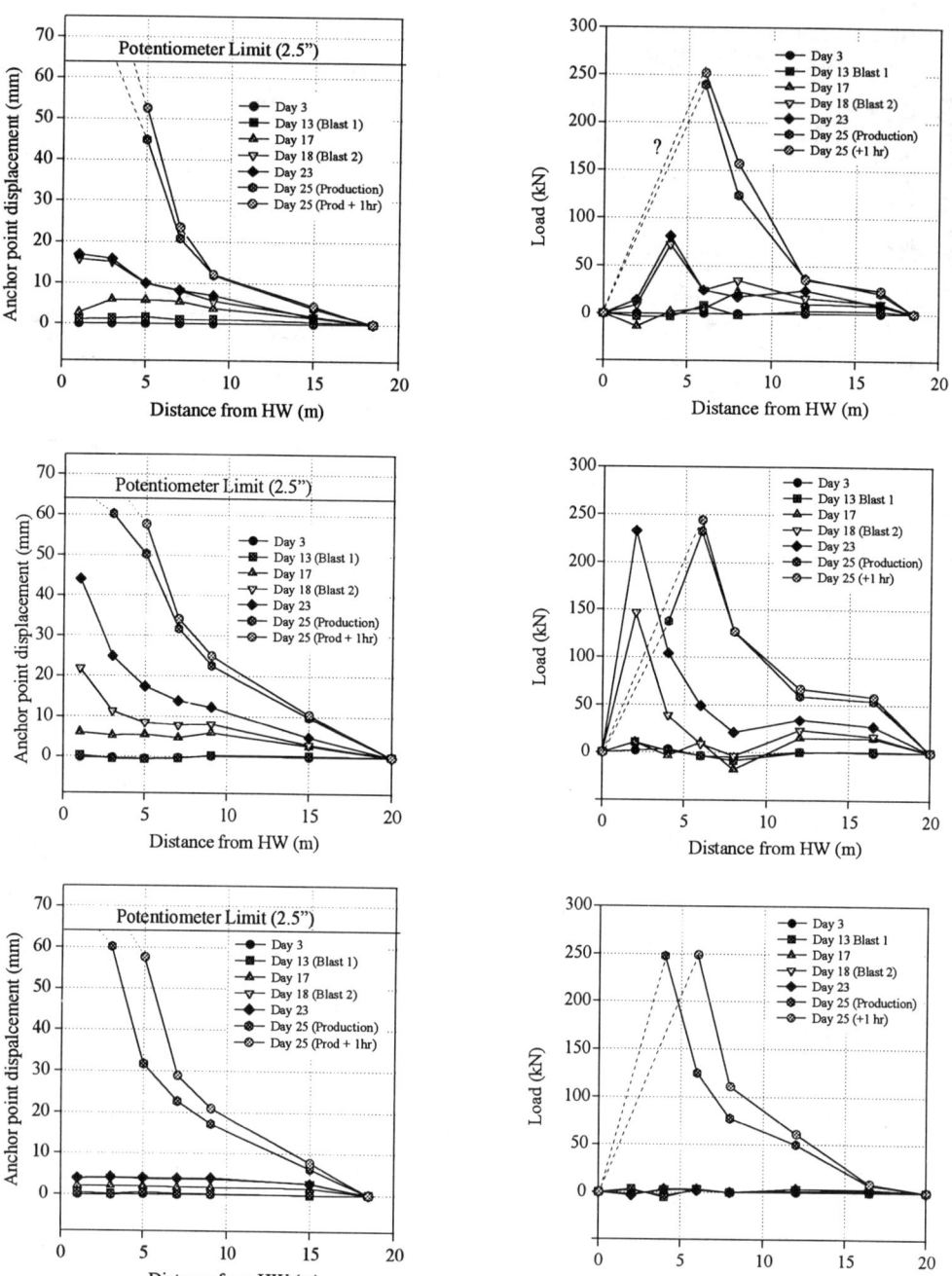

Figure 8: Displacement of the anchor points located at 1, 3, 5 7, 9, and 15m along the cable during the experiment (153-10, 153-12 and 154-10).

Figure 9. Load distribution along the SMART cables.

Based on an initial underground experiment with the instrument the following conclusions can be drawn.

From a technical perspective:

1. They performed equally well when the head was immersed in the grout as when it was left exposed.
2. Following initial installation, the instrument readout remained steady in the 2 weeks preceding mining. They showed no tendency to drift and no effects due to the exothermic reactions associated with curing of the cement grout.
3. The response of the cables was successfully monitored using both a hand-held unit and a data acquisition system. The advantage of an electronic readout was demonstrated in that the response of the cables immediately following a blast was recorded.
4. The anchor point displacements were higher than expected, primarily because of the length of the cables (around 20m), and the high rock mass displacements in the hanging wall (over 200mm). The range of the potentiometers (63mm) was exceeded, and a longer range will be used in future.

From an operational perspective:

1. The cables were quickly and efficiently installed. using the routine procedure.
2. The response to mining was predictable: cable bolts closer to the excavations exhibited higher loads.
3. The results indicated that the cables were fully grouted during installation (sometimes a concern) and that very high loads (exceeding the yield point of the steel, though not the strain required for rupture) were mobilized. There is little doubt that the cable bolts played an important role in stabilizing a H/W subjected to very high rock mass deformations.
4. Detailed response to the blast indicated sporadic loading of the cables in the hour following the blast. The rate of these events decreased with time as stability was progressively attained.
5. Visual observations combined with the CMS survey indicated that the SMART cables behaved identically to the regular cable bolts (*i.e.* no better, no worse).

ACKNOWLEDGEMENTS

This work is funded from both a joint industry initiative between Barrick Gold, Inco and HBMS and Queen's University combined with MIROC (now CAMIRO) and with matching NSERC CRD funding. The authors are indebted to P.Gauthier for his support and permission to publish this research.

REFERENCES

Goris, J.M., Duan, F. and Pfarr, J. 1990. Evaluation of cable supports at Homestake mine.*CIM Bulletin*, 84, 146-150.

Hutchinson, D.J. and Diederichs, M.S. 1996. Cable bolting in Underground Mines. BiTech. Publishing Ltd. Richmond, B.C., Canada.

Hutchinson, D.J. 1992. A field investigation of cable bolt reinforcement of open stopes at Ansil mine. Ph.D. thesis. University of Toronto.

Kaiser, P.K. Yazici, S and Nose, J. 1992. Effect of stress change on the bond strength of fully grouted cables. *Int. J. Rock. Mech. Min. Sci & Geomech. Abstr.* **29(3)**, 293-306.

Maloney S., Fearon R., Nose J., and Kaiser P.K. 1992. Investigations into the effect of stress change on support capacity. *Rock Support* (eds. Kaiser, P.K. and McCreath, D) A.A. Balkema, Rotterdam.

MacSporran G.R., Bawden W.F., Hyett A.J., Hutchinson D.J., and Kaiser P.K. 1992. An empirical method for the analysis of failed cable bolted ground: Research in Progress. 94th CIM AGM, Montreal.

Windsor, C.R. and Thompson A.G. 1993. Rock Instrumentation - evelopments and case studies from Australia. *Comprehensive Rock Engineering. Principles, Practises and Projects.* (ed. Hudson, J.A). Pergammon Press, **5**, 193-225, Oxford.

Application of boundary element method for displacement back analysis of tunnels in viscoelastic rock mass*

Lin Xue & Rong Luo
Qingdao Institute of Architecture and Engineering, People's Republic of China

ABSTRACT: The viscoelastic distinguish model and parameters back analytical method is proposed in this paper. The theory is clear, the method is simple. The back analysis theory and the boundary element method is combined with. And the engineering example of displacement back analysis is given.

KEY WORDS: distinguish model, parameters back analysis, analytical method, boundary element method, engineering example

1 INTRODUCTION

The constitutive relationship of rock is very important in the rock engineering design and construction. According the displacement data obtained by the construction site, the distinguish rock mechanical model is foundation of mechanical parameters back analysis.

Recently, the creep compliance back analysis of continuity mechanics in rock develops in theory[1,5]. That is: the creep compliance back analysis theory of continuity mechanics in rock is established, the distinguish rock mechanical model and parameters back analysis is obtained by the displacement monitoring data in truncated viscoelastic mechanical model, in the mechanics[1,2]. In the mathematics, to study viscoelastic plane problem using complex variable, the new theory formula[1] is used for viscoelastic boundary displacement analysis for the arbitrary hole.

According to the research development of the creep compliance back analysis and the engineering application, we must solve the combination problem of back analytical method and numerical method.

In this paper, the combination method of the creep compliance back analysis and boundary element method for the viscoelastic is given and it is used in the engineering application successfully.

2 TWO ANALYTICAL METHODS FOR THE VISCOELASTIC DISTINGUISH MODEL

Recently, the two analytical methods for tunnels in rock engineering and midde and small test pieces are found in viscoelastic distinguish mechanical model and parameters back analysis.

Firstly, five criterion theorems[1] of viscoelastic mechanical model of rock is used in plane strain problems. The inverting values of viscoelastic tunnels creep compliance and generalized creep compliance are obtained by different mechanical models of distinguish rock criterion theorem, and then, according to different mechanical model inverting formula the mechanical parameters of rock are calculated[4].

Furthermore, to study distinguish model and parameter back law, we find no relationship between diffevent viscoelastic model mechanical parameters are the inverting vesults in back problems. It has proved[3]:

Theorem 1. For kelvin model. H-K model, H/M model and M-K model, the back values after delayed time τ_d are no relationship with distinguish model.

Theorem 2. For kelvin model. H-K model and cascade kelvin model in M-K model, the back analysis of mechanical parameters is no relationship with the different model.

Theorem 3. The distinguish of H-K model and

---

* Supported by NSFC (No. 49372153)

H/M model is equal to the inverting result.

The generalization formula is shown in reference 3. The generalization formulas with two mechanical parameters of kelvin model. H-K model and cascade kelvin model in M-K model are shown in reference 3.

If the parameter back formula of M-K model is used as the parameter generalization formulas for kinds of different viscoelastic mechanical model, it has proved that the back result of parameter $\eta_{M-K,M}$ (the cohesional coefficient of cascade Maxwell model) is unfinite using H-K model and H/M model for viscoelastic body. The two back parameters are unfinite using kelvin model and cascade Maxwell model for viscoelastic body. That is[3]:

Theorem 4, the back values of parameters E_k, η_k are fix, if the parameter $\eta_{M-K,M}$ is ∞, then the distinguish models are kelvin model, H-K model and H/M model.

Theorem 5, the back values of parameters E_L and τ_d are fix, if the parameter Ee is ∞, then the distinguish model is kelvin model.

Here, E_k and η_k are two mechanical parameters of kelvin or cascade kelvin model in H-K model and M-K model. E_L is the long-term elastic modulus. Ee is the instantaneous elastic modulus. These parameters are shown in reference 1,3,5.

Now, the distinguish problems of rock viscoelastic mechanical model has given more attention in the world. We explain again: it is no necessary using all kinds of different mechanical model for the inverting problems of viscoelastic body. We think that only model is Burgers model (M-K model) and other models are special model of Burgers model, and then the problem is simple[2][3]. Now we proposed the second analytical method for distinguish model and parameter back analysis. The resolution criterion and the resolution formula are given. This method is used in plane strain problem and plane stress problem and also used in one way or multiway state of stress in site test or lab. The viscoelastic mechanical model is distinguished by the inverting value of creep compliance and the mechanical parameters are also obtained. This method is simple and it is very easy used in engineering.

3 APPLICATION OF BOUNDARY ELEMENT METHOD FOR DISPLACEMENT OF TUNNELS IN VISCOELASTIC BODY

A new method is proposed to solve the displacement of tunnels in viscoelastic body with boundary method[6]. For viscoelastic plane problem, the complex variable functions of displacement are[1][6]:

$$u(t) + iv(t) = X_1(z)J_1(t) + X_2(z)J_2(t) \quad (1)$$

$J_1(t)$ and $J_2(t)$ are creep compliance and generalization creep compliance. $X_1(z)$ and $X_2(z)$ is the functions of complex variable.

For the elastic plane problem, the complex variable functions of displacement are

$$u + iv = \frac{1}{2G}X_1(z) + \frac{1}{2(3K+G)}X_2(z) \quad (2)$$

G is the modulus of shearing, K is the bulk modulus.

The analytical function $X_1(z)$ and $X_2(z)$ is relationship with the boundary conditions of stress geometry, size of zone, the displacement of z, it is no relationship with the mechanical parameters. So the analytical function $X_1(z)$ is same with $X_2(z)$ if these factors are same in elastic plane problem and viscoelastic plane problem.

To use boundary element method for elastic plane problem, the displacement $u_x(p)$ and $u_y(p)$ for the point p of tunnels are calculated. If the point z in equation (2) is same as the point p. So the point p of boundary element mumeral method is

$$u_x(p) + iu_y(p) = \frac{1}{2G}X_1(z) + \frac{1}{2(3K+G)}X_2(z) \quad (3)$$

We select two group of mechanical parameters G_1, k_1 and G_2, k_2 under allowance value, the displacement of point p is get from boundary element method, then the system of linear equations is obtained from equation (3):

$$\left.\begin{array}{l} a_i Q_{1x}(z) + b_i Q_{2x}(z) = u_{xi}(p) \\ a_i Q_{1y}(z) + b_i Q_{2y}(z) = u_{yi}(p) \end{array}\right\} \quad (4)$$

Where, $Q_{ix}(z)$ and $Q_{iy}(z)$ is the generalized stress of the point z, i=1,2 and $Q_{ix}(z)=\text{Re}Xi(z)$; $Q_{iy}(z) = 1mX_i(z)$, $a_i = \frac{1}{2Gi}, b_i = \frac{1}{2(3K_i+G_i)}, i=1,2$

The numerical results of analytical functrons are get from the equation(4)

$$\left.\begin{array}{l} X_1(p) = Q_{1x}(p) + iQ_{1y}(p) \\ X_2(p) = Q_{2x}(p) + iQ_{2y}(p) \end{array}\right\} \quad (5)$$

Where, $Q_{ix}(p), Q_{1y}(p), Q_{2x}(p), Q_{2y}(p)$ is the numerical result of generalized stress.

For example, infinitesimal zone hole under homogeneous stress boundary conditions is the elastic plane problem, the error between numerical result and theory result is under 2%.

The displacement result of viscoelastic plane problem under arbitrary hole and arbitrary boundary conditions from the equation (1) and the equation (5) is obtained

$$u_x(p,t) + iu_y(p,t) = X_1(p)J_1(t) + X_2(p)J_2(t) \quad (6)$$

Where, $u_x(p,t)$ and $u_y(p,t)$ is the viscoelastic displacement compoment of the point p (or point z) at time t

The two analytical functions depend on numerical result in the equation (6); For the tunnels analysis, the new method is simple because of ignoring Laplace transform of boundary element method. So we call this method in the equation (6) for viscoelastic plane problem as semi-analytical formula.

4 THE SEMI-ANALYTICAL METHOD OF DISPLACEMENT ANALYSIS

For arbitrary hole tunnels, the displacement $u_x(p,t)$ and $u_y(p,t)$ of the point p (or z) in the tunnels are get from measurement, then the system of linear equations are obtained from the equation (b).

$$\left. \begin{array}{l} Q_{1x}(p)J_1(t_k) + Q_{2x}(p)J_2(t_k) = u_x(p,t_k) \\ Q_{1y}(p)J_1(t_k) + Q_{2y}(p)J_2(t_k) = u_y(p,t_k) \end{array} \right\} \quad (7)$$

Where, $Q_{1x}(p), Q_{2x}(p), Q_{1y}(p), Q_{2y}(p)$ is the numerical results of the generalized stress of the point p.

The creep compliance $J_1(t_k)$ and the generalized creep compliance $J_2(t_k)$ are get from equation (7) at any time $t_k, k=1,2,\cdots,n$. So the mechanical model of distinguish tunnels and back mechanical parameters are get from one of the two analytical method above.

5 ENGINEERING APPLICATION

There is one metre radius hole of the tunnels, it is bonaogeneous continuous isotropic viscoelastic body. The geothermal stress $\sigma_x = 2.5$ MPa, $\sigma_y = 5$ MPa, $u = 0.25$. The measurement results are obtained from reference 7.

The generalized stress get from different methods are shown in table 1, the creep comliance results are shown in table 2.

The distinguish mechanical model of rock and back mechanical parameters are obtained using resolution criterion of the viscoelastic mechanical model[2]. Here the inverting formula of the mechanical parame-

Table 1. Generalized stress analysis of measument points of circle diversion tunnel

coordinate of point		boundary element method		theory method	
r (m)	0 (degree)	$Q_{1r}(p)$ MPa·m	$Q_{2r}(p)$ MPa·m	$Q_{1r}(z)$ MPa·m	$Q_{2r}(z)$ MPa·m
1.5	90°	1.185	2.463	1.204	2.5
2.5	30°	2.062	−2.156	2.085	−2.25
1.5	120°	1.219	1.209	1.227	1.25
1.5	0°	1.282	−2.440	1.296	−2.5
1.15	0°	0.593	−0.947	0.591	−0.978

Table 2. The inverting valve of the creep compliance of rock

time	boundary element analytical method	analytical method
days	$J_1(t_k) \times 10^{-4} \frac{1}{\text{MPa}}$	$J_1(t_k) \times 10^{-4} \frac{1}{\text{MPa}}$
10	1.408142	1.393189
20	1.43193	1.416711
30	1.463038	1.447476
40	1.487775	1.471947
60	1.524902	1.508614
80	1.556760	1.540215

ters is algebra formula. It shows the rheological law of the rock. When the engineering application, the results must be considered as possible[2]. Generally, the inverting mechanical model and the inverting mechanical parameters can be accepted using actual displacement at stationary creep stage.

The actual displacement in reference 7 is at stationary creep stage after 20 days. The mechanical model and mechanical parameters can be determined with the inverting value of the creep compliance after 20 days, 40 days, 60 days, 80 days. The results are shown in Table 3.

From the table 3, the four mechanical parameters are obtained, so it is Burgers Body.

The effect of the inverting results causing by measurement error of displacement will be discussed in another paper.

Table 3. The inverting vesults after 20days, 40days, 60days, 80days

mechanical parameter	boundary element analytical method	analytical method
E_1 MPa	7636	7716
E_2 MPa	77635	78668
η_1 MPa·d	6715261	6784733
η_2 MPa·d	1224777	1245109
τ_d d	15.77	15.83

6 CONCLUSION

A new method combined with the boundary element method and the creep compliance method is proposed in this paper for elastic plane problem. This method is simple. First, the system linear equations in composed of the generalized stress from the boundary element method and the creep compliance and the generalized creep compliance. Second, using the resolution theorem of viscoelastic mechanical model, the distinguish mechanical model of rock and back mechanical parameters are obtained. This new method solves the viscoelastic distinguish model of complex hole on the tunnels and the inverting mechanical parameters problem. So the inverting theory of the creep compliance for continuum rock has wide use in the future.

REFERENCES

[1] Xue Lin, Theorem of Identification for Viscoelastic Model of the Rock Mass and Applications, J. of Rock-soil Engineering (China), 1994, Vol,16, No5, 1~10

[2] Xue Lin, Criteria for Mechanical Model of Viscoelastic Body and Their Applications in Rock Tests, Rock and Soil Mechanis (China), 1996, Vol,17, No1, 9~16

[3] Xue Lin, Study on Viscoelastic Machanical Model Recognition of Rock Mass and Methods of Parameter Inversion Analysis, Journal of Engineering Geology (China), 1995, Vol,2, No1, 70~77

[4] Xue Lin, Yang zhifa, The Analytical method of Viscoelustic Mechanical Parametevs of Rock, Annual Opening lab Report of Chinese Academy of Geonomy Sciences, Earthquake Press, 1993, 71~85

[5] Xue Lin, Displacement Theory of Rock Mass Mechanics and Their Engineering Applications, Beijing, Chinese Science Technology press, 1996, 617~623

[6] Xue Lin, Wang Rong, The New Method for Viscoelastic Displacement of the Tunnels using Boundary Element Method, Engineering Mechanics, Tsinghua University Press, 1996, 261~265

[7] Yuan Yong, et al, The Inverting Distinguish Theory of Rock Constitutive Model, Computer Method on Rock Mechanics and Engineering Application, Wuhan Survey and Map University, Wuhan, Vol.1 337~343

Limitations of ubiquitous joint models

Ethan M. Dawson
Dames and Moore, Los Angeles, Calif., USA

Yeon-Jun Park
Department of Civil Engineering, University of Suwon, Korea

ABSTRACT: Jointed rock is often modeled using ubiquitous joint models, anisotropic plasticity models with yield conditions that simulate slip along joint sets. In this paper, a ubiquitous joint model is derived for a rock mass cut by two sets of continuous joints. The model is used to compute the bearing capacity of a footing resting on jointed rock. Comparison to a series of Distinct Element simulations with different joint spacings, suggests that ubiquitous joint models are only appropriate when the joint spacing is small.

1 INTRODUCTION

Equivalent continuum models for jointed rock, often referred to as compliant joint models or ubiquitous joint models, are a commonly used tool in rock mechanics. In equivalent continuum models, a discontinuous rock mass is treated as a homogeneous material with bulk properties that capture the overall behavior of the system of blocks and joints. This is in contrast to discrete models, such as Distinct Element models, in which each individual joint and block in the rock mass is simulated explicitly.

Elastoplastic equivalent continuum models. which account for failure and slip along joints, have been developed by Morland (1974), Zienkiewicz and Pande (1977), Thomas (1982), Blanford et al. (1987), Chen (1989) and others. These are essentially anisotropic plasticity models. The material is assumed to contain an infinite number of slip planes at predetermined orientations: the orientations of the joints. Plastic slip on joints is usually assumed to be governed by the Coulomb failure criterion

$$|\sigma_s| \leq -\sigma_n \tan \Phi + c \quad (1)$$

where Φ is the joint friction angle and c is the joint cohesion. Note that tensile stresses are positive. Anisotropic, plastic yield conditions are then derived from the conditions for slip on the joint planes.

The fundamental assumption in these models is that the stress acting on the joints is the same as the stress acting on a plane in the continuum model at the same orientation as the joints. In other words, the normal stress σ_n and the shear stress σ_s acting on a joint are given by

$$\sigma_n = \sigma_{ij} n_j n_i \quad \sigma_s = \sigma_{ij} n_j t_i \quad (2)$$

where n_i and t_i are the components of the unit normal and unit tangent vector to the joint plane and σ_{ij} is the stress tensor for the equivalent continuum model. These normal and shear stresses can then be used in the joint yield condition (e.g. (1)) to produce yield conditions for the equivalent continuum model.

Equivalent continuum models are quite accurate for homogenous stress states. In other words, they closely approximate the behavior of discrete models employing the same constitutive assumptions for the joints (see for example Chen (1989)). However, for problems involving non-uniform stress fields, the behavior of discrete and continuum models can be quite different. Equivalent continuum models predict greater displacements (Petney and Blanford, 1989), collapse loads that are too low (Besdo, 1985; Dawson and Cundall, 1992, 1996; Dawson, 1995) and incorrect failure mechanisms (Cundall and Fairhurst, 1986).

One reason for the poor performance of ubiquitous joint models is that models based on the simple continuum cannot account for the bending moments transmitted between neighboring blocks

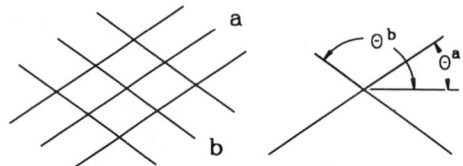

Figure 1: Rock mass with two sets of joints at angles θ^a and θ^b from the x_1-axis.

in a rock mass. This issue has been addressed in studies by Besdo (1985), Mühlhaus (1989,1990), Dawson and Cundall (1992, 1996), Dawson (1995), and Adhikary et al. (1995) by developing ubiquitous joint models based on the micropolar (Cosserat) continuum.

In this paper we examine another reason for the differences between discrete and continuum models. A ubiquitous joint model for a rock mass with two joint sets is applied to determine the bearing capacity of a footing. Numerical simulations and a limit analysis upper bound solution are compared to Distinct Element simulations done for various joint spacings. The results show that since ubiquitous joint models are free to fail along the most critical failure surface, they usually predict lower collapse loads than discrete models, which can only fail at pre-existing joint locations.

2. UBIQUITOUS JOINT MODEL

We present here a ubiquitous joint, anisotropic plasticity model for a rock mass cut by two sets of joints at arbitrary orientations, as shown in Figure 1. We label the two joint sets 'a' and 'b'. The orientations of the joint sets are described by the angles θ^a and θ^b between the joint planes and the positive x_1-axis. The model allows for plastic slip along joints and plastic tensile failure of the joints, but does not allow for plastic failure of the intact rock between the joints.

The ubiquitous joint model is elastic–perfectly plastic and is based on the conventional flow theory of plasticity. The stresses and strains for the plane strain model are written in the vector notation:

$$\boldsymbol{\sigma} = \begin{bmatrix} \sigma_{11} \\ \sigma_{22} \\ \sigma_{12} \end{bmatrix} \qquad \boldsymbol{\epsilon} = \begin{bmatrix} \epsilon_{11} \\ \epsilon_{22} \\ \gamma_{12} \end{bmatrix} \qquad (3)$$

where $\gamma_{12} = u_{1,2} + u_{2,1}$, with displacement components u_i.

As usual, the strain rate is decomposed into an elastic part and a plastic part

$$\dot{\boldsymbol{\epsilon}} = \dot{\boldsymbol{\epsilon}}^e + \dot{\boldsymbol{\epsilon}}^p \qquad (4)$$

and the stress rate $\dot{\boldsymbol{\sigma}}$ is related to the elastic strain rate by

$$\dot{\boldsymbol{\sigma}} = \boldsymbol{D}\dot{\boldsymbol{\epsilon}}^e = \boldsymbol{D}\left(\dot{\boldsymbol{\epsilon}} - \dot{\boldsymbol{\epsilon}}^p\right) \qquad (5)$$

where $\boldsymbol{D}$ is a matrix containing the elastic constants.

To specify the plastic strain rates, a set of yield conditions along with a set of plastic potentials is required. We describe plastic failure of the jointed material with four yield conditions f_i $(i = 1 \ldots 4)$. Corresponding to each yield condition is a plastic potential function g_i. The plastic strain rate is expressed in terms of the gradients of these plastic potentials.

$$\dot{\boldsymbol{\epsilon}}^p = \sum_i \lambda_i \frac{\partial g_i}{\partial \boldsymbol{\sigma}} \qquad (6)$$

where

$$\lambda_i = \begin{cases} \geq 0 & \text{if } f_i = 0 \\ = 0 & \text{if } f_i < 0 \text{ or } (f_i = 0 \text{ and } \dot{f}_i < 0) \end{cases} \qquad (7)$$

The condition $(f_i = 0$ and $\dot{f}_i < 0)$ implies that the material is at the yield surface but is unloading elastically.

We number the yield conditions and plastic potentials as follows:

f_1, g_1: Sliding on joint set a.
f_2, g_2: Tension on joint set a.
f_3, g_3: Sliding on joint set b.
f_4, g_4: Tension on joint set b.

The yield conditions are simply the conditions for Coulomb sliding or tensile failure of the joints and have the form:

$$f_1 = |\sigma_s^a| + \sigma_n^a \tan \Phi^a - c^a = 0 \qquad (8)$$
$$f_2 = \sigma_n^a - T^a = 0 \qquad (9)$$
$$f_3 = |\sigma_s^b| + \sigma_n^b \tan \Phi^b - c^b = 0 \qquad (10)$$
$$f_4 = \sigma_n^b - T^b = 0 \qquad (11)$$

where $\sigma_n = \sigma_{lk} n_l n_k$ is the normal stress acting on a plane in the continuum model at the same orientation as the joints and $\sigma_s = \sigma_{lk} n_l t_k$ is the shear stress, while T^a and T^b are the tensile strengths of joint sets 'a' and 'b', respectively, and c^a and c^b are the shear strengths of joint sets 'a' and 'b'.

For tensile failure we assume an associated flow rule, but for sliding failure we allow for a non-associated flow rule in which the joint dilation angle, ψ^a or ψ^b, can be less than the joint friction angle Φ^a or Φ^b. The plastic potentials have the form:

$$g_1 = |\sigma_s^a| + \sigma_n^a \tan \psi^a \quad (12)$$
$$g_2 = \sigma_n^a \quad (13)$$
$$g_3 = |\sigma_s^b| + \sigma_n^b \tan \psi^b \quad (14)$$
$$g_4 = \sigma_n^b \quad (15)$$

3. NUMERICAL IMPLEMENTATION

The elastoplastic model described above has been implemented in the explicit-finite-difference code, FLAC (Itasca Consulting Group, 1993) as a user-defined consitutive model using FLAC's built-in FISH macro language.

Although it is a finite difference code, FLAC is quite similar to a finite element code. For a given set of element shape functions, the set of algebraic equations solved by FLAC is identical to that solved in the finite element method. However, in FLAC, this set of equations is solved using dynamic relaxation (Otter, 1966). Dynamic relaxation is an explicit time marching procedure in which the full dynamic equations of motion are integrated step by step. Static solutions are obtained by including damping terms which gradually remove kinetic energy from the system.

4. BEARING CAPACITY PROBLEM

The ubquitous joint model is tested by comparing it to a discrete model for determining the bearing capacity, P, of a footing of width, b, resting on jointed rock. The rock mass is cut by two sets of joints with spacing s: one set dipping at an angle α and the other dipping at an angle $-\alpha$. For simplicity we assume that the shear strength of the joints is due entirely to a cohesion c and that the joint friction angle Φ is equal to zero. Distinct Element models of the problem are shown in Figures 5 and 6.

4.1 Ubiquitous Joint Numerical Simulation

For the numerical simulation using FLAC, we take advantage of the symmetry of the problem and

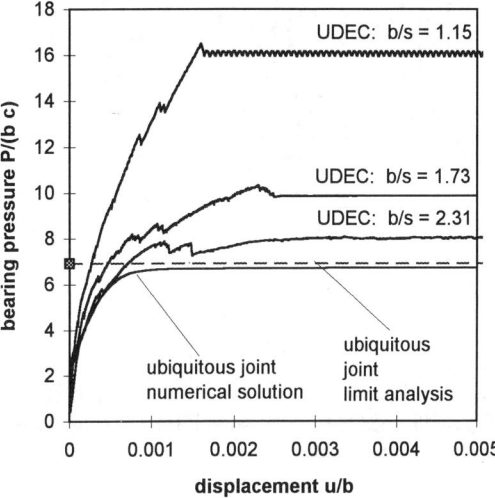

Figure 2: Bearing pressure versus displacement for Distinct Element simulation, ubiquitous joint numerical simulation and ubiquitous joint upper bound solution.

model only the right half of the problem. Along the left edge of the mesh, horizontal displacements are fixed, while along the bottom and right edge of the mesh, both degrees of freedom are fixed. The bearing capacity problem is simulated by giving a small velocity to the footing nodes and summing the vertical reaction on these nodes as they move downward. Force versus displacement for the ubiquitous joint model is shown in Figure 2.

The velocity field for the ubiquitous joint model (Figure 3) shows that a well defined failure mechanism develops, similar to the typical bearing capacity failure mechanism for cohesive soils. This is rather surprising because it requires shearing along a horizontal plane, not coincident with the joint planes. However, Distinct Element simulations presented below (see Figures 5 and 6) fail by a similar mechanism.

4.2 Limit Analysis Kinematic Solution

To verify the results of the ubiquitous joint numerical simulation, we derive a limit analysis solution to the bearing capacity problem. We can derive an upper bound solution by computing the energy dissipated by any kinematically admissible velocity field. The numerical solution using FLAC suggests

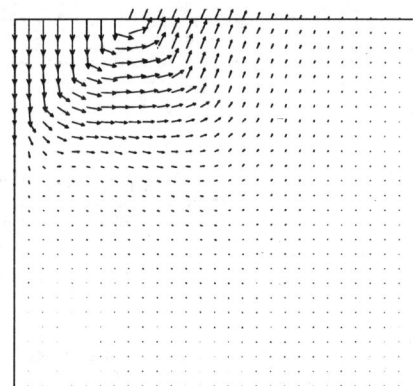

Figure 3: Velocity field at failure for ubiquitous joint numerical simulation

a velocity field composed of rigid blocks separated by plane velocity discontinuities as shown in Figure. 4. This velocity field contains discontinuities aligned with the joint planes as well as two horizontal velocity discontinuities which are not aligned with the joint planes.

The energy dissipated by a velocity discontinuity aligned with a joint set can be derived by considering a rotated coordinate system with the x_1-axis parallel to the joint plane. Using joint set 'a' for example, we then have $\theta^a = 0$. With the dilation angle ψ^a set equal to zero, the flow rule can be written in matrix form as

$$\dot{\boldsymbol{\epsilon}}^p = \begin{bmatrix} \epsilon_{11}^p \\ \epsilon_{22}^p \\ \gamma_{12}^p \end{bmatrix} = \lambda_1 \begin{bmatrix} 0 \\ 0 \\ \text{sign}(\sigma_s^a) \end{bmatrix} \quad (16)$$

where $\text{sign}(\sigma_s^a)$ is the sign of the shear stress acting on joint set 'a'. In the rotated coordinate system, the shear stress acting on the joint plane is simply σ_{12}. If the friction angle, Φ^a, is equal to zero, then $\sigma_{12} = c^a$. Thus the rate of work dissipated per unit volume, $\dot{d}$, is given by

$$\dot{d} = \dot{\gamma}_{12}\,\sigma_{12} = |\dot{\gamma}_{12}\,c^a| \quad (17)$$

and for a shear velocity discontinuity aligned with the joint plane, the rate of work dissipated per unit area is

$$\dot{d} = |[\dot{u}_s]|\,c^a \quad (18)$$

We next compute the rate of energy dissipation for a horizontal velocity discontinuity, demonstrating that the flow rule (6) for the ubiquitous joint model allows for velocity discontinuities along

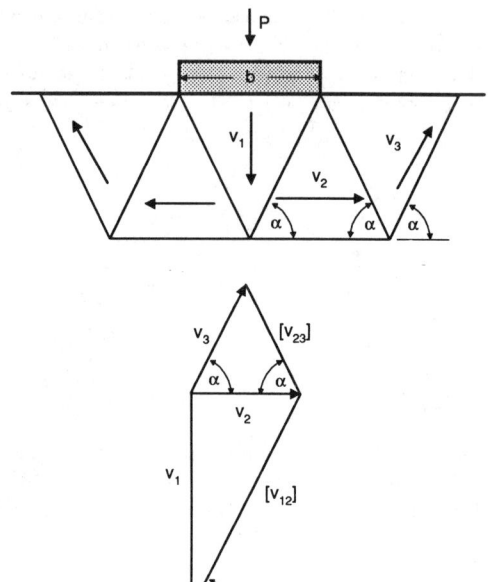

Figure 4: Kinematic mechanism for upper bound solution and corresponding hodograph.

planes not aligned with the joint planes. The flow rule for simultaneous sliding along two joint sets can be written

$$\dot{\boldsymbol{\epsilon}}^p = \lambda_1\,\mathbf{g}_1 + \lambda_3\,\mathbf{g}_2 \quad (19)$$

where, for $\psi^a = \psi^b = 0$,

$$\mathbf{g}_1 = \begin{bmatrix} -\text{sign}(\sigma_s^a)\sin\theta^a\cos\theta^a \\ \text{sign}(\sigma_s^a)\sin\theta^a\cos\theta^a \\ \text{sign}(\sigma_s^a)\left[\cos^2\theta^a - \sin^2\theta^a\right] \end{bmatrix} \quad (20)$$

and

$$\mathbf{g}_3 = \begin{bmatrix} -\text{sign}(\sigma_s^b)\sin\theta^b\cos\theta^b \\ \text{sign}(\sigma_s^b)\sin\theta^b\cos\theta^b \\ \text{sign}(\sigma_s^b)\left[\cos^2\theta^b - \sin^2\theta^b\right] \end{bmatrix} \quad (21)$$

For the footing problem considered, $\theta^a = -\theta^b = \alpha$. Assuming that the sign of the shear stress acting on both joint planes is positive, the flow rule reduces to

$$\dot{\boldsymbol{\epsilon}}^p = (\lambda_1 + \lambda_3)\begin{bmatrix} 0 \\ 0 \\ \cos^2\alpha - \sin^2\alpha \end{bmatrix} \quad (22)$$

It can be shown that the rate of energy dissipation per unit volume for this plastic strain rate is

$$\dot{d} = c \left| \dot{\gamma}_{12} \left(\cos^2 \alpha - \sin^2 \alpha \right) \right| \qquad (23)$$

and that for a horizontal shear velocity discontinuity, the rate of energy dissipation per unit area is

$$\dot{d} = c \left| [\dot{u}_s] \left(\cos^2 \alpha - \sin^2 \alpha \right) \right| \qquad (24)$$

The total rate of energy dissipated by the kinematic mechanism shown in Figure. 4 is the sum of the energy dissipated along the velocity discontinuities. The block velocities and the velocity jumps at the discontinuities, as defined in Figure 4, are given by

$$v_2 = \frac{\cos \alpha}{\sin \alpha} v_1 \qquad (25)$$

$$v_3 = \frac{1}{2 \sin \alpha} v_1 \qquad (26)$$

$$[v_{12}] = \frac{1}{\sin \alpha} v_1 \qquad (27)$$

and

$$[v_{23}] = \frac{1}{2 \sin \alpha} v_1 \qquad (28)$$

Equating the rate of work done by the footing load, P, to the rate of work dissipated gives

$$P^u = 2bc \left(\frac{1}{\cos \alpha \sin \alpha} + \frac{1}{\left| \cos^2 \alpha - \sin^2 \alpha \right|} \frac{\cos \alpha}{\sin \alpha} \right) \qquad (29)$$

for $0 \leq \alpha \leq 90$. This solution is an upper bound to the true bearing capacity, P.

4.3 Distinct Element Simulation

Discrete simulations of the bearing capacity problem were performed for three different values of the ratio of the footing width b to the joint spacing s. The simulations were done with the Distinct Element program UDEC (Itasca Consulting Group, 1991), a two-dimensional numerical code for modeling assemblies of rigid or deformable blocks. For the bearing capacity problem, rigid blocks were used, with the corners of the blocks rounded to prevent interlocking after the joints become slightly offset during shearing.

Examples of the Distinct Element models are shown in Figures 5 and 6. To simulate the bearing capacity problem, displacement was fixed for blocks on the left and right sides of the model and

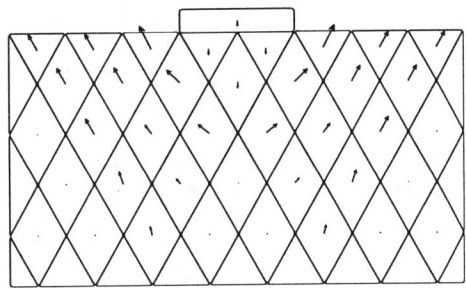

Figure 5: UDEC model of footing on jointed rock along with block velocity vectors for a footing width to joint spacing ratio of $b/s = 1.15$.

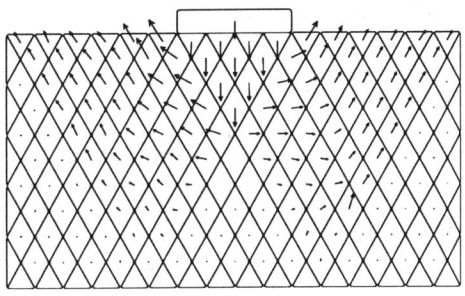

Figure 6: UDEC model of footing on jointed rock along with block velocity vectors for a footing width to joint spacing ratio of $b/s = 2.31$.

along the bottom of the model. The footing block was given a small fixed velocity and the reaction on the block was recorded as it moved downward. This reaction is plotted versus displacement in Figure. 2.

For the smallest ratio of footing width to joint spacing, $b/s = 1.15$, the bearing capacity for the Distinct Element model is more than twice that predicted by the ubiquitous joint model. However, as the ratio b/s becomes larger, the Distinct Element results become closer to the ubiquitous joint model results. For $b/s = 2.31$, the Distinct Element results are already within 15% of the ubiquitous joint model predictions.

5 CONCLUSION

The bearing capacity problem studied here illustrates one reason why ubiquitous joint models do not accurately represent the behavior of discrete models when the joint spacing is large. The situation is analogous to slope stability analysis where the shear stress is computed along many different potential failure surfaces in order to find the most critical failure surface.

For a discrete model, if the joint spacing is large, it is unlikely that a joint or set of joints will lie along the most crtical failure surface. Failure will take place, instead, along a less critical failure surface leading to a higher bearing capacity. As the joint spacing becomes smaller, it is more likely that a joint or set of joints will lie close to the critical failure surface. The ubiquitous joint model represents the extreme case where there are joints everywhere, so that failure can take place exactly along the crtical surface.

We can conclude that ubiquitous joint models are only appropriate when the joint spacing is small compared to the length scale of the problem being considered. Alternatively, the ubiquitous joint model can be thought of as representing a worst-case scenario in which joint planes just happen to be located along the most critical failure surface.

REFERENCES

Adhikary, D.P., A.V. Dyskin and R.J. Jewell 1995. Analysis of fracture and deformation processes during flexural toppling in foliated rock slopes, in Rossmanith, editor, *Mechanics of jointed and faulted rock*. 611–616. Balkema, Rotterdam.

Besdo, D. 1985. Inelastic behavior of plane frictionless block-systems described as Cosserat media. *Arch. Mech.*, 37(6):603–619.

Blanford, M.L., S.W. Key & J.D. Chieslar. A general 3-D model for a jointed rock mass. In C.S. Desai et al., editor, *Proceedings, Second International Conference on Constitutive Laws for Engineering Materials: Theory and Application*, pages 35–46, University of Arizona, Tucson, Arizona, Jan. 1987.

Chen, E.P., A computational model for jointed media with orthogonal sets of joints. *J. Appl. Mech. Trans. ASME*, 56:25–32, 1989.

Cundall, P.A. & C. Fairhurst. Correlation of discontinuum models with physical observations- an approach to the estimation of rock mass behavior. *Felsbau*, 4(4):197–202, 1986.

Dawson, E.M. and P.A. Cundall, Cosserat plasticity for modeling layered rock. in *Proceedings of the Fractured and Jointed Rock Masses Conference*, International Society for Rock Mechanics, 1992.

Dawson, E.M. 1995. *Micropolar Continuum Models for Jointed Rock*. Ph.D. thesis, University of Minnesota, Minneapolis, Minnesota.

Dawson, E.M. and P.A. Cundall 1996. Slope stability using micropolar plasticity, in M. Auberin et al., editor *Rock Mechanics, Tools and Techniques* Proceedings of the 2nd North American Rock Mechanics Symposium, Montreal, Quebec, pp. 551-558, 1996.

Itasca Consulting Group 1993. *FLAC (Fast Lagrangian Analysis of Continua) Version 3.2*. Itasca Consulting Group, Minneapolis.

Itasca Consulting Group 1991. *UDEC (Universal Distinct Element Code) Version ICG 1.7*. Itasca Consulting Group, Minneapolis.

Morland, L.W., Continuum model of regularly jointed mediums. *J. Geophys. Res.*, 79(2):357–362, Jan. 1974.

Mühlhaus, H.-B. 1989. Application of Cosserat theory in numerical simulation of limit load problems. *Ingenieur-Archiv*, 59:124–137.

Otter, J.R.H., A.C. Cassell, & R.E. Hobbs 1966. Dynamic relaxation. *Proc. Inst. Civ. Eng.*, 35:633–656.

Petney, S.V. & M.L. Blanford. *Understanding the Use of a Continuum Joint Model for the Analysis of an Excavated Drift*. Technical Report RSI Publ. No. 89-01, RE/SPEC Inc., Albuquerque, 1989.

Thomas, R.K., *A Continuum Description of Jointed Media*. Technical Report SAND81-2615, Sandia National Laboratories, Albuquerque, 1982.

Zienkiewicz, O.C. & Pande, G.N., time-dependent multilaminate model of rocks – a numerical study of deformation and failure of rock masses. *Int. J. Numer. Anal. Methods Geomech.*, 1:219-247, 1977.

Complex behaviour of the rock mass around the excavation face in large rock caverns

Y.N.Lee, Y.H.Suh, D.Y.Kim & K.S.Jue
Hyundai Engineering and Construction Co., Ltd, Seoul, Korea

ABSTRACT: This paper presents a complex deformation behaviour of rock mass around the excavation face in 30m high underground oil storage cavern. Changes in displacement were monitored using multipoint borehole extensometers(MPBX), installed before the excavation face passed the instrument station. Displacements, measured during the excavation of four 7.5m high sections, show a complex deformation behaviour due to blasting and initial stress relief due to excavation. Measured displacements were compared with the results of three-dimensional analysis to study the effect of stress relief due to several stages of excavation on the deformation behaviour. Dynamic analysis was also carried out to study the effect of blasting on the deformation of rock mass. Based on the results of these studies, the complex deformation behaviour of the rock mass monitored from the field instrumentation is explained.

1. INTRODUCTION

30m high underground cavern was excavated for the storage of oil in southern part of Korea. During the excavation work, field tests were carried out to determine the deformation modulus of the rock mass and the initial stresses at the site. A monitoring system was provided for observing and recording the behaviour of the rock mass in response to the excavation process. The results obtained from the instrument system were analyzed to understand the complete history of deformation behaviour. Numerical analysis was carried out to study the effects of stress relief due to excavation and blasting on the deformation of the rock mass, and the results of this analysis were compared with the measured results.

2. PROJECT DESCRIPTION

Major underground oil storage facility consists of six storage cavern, construction tunnels, two shafts and water curtain tunnels, as shown in Figure 1. Six storage caverns, to be left unlined at EL.(-)30m - EL.(-)60m, are aligned parallel to each other in the direction of N80°W and each of these caverns is horseshoe-shaped and its dimension is 18m wide, 30m high, and 400 - 600m long. The construction tunnel is also horseshoe-shaped and the dimension of 8m wide and 7.5m high. The tunnel elevation changes from EL.(+)10.0m at the entry to EL.(-)60.0m at the bottom level of cavern with about 12 % slope in between these two levels. The water curtain tunnels at EL.(-)1.0m are horseshoe-shaped, 4.5m wide and 5m high.

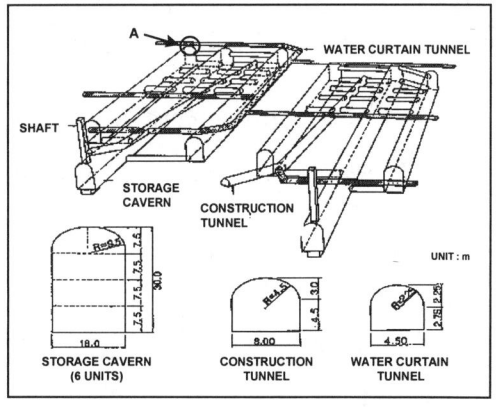

Figure 1. Schematic view of the underground oil storage facility.

The bedrock at the site is metamorphosed andesite tuff of the Late Cretaceous Period. The tuff is dark grey colored and very tight and has welding texture at the boundary of rock fragments. There are three joint sets(N70 - 80°E/70 - 80°NW, N10°E/70 - 80°SE and N45°W/20°SW) and random joints in the bedrock. These joints are in general tightly-healed, but locally coated or filled with calcite.

The mechanical properties of the intact rock obtained from site investigation are shown in Table 1. The average uniaxial compressive strength is about 250MPa and the average modulus of elasticity is 83GPa, indicating that the tuff can be classified as rock having very high strength and medium modulus ratio(Deere and Miller, 1966).

Table 1. The mechanical properties of intact rock.

Property	No. of samples tested	Average value	Range
Uniaxial compressive strength(MPa)	12	250	215-331
Tensile strength(MPa)	12	22	16-31
Elastic modulus (GPa)	12	83	70-97
Poisson's ratio	12	0.23	0.21-0.25
Internal friction angle(°)	4	51.5	49-54
Cohesion(MPa)	4	53	48-59

3. EXCAVATION AND SUPPORT

The excavation of storage caverns was carried out in four 7.5m high sections(gallery, bench-1, bench-2, and bench-3) using the drill and blast method with 3.4m face advance per round. The excavated caverns were supported by a combination of rock bolt(slack bolt type) and shotcrete, and supporting work started at the section 50m away from cavern face. For the installation of rock bolt, 3 to 7m long rebars, 25mm in diameter, were inserted into the predrilled hole filled with cement paste plus quick-setting and expansion admixtures. A random bolting pattern was adopted where the excavated rock mass showed sound quality, while systematic pattern bolting was used for the area where the rock mass was of poor quality. The spacing of rock bolts varied according to the quality of rock, but a square grid of 1.5m x 1.5m spacing was mainly adopted for the support of the crown.

4. MONITORING

4.1 *Monitoring instruments*

Displacements and stress changes during excavation were monitored using MPBX and vibrating wire stressmeters, before the excavation face of the cavern passed the measuring station. The layout section of the instrumentation is shown in Figure 2. The monitoring of displacements at the depths of 3m, 7m, 15.7m, and 26.4m above the cavern crown, and the stress changes at the depths of 3.5m 10.6m and 23.15m above the crown have been carried out, in a vertical hole from a water curtain tunnel. A complete history of displacement was monitored during excavations of gallery, bench-1, bench-2 and bench-3.

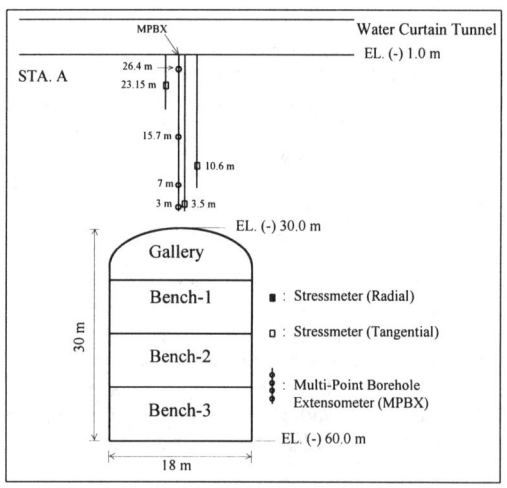

Figure 2. Layout section showing instrumentation details.

4.2 *Monitoring results*

The displacements above the crown are plotted against the distance between the measuring station and the excavation face in Figure 3(where D is the width of the cavern and x is

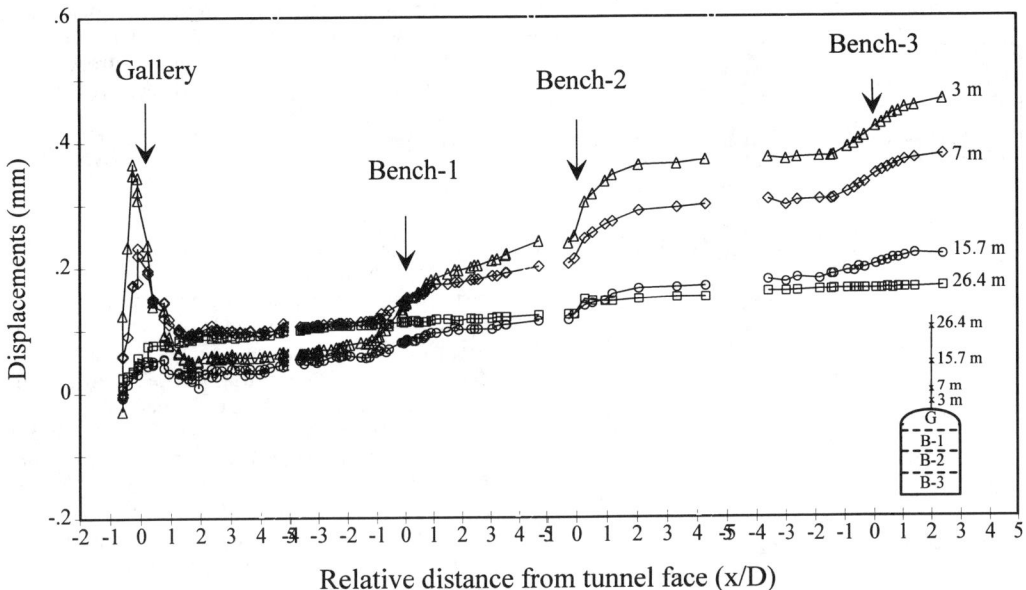

Figure 3. Measured displacements above the crown during excavation.

the distance between the measuring station and the excavation face). From the results shown in this Figure, the following observations may be made :

(1) A small amount of upward displacement less than 1mm was measured at the end of bench-3 excavation, implying that the bedrock is competent.

(2) In gallery excavation, upward movement was recorded at 3m and 7m above the crown as the excavation face approached near the measuring station. Once the cavern face passed the instrument station, the crown moved downward sharply and the movement rapidly converged soon after the excavation face passed the instrument station(x/D of 1 to 2).

(3) As the excavation of bench-1, bench-2 and bench-3 was progressively made, the cavern crown continuously moved upward and, at the end of bench-3 excavation, ended above the original position before excavation.

As noted above, a complex deformation behaviour of the rock mass above the crown was observed during excavation. This behaviour may be attributed to the redistribution of initial stresses and blasting near the instrument station.

5. NUMERICAL ANALYSIS

In attempt to investigate the complex deformation behaviour of the rock mass mentioned above, static analysis was carried out to study the effect of stress relief due to excavation on the deformation behaviour and dynamic analysis was carried out to study the effect of blasting on the behaviour of rock mass.

5.1 Static Analysis

To study the three-dimensional behaviour of rock mass around the cavern, numerical analysis was carried out using the commercially-available program 3DFLAC. In Figure 4, a three-dimensional model(250m in height, 100 m in width and 150m in length) used in this analysis is shown. Some of input data used in this analysis were determined from laboratory tests and summarized in Table 1. The deformation modulus of the rock mass was determined from the plate loading test at the site and details may be found in Lee et al.(1996). Deformation moduli determined from this test range from 37GPa to 43GPa, with an average value of 40GPa, which was used in the analysis as a representative modulus of the rock mass.

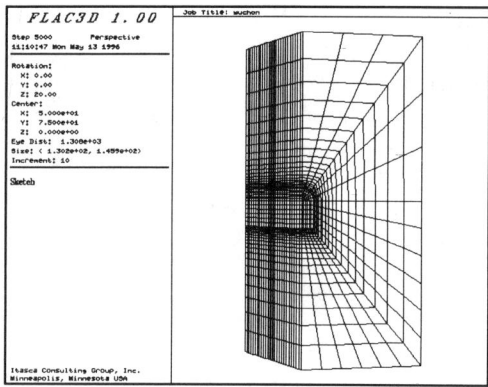

Figure 4. Numerical model for static analysis.

The initial stresses of the rock mass were measured from the overcoring method with USBM deformation gauge(Lee et al., 1995). The average ratios of measured horizontal to vertical initial stresses are 3 for the direction parallel to the cavern axis and 2 for the direction perpendicular to the cavern axis. The vertical stress at the cavern level was estimated to be between 5 to 7MPa, depending on the overburden depth above the cavern level.

Calculated displacements corresponding to different stages of excavation are plotted against relative distance of cavern face to the instrument station in Figure 5. From the results shown in this Figure, it can be noted that a small amount of upward displacement less than 1mm is recorded at the end of bench-3 excavation, which is comparable to the measured one. The overall trend of displacement during 4 stages of excavation is similar to the measured one. It is also interesting to observe that the cavern crown moved downward during gallery excavation and upward upon subsequent excavation of benches, as measured in the field. The upward movement of crown may be attributed to the redistribution of stresses and changes in the shape of excavation at different stages of excavation.

5.2 Dynamic Analysis

To investigate the upward movement of crown due to the excavation by blasting near the instrument station, dynamic analysis using FLAC was carried out to simulate the excavation by drilled and blasting. Two dimensional axisymmetric model, having the size of 300 m in width and 200m in length, was used for the analysis as shown in Figure 6. Blasthole was installed in front of the tunnel face to calculate the deformation behaviour above the crown due to blasting. Analyzed shape of excavated section was assumed to be a circular tunnel, with diameter of 12.6m, equivalent to the cross section area of gallery.

The mechanical properties used in the analysis are same as the ones used in static analysis(see Table 1) and initial stress condition was assumed to be a hydrostatic stress condition($\sigma_x = \sigma_y = \sigma_z = 6.75$MPa). A damping coefficient and a natural frequency used are 5% and 25Hz respectively. Viscous boundaries were used in order to avoid a reflection of propagating waves back into the model. The element size was adjusted in order to avoid errors being introduced from spurious reflections of waves propagating in a graded mesh.

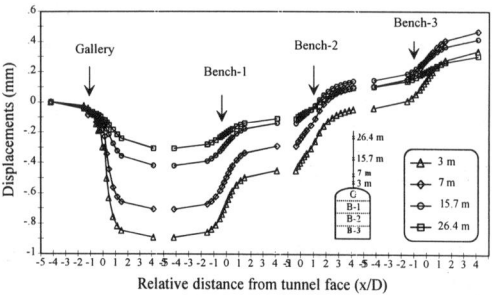

Figure 5. Calculated displacements above the crown during excavation(static analysis).

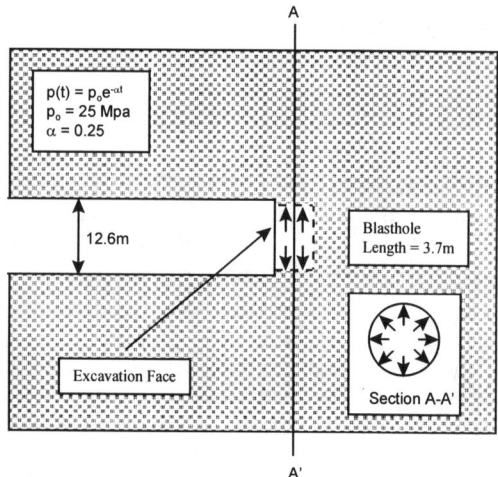

Figure 6. Schematic view of dynamic analysis.

Being difficult to obtain a precise description of pressure-time history for blast loading, an empirical equation of a negative exponential function of time suggested by Stecher et al.(1981) was adopted to simulate a variety of explosive-induced loads applied to blasthole wall. This function can be given in equation (1) :

$$P(t) = P_o e^{-at} \quad (0.25 \leq a \leq 0.5) \quad (1)$$

where P(t) is the pressure acting at any point on the blasthole wall at a given instant of time, a is a experimental constant defining the particular form of the function, P_o is the peak pressure. To determine the peak pressure, following empirical equations may be used(Yoon, 1992) :

$$P_o = \frac{P_D}{2} \quad (2)$$

$$P_D = 0.000424 \, V_e^2 \rho_e (1 - 0.543 \rho_e + 0.193 \rho_e) \quad (3)$$

where P_D is the detonation pressure, V_e is the velocity of detonation, ρ_e is the density of explosives. Table 2 summarizes the input data to simulate the explosive induced loads. Considering the distance between blastholes and the loads acting on rock mass by blasting, a peak pressure of 25MPa was adopted. All the input data to simulate the explosive-induced loads are summarized in Table 2.

Table 2. The input data to simulate the explosive induced loads.

Input	Value
Peak pressure(MPa)	25
Experimental constant(a)	0.25
Dynamic time(sec)	0.5

Calculated displacements above the crown considering the effect of blasting and stress relief are presented in Figure 7. When the tunnel face approaches to the instrument station, the trend of upward movement is observed as in the case of field measurement. When the cavern face approaches to instrument station, the movement of cavern crown is influenced by both the stress relief and blasting effect. As noted in Figure 5, stress relief near cavern face causes the crown to move downward, but blasting near cavern face make the crown move upward. However, the effect of blasting on the movement of crown is more dominant than that of stress relief, resulting in the upward movement of crown. Maximum upward displacement was recorded just before the tunnel face passed the instrument station, and, upon cavern face passing this station, downward displacements were occurred due to the dominant effect of stress relief.

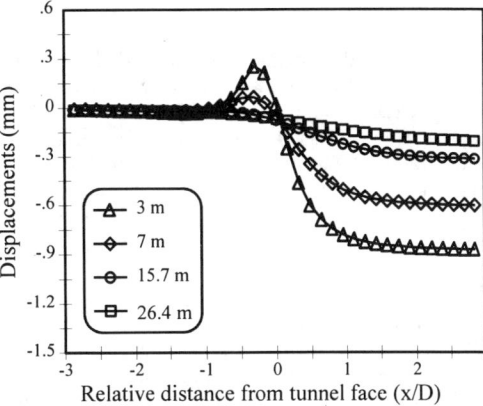

Figure 7. Calculated displacements above crown, considering the effects of blasting and unloading due to excavation.

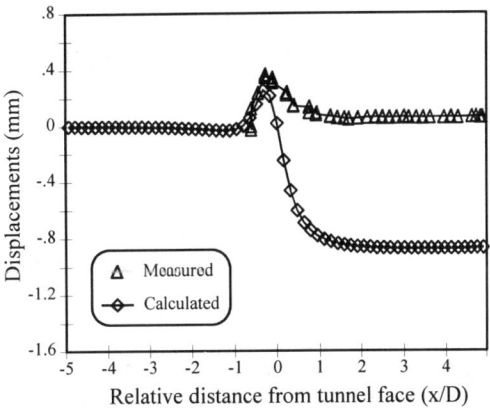

Figure 8. Comparison of displacements at 3m above the crown.

The results of this analysis are compared with the measured data for 3m above crown in Figure 8. It can be noted that the trend of upward movement near cavern face is observed from both the measurement and the analysis, indicating that the upward movement near the face is mainly due to the effect of blasting.

CONCLUSION

A complete deformation history of 30m high underground oil storage cavern was monitored during excavation using MPBX installed before cavern face passed instrument station. To understand the monitored behaviour of rock mass near cavern face, 3D static analysis and 2D dynamic analysis were performed. From the results of field measurement and numerical analyses, the following conclusions may be drawn :

(1) The deformation behaviour of the rock mass at cavern crown was influenced by both blasting and stress relief during excavation.

(2) When the excavation face of gallery approaches to the instrument station, the effect of blasting is dominant, while the effect of stress relief is governing the deformation behaviour during the subsequent excavation of 3 benches.

(3) Changes in shape of the excavated section during four stages of excavation and corresponding stress redistribution cause the crown to move downward during the gallery excavation and upward for the subsequent excavation of 3 benches.

REFERENCES

Deere, D. U. & Miller, R. P. 1966. Engineering classification and index properties for intact rock. *Air Force Weapons Lab. Rep. AFWL-TR-65-16*, Kirkland, New Mexico.
Lee, Y. N., Suh, Y. H. and Kim, D. Y. 1996. Deformability of metamorphosed andesitic tuff from plate loading test. *Proc. of the 2nd North American Rock Mechanics Symposium*: NARMS '96, Montreal, Canada, 1573-1579.
Lee, Y. N., Suh, Y. H., Kim, D. Y. & Nam, H. K. 1995. Three-dimensional behaviour of large rock caverns. *Proc. of 8th Int. Congress on Rock Mechanics*, V. 2: 505-508.
Stecher, Fourney 1981. Prediction of crack motion from detonation in brittle material. *Int. J. Rock. Mech. Min. Sci. & Geomech. Abstr.*, 18:1, 23-33.
Yoon, J. S. 1992. New blasting technique. 歐美書舘.

Modelling techniques for safety evaluation: Slope stability and landslides

, Rotterdam, ISBN 90 5410 910 6

Landslide hazards and stability analysis of coastal cliff regions of Bangladesh

Md. Hamidur Rahman
Department of Geology and Mining, Rajshahi University, Bangladesh

ABSTRACT : In this paper an attempt has been made to illustrate the problem of landslide and slope stability of heterogeneous coastal cliff of Cox's Bazar area, Bangladesh. The calculation of the safety factor of the slope of the coastal cliff were done by finite element method (FEM) and by the widely used traditional methods. The values of the safety factor indicates that the slope is stable but the coastal cliff slope has a history of failures in different places every year. Undisturbed monolithic soil samples were collected for different laboratory tests. The soil parameters which were used for the slope stability calculations were obtained by Unconfined compression, Triaxial compression and Direct shear box tests performed in the laboratory. The results of the study includes the distribution of vertical stress, horizontal stress, shear stress, maximum shear stress, principal stresses, displacements, and the safety factor (SF) of the coastal cliff slope. The satisfying results of the slope stability analysis of the coastal cliff lies mainly in a proper engineering geological recognition of the slope and evaluation of the physico-mechanical parameters of soils including their natural changeability.

1 INTRODUCTION

The studied area lies between latitude 21°12′ N to 21° 30′ N and longitude 9°20′ E to 92° 05′ E within Cox's Bazar Sadar (Figure 1). Physiographically the area is situated on the low hill ranges between Banghkhali river on the east and Bay of Bengal on the west with a long open beach towards the sea.

Bangladesh occupies a major part of the Bengal Basin. The area under investigation belongs to the folded flank, representing the Cox's Bazar coastal cliff section. In the cliff section, the Neogene sediments are well developed but many of these are covered with piedmont, floodplain and marsh deposits of Recent age. In the stratigraphic order the sequences are Bokabil Formation, Tipam Sandstone Formation and Duptila Formation.

The Cox's Bazar coastal cliff section is continuously facing different types of erosion and weathering which give rise to the shallow

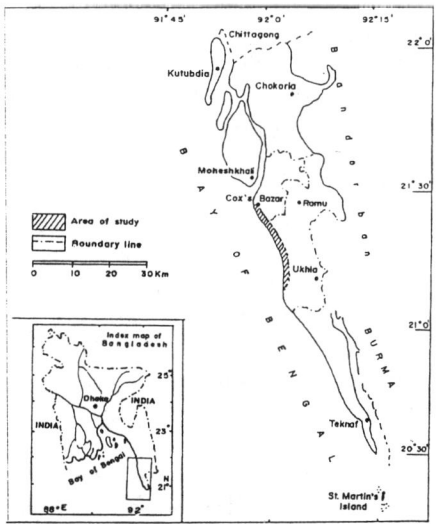

Figure 1. Map showing location of the study area.

failure of the cliff section. The stability of the coastal cliff depends on various interconnected geologic and other natural and artificial factors and conditions.

The sliding phenomena of the soil masses generally causes great damage in forest, agriculture, dam, embankment, building structures, etc. Since such types of natural disaster gradually unbalancing the environmental equilibrium and for this reason attention has been made in the international forum about the problem.

The analysis of stability of the coastal cliff slope has been done for the safety factor by means of finite element method (FEM) and other traditional methods. The stability of earth coastal cliff slopes should be very thoroughly analyzed since their failure may lead to loss of human life as well as colossal economic loss. A thorough knowledge of shear strength and related properties of soil is essential to design the slope and height of any earth structures.

For accurate slope stability analysis, the division of the soil massive should be taken into consideration. Slope with a complicated geological structure, the situation of the individual structural units of soil massive differing with the deformability parameters has a considerable influence on the character of the distribution and value of stresses, displacements and stability of slopes (Rahman et al, 1987). It was found that the investigated coastal cliff slope is composed of heterogeneous soil massives. The physico-mechanical parameters which was used for the calculation (Table 1) were obtained by Unconfined compression, Triaxial compression and Direct shear box tests performed in the laboratory.

The size and way in which the slope failure takes place depends on many factors. The presence of surface of weakness is also an important factor among others. Recently few minor and major landslides has been observed in the different parts of the Cox's Bazar coastal cliff section. The developing landslide movements in different parts of the coastal cliff are the effects of local instability of escarps mainly by the processes of weathering, water saturation and excess water waves, etc.

Table 1: Parameters of the Sediments Composed the Cox's Bazar Coastal Cliff Slope.

	Soft Soils	Medium Soils	Hard soils
Modulus of Deformation E_o (MPa)	6.0	10.0	15.0
Cohesion c(MPa)	0.02	0.03	0.04
Internal friction Co-efficient	0.176	0.194	0.230
Poisson's ratio v	0.40	0.35	0.30
Bulk Density γ KN/m^3	19.0	20.0	21.0

Slumping is also the main type of mass movement that is endangering the coastal cliff slope in different parts. Thus the prediction of possible landslide processes lies mainly in the calculation analysis taking into account the properties of soil strength and the state of stresses in the slope. Both the state of stress and the strength of the soil depends on many natural and artificial factors. The clay materials show an appreciable change in strength when these come in contact with water (Hossain K.M. et al 1986, Rahman M.H. 1987).

The section in relation to the original structure and the contour of the slope has been very slightly simplified.

The slope has been divided into triangular elements because of the impossibility of taking into account little inserts and soil lenses in the model. The section through the slope has been divided into 797 triangular elements, 449 nodal points and 59 boundary points (Figure 2), the existing instructions concerning the range of boundaries of the limitless medium have been taken into account (Desai C.S. et al 1972).

The model is burdened with the mass force (the weight of the model itself) and in the areas of expected concentrations of stresses, a thicker net of elements were applied. The computations were done by means of the computer program (Wang F.D. et al 1972).

By using the possibilities of the program, the drawings of the model of the slope were presented together with the isolines of all stresses and direction of the maximum principal stresses. Compressive stresses are designated (-) and tensile ones (+).

The satisfying results of analysis of stability of the Cox's Bazar coastal cliff slope lies mainly in a proper engineering-geological recognition of the slope and evaluation of the physico-mechanical parameters of soil massives including their natural changeability (Rahman M.H. et al 1987).

2 RESULTS OF INVESTIGATIONS

2.1 Vertical stress

The isolines of the vertical stress run almost parallel to the slope and the value of stress increases together with the depth. The maximum value of vertical stress is (-) 1.20 Mpa occurs at the bottom of the slope. The vertical stresses in the whole model have a compression character (Figure 3). The values of vertical stress and partly their distribution may vary depending on the order of the arrangement of the soil layers in the slope model.

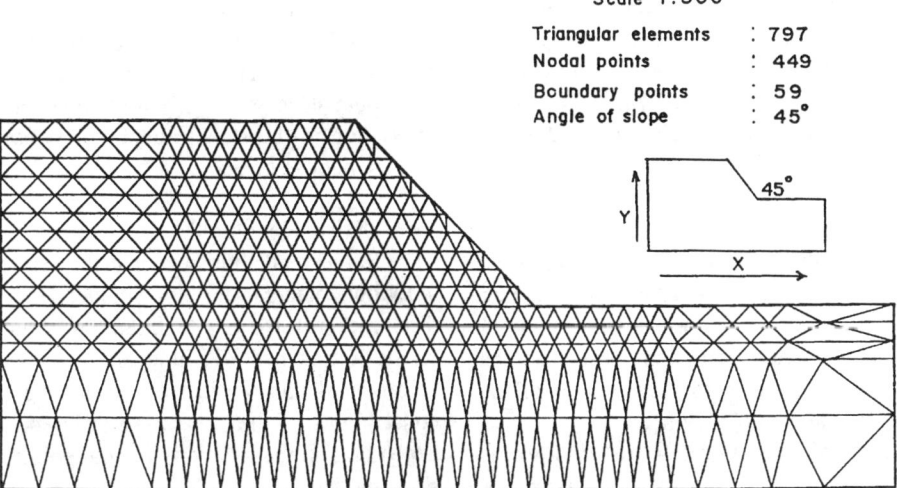

Figure 2. Model of slope divided by Triangular elements.

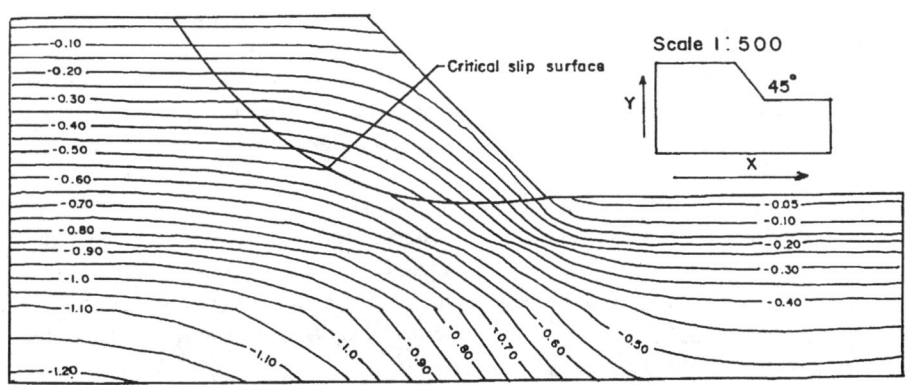

Figure 3. Vertical Stress Contour Plot.   C.I.=0.05MPa

2.2 Horizontal stress

The horizontal stress both in homogeneous and in the heterogeneous slope have a more complicated distribution than in the case of the vertical ones. The nature of stress depends on the types of soils (Rahman M.H. 1990).

In the model of heterogeneous slope the values and division of horizontal stresses are more complex and generally depends on the mutual arrangements of soil layers. Higher values of horizontal stress of about (-) 0.38 Mpa occurs at the bottom of the slope. Horizontal stresses have compression characters only. In the contact surface of different layers the values of horizontal stress becomes zero.

In the forefield and subsoil of the heterogeneous slope, the horizontal stress largely depends on the kind and order of arrangement of the layers of soil in the slope above the subsoil as to their kind and value (Rahman M.H. 1990).

2.3 Shear stress

The isolines of shear stress in the model of the heterogeneous slope are arranged in a manner similar to concentric ones, where the maximum values are obtained in the neighborhood of the foot of the slope. The maximum values of the shear stress is about (+) 0.12 Mpa.

The values of shear stress decreases towards the forefield of homogeneous slope. The distribution of shear stress in heterogeneous slope model is more complicated than in homogeneous slope model.

2.4 Principal stress

Maximum and minimum principal stresses are derivatives of horizontal, vertical and shear stresses. The isolines of the stresses and their values are analogous. The direction of maximum principal stress and their isoline diagram shows the range of a considerable disturbance of stress distribution in the neighborhood of the slope in relation to the geostatic state. The maximum value is about (-) 1.28 Mpa occurs at the forefield of the slope. The direction of maximum principal stress is almost parallel to the slope's inclination. Inside the slope they become almost vertical and in the forefield they are almost horizontal, particularly at the surface area of the forefield of the slope.

2.5 Maximum Shear Stress

The isolines of maximum shear stress for heterogeneous slope are greater at a comparable depth in relation to the areas inside the slope. The maximum shear stress reach maximum values of about (+) 0.51 Mpa. The isolines of the stress are almost parallel to the slope inclination. Greater values of maximum shear stress and their increased height gradient appear at the foot of the slope. In the slope model there appears an increased stress gradient in the area where the soil layers meet. In heterogeneous slope model the arrangement of soil layers are not consequent. The distribution and value of maximum shear stress have traditional character in relation to the slope model of a consequent arrangement of layers.

2.6 Displacements

The values and direction of displacements of individual points of the slope model depends on the location of the points in the model and also on the arrangement of soil layers.

Horizontal and vertical displacements of the nodal points of the slope have different values depending on the location of nodal points in the slope model.

Consequently decreasing strength downwards the slope. The displacements are more complicated. The vertical displacements in the forefield of the slope are (+) toe failure of the subsoil but they are much smaller than in the case of homogeneous slopes made of weak soil. Maximum vertical displacements which are (-) subsidence appears at the back of the slope from the upper surface of the slope model (Figure 4).

The general direction of horizontal displace- -ments in the model is towards the slope's inclination and the forefield the direction of vertical displacements downwards of the slope model.

The maximum horizontal displacements about

2.2 cm appears lower middle part of the slope model. Similarly as in the case of maximum vertical displacement is about 8 cm appears in the upper part of the slope model.

2.7 Safety factor of the coastal cliff slope

The analysis of the plane state deformation have been done with the help of finite element method (FEM) to define the stress area in the model of the slope assuming circular cylindrical slip surface along which horizontal stress (σ_x), vertical stress (σ_y) and tangent stress (τ_{xy}). Normal stress (σ_n), and shear stress (τ_{nm}) are designed at every point on the slip surface from the following equations:

$$\sigma_n = 0.5(\sigma_x + \sigma_y) + 0.5(\sigma_x - \sigma_y)\cos 2\theta + \tau_{xy}\sin 2\theta \quad \ldots (1)$$

$$\tau_{nm} = \tau_{xy}\cos 2\theta - 0.5(\sigma_x - \sigma_y)\sin 2\theta \quad \ldots (2)$$

The safety factor (SF) is obtained from the proportion of shear resistance (S_R) from Coulomb equation to shear stress τ_{nm} at every point of slip surface.

$$SF = \frac{\sum S_R \cdot dl}{\sum \tau_{nm}} = \frac{\sum (\sigma_n \cdot tg + c) \cdot dl}{\sum \tau_{nm} \cdot dl} \quad \ldots (3)$$

where dl is a unit length of slip surface.

Slope stability calculations were done

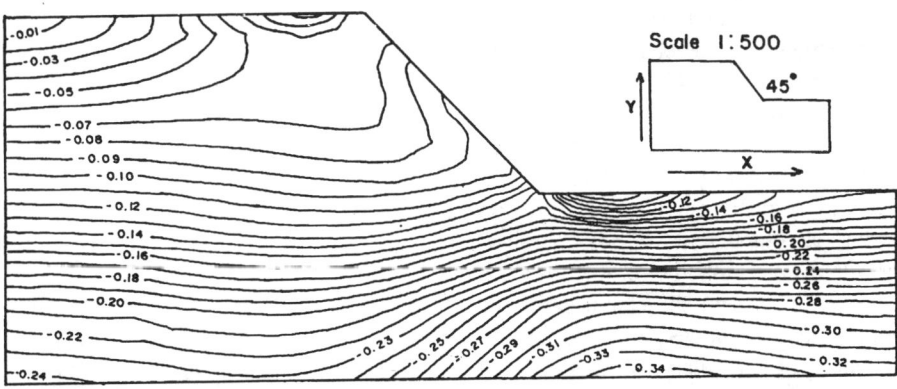

Figure 4. Horizontal Stress Contour Plot.

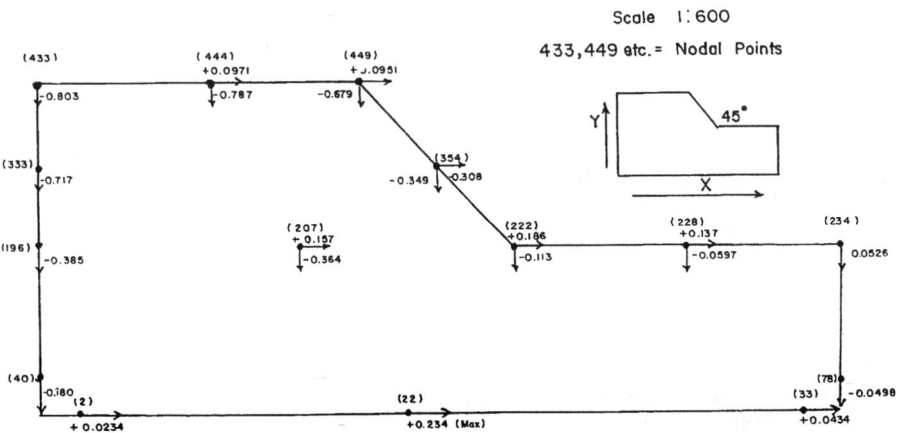

Figure 5. Displacement of some points of slope model of Cox's Bazar coastal cliff.

by finite element method (FEM) assuming the circular cylindrical slip surface have indicated that the safety factor (SF) is 1.43 of the Cox's Bazar coastal cliff slope.

Comparative calculations of the safety factor(SF) for the critical slip surface as in the FEM with Fellenius method SF is 1.30 and by Bishop's method SF is 1.34. It was observed that particularly Fellenius method gives more reliable results which is also stressed by other authors.

3 CONCLUSIONS

The stability of earth slopes should be very thoroughly analyzed since their failure may lead to loss of human life as well as colossal economic loss. A very detailed knowledge of shear strength and related parameters of soil masses are essential to design the slope and height of the earth structures.

For the sake of accurate slope stability analysis, the division of soil masses should be taken into consideration. Slope with complicated geological structures, the situation of individual structural units of soil massives differing with the deformability properties has a considerable influence on the character of the distribution and the values of stresses and displacements which also influences the stability of slopes.

4 REFERENCES

Hossain, K.M. and Dutta D.K. (1986) Slope stability problem of the Chittagong University campus, *Dhaka University Studies*, Part B/1, pp. 77-88, 1986.

Rahman, M.H. (1987) Some important remarks on slope stability, *Rajshahi University Studies*, Part B, **Vol-15**, pp.117-122.

Rahman, M.H. (1990) Analysis of slope stability problem of the heterogeneous slope by finite element method. *Proc. of 6th International Congress of IAEG*, Amsterdam, The Netherlands, August 6-10, 1990, pp.2279-2285.

Rahman M.H. and Rybicki S. (1987) Analysis of stresses, displacements and stability of heterogeneous soil slopes on the example of open pit sulphur mine at Machow, Poland, with the use of finite element method. *Proc of National Conference on Soil Mechanics and Foundation Engineering*, **Vol-52**, No(24), pp.287-292, Poland 1987.

Wang F.D., Sun F.C. and Rophan D.M. (1972) Computer program for pit slope stability analysis by finite element stress analysis and limiting equilibrium method. Report of investigations, #7685, Bureau of Mines, U.S. Department of the Interor.

Safe and economical rockfall protection barriers

W. Gerber
WSL Swiss Federal Institute for Forest, Snow and Landscape Research, Section Natural Hazards, Birmensdorf, Switzerland

B. Haller
GEOBRUGG Fatzer AG, Romanshorn, Switzerland

ABSTRACT: Flexible wire net systems are an integral part of rockfall mitigation measures. In Switzerland they are often used to protect transportation lines as well as residential areas. A proper site analysis and the choice of the appropriate rockfall protection system are the key components to a safe and economic rockfall hazard analysis. Although wire nets have been well known as rockfall protection systems since the early sixties, no field testing was conducted before 1989 and little information was recorded on the capabilities of these systems. Extensive field testing of rockfall protection systems was carried out on three different locations in California and in Switzerland from 1989 till 1996. Many different designs were tested and improved with the goal to maximize safety and minimize maintenance needs. A result of the tests is an improved and advanced generation of ring net barriers which are capable of stopping rockfall energies in the range from 40 kJ to 2350 kJ. Rockfall impact energies were analysed and the response and behaviour of the systems studied. The detailed analysis of rock deceleration in the barrier allows the calculation of the distribution of the impact forces in the system components. Future measurements of anchor forces provide basic data for the dimensioning of systems for inter- and extrapolation.

1 INTRODUCTION

This article presents an analysis of experimental data from rockfall experiments which were conducted over a ten year period in Switzerland. The tests were a co-operation between a State Research Institution (WSL) and a private company (GEOBRUGG). The WSL was primarily interested in the influence on defensive structures to protect transportation lines and mountain communities from rockfalls. In this paper rockfall speed and the quality of the protective wire net constructions are discussed.

The primary design factors for wire net systems are the rock size and speed, moreover, the rock's kinetic energy. The purpose of the wire net systems is to transfer the rock's kinetic energy into deformation energy in the wire net system. To stop the rock, the kinetic energy of the rock must be smaller than the maximum possible inner deformation energy of the construction. During the tests the deformation energy was increased several times by usage of improved net components and construction changes. For example, in 1985 rocks with kinetic energies of 250 kJ could be stopped. Nowadays ring net constructions capable of withstanding more than 2000 kJ are possible. Field tests proved to be the sole method to further develop advanced rockfall protection barriers.

In the following the rockfall test methods and results are discussed. For the first time the impact forces for large single falling rocks have been experimentally determined. Afterwards, the newest construction of rockfall barriers with ring net design are presented in detail.

2 EXPERIMENTS ON ROCKFALL

2.1 *Protection measures before 1981*

60% of the surface area of Switzerland lies in the Alps. Rockfall originating in the mountains endangers a large number of people on travelled ways and in buildings. Therefore protective measures against rockfall have started at an early date. The systems erected at that time mainly consisted of a combination of wood and steel. Such structures are only effective if a relatively small amount of energy is involved, as their deformation capacity is limited. Other constructions, for example with wire mesh, can sustain greater deformation. However, as the strength of such systems is also limited, the internal energy absorbing capacity is only slightly enhanced. Constructions of wire rope mesh display a greater energy absorbing capacity with a concomitant increase in strength. Such systems were subjected to

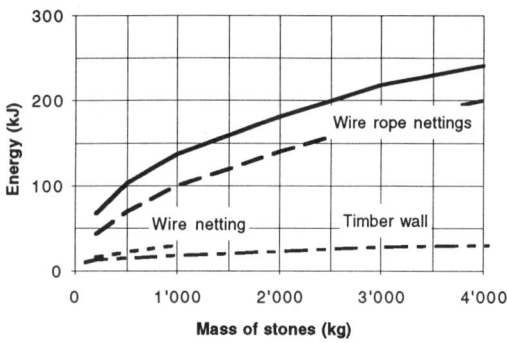

Fig. 1: Maximum energy absorbing capacity of different protective systems.

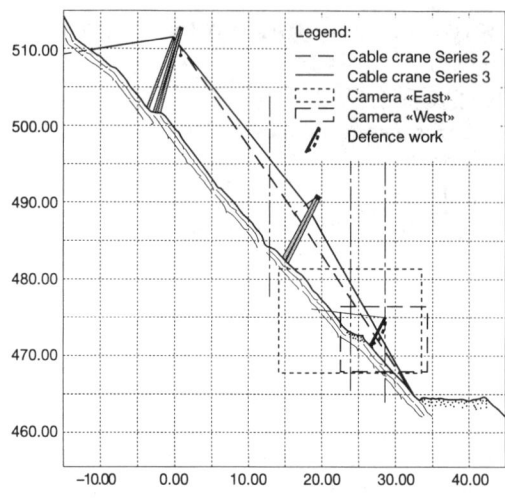

Fig. 2: Test site at Beckenried, longitudinal section with cable crane.

tests with stones in free-fall even before 1981. Figure 1 gives an overview of the maximum internal energy absorbing capacity of the protective structures tested at that time (HEIERLI et al. 1981). It can be seen that timber constructions display an internal energy absorbing capacity of 20 kJ when tested with a mass of stones of around 1000 kg. Stuctures of wire mesh withstand only slightly more impact energy. At the time of testing, constructions of wire rope net dissipated most energy: it was possible to test them with a mass of rock of 4000 kg, under which conditions they were able to dissipate 250 kJ of energy.

2.2 *Experiments on rockfall 1988-96*

A spectacular research project was started in 1989 by CALTRANS, the California Department of Transportation, Division of New Technologies, to test a wire rope net system under real conditions (SMITH, DUFFI 1990). An ambitious research project as a confirmation of the CALTRANS-Tests was then started in Oberbuchsiten in Switzerland. In these tests systems were developed to stop rocks with kinetic energies of about 1000 kJ (DUFFI 1992).

A next test series with the possibility for more sophisticated data collection and for repeatable tests followed in Beckenried (GERBER, BÖLL 1993). The terrain at Beckenried comprises mainly a stable rock surface with a slope of 45°. A cable crane serves to transport stones to the top, and the present facility is also used to project stones directly against the protection barrier. We filmed the stones as they travelled with a fast-frame camera (54 frames per second) from two different positions, depending on the aim of the experiment. The camera was placed at the site "East" to film the movement of the stones over the slope and at the site "West" to film the stones hitting the barrier. Figure 2 shows a profile of the test site with the fields covered from the two camera locations.

Test series: In Series 1, rocks were set in motion over the surface of the slope from a point about 30 vertical metres above the barrier, whence they bounced downhill and hit the barrier at some point or other. This series, mainly conducted at the beginning of the experiments, provided valuable information on the way the rocks moved over the steep slope. The aim of Series 2 was to obtain higher levels of energy. In 1991 a new cable crane was erected and used to set in free-fall rocks suspended from the bearer cable; the stones were released in full flight and hit the ground before rolling or bouncing into the net. In Series 3, the rocks were projected directly at the barrier without hitting the ground beforehand. This made it possible to calculate the energy of the rocks in advance and to aim the stones at specific points of the barrier.

Data collection: Before each test, the current condition of the barrier was marked and recorded. This enabled us to determine which elements reacted and where energy had been dissipated. We also recorded data such as the distance of the rock from the camera, slope of the area of impact etc. Most of the filming was done from the site "East". Only during 1993, when the deceleration path of the rocks and their movement in the barrier were being filmed, the camera was located at the site "West".

2.3 *Evaluation methods*

For the evaluation of the films "East", each photograph was traced with a special device to produce a diagram of the trajectory of each stone. Given the scale of the photographs, it was possible to calculate

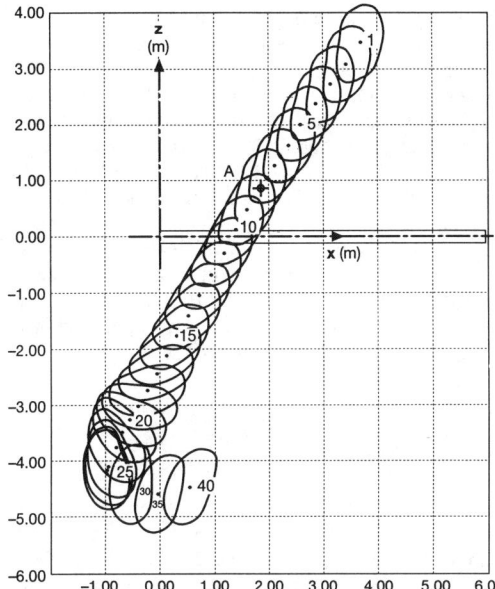

Fig. 3: Record of trajectory used to analyse test no. 138

3.1 Velocity of the stones

For each test, the velocity of the stone at the moment of impact against the barrier was determined. In Figure 4 it can be seen that the velocity depended on the test series. In Series 1 the stones reached the barrier after repeated contact with the ground. The broad scatter is a logical consequence of this. The mean velocity was 15 m/s. From the results of this series we were also able to study conditions on impact and thus obtain valuable information on the determinative parameters. In Series 2 there were no great changes in velocity although the stones hit the ground only once or twice before impact against the barrier. Much more energy was dissipated through contact with the ground. In Series 3 it was possible to keep the velocity constant, at 26.5 m/s and thence to compute the expected energy beforehand.

the mathematical function of each trajectory and thus the velocity of each stone at any given moment. From this it was also possible to compute the kinetic energy. The position of the stones was also traced from the films "West". In this case, however, it was not possible to compute the velocity as a mathematical function, as the stones were continually being retarded. For this reason the velocity of each stone at any given moment was computed as the difference between its position in two consecutive frames. For technical reasons concerning measurement, velocities vary considerably. Nevertheless we used them as basic data for further calculations. The relationship between velocity and time allowed calculation of the retardation, and from that and the mass of each stone it was possible to compute the effective forces. Figure 3 shows the trajectory of the rock in Test No. 138 (mass 2170 kg) in the individual frames. The time difference between the separate frames is 1/54 second. From frame no. 1 to frame no. 8 the stone is in free flight. After frame no. 8 the stone is retarded (point A). Here the maximum velocity is 25 m/s. The subsequent frames up to no. 26 were used to analyse of the braking process.

3 RESULTS AND DISCUSSION

Only the most important results are discussed, and for reasons of space only some of these are shown in the diagrams. Some 180 tests were conducted.

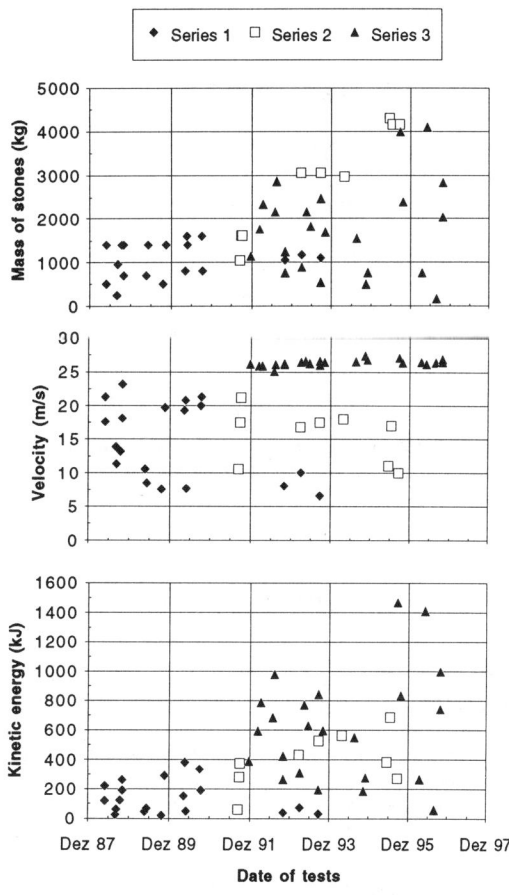

Fig. 4: Masses, velocities and energies of rocks crossing the plane defined by the fence posts.

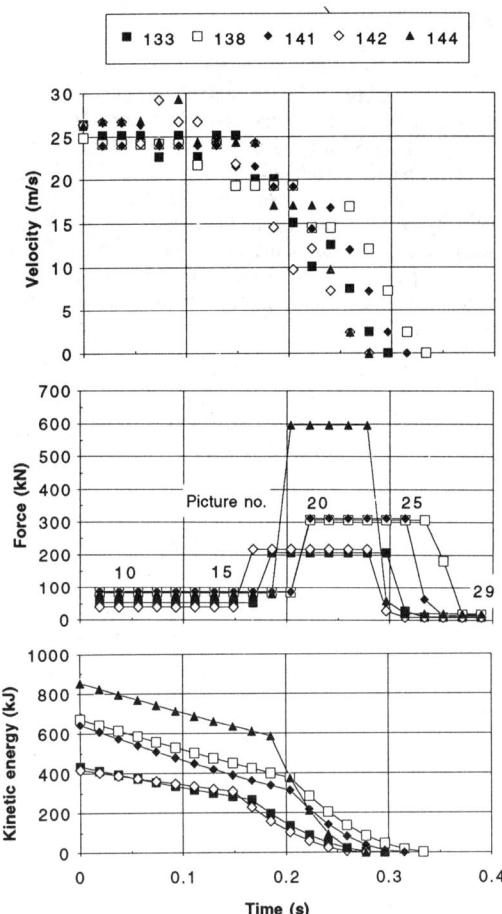

Fig. 5: Velocities, forces and energy dissipation in the protective system.

3.2 Kinetic energy

Up to the end of 1990 it was only possible to increase the impact energy on the barriers to 400 kJ. Only with the introduction of the new cable crane was it at all possible to produce higher levels of energy. Over the following 3 years barriers capable of absorbing a kinetic energy of 1000 kJ were designed. The development work was continued up to 1996, by which time the value had increased to 1500 kJ. This development can also be traced in Figure 4.

3.3 Velocities, forces and energy absorption

The deceleration paths of the rocks during some of the tests were drawn and the velocities computed. Figure 5 shows the velocities during the retarding phase. The maximum velocity at the beginning of retardation was 27 m/s. In the first half of the retarding phase, the velocity was reduced by only some 5-8 m/s. Not until the second half was the greater part of the energy dissipated. That means that only during the second half of the phase the strong forces necessary to halt the rock come into play. For example by test no. 138 the force was 300 kN. It acted for 0.15 seconds before the rock halted. In another test a force of as much as 600 kN was necessary. Once the force has been calculated, the energy dissipation within the barrier can be traced. Very interesting detailed studies on the relationship of forces evolved, when in one test a support in the barrier was deliberately struck (BÖLL 1995).

4. CHARACTERISTICS OF ADVANCED ROCKFALL PROTECTION BARRIERS

Rockfall protection barriers ideally absorb the energy from a rockfall safely, within the elastic limits and without the need for maintenance after the impact, independent from the point of impact. The developments of barriers from the past years using the results of field tests and its analysis has led towards this ideal behaviour. The design, layout and anchoring of a rockfall barrier considers easy installation, taking into account that such installations mostly have to be realised in difficult, steep and remote terrain. Lightweight parts, a minimum of anchors and quick erection are important aspects. Figure 6 shows the principal design of an advanced rockfall protection barrier with ringnets as tested in Beckenried.

4.1 Nets

The nets are the major component in the system. They are often the first part to be impacted and must transfer the forces to the structure, that is to support ropes, suspensions, posts and finally to the anchors. Ringnets, arranged like Olympic Rings, made of high grade steel wires have proved to be the most effective in their energy absorption capacity. This type of net has a very high internal elastic energy absorbing capacity and therefore is maintenance - free in a large range. Also trees can be stopped safely with ringnets. These ringnets can be fabricated with Al/Zn coated wires to improve the corrosion protection for a long lifetime.

4.2 Brake devices and support-ropes

The impact load has to be transferred from the net into the support ropes. It is essential that the support ropes are designed to react uniformly independent from the point of impact in the net. A special double support rope design with incorporated energy ab

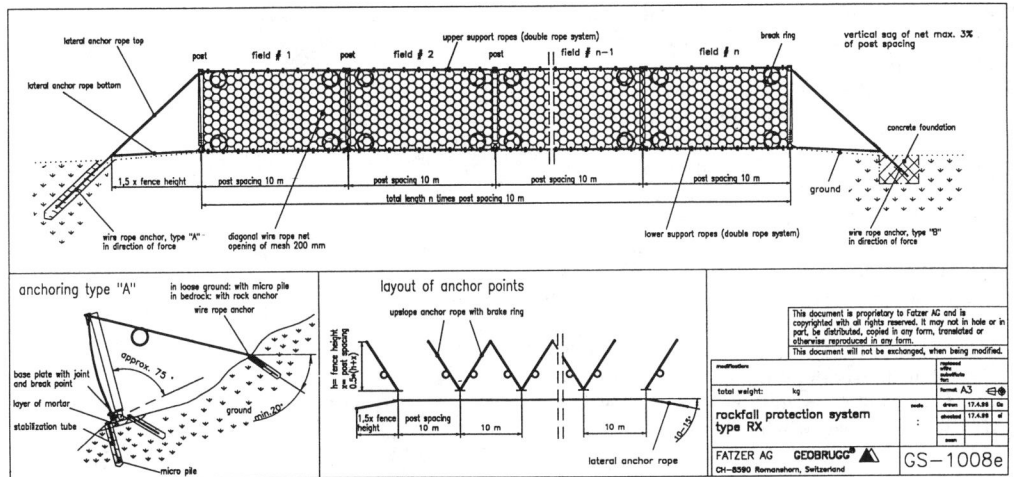

Fig. 6: Layout of an advanced GEOBRUGG RX-Type Rockfall Barrier

sorption devices at specified locations proved to provide the optimum balance between energy dissipation, net sagging and maintenance. The brake ring used allows a multiple impacting of the system without destroying the ropes or reducing their break strength, which is an important safety factor. The new support-rope design increased the flexibility of the system at the connections with the posts considerably without increasing maintenance needs.

4.3 Posts and anchors

The principal function of the posts is to support the net. In case of an impact to the net the post must not collapse to keep the net upright. A minimum size of the used wide flange (HEB) -profiles has been evaluated for each net height. In case of direct impacts with the posts near the baseplate a rated break device at the connection of the post to the baseplate allows the post to separate from the baseplate and therefore to protect the anchors of the baseplate from being destroyed. The anchoring of the lateral and upslope anchor ropes is provided with wire rope anchors. Wire rope anchors with double sleeve are the best solution to absorb high impact forces, especially if the axis of the anchor is not exactly in line with the one of the anchor-ropes.

tem. This represents therefore the worst case. Procedures with ground contact do not allow to determine the energy taken by the system, since a considerable part can be absorbed by the ground.

Through such experiments the WSL can improve their knowledge on the movement of stones over terrain and also during the retarding phase. The calculation of the forces acting on the stone, however, only provides a starting point for further experiments. What must be done now is to distribute the forces over the net and their supporting elements, and lead them into the individual anchors. Sophisticated measurement of the forces at the anchor points should be taken to verify the calculations. The aim is to improve the basics for dimensioning protection barriers and their components.

On the basis of these experimental results industry could develop protection structures in which the balance between all the components is maintained under varying conditions. Balanced flexibility means: no very rigid components, which lead to greater forces, but also no components which are too flexible, as these reduce the effectiveness of the construction and therefore lead to increased maintenance work. Such balancing also means that none of the components should be over-dimensioned; only when this is achieved the erection of protective systems will become financially worthwhile for the client.

5 OUTLOOK, CONCLUDING REMARKS

The tests shows different cases of stopping rocks in a protection barrier. The effect that no ground contact occurs before the rock is fully decelerated means, that the entire energy has to be dissipated by the sys-

6 REFERENCES

BÖLL, A., 1995: Tragsicherheit von Stahlstützen in Steinschlagverbauungen. In: Schweizer Ingenieur und Architekt SI+A Heft Nr.45, Verlags-AG der akademischen Vereine, Zürich.

DUFFI, J.D., 1992: Field tests of flexible rockfall barriers, Report for GEOBRUGG FATZER AG, Romanshorn. 135 S.

GERBER, W., BÖLL A., 1993: Massnahmen zum Schutz gegen Rutschungen und Steinschlag. In: Eidgenössische Forschungsanstalt für Wald, Schnee und Landschaft (Hrsg.): Naturgefahren, Forum für Wissen 1993: S. 33-38.

HEIERLI, W., MERK, A., TEMPERLI, A., 1981: Schutz gegen Steinschlag. Forschungsarbeit 6/80 auf Antrag der Vereinigung Schweizerischer Strassenfachleute (VSS). Bundesamt für Strassenbau, Bern. 120 S.

SMITH D.D. , J.D. DUFFI, 1990: Field tests and evaluation of rockfall restraining nets. Report No. CA/TL - 90/05. California Departement of Transportation, Sacramento. 138 S.

Study on prediction of landslide of rock slopes

Xiurun Ge, Dongjun Xu, Xianrong Gu, Congxin Chen & Yongsheng Shi
Institute of Rock and Soil Mechanics, The Chinese Academy of Sciences, Wuhan, People's Republic of China

ABSTRACT: This paper describes the study of prediction of landslide of rock slopes, i.e., based upon the displacement monitoring data of slopes of ages, send out a landslide warning by taking the displacement rate where displacement-time curve enters from an increasing stage of constant rate into accelerating stage as the critical displacement rate of landslide; using analysing method of time sequences, predict development tendency of displacements in coming years according to the displacement law in the past years; and predict the time of landslide by taking the displacement rate of landslide determined from landslide case histories at home and abroad as the criterion of landslide time. Prediction of both time and displacement rate of landslide one year before schedule is well coindcident with the real situation.

1 INTRODUCTION

This study has been carried out in Daye Iron Mining — the biggest deep concave open iron mining in China. The slope of the Mining, composed of diorite, has been lowered at present from $\triangledown 241m$ to $\triangledown -120m$. Since 1990 when landslide of A_1 (6000m³) took place between elevations of $\triangledown 84m \sim \triangledown 12m$, a 60m wide slope rockmass located in the hanger of faults F_9 and $F_9{'}$ between elevations of $\triangledown 180m \sim \triangledown 48$ had constantly moved towards sliding body A_1 and finally, a huge landslide of 9000m³ in volume resulted in March — July of 1996.

This paper will describe the displacement monitoring results of the slope from 1990 till July of 1996 and the research results about slope landslide, i.e., based upon the monitoring results, predict the coming slope displacement using the method of determining the critical displacement rate of landslide at which the displacement turns from a state of constant rate increasing to a state of accelerating increment and the method of time-sequence analysis and thus determine the criterion of landslide rate to predict landslide time. The *in-situ* measured data of recent seven years have indicated that the predicted critical displacement rate of landslide and the predicted displacement rate in the accelerating stage of landslide and the landslide time are coincident with the real situation.

2 GEOLOGICAL ENVIRONMENTS OF LANDSLIDE AREAS AND FACTORS INDUCING LANDSLIDE

Shown in Fig. 1 is the engineering geological plan of interest. The slope is mainly composed of diorite and its unstable areas are most concentrated in the scope of the hanger of faults F_9 and $F_9{'}$ with a width of 50~70m. This area has the lithology of diorite of δ_5^{2-1} strongly altered into kaolinite and chlorite, belonging to the weak rock type of fractured and loosen structure.

The sliding body is bounded on the top by fault F_9 with a strike of $N \angle 67° \sim 81°W/SW \angle 58° \sim 79°$ and on the east by fault $F_9{'}$ with a strike of $N \angle 65°W/SW \angle 57°$. The fracture structure of $N \angle 48°/NW \angle 71°$ developed inside the sliding body makes the body possess the potential energy which tends to force it to topple over towards the direction of SE 120°.

The strike of the slope in this area is $N \angle 75° E/SE \angle 43°$, the global slope angle is $43° \sim 45°$ above $\triangledown \pm 0$ and $50° \sim 56°$ below it, the mining bench has a slope angle of $60° \sim 70°$.

The above unfavourable lithology and fracture structure comprise the engineering condition making the slope move southeastwards (120°). The rainwater has for ages firstly infiltrated into the slope and then into the deep of the rockmass along the toppling-over deformed fractures of NE direction, and it has gradually weakened the rockmass strength and exerted a certain sliding-down force on the rockmass. The water in the rainy seasons made displacements increasingly develop obviously. That specially strong precipitation in 1996 did add a sliding-down force to the underground water then on the slope, which had been in the state of constant rate rehology, to induce and accelerate the slope's displacement. In March of 1996, it rained almost everyday with a rainfall of up to 197.0 mm per day. In 13th of March, a transversal gap with a width of 2000mm appeared in the elevations of ▽150 and ▽132, as a result, the slope displacement of the whole unstable area started accelerating; in 18th of April with a rainfall of 108.7mm, the vertical and horizontal cracks of the area developed downwards and the slope displacement aggravated further; going into June, the rainfall reached 274.3mm within three days of 3rd, 4th and 5th and the months rainfall was 609.2mm with an average of 91.4mm per day. In 2nd of July, its rainfall was 117.1mm, resulting finally in a huge landslide.

From the above, it can be considered that strong precipitation is the factor to induce the landslide of the slope which has been in a state of ultimate equilibrium.

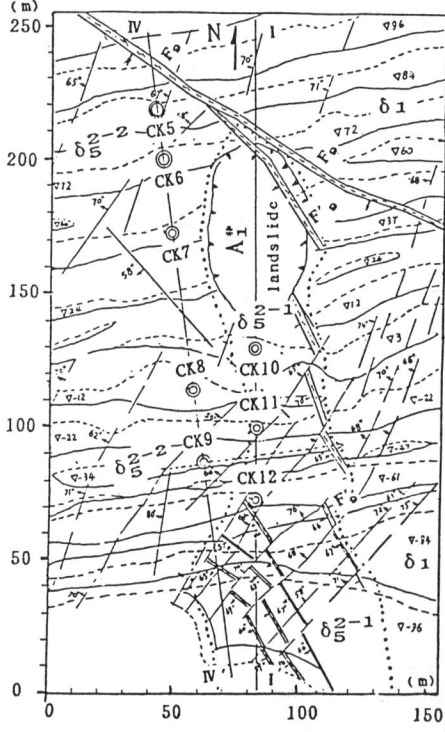

Figure 1. Engineering geological profile

3 MONITORING RESULTS OF SLOPE DISPLACEMENTS

Listed in Table 1 are the displacement monitoring results obtained using level probe and altometer at 31 measuring points in the sliding area, including horizontal displacement (L), its azimuth (θ) and annully average displacement rate, V, (mm/day). Fig. 2 shows the horizontal displacement rates (mm/day in unit) of all measuring points in the sliding body in 7th of June, 1996 and the displacement azimuths, θ, all plotted according to Table 1.

Table 1. Monitoring results of slope displacement

No. of Points	1993		1994		1995		1996.6.7	
	L	V	L	V	L	V	L	V
180—12	85.2	0.11	148.6	0.23	265.0	0.40	682.0	3.28
156—11	191.6	0.28	457.0	0.95	488.0	0.22		
156—12	265.0	0.41	571.0	1.15	925.0	1.17	1876.0	11.4
132—11	253.6	0.37	492.6	0.84	731.0	0.81	1156.0	4.74
132—11	65.8	0.065	154.0	0.41	395.0	0.79	787.0	4.16
108—8	280.0	0.34	526.5	0.77	780.0	0.66	1125.0	3.95
108—9	191.0	0.21	369.0	0.55	517.0	0.46	810.0	3.47
108—10	28.0	0.35	526.5	0.77	771.0	0.59	841.0	0.77
96—8	722.0	0.57	971.0	0.73	1213.0	0.65	2417.0	7.5
96—9	413.0	0.35	566.0	0.46	702.0	0.43	1630.0	7.1
84—10	446.0	0.56	726.0	0.75	936.0	1.05	1780.0	6.4
84—11	343.0	0.42	549.0	0.51	732.0	0.51	1615.0	6.75
72—10X	473.0	0.67	687.0	0.72	887.0	0.62	1917.0	7.93
72—11	458.0	0.48	642.0	0.53	811.0	0.55	1607.0	6.16
72—11X	269.0	0.34	401.0	0.35	529.0	0.42	1129.0	4.6
48—8—1	243.0	0.46	463.0	0.60	647.0	0.53	1532.0	6.82
48—8X	323.0	0.44	477.0	0.53	618.0	0.43	1219.0	5.14
48—9X	294.0	0.39	423.0	0.40	546.0	0.37	1112.0	7.10
48—10	202.0	0.20	269.0	0.20	334.0	0.19	630.0	2.25
平均		0.369		0.60		0.57		5.51

L: Horizontal displacement (mm)
V: Rate of horizontal displacement (mm/day)

Fig3 gives the relationship of horizontal displacement L with time and precipitation of six typical points.

It can be seen from the above monitoring results that the area around the nineteen measuring points within the elevations of ▽48m～▽180m *was* in a state of constant rate rehology in 1994 and 1995. The average displacement rate of those nineteen points remained 0.60mm/day and 0.57 mm/day respectively. In the early 1996, the area entered into the stage of accelerated rehology. The landslide started from the upper of ▽48m and afterwards gradually moved

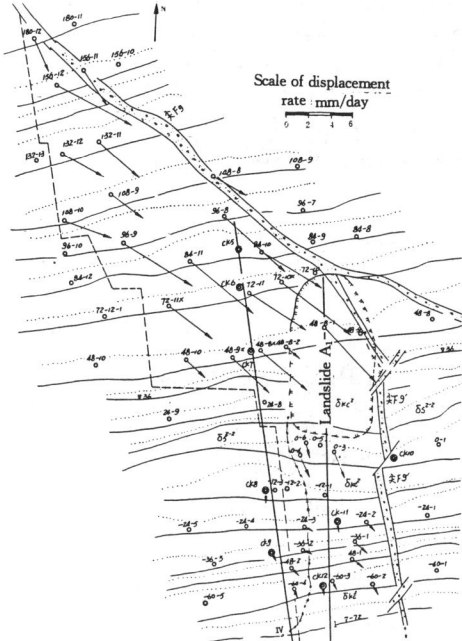

Figure 2. Velocity and azimath of slope displacement

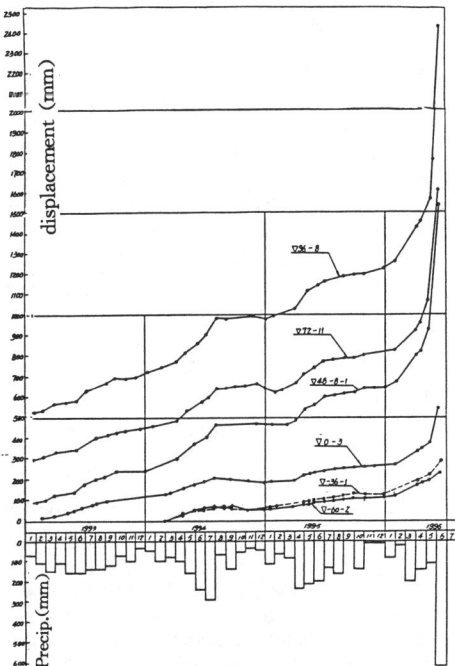

Figure 3. The relationship of holizontal displacement versus time and precip.

towards sliding body A_1 and then developed to the lower body A_1. The displacement direction of the sliding body was firstly SE∠120° at the elevation of ▽48m and then gradually deflected towards SE∠150°~160° at the lower of sliding body A_1. Almost all annully average displacement rate of each measuring point at the upper ▽48m reached 0.5mm/day in 1995, this value can be taken as the critical displacement of landslide.

4 PREDICTION OF SLIDING TIME

4.1 *Determination of critical displacement rate for landslide*

As shown in Table 1 and Fig. 3, by analysing the increasing law year by year of the annual average displacement rate of each measuring point, we can define the initial increasing stage, constant stage and accelerating stage of the slope displacement.

Providing the time can be determined accurately when the slope displacement turns from the constant stage into the accelerating stage, we may precisely make a prediction of landslide and take preventive measure in time.

The measuring data show that even for the stage of constant rate displacement, the slope displacement increased more rapidly during rainy seasons and became smaller during dry seasons. Accordingly, it is of prime importance to plot the relationship curve of displacement to time and to precipitation in time in order to find the changing law of displacement variation with time.

Considering the monitoring data of those nineteen measuring points between elevations of ▽108m~▽48m, we think that the years of 1994 and 1995 were the constant rate stage of the slope displacement, during which the nineteen points had annual average displacement rates of 0.60mm/day and 0.57mm/day respectively. In those two years, there were thirteen points having a displacement rate of beyond 0.5mm/day and five points having a rate of beyond 0.4mm/day. In the early 1996, the displacement rates of those nineteen points increased rapidly with an average value of up to 5.51mm/day in 7th of June, 1996.

Based upon the above, the critical displacement rate of landslide at which the slope displacement turns from the constant rate stage into the accelerating stage is taken as 0.50mm/day.

4.2 Predict landslide time using method of time sequences

In the early 1995, based upon the displacement monitoring data of the slope from 1990~1994, we predicted the slope displacement value for each 15-day long time interval in the years of 1995 and 1996 using method of time sequence and then predicted landslide time according to the criterion of landslide velocity.

(1) Analysing method of time sequences

So-called analysing method of time sequences is mainly aimed at relative stable time sequences to establish model and to make prediction. Providing a time sequence, $\{X_t\}$, meets the condition of $EX_t^2 < \infty$ and

$$Ex_t = \mu \quad (\mu = \text{constant}) \quad (1)$$
$$\gamma_{ts} = \gamma_{t-s} \quad (2)$$

then $\{X_t\}$ is defined as a wide stable time sequence.

For a stable, normal and 0-mean-value time sequence of $\{X_t\}$, if the value of x_t is not only related to those of $X_{t-1}, X_{t-2}, \cdots, X_{t-n}$ in the previous n steps but also to the interference of $a_{t-1}, a_{t-2}, \cdots, a_{t-m}$ in the previous m steps, then a model of ARMA can be obtained as follows according to the concept of multiple regression:

$$X_t = \sum_{i=1}^{n} \varphi_i X_{t-1} - \sum_{j=1}^{m} \theta_j a_{t-j} + a_t,$$
$$a_t \sim NID(0, \sigma_2^2) \quad (3)$$

The model expressed by (3) is called n-order self regression and m-order sliding average model, written as ARMA (n,m). Especially, given $\theta_j = 0$ in eq. (3), we have n-order self regression model, AR (m); whereas given $\varphi_i = 0$, then we have m-order sliding average model of MA (m).

The time-sequence model is a kind of parameter model, when established, its main purpose is to determine the parameter of the model. In addition, it includes checking and processing of data, selection of model-establishing scheme, checking of model adaptation and optimization of parameters.

In the case of unstable time-sequence for which it is no way to establish the model of ARMA, we adopt an extracting method of tending terms, i.e., picking-up the tendency of determinacy from an unstable time-sequence and express it by a clear function and then establish model of ARMA to the residual difference sequence and finally extract the tendency term using Marquardt method.

(2) Superposition of slope displacements and time-sequence model

A slope is affected by the condition of itself and various environmental factors, its displacements exhibit characters of determinacy and randomness. So, the overall displacement of the slope can be expressed as

$$X_t = d_t + \varepsilon_t \quad (4)$$

where d_t is the diterminative part of the displacement,

$$d_t = \sum_{i=1}^{p} \beta_i e^{a_{it}} + \sum_{j=1}^{q} e^{\gamma_{jt}} \left[A_j \cos \frac{2\pi t}{T_j} + B_j \sin \frac{2\pi t}{T_j} \right] \quad (5)$$

and ε_t is the stable random part of the slope displacement.

The first term of eq. (5) represents the displacement caused by its own structure condition and the second the displacement caused by the environment factors.

From eq.s of (3),(4) and (5), we can get a general model of the slope displacemnt in the form of

$$X_t = \sum_{i=1}^{p} \beta_i e^{a_{it}} + \sum_{i=1}^{q} e^{\gamma_{it}} \left[A_j \cos \frac{2\pi t}{T_j} + B_j \sin \frac{2\pi t}{T_j} \right]$$
$$+ \sum_{j=1}^{n} \varphi_j \varepsilon_{t-1} - \sum_{j=1}^{m} \theta_j a_{t-j} + a_t \quad (6)$$

eq. (6) defines a model of superposition and time-sequence for the slope displacement.

(3) Predicting results of slope displacements

According to the above analysing method of time sequence and the model of superposition and time-sequence for the slope displacement, we selected seven points in the large deformation area to predict the displacement of the years of 1995 and 1996 according to the monitoring data of 1990~1994. Table 2 lists the predicting results of June and July of 1996, in which S_x, S_y, X_z are respectively the horizontal and vertical components of the displacements along coordinate axes of x, y, z, δ_x, δ_y and δ_z are respectively the displacement increments along axes of x, y, z in the predicting time interval (15 days), $S = \sqrt{\delta_x^2 + \delta_y^2 + \delta_z^2}$, V is the displacement rate (mm/day) in the predicting time interval.

(4) Criterion of displacement of landslide and prediction of landslide time

According to the knowledge of the known information, researchers both at home and abroad suggest that a displacement rate value be used as the criterion for landslide. Although the suggestion is rational, the suggested rates differ somewhat from each other, for example, 100, 14.4, 24 mm/day etc.. These criteria or critical values have all been suggested specially from respective gological enviroment, hydrology, meteorology and topography of slopes, therefore no common standard can be made.

Table 2. Predicted results of slope displacement

No. of points	Prediction of date (1996 y.)	Predicted displacement value (mm)				displacement rate V (mm/day)
		δ_x	δ_y	δ_z	$S = \sqrt{\delta x^2 + y^2 + \delta z^2}$	
96—8	6.5,	−12	59	−6	60	4.0
	6.20,	−62	56	−6	84	5.6
	7.5,	−80	51	−7	95	6.3
84—10	6.5,	−37	51	−23	67	4.5
	6.20,	−44	49	−23	70	4.7
	7.5,	−46	45	−23	68	4.5
84—11	6.5,	−40	52	−14	67	4.5
	6.20,	−53	49	−25	26	5.0
	7.5,	−53	43	−30	75	5.0
48—8—1	6.5,	−57	86	−15	104	6.9
	6.20,	−72	108	−14	131	8.7
	9.5,	−81	119	−19	146	9.7
48—8X	6.5,	−32	32	7	46	3.1
	6.20,	−34	30	8	46	3.1
	7.5,	−34	30	8	46	3.1
72—10X	6.5,	−11	41	−14	45	3.0
	6.20,	−66	52	−42	93	6.2
	7.5,	−68	53	−62	106	7.1
72—11	6.5,	−22	34	−16	44	2.9
	6.20,	−22	39	−15	47	3.1

Table 3. The monitoring value of land slide displacement (mm)

No. of points	monitoring date (1996 y.)	Horizontal displacement δ_L	vetical displacement δ_H	$S = \sqrt{\delta_L^2 + \delta_H^2}$	displacement rate (mm/day)
96—8	4.16	17	18.7	25.0	1.78
	5.09	117	2.5	117.0	5.08
	6.07	858	96.3	863	30.8
84—10	4.16	0	44.7	44.7	3.17
	5.09	145	6.0	145.1	6.30
	6.07	476	224.4	526.2	18.80
84—11	4.16	9	24.4	26.0	1.86
	5.09	116	3.2	116.0	5.04
	6.07	590	123.6	602.8	21.53
48—8—1	4.16	12.0	28.3	30.7	2.19
	5.09	112.0	12.0	112.6	4.9
	6.09	605.0	164.5	626.9	22.39
48—8X	4.16	10.0	15.1	18.11	1.29
	5.09	81.0	4.6	81.13	3.52
	6.07	455.0	82.8	462.47	16.51
72—10X	4.16	21.0	46.8	51.30	3.66
	5.09	125.0	20.6	126.68	5.51
	6.07	701.0	287.3	287.39	16.26
72—11	4.16	0	33.0	33.0	13.6
	5.09	111.0	13.3	111.79	4.86
	6.07	543.0	201.1	579.0	20.68

In accordance with the slope structure, engineering geology, hydro-meteorological condition of the Mining and based upon the monitoring data obtained from those nineteen points in the sliding area, we analysed the increasing law of the predicted displacement rates year by year of seven measuring points and selected the value of 5.0 mm/day as the displacement rate during landslide and took the time interval in which the predicted displacement of most points had reached or exceeded the value of 5.0mm/day as the time when landslide would occur and then sent out warning of landslide.

The predicting result indicated in June of 1995 that a landslide in the area would take place around 20th of June, 1996 with a rate of 5.0 mm/day.

(5) Comparison between predicted and measured displacements of landslide

The real date of landslide in the area was 1st of July, 1996, the accelerating stage (Fig. 3) of the sliding body displacement started in March and the rapidest time-interval of displacement increment was 16th of April to 9th of May. The monitoring continued to 7th of June. Table 3 lists the measured displacements of those seven measuring points for comparison.

The predicted and date of landslide displacement rate were 5.0mm/day and 20th of June respectively, whereas the actual displacement rate was 5.0mm/day and the date of landslide was 9th of May—more than one month earlier than the predicted time. And from Fig. 3, it can be known that the rapid acceleration of the displacement of landslide took place during March to 6th and 7th of June of 1996. It seems that which of the time-inievtals during landslide process should be selected as the criterion of landslide remains to be studied. Nevertheless, predicting landslide time using the analysing method of time-sequence based upon long-term monitoring data is feasible. The above comparison between predicted and measured data shows that the prediction of landslide of the slope was accordant with what it was and succeful.

5 CONCLUSION

The study carried out in Daye Iron Mining, China about the displacement monitoring and prediction of landslide of the Mining slope in a long-term unstable state has played a role of guiding safe production of the Mining. This study enables us to have knowledge that apart from the engineering geological conditions of unfavourable lithology and fracture structure and the slope structure such as strikes, the most important affecting factor to cause slope displacement and sliding is precipitation strength. Long time and continuous precipitation and heavy raining in short time is the factor inducing a slope which has been in a state of ultimate equilibrium to enter into a landslide stage. As concerns prediction of slope landslide, it seems the most valuable to predict the critical value of the displacement rate of a slope where the slope turns from the state of constant rate displacement to that of accelerating displacement. The reason is that precise prediction of slope landslide provides the engineer with enough time to take reinforcing and other preventive measure against landslide. The suggested critical displacement rate of 0.50mm/day is practical to the Mining slope of diorite.

It is feasible to predict the coming displacement of a slope using various theories and methods and then to predict the landslide time according to displacement rate of the landslide. This study has valuably taken the value of 5.0~10.0mm/day as the landslide rate of the rocky slope.

REFERENCES

Yongsheng Shi, Dongjun Xu. Application of time series analysis in slope displacement prediction. *Rock and Soil Mechanics*. Vol.16, No.4:1-7.

Dongjun Xu et al. 1996. Study of stability of rocky slope under high displacement. *Proc. of the Korea-Japan Joint Symp. on Rock Engineering*. 18—20 July. Seoul. Korea, pp.445-449.

ate the
Reliability-based analysis for rock slopes considering multi-failure modes

In-Mo Lee
Department of Civil Engineering, Korea University, Korea

Myung-Jae Lee
Institute of Construction Technology, Kyongho Engineering Co., Korea

ABSTRACT : This paper presents the results of sensitivity analysis performed on rock slopes to verify a newly developed reliability-based model considering uncertainties of discontinuities and multi-failure modes - plane, wedge, and toppling. The parameters that are needed for sensitivity analysis are the variability of discontinuity properties (orientation and strength of discontinuities), the loading conditions, and the rock slope geometry. The variability in orientation and friction angle of discontinuities, which can not be considered in the deterministic analysis, has a great influence on the rock slope stability. The stability of rock slopes including multi-failure modes is more influenced by the selection of dip direction of cutting rock face than any other design variables. The case study shows that the developed reliability-based analysis model can reasonably assess the stability of rock slope.

1 INTRODUCTION

The stability condition of rock slopes is greatly affected by the geometry and strength parameters of discontinuities in rock masses. Rock slopes involving movement of rock blocks along discontinuities are failed by the combinations of the plane, wedge, and toppling modes.

The kinematic analysis considering three basic failure modes is widely available in design practice (Hoek and Bray, 1981; Leung and Kheok, 1987). However, the stability analysis considering three basic failure modes has hardly been applied in the research and design practice due to difficult task of combining the translational mode (plane and wedge failure) and the rotational mode (toppling failure). Only the stability analysis of the rock slope considering a single failure mode - translational or rotational - has been studied.

In rock mechanics, practically all the parameters such as the joint set characteristics, the rock strength properties, and the loading conditions are always subject to a degree of uncertainty. Therefore, a study is needed to develop a probabilistic analysis tool of rock slopes considering multi-failure modes and uncertainty of discontinuities under several loading conditions.

2 STABILITY ANALYSIS OF ROCK SLOPES

2.1 Failure modes of rock slopes

The analysis of rock slope stability is fundamentally a two-part process: the kinematic analysis and the stability analysis. The first step is to analyze the discontinuities of rock mass to determine whether the orientation of the discontinuities could result in instability of the slope.

Once it has been known that a kinematically possible failure mode is present, the second step requires a limit equilibrium stability analysis to compare the resisting forces with the forces causing failure. Not only the resistance forces including friction and cohesion of discontinuities but also resultant forces including gravity forces, water pressures, seismic forces and support forces from the rock bolts must be reasonably considered for the stability analysis of three basic failure modes.

2.2 Plane and wedge failures

The method of vector analysis provides relatively simple formulations for all the quantities related to block morphology including the volume of each joint block, the areas of block faces, the positions of its vertexes, and the positions and attitudes of its faces and edges. The use of vectors also permits kinematic and static equilibrium analysis of key blocks under self-weight, water pressure, seismic force, and support pressures.

From a rock slope design perspective, the most important characteristic of a discontinuity is its orientation, which is best defined by two parameters; dip(α) and dip direction(β). The unit normal vector of the discontinuity plane, $\hat{n}$, is given as

$$\hat{n} = (\sin\alpha\sin\beta, \sin\alpha\cos\beta, \cos\alpha) \quad (1)$$

2.2.1 *Kinematic analysis*

Three major components of block theory are as follows (Goodman and Shi, 1985):
(1) Finiteness analysis determines whether the rock joints and excavation surfaces combine to an isolated rock block and separate from the rest of the rock mass.
(2) Removabilty analysis determines whether the isolated block has a shape that allows the rock block to move into the excavation without movement of any other parts of the rock mass. The steps of finiteness and removability analyses are commonly entitled as kinematic analysis.
(3) Stability analysis determines whether driving forces acting on the block are sufficient to overcome resisting forces.

2.2.2 *Stability analysis*

If the kinematic analysis indicates that requisite discontinuities conditions are present, the rock slope stability for plane and wedge failures must be evaluated by a limit equilibrium analysis, which considers the friction force and cohesion strength along the failure surface and the resultant force. Vector analysis facilitates the analysis of block stability under gravity force, water pressure, seismic force, support force, friction, and cohesion.

The condition of equilibrium for a potential or real key block B, described in Figure 1, is given as

$$\vec{r} + \sum_i N_i \hat{v}_i - T\hat{s} = 0 \quad (2)$$

where, N_i = normal reaction force of joint plane i, $\hat{v}_i$ = unit vector normal to joint plane i, T = resultant of the tangential frictional force, $\hat{s}$ = unit vector of sliding direction, and $\vec{r}$ = resultant force.

The resultant ($\vec{r}$) of all forces, including gravity force, water pressure, seismic force, and support force from the rock bolts is given as

$$\vec{r} = \vec{W} + \vec{U} + \vec{F}_d + \vec{Q} \quad (3)$$

where, $\vec{W}$ = gravity force, $\vec{U}$ = hydrostatic force, $\vec{F}_d$ = seismic force, and $\vec{Q}$ = support force.

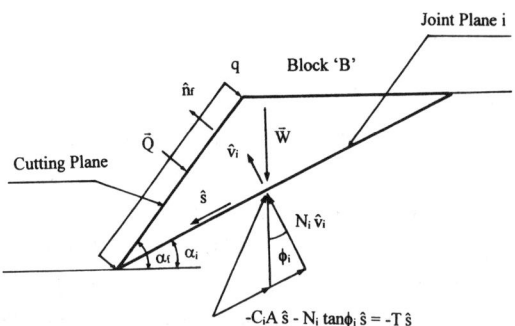

$\hat{n}_f$: unit vector of cutting plane
α_f : dip angle of cutting plane
α_i : dip angle of joint plane i
C_i : cohesion of joint plane i
ϕ_i : friction angle of joint plane i
q : uniformly distributed stabilizing pressure (normal to cutting plane)

Figure 1. Limit equilibrium on block B

2.3 *Stability analysis for toppling failure*

Toppling failures most commonly occur in rock masses which contain a large number of slabs or columns formed by a set of fractures that strike approximately parallel to the slope face and dip steeply into the face. Toppling failures are characterized by significant horizontal movements at the crest and very little movement at the toe. To accommodate this differential movement between the toe and the crest, interlayer movement must occur. Thus, the shear strength between layers is crucial to the stability of slopes structurally susceptible to toppling.

An analysis method of toppling failure presented by Goodman and Bray(1976) provides the fundamentals for the stability of rock slopes susceptible to toppling. This study is concerned with the toppling failure under water pressure, seismic, external, and gravitational loading conditions.

A numerical analysis method for toppling analysis is developed in this study, modifying and combining analytical method by Goodman and Bray(1976) and numerical method by Zanbak(1983).

2.3.1 Kinematic Analysis

The necessary kinematic conditions for toppling failure can be summarized into two parts: the one is the direction of toppling plane and rock slope face; and the other is the dip angle of joint plane and rock slope face and the friction angle of the joint plane. The strike of the joint plane must be approximately parallel to the rock slope face. The dip of the joint plane must be directed into the rock slope face. In order for interlayer slip to occur, the normal to the toppling plane must have a plunge less than the inclination of the slope face and less than the friction angle of the surface. The condition can be formulated as (Goodman and Bray, 1976)

$$(90° - \alpha_A) \leq (\alpha_1 - \phi_A) \quad (4)$$

where, α_A = dip angle of discontinuity A, ϕ_A = friction angle of discontinuity A, and α_1 = cutting angle of rock slope.

2.3.2 Stability Analysis

Figure 2 shows a cross section of the rock slope with a system of blocks on a stepped failure surface. Two discontinuity systems A and B with 100% persistence are assumed to be present in the rock mass.

For the resulting geometry of rock blocks in the slope, resting on the estimated stepped failure surface, the n blocks are numbered in sequence starting with one at the toe shown in Figure 2. For the (i)th block shown in Figure 3, the force P(i) transferred to the (i-1)th block is calculated from limiting equilibrium condition. When considering the limiting equilibrium condition of a typical block (say i, on Figure 2), the following additional forces are acted on the block :
- the weight W(i) ;
- the shear forces T(i), T(i+1) on the sides and R(i) on the base ; these forces are related to the friction angle ϕ_A, at the sides, and ϕ_B, at the base ;
- the hydrostatic forces $U_l(i)$, $U_r(i)$ at the sides and $U_b(i)$ at the base, resulting from the water pressure distribution;
- the seismic force $F_D(i)$, which is as usual given by K×W(i), with K being the seismic coefficient ;
- the force P(i+1), transferred from the (i+1)th block, taken as normal to the side and applied at point B ; and
- the force Q(i) resulting from a stabilizing pressure distribution.

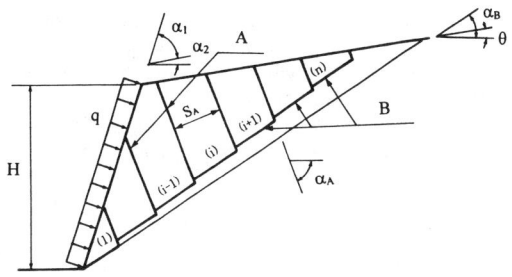

H : slope height
q : uniformly distributed stabilizing pressure
α_1, α_2 : cutting angle of slope and upper slope surface
α_A, α_B : dip angle of discontinuity A and B
S_A : spacing of discontinuity system A
θ : step angle

Figure 2. Idealized slope generating with a postulated failure surface

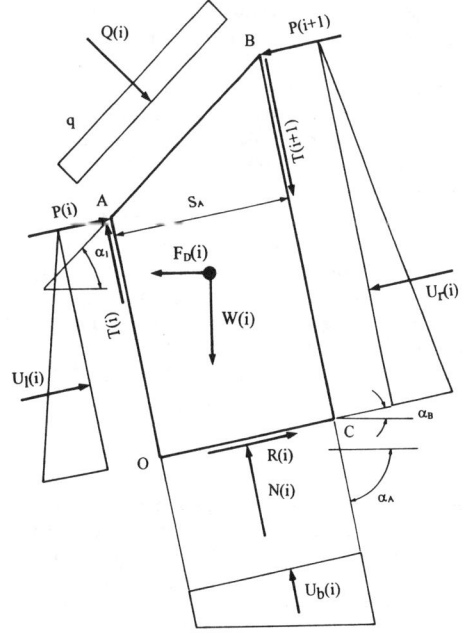

Figure 3. Forces applied to slice (i)

The force P(i) is assumed to be normal to the side of the (i)th block and applied at discontinuity A. The greater one between the calculated $P_t(i)$, toppling resistance of the (i)th block, and $P_s(i)$, shesr resistance of the (i)th block, will be the P(i) force exerted on the (i-1)th column. The overall stability of the toppling slopes is defined by the P_1 force exerted on the first column at the toe (Figure 2). The slope is considered unstable if the resultant P_1 force applied on the toe column is greater than zero. When $P_1 > 0$, then magnitude of P_1 is the retaining force in the base plane obtained from the limit equilibrium condition.

3 RELIABILITY-BASED ANALYSIS

3.1 General

The failures of rock masses are mainly controlled by discontinuities in the rock mass. Therefore, it is important to know features of discontinuities. These features are classified into three groups: (1) orientations of discontinuities; (2) strength properties of discontinuities; and (3) size and spacing of discontinuities.

In this study, the reliability-based analysis of rock slopes is accomplished by the Monte Carlo simulation based on independently generated random numbers. Some basic features and assumptions are given below:
- The kinematic analysis and the stability analysis for plane and wedge failures of rock slopes are accomplished by block theory.
- The kinematic analysis and the stability analysis for toppling failures of rock slopes are performed by a newly developed 2-D numerical method.
- The orientations of discontinuities are random variables assumed to be Fisher's distribution assessed by the clustering technique.
- The friction angles of discontinuities are random variables assumed to be normal distribution.
- This study does not consider any time related aspects such as creep or weathering.
- Correlations between parameters are neglected.

3.2 Evaluation of failure probability

In probabilistic reliability analysis, the safety margin (SM) of a rock slope is defined as the difference between the force resisting the sliding down the plane (R) and the force causing the sliding to occur (L). Failure is defined by the event SM<0. The probability of this event is:
(1) for plane and wedge failures,

$$P_f = P(SM = (R-L) \leq 0); \qquad (5)$$

(2) and for toppling failure,

$$P_f = P(SM = -P_1 \leq 0) \qquad (6)$$

where P_f is the probability of failure.
By generating numerous combinational sets of random variables, one can estimate the relative chance of each failure mode, P_{fp}, P_{fw}, and P_{ft}, and finally P_f:

$$P_{fp} = \frac{N_p}{N}, \quad P_{fw} = \frac{N_w}{N}, \quad P_{ft} = \frac{N_t}{N}, \quad P_f = \frac{N_f}{N} \qquad (7)$$

where, P_{fp} = probability of plane failure, P_{fw} = probability of wedge failure, P_{ft} = probability of toppling failure, P_f = probability of overall failure, N_p = number of plane failure, N_w = number of wedge failure, N_t = number of toppling failure, N_f = number of overall failure irrespective of failure modes = $N_p \cup N_w \cup N_t$, and N = total number of sets analyzed.

4 APPLICATION OF THE PROPOSED MODEL

4.1 General

The reliability analysis of rock slope stability using the proposed model was performed for a hypothetical site. This site is assumed to have the system of four discontinuity sets as shown in Table 1.

Table 1. Summary of input parameters

	Orientation			Friction Angle		S (m)
	E(α)	E(β)	K	E(ϕ)	σ(ϕ)	
1	35°	20°	100	30°	3°	3
2	15°	125°	100	30°	3°	3
3	60°	220°	100	30°	3°	3
4	75°	300°	100	30°	3°	3

The four discontinuity sets are numbered 1, 2, 3, and 4. Their mean values of dips and dip directions are shown in Table 1. Fisher's constants are assumed to be 100. The friction angles of discontinuities are; mean value of 30° and standard deviation of 3°. The spacing of discontinuities is considered to be 3m. The cohesion of discontinuities is taken to be zero.

The available region of design variables are completely surveyed to evaluate the probability of failures with the following conditions:
- dip direction of rock slope ; β = 0 ~ 360°
- cutting angle of rock slope ; α = 30° ~ 90°
- support pressure ; q = 0 ~ 10t/m².

The other parameters used for the study are :
- height of rock slope ; H = 30m
- unit weight of rock mass ; γ = 2.5t/m³
- angle of upper slope surface ; α₂ = 0.

4.2 Results and discussion

The stability of rock slope including plane, wedge, and toppling failure modes was found to be more influenced by the selection of dip direction of cutting rock face than any other design variables as shown Figure 4. The toppling failure is more sensitive to the variation of cutting angle of rock slope than any other failure modes as shown Figure 5. Figure 6 shows that both of plane and wedge failures are more sensitive to the variation of support pressure than toppling failure is.

Figure 7 shows that the failure probability is highly dependent on the Fisher's constant(dispersion of discontinuity orientation). Figure 8 also shows that the increase of coefficient of variation of friction angle of discontinuities gives an increase of failure probability. It means that variability in orientations and friction angles of discontinuity set can lead to failures which would not be predicted by just performing deterministic analysis using mean orientations and mean friction angles.

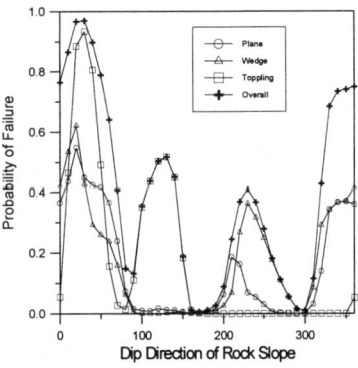

Figure 4. P_f versus β_f ($\alpha_f = 70°$ and $q = 0$ t/m^2)

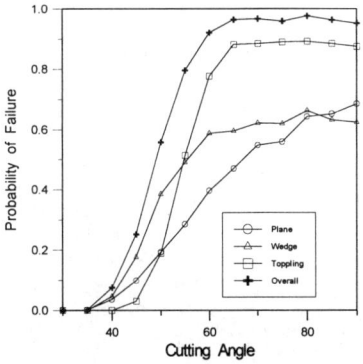

Figure 5. P_f versus α_f ($\beta_f = 20°$ and $q = 0$ t/m^2)

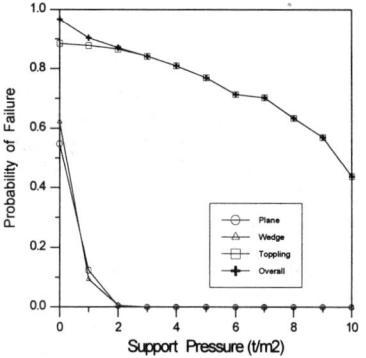

Figure 6. P_f versus q ($\beta_f = 20°$ and $\alpha_f = 70°$)

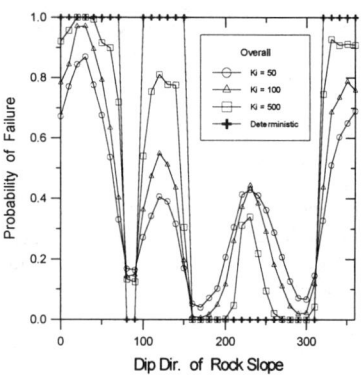

Figure 7. P_f versus α_f with the variation of K

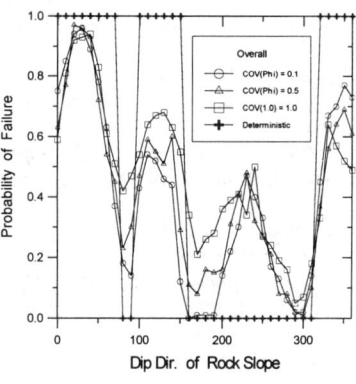

Figure 8. P_f versus α_f with the variation of coefficient of variation of ϕ

5 CONCLUSIONS

A reasonable assessment of the rock slope stability will be made by evaluating the multi-failures modes and considering uncertainties of discontinuity characteristics. This study was performed to provide a new numerical model of reliability-based analysis for rock slope stability.

For evaluating the sensitivity of each design variable on the rock stability, a case study was performed. The sensitivity analysis reveals that the variability in orientations and friction angles of discontinuity, which can not be considered in the deterministic analysis, has a great influence on the rock slope stability.

The stability of rock slope including plane, wedge, and toppling failure modes is more influenced by the selection of dip direction of cutting rock face than any other design variables. The case study shows that the developed reliability-based analysis model can reasonably assess the stability of rock slope.

ACKNOWLEDGEMENT

The research reported here is part of the project entitled "Reliability-based optimization for rock slopes", supported by the Korea Science and Engineering Foundation (KOSEF). This support is greatfully acknowledged.

REFERENCES

Goodman, R. E. and Bray, J. W. (1976), Toppling of rock slopes, *ASCE Specialty Conference on Rock Engineering for Foundations and Slopes*, Vol.2, pp. 201-234.

Goodman, R. E. and Shi, Gen-Hua (1985), *Block theory and its application to rock engineering*, Prentice-Hall, Inc.

Hoek, E. and Bray, J. W. (1981), *Rock slope engineering*, The Institution of Mining and Metallurgy.

Leung, C. F. and Kheok, S. C. (1987), Computer aided design of rock slope stability, *Rock Mechanics and Rock Engineering* 20, pp.111-122.

Zanbak, C. (1983), Design charts for rock slopes susceptible toppling, *ASCE, Journal of Geotechnical Engineering*, Vol.109, No.8, pp. 1039-1062.

Statnamic test for estimating the bearing capacities of rock socketed piles

J. H. Kim & M. M. Kim
Department of Civil Engineering, Seoul National University, Korea

S. H. Lee
Institute of Technology, Engineering and Construction Group, Sam Sung Co., Korea

ABSTRACT: As the structures became massive in recent years, large diametered cast in-situ concrete piles socketed on rock are being used increasingly as a pile foundation. However, bearing capacities of rock socketed piles vary with types of rock and its weathering status. Thus, it is essential to verify the capacity of the pile. In the case of large diametered rock socketed piles, it is difficult to carry out static load test due to its high costs and excessive load capacity. Statnamic load test is developed to take the place of conventional load tests. In this paper, by comparing the bearing capacities obtained from Statnamic testing with those of empirical methods on rock socketed piles; the problems in design of rock socketed piles are investigated. The empirically estimated bearing capacity is in large variation. Therefore, it is essential to confirm design load from preliminary pile load tests. The settlement of the Statnamic test results are not enough to analyze the ultimate capacity. Therefore, as an interim conclusion, it is safe to interpret the extrapolated load-settlement curve by polynomial functions.

1. ESTIMATION OF BEARING CAPACITIES OF ROCK SOCKETED PILES

Rock-socketed piles are designed to support compressive load through shaft resistance or the point resistance, or both. Shaft resistance of rock-socketed pile are related to the uniaxial compressive strength by many researchers, for instance, Rosenberg & Jouneaux(1976), Horvath & Kenny(1979), William & Pell(1981), Rowe & Armitage(1984), etc. The point resistance is calculated by means of uniaxial compressive strength & RQD or empirical estimation methods, etc.

2. TEST PILE PROPERTIES AND SUBSURFACE CONDITIONS

The test pile is a 1.5m diameter RCD bored pile that is constructed as a foundation of the S. Grand Bridge. Fig. 1 shows the arrangement of piles and the location of the test pile. The test pile is cased into EL-17.53m and installed into a depth of EL-19.64m. Subsurface consists of fill, weathered soil, weathered rock, soft rock with increasing depth.

The weathered zone in this subsurface is much more disturbed than the common weathered zone and this zone is not weathered sequentially because of folds and faults. There are 0.2 ~ 1.0m depth soft rocks of schist and gneiss or quartzite vein in the weathered zone. As N-values varies from 10 ~ 50 or more(50/14cm), this zone is classified from medium to very dense. The soft rock of schist is bedded under the weathered zone and TCR, RQD are 0 ~ 99%, 0 ~ 40% respectively in this soft rock stratum.

From the slime observed during RCD construction, the test pile is thought to be socketed about 2m in soft rock(quartzite) under EL-17.53m. However from the additional site investigations adjacent to the test pile, the weathered soil or the weathered rock is found to be located under a 1.5m socketed portion. Fig. 1 shows the construction depth of the test pile and the results of site investigation(B-1, B-2) before construction.

3. STATNAMIC LOAD TEST AND RESULTS

3.1 *Principle*

During the Statnamic testing, the load on the pile top is obtained by launching reaction mass which is moved upward by fast expanding high pressure gases. The reaction of launched mass that is accelerated with 20g(g is the gravity acceleration) upward pushes the pile into the ground with 1g. So, the weight of the reaction mass required is only 5% of the maximum design load.

The duration of loading is about 100ms ~ 150ms

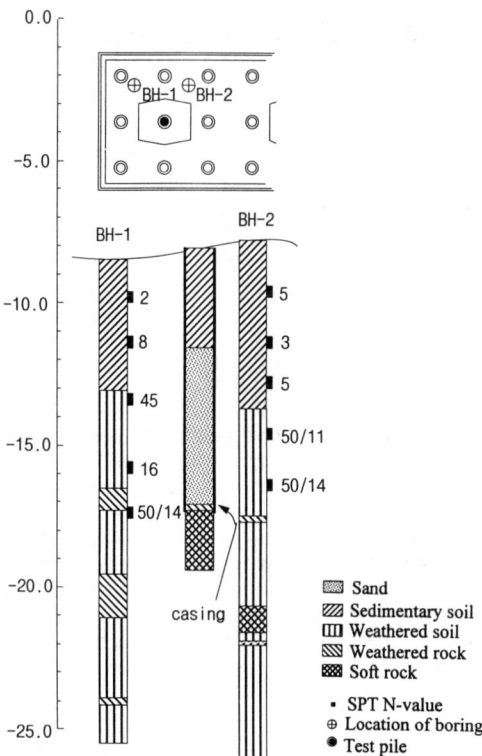

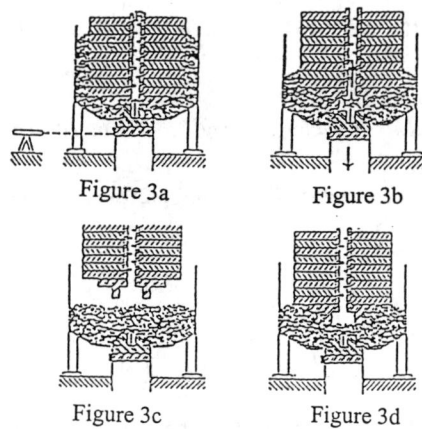

Figure 3. Statnamic setup and procedure

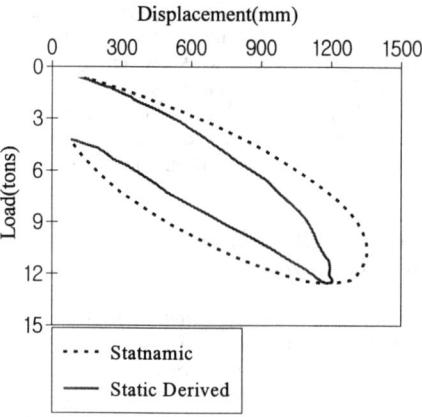

Figure 4. The results of the Statnamic load test(load-settlement curve)

Figure 1. Depth of test pile and results of site investigation

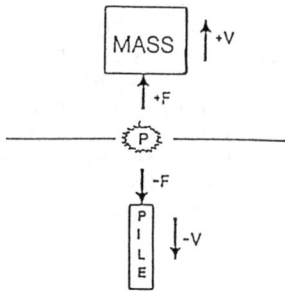

Figure 2. Schematic of Statnamic method

and is 50 times of the loading period of dynamic load tests. Fig. 2, 3 each shows the schematic of Statnamic method and Statnamic setup and procedure.

3.2 *Result and interpretation of Statnamic test*

Fig. 4 shows the results of Statnamic test. The Unloading point method is used to derive static load-displacement relations from the results of Statnamic test. In recent studies, comparing the results of static tests with those of Statnamic tests shows that the unloading point method is reliable for the estimation of static load-displacement relations(Middendorp et al.(1992), Matsumoto et al.(1994), Hovarth et al.(1993)).

From the test results maximum load is 1200ton, and maximum displacement is 12mm. The interpretation methods used are Chin's method, Vander Veen's method, Mazurkiewicz's method, Davisson's method, Modified Davisson's method, De Beer's method, Fuller & Hoy's method and Butler & Hoy's method. Among these methods, for Davisson's method, Modified Davisson's method, De Beer's method, Fuller & Hoy's method and Butler & Hoy's method, there is a need to extend the load-displacement curve for interpretation. Therefore the curve is extrapolated by 1. Hyperbolic functions and 2. Polynomial functions. From chin's method, Vander Veen's method, Mazurkiewicz's method the average allowable load is 615ton. The average allowable load interpreted from extrapolating the curve by hyperbolic functions is 829ton, and that from extrapolating the curve by polynomial functions is 572ton.

The results of the analysis from the curve extrapolated by hyperbolic functions are 30% higher than those from Chin's method, Vander Veen's method and Mazurkiewicz's method in which the extrapolation is not needed. The results of analysis from the curve extrapolated by polynomial functions are 7% lower than those from the same methods. Therefore, if it is necessary to extrapolate the load-displacement curve for interpretation, it is conservative to apply the polynomial functions in extrapolation.

In carrying out Statnamic test, it is difficult to predict the maximum displacement that is occurring during testing. If the displacement hasn't occurred enough, it may be cumbersome to interpret the results of Statnamic tests.

4. COMPARISONS OF THE BEARING CAPACITIES BASED ON CURRENT ESTIMATION-METHODS WITH THE INTERPRETATION RESULTS OF STATNAMIC TEST

When judged by both the accurate results of site investigations near test pile and the inspection of slimes in RCD construction, the test pile is socketed in shallow-mica stratum. Rock-socketed length is approximately 1.5m. As there is a possibility that the stratum right below the test pile-toe is weathered soil, it is assumed that the toe resistance is negligible. Shaft resistance around soil with casing is also neglected. Only shaft resistance of rock socket, based on several empirical methods, is compared to the interpretation results of Statnamic test.

Fig. 5. presents design load(742ton), the bearing capacities estimated, and the interpretation results of Statnamic test. When the shaft resistances of rock socket are calculated using the various empirical methods, the highest value(1200ton) is estimated as 4 times of the lowest value(300ton). As good luck would have it; the average value(745ton) is almost equal to the design value.

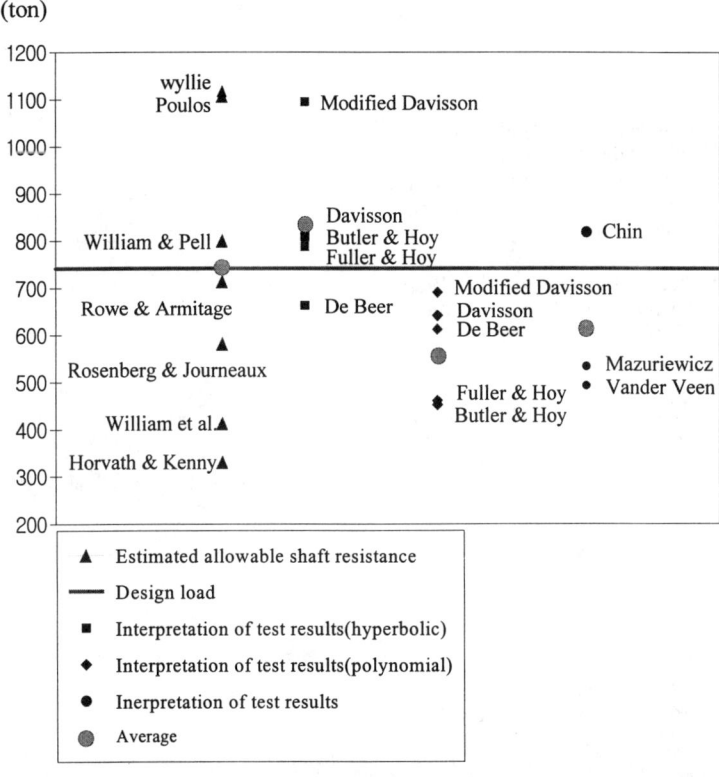

Figure 5. A comparison between the bearing capacity based on current estimation methods and the interpretation results of Statnamic test.

5. CONCLUSION

1. When various methods of estimation on the shaft resistance of rock socket are applied to the test pile, there is 4 times' variation of the lowest value between the lowest value (300ton) and the highest value (1200ton). Therefore, in case of constructing important structures, the bearing capacity of piles must be confirmed from preliminary pile load tests. If there is no preliminary pile load test, it is desirable to take the conservative value among the empirical predictions.

2. When there is a need extend load-settlement curve in interpreting the results of Statnamic test, it is found to be conservative to extrapolate the curve by polynomial functions for this particular test result. The results of analysis from the curve extrapolated by hyperbolic functions are 30% higher than those from the interpretation methods in which the extrapolation is not needed. The results of analysis from the curve extrapolated by polynomial functions are 7% lower than those from the same methods.

3. Large capacity(3000ton) Statnamic testing can be done without high costs and long hours. However, if the displacement hasn't occurred enough during the test, it could have probably failed to get the ultimate capacity from the Statnamic test results.

REFERENCES

Horvath, R.G., Bermingham, P., Middendorp, P. 1993. The Equilibrium Point Method of Analysis for the Statnamic Loading Test with Supporting Case Histories. *Proc. Of the Deep Foundations Conference*: 18-20. Pittsburgh.

Horvath, R.G., Schebech. D., Anderson. M. 1989. Load-Displacement behaviour of socketed piers-Hamilton General Hospital. *Can. Geotech. J. 26*: 260-268.

Middendorp, P. 1993. First experiences with statnamic load testing of foundation piles in Europe. *Proc. 2nd Int. Geotech. Seminar. Deep Foundations on Bored and Auger Piles*: 265-272. Ghent Univ., Belgium.

Middendorp, P., Bermingham, P. and Kuiper, B. 1992. Statnamic Load testing of foundation piles. *Proc. 4th Int. Conf. Appl. Stress-wave Theory to Piles*: 585-588. The Hague.

M.J. Tomlinson 1981. *Pile Design and Construction Practice, 3rd ed.*: 356-369. London; A View Point Publication.

Pells, P.J.N., Rowd, R.K. and Turner, R.M. 1980. An experimental investigation into side shear for socketed piles in sandstone. *Proc. Int. Conf. On Structural Foundations on Rock*: 291-302. Sydney.

Prakash, S. & Sharma, H.D. 1990. *Pile foundations in engineering practice*. New York; John Wiley & Sons.

A.Matsumoto, M.Tsuzuki 1994. Statnamic tests on steel pipe piles driven in a soft rock. *Proc. Int. Conf. On Design and Construction of Deep Foundations.*: 586-660. Orlando.

Putting forward a new concept of grouting engineering – Groutable period

Xinghua Wang & Quqing Gao
Institute of Geotechnical Engineering, Southwest Jiaotong University, People's Republic of China

ABSTRACT: The paper states some special properties of clay-hardening grouts (CHG). According to the properties, the concept, measuring instrument, and measuring standard of the groutable period (GP) have been first put forward. At last, a practical grouting engineering is set to confirm the suitability and soundness of the measuring standard of the GP.

1 INTRODUCTION

During grouting, according to different purposes and conditions of grouting, using grouts should be different. Often used grouts are chemical and cement based grouts. Because properties of the two kinds of grouts are quietly different, different grouting technology must be used when different grouts are used. To protect chemical grouts, for example; sodium silicate based grouts, from gelling in grouting tube or pipe, the grouting technology of dual pipes and double packer injection or 'two-shot' injection must be used. If gelling time of the grouts is longer, the technology of 'one-shot' injection can be used, such as normal cement based grouts. The reason that dual pipes and double packer injection method have to be used is that as soon as chemical grouts and/or sodium silicate meet water or another ingredient (reactant), chemical reaction will take place, causing the mass to solidify or gel, the gelling time is very short and strength of the solidified grouts increases very fast. But the gelling and final setting times of cement based grouts are longer and its strength increase slow. So in the gelling time it could be grouted with simple pipe injection method. The strength of cement based grouts increases, even though, slower than that chemical grouts dose, it is more relatively, such as compressive strength of the grouts, one day, may be to some MPa. To protect the grouts from gelling in grouting tube a concept about time has been put forward, pumpable period. In the time, pumpable period, grouts can flow in grouting pipe and does not gel or solidify. When time surpasses the pumpable period, the technology of 'one-shot' injection will not be able to be used, otherwise, the grout will gel or solidify in the tube. Also after pumpable period, the grouts injected in rock or soil will solidify or gel, strength of the grouts will increase very fast. So newer injected grouts can not push them through the gelled or solidified grout in rock or soil by pump pressure and grouting can not be carried out repeatedly. It can only push it through the space between gelled grouts and fracture wall and fill the fissure that is not filled in the first grouting and second grouting may be carried out. So concept of pumpable period is a very important property of grouts. It is widely used in practical grouting engineering.

2 MEASURING TECHNIQUES

2.1 Measuring instrument

Measurement of plastic strength (Ps) of CHG is carried out with an improved Vicat that the needle is replaced by a circular cone. The cone being on the surface of measured material will sink into the surface h depth (see Fig. 1) with weight of the cone.

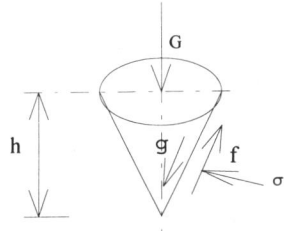

Fig. 1 Sketch of Calculating Ps

When the cone to be balance will stop to sink, shear stress of the surface of the cone is known as the plastic strength of the CHG.

$$Ps = \frac{Ka \times G}{h^2}$$

where: $Ka = \cos^2(\alpha/2) \times \ctg(\alpha/2)/\pi$
α: angle of the cone (°).
h: sinking depth of the cone (mm).
G: total mass of the cone (g).

2.2 *Experiment materials*

The used clay is a washed kaolinite clay produced by Donghu Kaolinite Mine, Hunan, P.R. China. The used cement is a common silicate cement (425#) manufactured by Xiangxiang Cement Factory, Hunan, P.R. China.

2.3 *Compounding the grouts*

The grouts are compounded in some programs. First, clay is immersed in the water two days. Second, the mud is stirred with stirrer one day. Third, cement is added into the mud in some proportion, proportion of weight of cement to volume of mud. At last, reactant is added into the cement-clay grouts in the some proportion; proportion of weight of reactant to volume of cement-clay grouts.

Table 1. Gelling time of CHG

Viscosity of mud*	Specific gravity	Cement content (%)	Content of Reactant (%)						
			1	2	3	4	5	6	7
36.5"	1.355	5	75'	26"	51"	1'35"	4'54"	7'27"	70'
		10	31'	11"	16"	34"	36.8"	1'5"	1'50"
		15	20'	14"	12"	18.5"	28"	32"	45"
58.3"	1.36	5	54'		45"		2'6"		45'30"
		10	29'		20"		55"		2'5"
		15	15'		14"		25"		52"
87.5"	1.425	5	30'	19"	45"	1'50"	2'4"	5'50"	21'
		10	20'	12.3"	12.5"	31.5"	56"	1'20"	2'5"
		15	14'	7.5"	10"	13"	15"	23"	45"

*: Viscosity is measured with Marsh-Cone viscometer.

Table 2. Plastic strength (Ps) of CHG   × kPa

No.	Viscosity of mud	Specific gravity	Cement content (%)	Reactant content (%)	30 min.	60 min.	4 hour	1 day	4 day	10 day
41	30"	1.34	10	1	0.165	0.165	0.324	8.681	85.332	111.98
43	30"	1.34	10	3	0.314	0.343	1.04	64.08	167.82	325.70
44	30"	1.34	10	5	0.705	0.845	2.499	53.07	167.82	498.73
45	30"	1.34	10	10	8.838	11.898	19.005	74.459	1383.6	1483.8
47	30"	1.34	15	1	0.176	0.22	0.526	14.997	266.57	546.18
49	30"	1.34	15	3	0.379	0.466	4.723	82.275	546.18	813.67
50	30"	1.34	15	5	0.637	0.787	14.026	276.48	817.87	937.94
51	30"	1.34	15	10	7.128	9.183	40.234	636.29	1181.4	1633.9
52	40.5"	1.38	10	1	0.283	0.494	1.707	16.194	66.643	338.77
53	40.5"	1.38	10	3	0.729	0.787	3.969	114.54	251.52	475.29
54	40.5"	1.38	10	5	2.17	2.789	3.969	192.71	451.79	457.97
55	40.5"	1.38	15	1	0.376	0.376	1.998	53.079	190.96	
56	40.5"	1.38	15	3	0.633	0.845	14.046	128.49	332.37	564.01
57	40.5"	1.38	15	5	1.211	1.473	40.234	328.93	757.06	757.06

3 BEHAVIORS OF CLAY--HARDENING GROUTS (CHG)

Because of the advantages of clay-hardening grouts (CHG): high water assimilation; high efficiency of sealing water; good rheological behavior; good dilution resistance what can be grouted under condition of flowing underground water and get success; and very low cost, it is being paid a good deal of attention. One of important behaviors of CHG is that it can be grouted repeatedly. If because of some reasons effect of the grouting is not very good, CHG can be grouted repeatedly. And it will fill the fissure and space that are left behind by the last grouting of CHG or push it through the semi-gelling or semi-solidifying stone of CHG and continue to fill the fracture and fissures of stratum. Efficiency of sealing water will get increasing. It is because the gelling time of CHG can be adjusted from some seconds to some dozens minutes (see table 1) and final setting time is very long; dozens hours to some days. The early plastic strength of CHG increases very slowly (see table 2).

The increase rate of plastic strength of CHG is not fast in the first hours. After 4 hours increasing rate of Ps is more. But the value is less relatively in the first days, less one MPa. After 4 days the value increases more fast than the value dose in the first days. The value at 10 day age may be some, dozens, and even hundreds times more than that at one day age. Amount of the value is with relation to the composition of CHG.

Because of the specialty that the final setting time of CHG is longer and the early strength of CHG is lower, the grouts can be grouted repeatedly to overcome the defect left behind by the CHG grouted only one. But how lone can CHG be grouted repeatedly? It is evident that the time range that CHG can be grouted repeatedly would not be forever. There is a range of time that it can be grouted repeatedly. When the time CHG injected in stratum overruns the range of time, CHG will not be grouted repeatedly also.

4 CONCEPT OF GROUTABLE PERIOD (GP)

In the time range CHG can be grouted repeatedly. But the time range is limit. According to author's experience the time range will be varied with the composition of CHG. To use correctly CHG it is very important that the time range in that CHG can be grouted repeatedly should be got a firm grasp. To be used in grouting engineering concept of groutable period (GP) should have been put forward in accordance with the concept of the pumpable period.

As soon as adding reactant CHG will gel and lose its fluidity very fast. But under action of out-force (pump pressure) it can still move. To move CHG the out-force (pump pressure) must surpass the coherence of the grouts. Plastic strength (Ps) of the grouts represents just measurement of the coherence and ability that solidified stone of CHG resists shearing.

To get a quantitative concept Ps is used as a measurement of the groutable period, i.e. when CHG is solidified to some extent and the Ps surpasses a value, CHG can not be grouted repeatedly. So the time range from that the reactant is added into the grouts to time that Ps of CHG surpasses a certain value and CHG can not be grouted repeatedly is defined as Groutable Period (GP) of CHG.

To define GP critical value of Ps must be ensured first. It is suitable that how much the critical Ps value is for the special grouts: CHG. According to the research results of out- or in-door experiment and practical grouting engineering, 50kPa is fixed for the critical value of Ps temporarily. Accurate definition of GP is that the time range from that the reactant is added into the grouts to time that Ps of CHG surpasses 50kPa and CHG can not be grouted repeatedly.

To measure GP of CHG, the improved Vicat was standardlized: the cone angle is 30°; the total mass of the cone and the rod is 300g. When the depth that the cone sinks is larger than 8mm, CHG can be grouted repeatedly. Using the standardlized instrument the definition of GP is the time range from adding reactant into CHG to time that the depth that the cone sinks is equal and/or larger than 8mm.

5 EXAMPLE OF GROUTING

Chengmenshan Copper Mine, Jiujiang, P.R. China, is a large muti-metal mine. Hydrogeological condition of the mine is very complex. Most deposit of the mine is under water level of Chengmeng Lake. The grouting engineering was located on a large gradient, large fault zone at the east of the mine. In the area, Karst proportion is high (above 15%); coefficient of water transmissibility is high (large to 830m^2/d). The distance between two grouting holes is 14.6M. Before grouting, LU numbers of the grouting sectors of the holes are 5.736 and 8.603 separately.

The depth of the No.1 hole is 150.75M. The grouting sector is 42.6M-150.75M. The grouting method is downstage without packer and/or downstage with packer. Total grouting number is 20. Clay mud 589.48M^3 was grouted in the stratum. Grouting pressure was raised to 0.8 MPa from 0.0. LU number was decreased to 0.00505. The depth of the No.2 hole is 150.55M. Grouting sector is 52.4M-125.5M. Grouting number is 5. Clay mud 235.03M^3 was grouted into the stratum. Grouting method was downstage without packer. Grouting pressure was raised to 1.2MPa from 0.0. LU number was decreased to 0.00775.

After completing the grouting, a checking hole was drilled between the holes. During drilling the checking hole, more integral cementing stone of CHG with sand in the fissure have been got out from the hole, which proves that curtain formed by CHG has been linked into whole between the holes, and having formed an anti-water curtain. And it also proves that diffused distance of CHG in the fissure is over 8M. Through pumping water experiments, LU number of the checking hole is 0.232. It is lower than the 0.5 that the designing institute required. It is 1/31 of average LU numbers of the two holes. It proves that antipermeability of the curtain is good. But its cost is only 1/3 of cement grouts.

During grouting, especially No.1 hole, CHG was discontinuously grouted repeatedly. Grouting interval is about two days usually, the most is about five and half days. At the first of each grouting, original pressure at the mouth of the holes is high to 0.5~ 2.0MPa, and amount of injected grouts is less. But the status of high original pressure and low injected amount had not been kept for a longer time. As soon as new grouting CHG wedges in the stone that is in semi-solidifying, the pressure will be decreased, some times, even though, to 0.0MPa and amount injected will increase great. The original pressure of each grouting is relative with composition of CHG injected, the amount of injected grouts, and time interval between groutings. The more the injected amount and the time interval are, the more the original pressure.

6 CONCLUSIONS

Some conclusions may be got from above analysis:

1. The concept of groutable period (GP) is a new concept of grouting engineering. It provide a theory bass for scientific grouting henceforth.

2. The submitted concept, measuring instrument and standard of GP for CHG are credible and correct by validation of practical grouting engineering. Because the concept and measuring standard of GP are only suitable for the special grouts; CHG, whether it will be suitable for other grouts should be proved and confirmed by practical grouting engineering.

REFERENCE:

Arvind V. Shroff; Dhananjay L. Shah, 1993. *Grouting Technology in Tunnelling and Dam Construction.* Rotterdam: A. A. Balkema Publisher.

Wang, Xinghua; Zeng, Xiangxi 1995. Study of method determining fluidity of Clay-based grouts. *Mining and Metallurgical Engineering.* 13(3): 182-184

Wang, Xinghua, 1995 Study of rheological properties and grouting technology of clay-cement (solidifying) grouts. Ph. D. Thesis, *Central South University of Technology*, Changsha, P. R. China.

Wang, Xinghua, 1996. Study and use of clay-hardening grouts. *Journal of Railway Engineering Society (supplement).* 12;405-409.

Wang, Xinghua 1996. A new technique of protection from underground water--clay hardening grouting method. *Geological Exploration for Non- Ferrous Metals.* 5(1): 62-64.

Wang, Xinghua. 1997. Effect factors and increasing law of plastic strength (Ps) of clay hardening grouts. *The Chinese Journal of Nonferrous Metals.* 7(1): 65- 68.

Wang, Xinghua, 1997. Study of a new cheap grouting materials--clay hardening grouts. *Proceedings of the International Symposium on Rock Mechanics and Environmental Geotechnology -RMEG'97, Chongqing, P.R. China, April 1-4, 1997.*

Wang, Xinghua 1997. *Application of Clay hardening Grouts in Underground Engineering.* Beijing, Chinese Railway Publisher.

Yonekura, R 1994. Recent chemical grout engineering for underground construction. *Proceedings of the International Symposium on Anchoring and Grouting Techniques.* 97-113. Econarc Communication (GROUP) Ltd., Hong Kong.

Implementation of safety measure by two dimensional and three dimensional stability analysis for a highly bedded rock slope

S.K.Chung, K.C.Han, S.O.Choi, C.Sunwoo, H.S.Shin & Y.Park
Korea Institute of Geology, Mining and Materials, Taejon, Korea

ABSTRACT: A stability analysis was carried out for a highly bedded rock slope in a Mesozoic sedimentary formation characterized by 3~4 major and 2~3 minor sets of discontinuity. One of the major joint sets is the bedding plane on which slickensides were imprinted proving that it is a plane of fault. The instability occurred even after the V-shape cutting of the original rock slope in an effort to improve the stability. The actual slope is figuring with two different small faces dipping toward 80° clockwise from true north on the right side and 345° on the left side, respectively. This paper describes the measurement of the displacement and the stability analysis by the stereographic projection on the one hand, and a stability analysis and validation of supporting systems with the FLAC 2D and FLAC 3D on the other hand, to obtain the results as follows:

1. The left side slope has a potential plane failure and a wedge failure, however the stability of the right side slope has improved having only a possibility of toppling failure in a relatively small area.
2. The bedding plane and the weight of the upper part of the slope were the principal parameters responsible for the instability.
3. Installation of the support such as cable bolts at several levels would improve the stability.
4. Unless the excavation of this upper part which has the priority to ensure the stability, a multiple monitoring system is imperatively recommended to observe the rock slope behavior and to implement the safety measure from the very beginning onset of the potential instability.

1 INTRODUCTION

A rock slope has been excavated on the left side of the Oongcheon River near the Boryeong Dam in Chungnam Province, just underneath of which the spillway has to be constructed. The geology of this area is composed of the sedimentary rocks belonging to the Daedong Group of the early Jurassic age. The rock masses are highly bedded and characterized by 3~4 major and 2~3 minor sets of discontinuity. One of the major joint sets is the bedding plane on which slickensides were imprinted proving that it is a plane of fault. The bedding plane having a very low shear strength is supposed as the most important plane of instability. The instability occurred even after the V-shape cutting of the original rock slope in an effort to improve the stability. The actual slope is figuring with two different small faces dipping toward 80° clockwise from true north on the right side and 345° on the left side, respectively. This study describes the mechanism of instability on the slope using the FLAC(Fast Lagrangian Analysis of Continua)-2D and FLAC-3D and a countermeasure appropriate to ensuring the long-term stability. A ubiquitous joint model was adopted to simulate the bedding plane in the sedimentary formation and the schistosity in the metamorphic formation during the numerical modeling. Excavation of the unstable rock masses and partial supporting system were analyzed as a countermeasure against the instability. Rock anchor, a kind of cable bolt, was taken into consideration for the method of the partial supporting system. A two dimensional and a three dimensional analysis were carried out for the stability analysis of the spillway slope of the Boryeong Dam. Two dimensional analysis was carried out followed by modeling a representative cross-section of the left side slope in order to identify the depth of the relaxed zone from the slope face. Three dimensional analysis was carried out by modeling both the left and the right side slope in a real scale. The failure behavior and the unstable area of the slope was evaluated by describing the failure mode on the actual slope with the basic data collected by the laboratory and in-situ measurement. Reinforcement and excavation of the unstable part of the slope was recommended and validated with the numerical simulation.

2 TWO DIMENSIONAL STABILITY ANALYSIS

A representative cross-section of the slope was modelled considering the tension cracks in order to validate the input data and to understand the mechanism of the slope failure. Two dimensional analysis was carried out for two cases such as;
1. When the slope was excavated with a slope angle of 1:1.4 as it was designed,
2. When the excavation of the lower part of the slope was stopped so as to keep the upper part of the slope with an angle of 1:1.4 and lower part 1:1.2.

The influence of the groundwater on the deformation of the slope and the slope failure was taken into consideration for both of the two cases.

Table 1. Mechanical properties of materials for the stability analysis.

Properties	Rock mass	Joint plane
density (kg/m^3)	2650	same
shear modulus (MPa)	1.96×10^2	same
bulk modulus (MPa)	2.14×10^2	same
friction angle (degree)	35	30
cohesion (MPa)	1.5×10^{-1}	5×10^{-3}
tensile strength (MPa)	5.44×10^{-2}	5×10^{-4}

2.1 *Excavation of the whole slope*

Table 1 shows the mechanical properties used during the stability analysis. All of the in-situ properties were estimated by the empirical criteria suggested by Hoek & Brown on the one hand and by Serafim & Pereira on the other hand considering the Rock Mass Rating value. Especially the bedding plane of the strata which has a dip of 32.5° similar to the slope was considered as a ubiquitous joint in the analyzed model. The model was divided into from 10,000 to

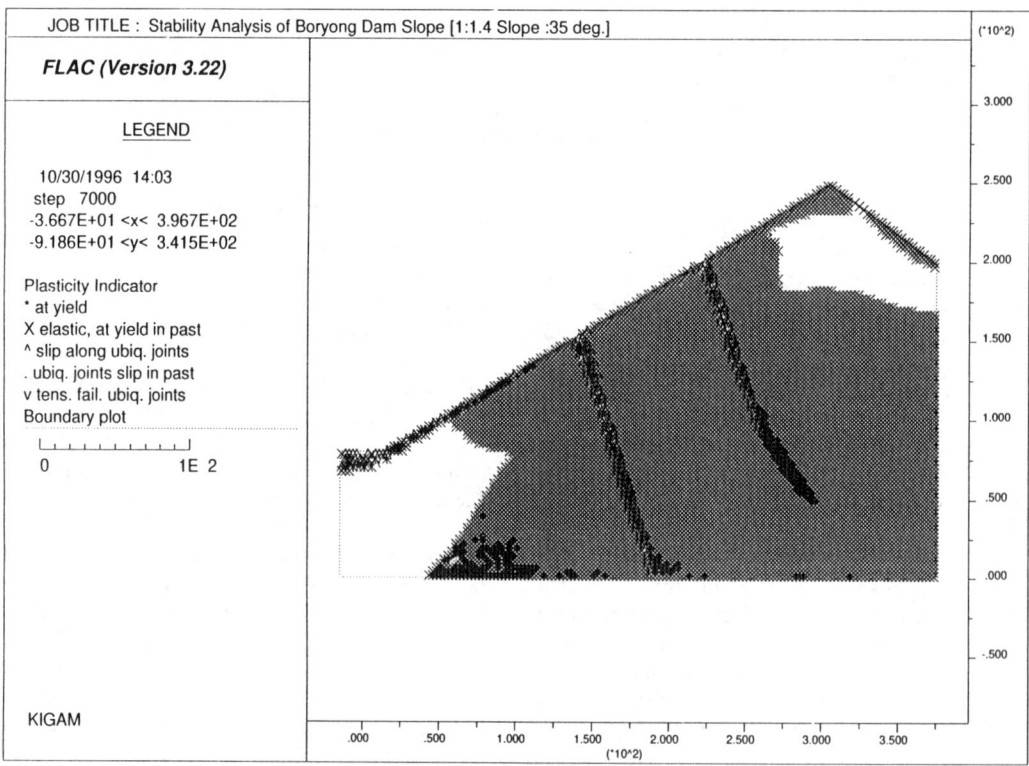

Figure 1. Distribution of the yielded area.

16,000 elements in order not to be affected by the boundary effect. The bottom of the model, 0 ML was fixed so as not to deform to the horizontal and vertical axis by taking the boundary effect into consideration and only the vertical deformation was allowed for the left and right boundary of the model. The first step of the simulation was performed by acting the gravity on the model, and for the second step, the load was transmitted into all the elements until the unbalanced force converged into a recommended value. The third step was performed by giving the tensile strength and cohesion of the materials and a large deformation was considered in order to represent a progressive failure mode.

The model composed of equilateral rectangular elements of 2.5m by 2.5m each side. Since the failure mechanism of the whole slope was the major concerns of the analysis, the benches of the slope were not taken into consideration.

Figure 1 shows the failure mode obtained by the analysis without presence of the groundwater. In the direction perpendicular to the bedding planes two failure surfaces were present from the slope face at the 200ML and 150ML due to the excavation of the lower part of the slope. The cracks developed on the surface of the slope below the 125ML imply a possibility of partial slope failure. An isotropic model for the rock mass will not yield such a failure mode. The result of the analysis gave the same failure mode even if the angle of the bedding plane varies from 22.5° to 30° showing that the failure does not take place along the weakness planes. This type of failure mode is described very well by the horizontal deformation in Figure 2.

Two discontiuities of deformation correspond to the failure surfaces in the Figure 1. The deformation vector is orienting toward the dip direction of the slope below 150ML. Once this failure mode is dominant, it is highly recommended to reinforce the slope lower than 79ML against a progressive failure and the lower part of the slope by rock anchor against the partial failure of the 125ML.

Figure 3 shows that two low vertical stresses developed in the deep part of the slope due to the

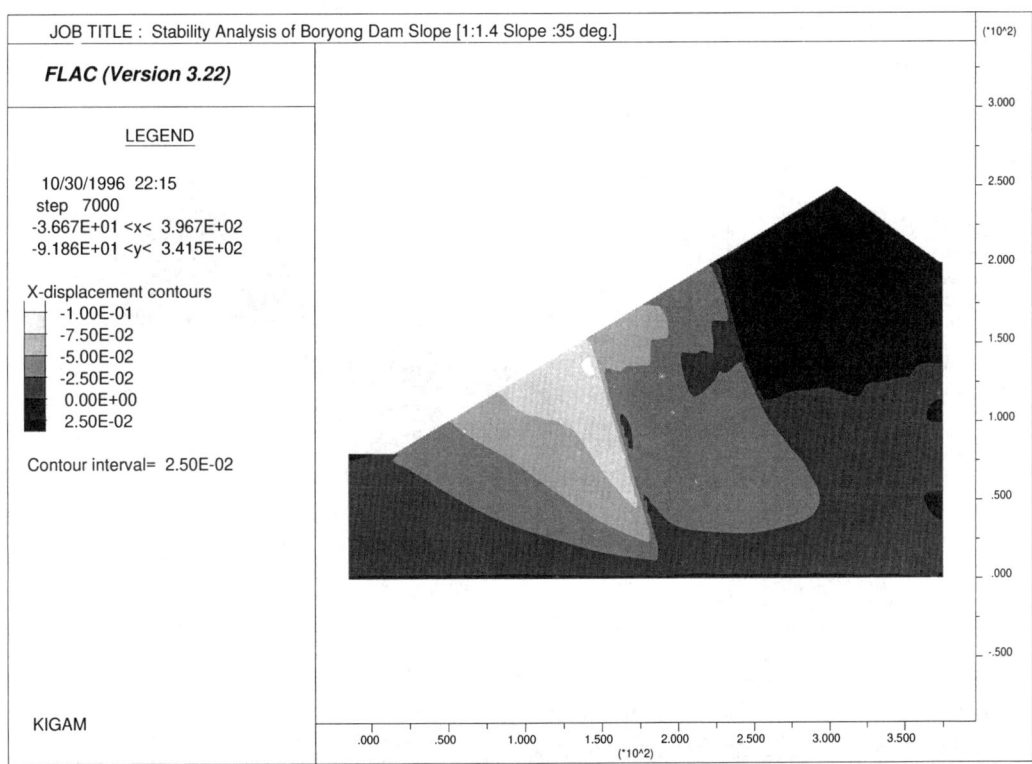

Figure 2. Distribution of the horizontal displacement

tension cracks. Some minor cracks also developed between two major failure surfaces indicating that some tension cracks can be formed between 200ML and 150ML.

In case when the slope is excavated as it was designed, two sets of major tension crack seems to develop in the direction perpendicular to the slope face and bedding plane. Three out of four sets of minor tension crack seems to develop between two major tension cracks. As the result of the failure along these major and minor tension cracks, the rock mass of the slope lower than 125ML is likely to move toward the toe of the slope.

The analysis performed considering the groundwater showed that the presence of the groundwater changed the position of two major failure surfaces toward the crest of the slope, emphasizing the magnitude of tension cracks and displacement of the lower part of the slope.

2.2 *Excavation of the upper part of the slope only*

When the excavation of the lower part of the slope was stopped, the result of the stability analysis shows that the slope was rather stable than when the slope was excavated with a slope angle of 1:1.4 as it was designed. However, the occurrence of the failure surfaces is concentrated on the lower part of the slope. An outstanding failure surface and a discontinuous crack were found in the upper part and in the lower part of the slope respectively(Figure 4.).

The horizontal displacement contour shows movement of the failed rock mass toward the dip direction of the slope in the upper part of the slope(Figure 5). Even though the slope seems to be more stable than the one excavated as it was designed due to confinement of the rock mass of the lower part, the slope should be reinforced with a proper design.

When the groundwater level is found at a depth from 30m to 100m below the slope surface, the horizontal displacement of the 150ML which is larger than that of the surrounding area may cause tension cracks followed by a rock burst. The tension cracks having

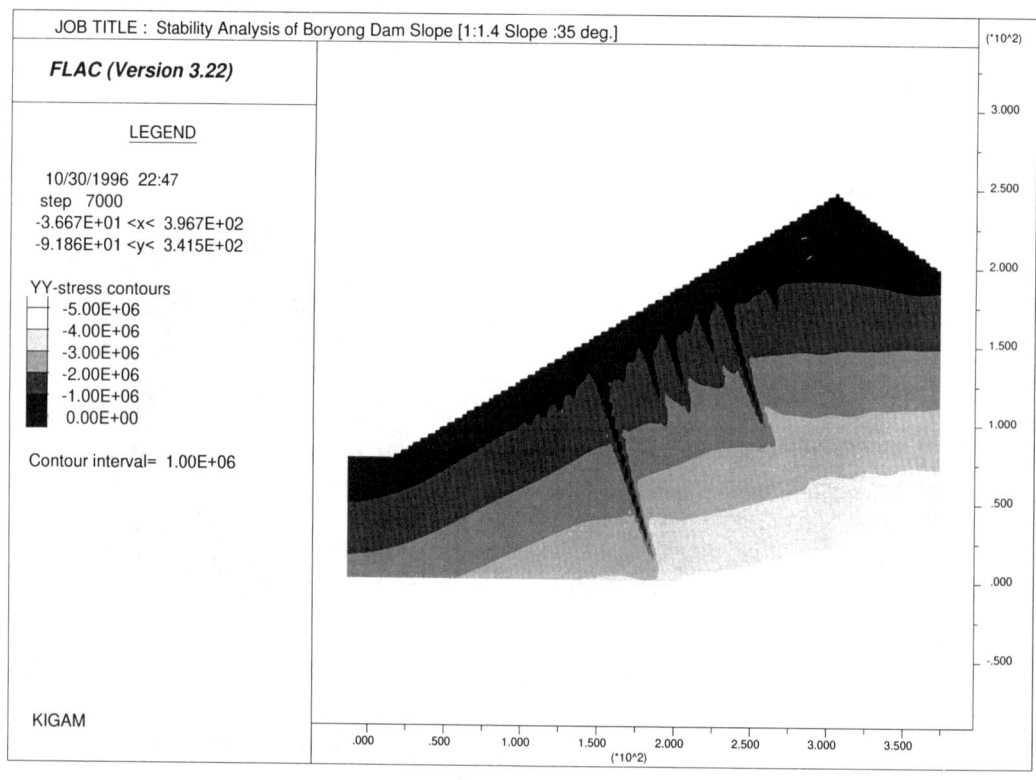

Figure 3. Distribution of vertical stresses.

an opposite dip direction to the dip direction of the slope coincide very well with those observed on the in-situ slope face. The stabilization of the displacement of the existing tension cracks does not necessarily guarantee the stability at 154ML of the slope. As the result of the stability analysis showed that the failure surface likely goes deep into the slope, the yielded area should not be left to prevent another failure of the slope.

3 THREE DIMENSIONAL STABILITY ANALYSIS

Three dimensional analysis with FLAC 3D was conducted to evaluate the stability of the excavated slope of the Boryeong Dam. The stability of the designed slope was analyzed first through the exact modeling of the whole slope without any reinforcement. In case when the result of the analysis shows instability of the slope, a reinforcement should be suggested to improve the stability of the slopes.

3.1 Generation of three dimensional model

The mechanical properties used during the stability analysis are the same as those of two dimensional analysis. The dip and the dip direction of the bedding plane are simply added to represent the highly bedded rock slope, as it is shown in the Table 2. The model in question is composed of about 32,000 blocks.

Table 2. Dip and dip direction of bedding plane for three dimensional analysis.

	Dip direction (degree)	Dip angle (degree)
Left side of slope	347.8	32.5
Right side of slope	332.4	42.0

3.2 Excavation of the whole slope

Figure 6 shows the block plot of state when the slope is excavated as it was designed. In this case, tension

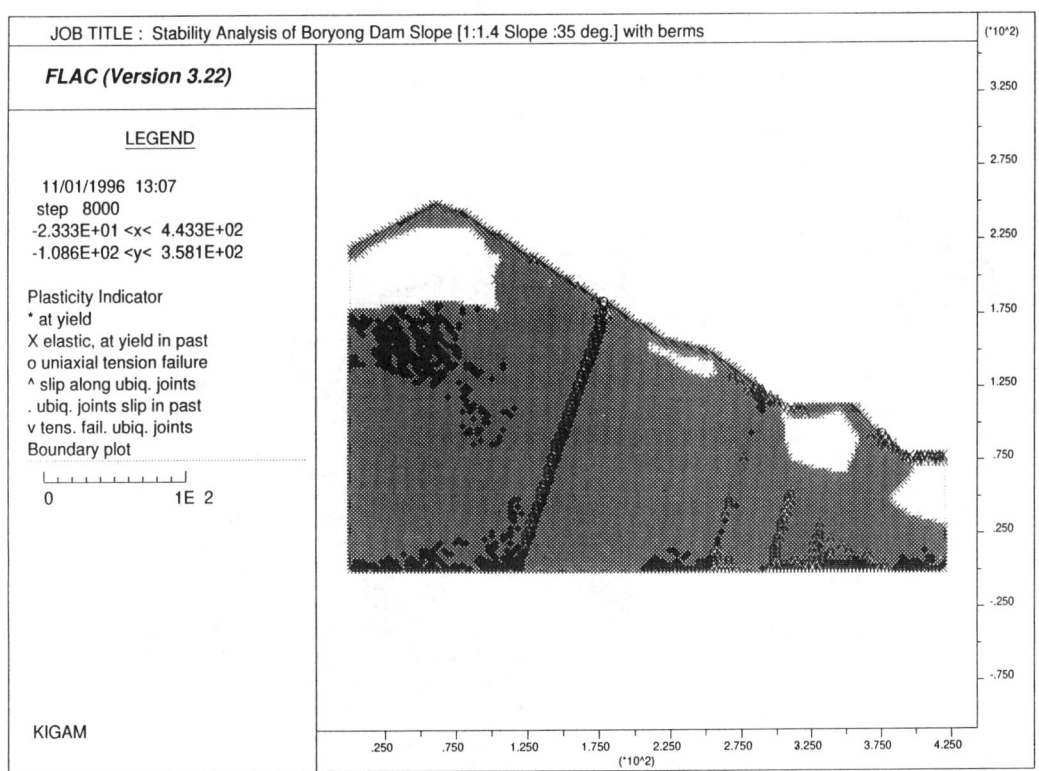

Figure 4. Distribution of the yielded area

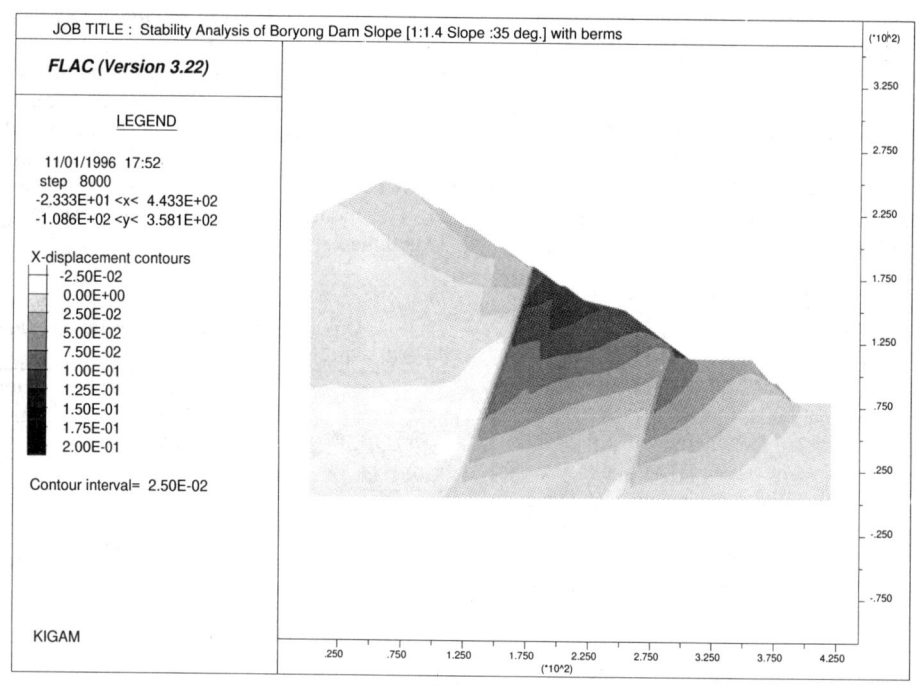

Figure 5. Distribution of the horizontal displacement

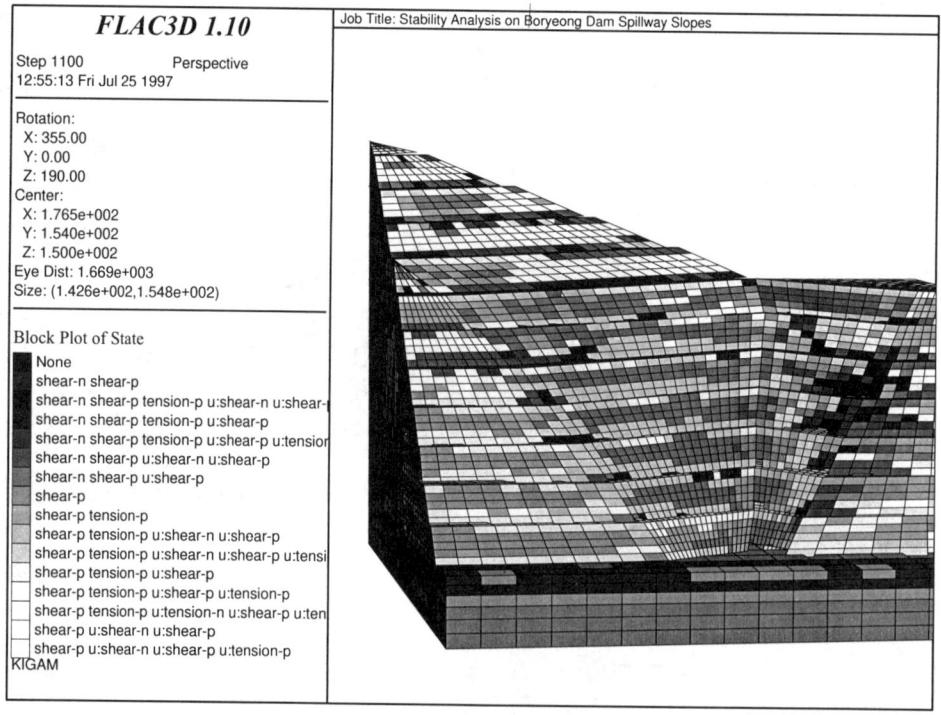

Figure 6. The block plot of state when the slopes are excavated in the proposed shape.

and shear cracks are developed in all over the left side of the slope, especially its failure patterns are concentrated in the region between 94ML and 184ML. This analytical result coincides with the actual slope. The results of the analysis, however, show that the active failure zones are generated without interruption in the upper part of the slope above 184ML, which may occur due to the instability of the lower part of the slope. The active failure zone above 184ML is likely to affect the stability of the lower part of the slope. The range of the yielded zone from the surface of the bench into the slope was also investigated through the sliced view on the left slope. The yielded zone below 184ML reached a depth equivalent to a bench height from the surface of the slope, meanwhile the whole area above 184ML of the slope face was in an active failure mode. Consequently, it was supposed that the reinforcement of zones above 184ML might affect the stability of the whole slope.

Moreover, the sharp edges of left and right slope had better have been removed in order to prevent the growth of the tension cracks.

4 COUNTERMEASURES AGAINST INSTABILITY

4.1 Two dimensional analysis

A two dimensional analysis was carried out considering a partial reinforcement system with cable bolts(Table 3.) for the 79ML, 124ML and 154ML, when the excavation of the lower part of the slope was stopped.

Three rows of vertical cable bolts of 50 m long at the 79ML, three rows of 30m long normal to the bench face of the 79ML, four rows of 20m long normal to the bench face of the 124ML and three rows of 20m long normal to the bench face of the 154ML were modelled.

The interval between two adjacent bolts and that between two adjacent rows was supposed 5m. Pretension could be as high as the tensile strength of the cable bolts in order to provide a better reinforcement by increasing the frictional force between bedding planes, however, it should not be higher than twenty tons when one takes the in-situ strength of the rock mass into consideration. The result of the analysis (Figure 6.) showed that the occurrence of the failure surfaces de. ased considerably at 79ML, 124ML and 154ML where the slope were reinforced with cable bolts. However, the presence of tension cracks

Table 3. Mechanical properties of the cable bolts

Diameter of Strand	$\Phi 12.7$mm
Area	$98.71\text{mm}^2 \times 7 = 691 \text{ mm}^2$
Diameter of Borehole	$\Phi 135$mm
Young's Modulus	200GPa
Tensile Strength	1.3MN
Yield Strength	1.1MN

were still existing, so it was hard to get rid of the potential failure of the slope only by reinforcement. The excavation of the upper part of the slope from the crest up to the 199ML which can create the tension cracks was suggested as an alternative.

The result of the analysis showed that the major tension cracks near 200ML became inactive and only minor tension cracks appeared and became stable on the crest of the slope.

4.2 Three dimensional analysis

Based on the results of the analysis for the slope excavated as it was designed without reinforcement, a new analysis was conducted assuming a partial reinforcement with cable bolts. Table 4 shows the mechanical properties of the cable bolt used for the stability analysis.

Table 4. The mechanical properties for the cable materials and the grout.

Cable area	$5 \times 10^{-4} \text{m}^2$
Cable modulus	98.6GPa
Cable ultimate tensile capacity	0.548MN
Grout compressive strength	20MPa
Grout shear modulus	9GPa

The numerical analysis was carried out firstly, for the case of partial reinforcement. Two rows of cable bolts of 20m long were installed in 80ML and 154ML, respectively, and a row of cable bolts of 30m long were installed at the bottom of the slope in the numerical analysis. The cable bolts were installed vertically in the bottom of the slope, meanwhile the others were installed 5~10° upward from the horizontal plane to maximize the effect of the support in 80ML and 154ML.

In this case, the active failure zones reduced considerably in the lower region of the whole slope comparing with the model without reinforcement. Nevertheless an instability was still existing in the upper part of 184ML.

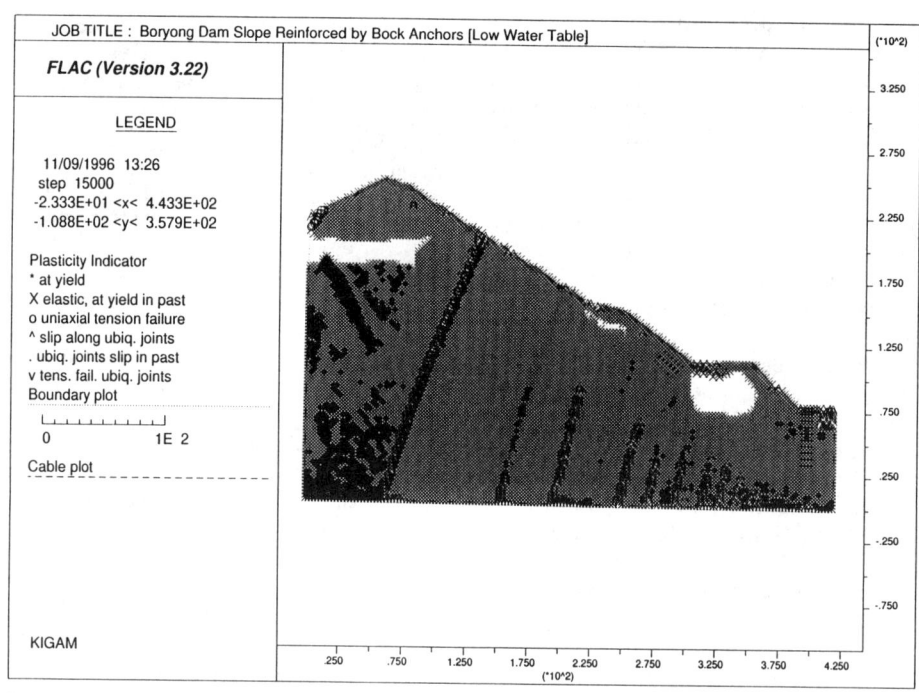

Figure 7. Distribution of the yielded area after reinforcement with cable bolts.

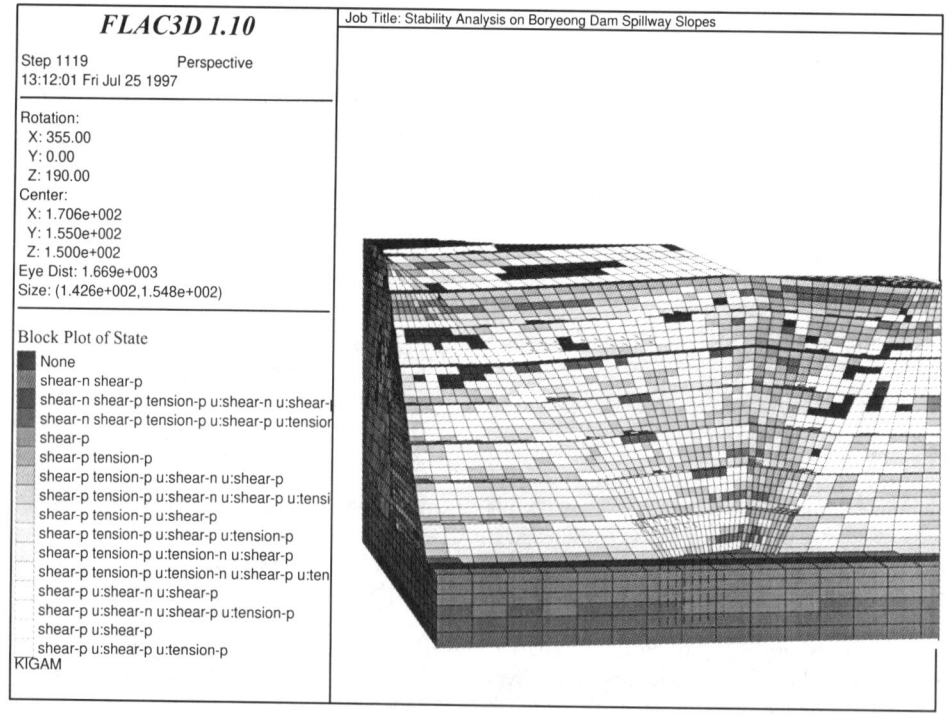

Figure 8. The block plot of state when the region above 184ML was removed and the partial support was installed.

Secondly, cable bolts were installed in the same manner as for the previous analysis, but the upper part of 184ML was removed, to have a more gentle slope. In this case, there was no active failure zones throughout the slope and the stability of the slope was guaranteed as it is shown in the Figure 8.

5 CONCLUSIONS

1. The left side slope has a potential plane failure and a wedge failure, however the stability of the right side slope has improved having only a possibility of toppling failure in a relatively small area.

2. The bedding plane and the weight of the upper part of the slope were the principal parameters responsible for the instability.

3. The stabilization of the displacement of the existing tension cracks does not necessarily guarantee the stability at 154ML of the slope. The yielded area should not be left to prevent another failure of the slope.

4. The yielded zone below 184ML reached a depth equivalent to a bench height from the surface of the slope, meanwhile the whole area above 184ML of the slope face was in an active failure mode.

5. Installation of the support such as cable bolts at several levels would improve the stability.

6. Unless the excavation of this upper part which has the priority to ensure the stability, a multiple monitoring system is imperatively recommended to observe the rock slope behavior and to implement the safety measure from the very beginning onset of the potential instability.

REFERENCES

Hoek, E. and E.T. Brown 1988. The Hoek-Brown failure criterion 1988 update, *Rock engineering for underground excavations*: 31-38. Toronto University of Toronto.

Serafim, J.L. and J.P. Pereira 1983. Considerations on geomechanics classification of Bieniawski, *Proc. Int. Symp. On Eng. Geol. and Underg. Constr.* 1. II31-42. LNFC. Lisboa.

FLAC(ver.3.2) User's Manual 1992. Vol.1,2,3. Itasca Consulting Group, Inc.

FLAC^{3D}(ver.1.1) User's Manual 1996. Vol.1,2,3. Itasca Consulting Group, Inc.

Evaluation of rock slope stability by analysis of discontinuities in Boryung damsite

C. Sunwoo, H. S. Shin, K. C. Han & S. K. Chung
Korea Institute of Geology, Mining and Materials (KIGAM), Taejon, Korea

ABSTRACT: The stability on the rock slope where failures had occurred in Boryung Dam site was evaluated using the stereographic projection techniques. SMR (Slope Mass Rating) approach which is suitable for preliminary assessment of rock slope stability in rock was also carried out for rating the slope rock mass. The 3~4 major discontinuity sets are distributed and all type of failures (plane, wedge and toppling failure) are present in overall slope faces. The dip of slope must be inclined lower than the minimum estimated angle of 26 degrees, otherwise the possibility of failures will always exist in rock slope.

1 GEOLOGY AND SCOPE OF SLOPE

As the Boryung dam for multi-purpose was constructed, the large scale rock slope (166m height and 350m width) was formed. The rocks in slope face are predominantly Jurassic sedimentary in origin but differ considerably in their properties. The interbedded sandstones, conglomerates and shales succession are present as three main kinds of rocks. The dykes several centimeters thick locally intruded these sedimentary rocks. And metamorphic gneiss is rarely present in eastern part. The regional dip of bedding is 30~45 degrees to the north with a consistent east-west strike. The rock mass is extensively jointed.

At first the large rock slope had been excavated with dip of 60°, but the middle part of slope face failed in shape of large wedge (20,000 m³) without any connection with rainfall and underground water. So the slope had been reexcavated by lowering the dip to 40 degrees (Table 1), but the failures also occurred in the same slope face. Plane and wedge failures and rock falls are common on the rock slopes in this area.

The 3~4 major discontinuity sets are distributed and all types of failures including the sliding of wedge along the intersection line of two or three discontinuity planes.

Core samples were selected from 12 boreholes to determine the rock properties and values of RQD were determined from 8 boreholes. The rock properties are given in Table 2. Though rock type is the same, the rock properties are locally very variable. Joint shear tests showed the friction angle of discontinuities is about 26.4°. The RQD for

Table 1. Dip direction and dip of slopes.

Slope		Dip direction	Dip
Left slope	L	22°	40°
	MID	35°	40°
Right slope	R	15°	40°

Table 2. Summary of rock properties.

		sandstone	shale	conglo-merate
density (kg/m³)		2663	2723	2667
wave velocity	P (m/s)	4510	5110	3970
	S (m/s)	2850	2990	2450
strength (MPa)	comp.	132	107	92
	tensile	9.5	11.0	8.0
Young's modulus (GPa)		33.13	31.89	27.24
Poisson's ratio		0.13	0.14	0.14
ϕ (°)		47.0	44.1	36.9

sandstones was 29%, 40% for conglomerates and 48% for shales respectively. These index present that the engineering quality of rock proposed by Deere (1968) is classified as a 'poor rock'.

2 DISTRIBUTION OF DISCONTINUITIES

Fig. 1 presents the panoramtic view of surveyed rock slope and rose diagrams illustrated the dip direction pattern of discontinuities in the slope. In the slope faces, three or four distinct sets of discontinuities are predominant (Table 3). The major set of discontinuities (ST1) shows the dip direction of 332° in the left side and 348° in the right side of slope.

The dip direction of discontinuity set ST1 corresponds to that of bedding and is similar to the orientation of the pattern of faults. This one appears to be the most significant for slope stability in this area. Most bedding planes are slickensided. Along these bedding plane, many plane and wedge failures are more likely to have occurred (Fig. 2 and Fig. 3).

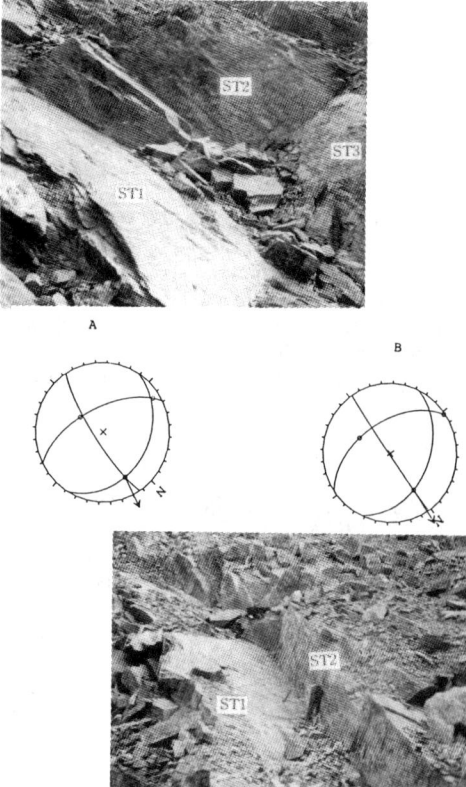

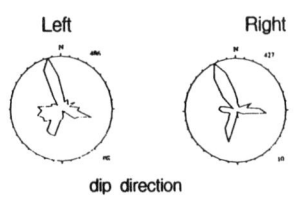

Fig. 1. Scope of rock slope and rose diagrams illustrated the dip direction pattern of discontinuities in Boryung Dam site.

Fig. 3. Wedge failures in the slope face.

Table 3. Summary of orientation, trace length and spacing for major discontinuities sets in overall slope.

		Dip direction (°)	Dip (°)	Trace length (m)	Spacing (m)
Left slope	ST1	348	33	43.8	0.25
	ST2	253	74	17.6	0.31
	ST3	103	79	15.1	0.35
	ST4	202	70	32.5	0.34
Right slope	ST1	332	42	25.6	0.20
	ST2	93	88	56.1	0.22
	ST3	200	67	15.2	0.21

Fig. 2. Bedding plane and plane failure.

The trend of dip direction and dip are gradually changed from right to left. The dip of bedding and discontinuities becomes steeper from left to right slope.

The mean discontinuity spacings are about 20cm~35cm (Table 3) and this spacing represents 'moderate' by the ISRM classification. The mean trace length of discontinuities ranges from 15m to 56m and represents very high persistence.

3 EVALUATION OF ROCK MASS

The parameter "discontinuity orientation" reflects on the significance of the various discontinuity sets present in a rock mass. The main set controls the stability of slopes. The geomechanics classification has widely applied in various types of engineering. Most of the applications have been in the field of tunnelling. Romana(1985, 1988) also applied rock mass classifications to the assessment of the stability of rock slopes. A factorial approach to rating adjustment for the discontinuity orientation parameter in Rock Mass Rating(RMR) system (Bieniawski, 1973, 1976, 1989) based on field data was developed.

The adjustment rating for discontinuities in rock slopes is product of three factors as follows: parallelism between the slope and the discontinuity strike(F_1), the discontinuity dip in the plane mode of failure (F_2) and the relationship between the slope angle and discontinuity dip(F_3). The adjustment factor(F_4) for the method of excavation depends on whether one deals with natural slope or one excavated by presplitting, smooth blasting, mechanical excavation or poor blasting.

The final Slope Mass Rating is of the form

$$SMR = RMR + (F_1 \times F_2 \times F_3) + F_4$$

Examination of rock conditions by RMR

Table 4. SMR value for each major discontinuity set (see Table 2).

	Left slope				Right slope		
	ST1	ST2	ST3	ST4	ST1	ST2	ST3
SMR	36.7	42.0	42.0	42.0	40.2	41.0	41.0

system reveals the rock mass 'fair rock'. The estimated friction angle for given rock masses that was calculated by RMR value is 26°.

As the wedge failures are no more dependent on RMR value than plane failures, the classification must be applied for each discontinuity system. The minor value of SMR is retrained for the slope.

Table 5. Tentative description for Slope Mass Rating (after Romana, 1993).

Class	SMR	Description	Stability	Predict failure
I	81~100	Very good	Very stable	None
II	61~80	Good	Stable	Some blocks
III	41~60	Fair	Partially stable	Some joints or many wedges
IV	21~40	Poor	Unstable	planar or large wedge
V	0~20	Very poor	Very unstable	Large planar or soil-like

In the classification estimates for each discontinuity set of the left slope, the minor SMR index showed 36.7 (Table 4). This index represents class IV and the slope 'unstable'. And the planar failures in some joints or big wedge failures are supposed (Table 5). The right slope with minor value of SMR 40.2 (Table 4) belongs to class III and represent the slope 'partially stable'. And the planar failures in some joints and many wedge failures are supposed (Table 5).

In broad sense, the support measurement such as systematic reinforced shotcrete or concrete should be taken to support the slope of class III or IV group.

4 ANALYSIS OF SLOPE STABILITY

In a majority of cases, the slope failures in rock masses are governed by discontinuities and develop across surfaces formed by one or several discontinuities. Basic failure modes are well known as plane, wedge, toppling and soil-type. The stability analysis on the slopes where the failures had taken place was carried out using the stereographic projection techniques.

The diagrams of Fig. 4 represent that mainly plane failures can occur in overall slope faces and partially toppling failures can occur.

BORYUNG LEFT JOINT (L SLOPE)
BORYUNG LEFT JOINT (M SLOPE)
BORYUNG RIGHT JOINT (R SLOPE)

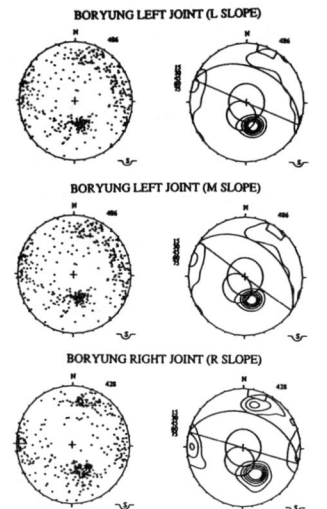

Fig. 4. Analysis of stability of overall slopes by stereographic projection method.

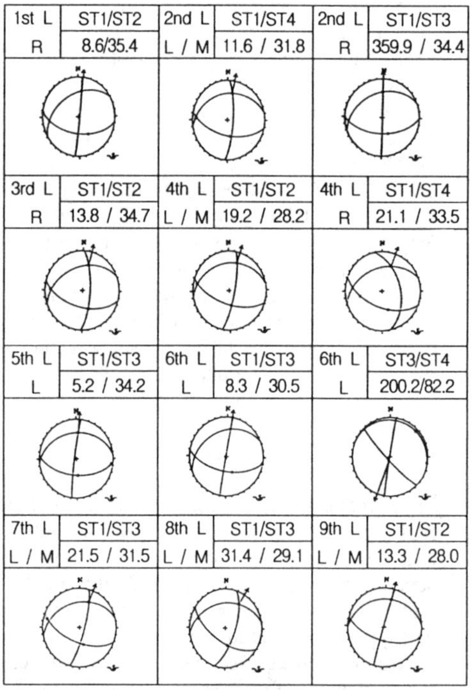

1st L R ST1/ST2 8.6/35.4
2nd L L / M ST1/ST4 11.6 / 31.8
2nd L R ST1/ST3 359.9 / 34.4
3rd L R ST1/ST2 13.8 / 34.7
4th L L / M ST1/ST2 19.2 / 28.2
4th L R ST1/ST4 21.1 / 33.5
5th L L ST1/ST3 5.2 / 34.2
6th L L ST1/ST3 8.3 / 30.5
6th L L ST3/ST4 200.2/82.2
7th L L / M ST1/ST3 21.5 / 31.5
8th L L / M ST1/ST3 31.4 / 29.1
9th L L / M ST1/ST2 13.3 / 28.0

Fig. 5. Potential wedge failures in each level of slope.

The sliding and toppling of wedge along the intersection line of discontinuity planes can also be presented in overall slope faces (Fig. 5). In each level of slope, the sliding possibility of wedges that will be formed by two discontinuity planes were analyzed. The estimated plunges of intersection lines formed by two discontinuity planes range from 28° to 34° in left slope and from 33° to 35° in right slope. The slope face in the right side is geometrically stabler than that in left side.

5 CONCLUSIONS

Though the rocks have strength of 100~130 MPa, rock slope was unstable with dip of 40°. The stability of surveyed rock slope appears to be controlled largely by the orientation and geometry of discontinuities within rock mass. The minimum plunge of potential wedge sliding by two discontinuity planes is 28° and the estimated friction angle for surveyed rock masses that was calculated by RMR value is 26°. In joint shear tests the friction angle of discontinuities indicated 26.4°. The internal friction angles of rocks vary between 37° and 47°. Therefore, we suggest that the minimum estimated angle for stabilizing the rock slope is 26°. The dip of slopes should be inclined lower than this angle. If the slopes could not be inclined lower than this angle, the support or stabilizaton measures such as systematic reinforced shotcrete or concrete should be considered to protect the slope from failures.

REFERENCES

Bieniawski, Z.T. 1973. Engineering classification of jointed rock mass, *Trans. South Afr. Ins. of Civ. Eng.*, Vol.15, No.12, pp.335~344.

Bieniawski, Z.T. 1976. Rock mass classification in rock engineering applications, *Proc. Symp. on Exploration for rock engineering*, Balkema, Rotterdam, Vol. 12, pp.97-106.

Bieniawski, Z.T. 1989. Engineering rock mass classifications, Wiley, New York. p.251.

Deere, D.U. 1968. Geological considerations, Rock mechanics in engineering practice, ed. R.G. Stagg and D.C. Zienkiewicz, Wiley, New York, pp.1-20.

Romana R.M. 1985. New adjustment ratings for application of Bieniawski classification to slopes, *Proc. Int. Symp. on the role of rock mechanics*, Zactecas, pp.49-53.

Romana R.M. 1988. Practice of SMR classification for slope appraisal, *Proc. Int. Symp. on Landslides*, Lausanne, Balkema, Rotterdam, pp.1227-1229.

Author index

Ahn, H.J. 727
Akagi, T. 643
Akutagawa, S. 853
Amano, S. 553
Aoi, T. 769
Aydan, Ö. 643

Baba, S. 789
Bae, G.J. 819
Baillargeon, M. 883
Barton, N. 547
Bawden, W.F. 875, 883

Cai, M. 587, 679
Cao, T. 697
Chang, K.M. 673
Chen, C. 921
Chen, W. 565
Chikahisa, H. 613
Cho, E.K. 595
Choi, K.S. 805
Choi, S.O. 703, 941
Christianson, M. 547
Chryssanthakis, P. 547
Chung, S.K. 739, 757, 941, 951

Dawson, E.M. 895
Dhar, B.B. 607
Duan, K. 601

Ekman, D. 619
El Tani, M. 667
Esaki, T. 857

Feng, Z. 697

Gao, Q. 937
Ge, X. 921
Gerber, W. 915
Ghosh, A.K. 869
Goel, R.K. 607
Gu, X. 921
Gupta, A.S. 661

Håkansson, U. 813
Haller, B. 915
Han, I.Y. 775

Han, K.C. 941, 951
Henning, J. 883
Hirakawa, Y. 763
Hirata, A. 789
Hong, S.J. 775
Hong, S.W. 819
Hyett, A.J. 875, 883

Inaba, T. 789

Jang, M.H. 715
Jee, W.R. 581
Jethwa, J.L. 607
Jiang, Y.J. 857
Jue, K.S. 901
Jung, H.K. 739, 863
Jung, S.L. 757
Jung, Y.B. 739

Kaneko, K. 769, 789
Kang, T. 775
Katsuyama, K. 691, 745
Kawamoto, T. 643
Kim, C.Y. 819
Kim, D.H. 739
Kim, D.Y. 901
Kim, H.Y. 731
Kim, J. 631
Kim, J.H. 933
Kim, J.W. 835
Kim, K.J. 673
Kim, M.H. 709
Kim, M.M. 933
Kim, S.H. 673
Kim, Y.S. 595
Kinashi, H. 553
Kiyama, T. 691
Koike, K. 769
Kong, G. 587
Krotov, N.V. 793
Kwaśniewski, M.A. 635
Kwon, B.D. 863
Kwon, K.S. 703
Kwon, S. 649

Lausch, P. 883
Lee, C.I. 779, 841

Lee, D.H. 649
Lee, H.K. 649, 709, 727, 739, 757, 805, 835
Lee, H.S. 709
Lee, I.M. 927
Lee, J. 775
Lee, K.S. 779
Lee, M.J. 927
Lee, S.H. 933
Lee, Y.K. 841
Lee, Y.N. 901
Li, C. 813
Li, H. 721
Li, X. 591
Liang, J. 587
Lim, T.J. 581
Liu, D. 559
Liu, T. 679
Lodus, E.V. 625
Lorig, L. 547
Luo, R. 891

Matsumoto, K. 613
Michihiro, K. 553
Miller, H.D.S. 649
Mochizuki, A. 763
Moosavi, M. 875

Nakagawa, K. 847
Nanda, A. 577

Obara, Y. 769
Ohmi, M. 769
Özçelik, Y. 683

Park, B.Y. 805
Park, C. 731
Park, K.O. 715
Park, Y. 731, 941
Park, Y.J. 895

Rahman, Md.H. 909
Rao, D.G. 869
Rao, K.S. 661
Rokahr, R. 825
Ruest, M. 883
Rutqvist, J. 619

Sakurai, S. 847, 853
Sengupta, S. 577
Seto, M. 691, 745
Shabarov, A.N. 793
Shen, J. 601
Shen, Z. 831
Shi, Y. 921
Shimizu, N. 847
Shimizu, Y. 643
Shin, H.S. 703, 819, 941, 951
Shiotani, T. 613
Sim, Y.K. 581
Sinha, A. 869
Sohn, K.I. 863
Staudtmeister, K. 825
Stephansson, O. 619, 631
Sugawara, K. 769
Suh, B.S. 863
Suh, Y.H. 547, 901
Sunwoo, C. 941, 951

Synn, J.H. 731

Tewatia, S.K. 785
Tsang, C.F. 619
Tsuchihara, H. 553
Tsutsui, M. 613

Uchita, Y. 763
Utagawa, M. 691, 745

Voznesensky, A.S. 751

Wan, H. 697
Wang, B. 565, 649, 697
Wang, J.A. 635
Wang, S. 591
Wang, X. 937

Xu, C. 697
Xu, D. 921

Xu, X. 697, 721
Xu, Z. 721, 831
Xue, L. 801, 891

Yang, H.S. 715
Yin, J. 697
Yokota, Y. 857
Yoon, Y.K. 655
Yoshioka, H. 553
Young, D.S. 571

Zhang, S. 697
Zhao, Y. 601
Zheng, Y. 559
Zhou, C. 679
Zhu, F. 801
Zhu, W. 565
Zoubkov, V.V. 793